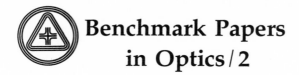

Benchmark Papers in Optics / 2

——————A *BENCHMARK* ® Books Series——————

INFRARED DETECTORS

Edited by
RICHARD D. HUDSON, Jr.
Hughes Aircraft Company
and
JACQUELINE WORDSWORTH HUDSON
Arjay Associates

Dowden, Hutchinson
& Ross, Inc.
Stroudsburg, Pennsylvania

Distributed by
HALSTED PRESS *A Division of John Wiley & Sons, Inc.*

To our mothers
  Marion A. Hudson
  Edith G. Wordsworth

Copyright © 1975 by **Dowden, Hutchinson & Ross, Inc.**
Benchmark Papers in Optics, Volume 2
Library of Congress Catalog Card Number: 75–4923
ISBN: 0–470–41837–0

77 76 75   1 2 3 4 5
Manufactured in the United States of America.

LIBRARY OF CONGRESS CATALOGING IN PUBLICATION DATA

Main entry under title:

Infrared detectors.

    (Benchmark papers in optics ; v. 2)
    Includes indexes.
    1. Infra-red detectors. I. Hudson, Richard D.,
1924-   II. Hudson, Jacqueline W.
TA1570.I53   621.36'2   75-4923
ISBN 0-470-41837-0

Exclusive Distributor: **Halsted Press**
a Division of John Wiley & Sons, Inc.

Benchmark Papers
in Optics

Series Editor: Stanley S. Ballard
University of Florida

Published Volumes and Volumes in Preparation

POLARIZED LIGHT
William Swindell
INFRARED DETECTORS
Richard D. Hudson, Jr., and Jacqueline Wordsworth Hudson
INTEGRATED OPTICS
John F. Ebersole and Charles M. Wolfe
REMOTE SENSING
Richard D. Hudson, Jr., and Jacqueline Wordsworth Hudson
LIGHT IN THE SEA
John E. Tyler

Acknowledgments
and Permissions

ACKNOWLEDGMENTS

THE OPTICAL SOCIETY OF AMERICA—*Journal of the Optical Society of America*
  Black Radiation Detector
  Factors of Merit for Radiation Detectors
  A Fast Superconducting Bolometer
  Improved Black Radiation Detector
  Low-Temperature Germanium Bolometer
  A New Classification System for Radiation Detectors
  Properties of Thermistor Infrared Detectors
  Proposal of the Detectivity D^{**} for Detectors Limited by Radiation Noise
  The Sensitivity of the Human Eye to Infra-Red Radiation
  The Ultimate Sensitivity of Radiation Detectors

PERMISSIONS

The following papers have been reprinted with the permission of the authors and copyright holders.

AMERICAN INSTITUTE OF PHYSICS
  Applied Physics Letters
    Heterodyne Performance of Mercury Doped Germanium
  The Review of Scientific Instruments
    The Design of Fast Thermopiles and the Ultimate Sensitivity of Thermal Detectors

INSTITUTE OF ELECTRICAL AND ELECTRONICS ENGINEERS, INC.
  Proceedings of the Institute of Electrical and Electronics Engineers
    Infrared Heterodyne Detection
    Photodetectors for Optical Communication Systems
  Proceedings of the Institute of Radio Engineers
    Bridges Across the Infrared–Radio Gap
    Description and Properties of Various Thermal Detectors
    Detectivity and Preamplifier Considerations for Indium Antimonide Photovoltaic Detectors
    Film-Type Infrared Photoconductors
    Fundamentals of Infrared Detectors
    The Measurement and Interpretation of Photodetector Parameters
    Noise in Radiation Detectors
    Phenomenological Description of the Response and Detecting Ability of Radiation Detectors
    Single-Crystal Infrared Detectors Based upon Instrinsic Absorption

INSTITUTE OF ELECTRONIC AND RADIO ENGINEERS—*The Radio and Electronic Engineer*
  High Performance Pyroelectric Detectors

THE INSTITUTE OF PHYSICS
  Journal of Scientific Instruments
    Solid State Devices for Infra-Red Detection
  Physics in Technology
    Modern Infrared Detectors

MICROFORMS INTERNATIONAL MARKETING CORPORATION—*Infrared Physics*

Advanced Design of Joule–Thomson Coolers for Infra-Red Detectors

Background Limited Photoconductive HgCdTe Detectors for Use in the 8–14 Micron Atmospheric Window

A Comparison of the Performance of Copper-Doped Germanium and Mercury-Doped Germanium Detectors

Photon Noise Limited D^* for Low Temperature Backgrounds and Long Wavelengths

THE OPTICAL SOCIETY OF AMERICA

Applied Optics

Closed Cycle Cryogenic Refrigerators as Integrated Cold Sources for Infrared Detectors

The Detectivity of Cryogenic Bolometers

The Detectivity of Infrared Photodetectors

Extrinsic Detectors

Indium Antimonide Submillimeter Photoconductive Detectors

Lead Selenide Detectors for Ambient Temperature Operation

Lead Selenide Detectors for Intermediate Temperature Operation

Photoconductive Indium Antimonide Detectors

Radiant Cooling

Solid-Backed Evaporated Thermopile Radiation Detectors

Journal of the Optical Society of America

Performance Characteristics of a New Low-Temperature Bolometer

PERGAMON PRESS LTD.—*Infrared Physics*

High Detectivity $Pb_x Sn_{1-x} Te$ Photovoltaic Diodes

Series Editor's Preface

Optics, pure and applied, comprises a very broad field indeed. Its roots are ancient; its modern applications are novel and exciting. A difficult and far-reaching task is hence faced by the series editor of the Benchmark Papers in Optics. He must consider the classical fields, so rich in history and so necessary for the understanding of all optical phenomena, and he must bring to the scientific audience current information on the burgeoning areas of applied optics. And what should be done in the several inter- or multidisciplinary fields that include optics?

For all subjects, the classical or benchmark papers must be included, be they a century or only a decade old, and in English or another language. And the papers must be placed in proper perspective by the commentaries written by knowledgeable and skillful volume editors. Thus the impact of the book will be satisfyingly greater than the sum of the original papers that are presented in facsimile. These volumes should be of value to scholars, teachers, applied physicists, engineers, and practical opticists. All this constitutes a courageous goal; we can only hope that it is being at least partially achieved.

This *Infrared Detectors* volume can be classified in the modern optics category. To be sure, the origin of infrared detectors dates back to 1800, but their modern realization is entwined intimately with the contemporary field of solid-state physics. The volume editors, Richard and Jacqueline Hudson, are to be commended for the thoroughness of their literature search, their wisdom in selecting the 38 papers to present, and the cogent and illuminating nature of their introductory remarks and their commentaries on the papers. The many bibliographic references, carefully classified, represent a major contribution in themselves. I believe that this book will prove to be most useful in enhancing our understanding of the science and art of the detection of radiant energy at infrared wavelengths.

Stanley S. Ballard

Preface

This book was written for those who use, produce, or simply wish to know more about infrared detectors. It contains a short history of the development of elemental (not imaging) detectors, a collection of 43 significant papers especially selected for this volume, explanatory commentary on each paper, and an extensive bibliography. We hope that it will prove to be a useful day-to-day reference for all workers in the field.

Our experience with detectors began with the World War II use of the infrared sniperscope and extends to the application of the newest tri-metal detectors. Our specialty is the design of infrared systems, and one of our stocks in trade is a continuously updated working knowledge of detectors. Since we are happiest when up to our elbows in hardware, we feel a deep kinship with those modern-day sorcerers who produce our all-important detectors. This, then, is our background and the background against which the book was written.

How does one select papers for a book like this from among the thousands that have been published? As in any field, only a few stand the test of time—those papers that we, and our colleagues in the detector-user's community, keep close at hand and refer to time and time again. They also are the papers that we have collected together in this volume. We have written a short history of detector development to introduce the collected papers. We think both old-timers and newcomers will enjoy this historical summary and the perspective it provides. We have grouped the papers in a way that seems logical to us; survey and tutorial papers first, followed by papers on specific types of detectors, and, finally, papers on cryogenics for detectors. We have written short commentaries about each paper. We describe what the paper contains, its significance, its relationship to other papers in this volume and, in some cases, offer a commendation to an author whom we feel to be particularly deserving. The experienced worker will find many of his favorite papers collected together for the first time. He may also find a classic or two that he has not yet encountered. Those who wish to learn more about detectors will find this an unusual tutorial text, because it offers the unique opportunity to learn directly from those most responsible for the advanced state of today's detector art.

The book is a particularly rich source of references on infrared detectors. The 43 papers contain nearly 800 citations to the literature. In our introduction and commentaries we have added nearly 300 more citations carefully selected to bring each paper up to date and further enhance its usefulness. In addition, there is, at the end of the introduction, a selected reading list of books devoted entirely, or in part, to infrared detectors. These references come from all over the world, and virtually every country with publications on infrared detectors is represented.

Good references are important to us but we feel even more strongly about their availability. We believe that citation of the *unpublished literature* (i.e., contractor's reports, internal company memoranda, classified reports, etc.) has no place in a book of this type. Our references are all in the *published literature,* and the reader can obtain them through public libraries.

Any book reflects the sum total of its authors' experiences. We are indebted to all who have contributed to these experiences. We regret that in the limited space available only a few key individuals can be named. We are deeply grateful to Stanley S. Ballard, the series editor and long-time "Mr. Infrared" in the United States, for his keen interest in our efforts, invaluable advice, and strong support; to Warren E. Mathews, who started the senior author on his writing career and continues to provide encouragement for both of us; to the authors who were kind enough to grant permission for reproduction of their papers and who, in many cases, took the time to offer valuable advice and suggestions; to A. S. Jerrems for his understanding, encouragement, and unflagging support of our writing efforts; to Betty Mathewes for her introduction, patient support, and capable handling of our correspondence; and to the Hughes Aircraft Company, and its subsidiary The Santa Barbara Research Center, for support of our efforts and the provision of a multitude of services that we might not otherwise have enjoyed.

Richard D. Hudson, Jr.
Jacqueline Wordsworth Hudson

Contents

I. CHARACTERISTICS OF CURRENTLY AVAILABLE INFRARED DETECTORS

II. FUNDAMENTALS OF INFRARED DETECTION

III. PHOTON DETECTORS

IV. THERMAL DETECTORS

V. ULTIMATE LIMITS ON THE PERFORMANCE OF DETECTORS

VI. TECHNIQUES FOR COOLING INFRARED DETECTORS

Contents by Author

Introduction

It all began with a thermometer!

> Thermometer No. 1 rose 7 degrees in 10 minutes by an exposure to
> the full red coloured rays. I drew back the stand . . . thermometer No. 1
> rose, in 16 minutes, $8\frac{3}{4}$ degrees when its centre was $\frac{1}{2}$ inch out of the visible
> rays.[1]

Thus wrote the astronomer Sir William Herschel in April 1800 as he announced
the discovery of what we now call the infrared portion of the spectrum. He was
looking for a better filter to protect his eyes while observing the sun. His first step was
to build a crude monochromator that used a thermometer as a detector so that he
could measure the distribution of energy in sunlight. His monochromator consisted
of a prism mounted in a hole in a window shade and tilted so that the resulting
spectrum fell on a table. Three thermometers, each on a stand, were arranged on the
table so that the bulb of one could be set to any desired point in the spectrum while
the other two, set off to the side, acted as controls. The temperature rise of the
exposed thermometer was proportional to the energy at that point in the spectrum.
The control thermometers made it possible to correct for changes in the ambient
temperature of the room during the many minutes required for an observation.
Later, Herschel used the same arrangement to measure the spectral transmission of
filters by noting the temperature rise at each point in the spectrum both with and
without the filter in front of the thermometer. It was not until four years later that
Young made the first measurements of the wavelength of light. Herschel divided the
visible spectrum into seven bands, designating each with the name of a common
color. Correspondingly, he divided the region beyond the red into six more bands,
designated simply by letters. We now know that the peak he found in the solar energy
distribution lies between 0.85 and 0.9 μm, a result explainable by the dispersion
characteristics of his prism and the fact that the bandpass of his monochromator/
detector varied with wavelength. Although others had used a thermometer to probe
a spectrum, Herschel was the first to have the curiosity to explore the region beyond
the red end of the spectrum. To him goes the credit for the discovery of the infrared
and to his thermometer goes credit for being the first infrared detector.

Herschel was born in 1738 in Hannover, Prussia. He was trained as a musician
and earned his living from music for many years. In 1773 he became interested in
astronomy and soon learned to make his own telescope mirrors because he could not
afford the commercially available ones and he was dissatisfied with their optical

quality. Within a few years he was making the finest mirrors in England. In 1780 he assured his place in history when, using one of his own telescopes, he discovered the planet Uranus. Two years later he was appointed Royal Astronomer to King George III. This appointment gave him some financial independence and the chance to devote his full energies to astronomy. He was a tireless observer and is generally regarded as one of the finest observational astronomers of all time. During 1800 Herschel published three papers [1–3] pertaining to his investigations of the spectrum [4, 5] but after finding his eye-protective filter (ink-tinted water in a small glass cell), he did no further work in the infrared. [1]

The thermometer that was used by Herschel has not survived to this day, so we must look to his papers for whatever description there may be of the first infrared detector. The thermometer, which Herschel referred to as Wilson's No. 1, was lent to him by Alexander Wilson, Professor of Practical Astronomy at the University of Glasgow. Where, and for what purpose, Wilson acquired the thermometer remains unknown. Since Wilson died in 1786, the thermometer must have been in Herschel's possession for quite some time. Wilson's No. 1 was a mercury-in-glass type with a ball of $\frac{1}{8}$-inch diameter that had been carefully blackened with lampblack. Herschel describes it as being of "exquisite sensibility" because of its small ball. It was fitted with a small ivory scale "properly disengaged from the ball." Its scale extended "only to 100" (probably degrees Fahrenheit). Herschel usually recorded its indications to $\frac{1}{2}$ degree and occasionally to $\frac{1}{8}$ degree. At times Herschel waited as long as 16 minutes before taking a reading. Unfortunately, this is all that we know about the first infrared detector.

The thermometer was the first of a trio of detectors that were to dominate the infrared detector field until the time of World War I. It remained unchallenged until the 1830s. Many sought to improve it, usually by developing smaller, faster-acting devices that were often equipped with scale-reading microscopes. Sir John Leslie, who was noted for his researches in the theory of heat radiation, developed a detector in 1818 that used the principle of the differential gas thermometer [6]. A similar principle was applied again in 1945 in the development of the Golay cell.

Development of the Thermopile

In retrospect, the end was in sight for the thermometer when Seebeck discovered the thermoelectric effect in 1821 and soon thereafter demonstrated the first thermocouple. A thermocouple consists of a pair of junctions between two dissimilar metals. If one of the junctions is warmed by, for instance, incident radiation, an electromotive force is established and a current will flow in an external circuit. Until about 1930 it was customary to measure the current with a sensitive galvanometer.

---

[1] Herschel called the new region the "thermometrical spectrum," the "invisible rays," and the "dark heat." He did not use the term infrared and it does not appear in the literature until the 1880s. The term does not seem to have been introduced by any particular person. It probably

In 1829 Nobili constructed the first thermopile by connecting a number of thermocouples in series. Macedonio Melloni, Professor and Director of the Institute of Physics at the University of Parma, recognized the significance of Nobili's development. In 1833 he helped him modify the design so as to make the thermopile suitable for measurements by radiation rather than by contact. The result, called the "thermo-multiplicateur" or, simply, Melloni's thermopile, used an array of bismuth and antimony blocks assembled so as to form a cube that was about 1 cm on a side. The blocks were connected so that the 38 pairs of junctions lay in opposite faces of the cube. Since the blocks were not inherently good absorbers of radiation, the junctions of one face were coated with lampblack. When this face was exposed to radiation, its junctions became the hot junctions and an emf was established between these and the cold junctions of the opposite face. The output of the thermopile was measured by a two-needle "multiplier," an early form of the astatic galvanometer. Melloni characterized his thermopile as being 40 times more sensitive and much quicker responding than the best available thermometers. He reported that it could detect the heat from a man at a distance of 25 to 30 ft [5, 7].

Melloni made detailed studies of the transparency of optical materials. He showed that conventional optical glass was of little value in the infrared since it transmitted only to about 2.5 μm. His most important discovery was that rock salt is remarkably transparent throughout much of the infrared. He made lenses and prisms of rock salt and used them in developing techniques that ultimately formed the basis for modern analytical infrared spectroscopy. For some of these studies he made long, narrow thermopiles from linear arrays of junctions. He did not seem to pursue this work, and the detectors were not particularly good. He made quantitative measurements of solar radiation and, from these, he noted day-to-day variations that he correctly attributed to the absorption by varying quantities of water vapor in the earth's atmosphere. He even proposed the word "thermochrose" for thermal radiation. Much of his work, and very complete descriptions of his detectors, appear in his book that was published in 1850 [8]. For nearly half a century his thermopile was the most sensitive detector available, clearly earning its place as the second member of the trio of outstanding detectors during the first 120 years of the infrared.

Development of the Bolometer

The third member of the trio, Langley's bolometer, appeared in 1880. Samuel Langley of the Allegheny Observatory and, later, the Smithsonian Institution, devoted most of his professional life to a study of the energy received from the sun and

---

appeared simply because of its obvious latin root. *Infra-* is a latin prefix meaning below or beneath, so the infrared region is the region below the red. The latin root is evident in the equivalent word in other languages, for example, *infrarouge* in French, *infrarot* in German, and *infrakrasnye* in Russian. Most dictionaries recognize infrared as an adjective, but workers in the field quite commonly use it also as a noun. In U.S. usage, the term has undergone the classical evolution from *infra red*, to *infra-red*, and finally to *infrared*.

its effects on the temperature of the earth [9]. At the time he began this work it was generally believed that the earth's atmosphere absorbed uniformly at all wavelengths, just as it appeared to do in the visible region. Langley's findings led him to question this assumption. He set out to measure the atmospheric transmission using the method, originated by Melloni, in which the sun serves as a source conveniently located outside of the atmosphere. Melloni had measured only the total transmission of the atmosphere, whereas Langley had to measure its variation with wavelength. Such spectral measurements required that the solar radiation be dispersed by a prism or diffraction grating so that the desired spectral interval could be selected by a narrow slit or an equally narrow detector. The thermopile had neither the right shape nor sufficient sensitivity for this application.

Langley's search for a more suitable detector led to his invention of the bolometer, a detector that makes use of the change in electrical resistance of certain materials when their temperature is changed. For his first bolometers Langley used two thin ribbons of platinum foil, connected so as to form two arms of a Wheatstone bridge. The strips, blackened on one side, were arranged so that one could be exposed to the radiation while the other was shielded from it. The radiation warmed the strip and caused a small change in its resistance. The change in resistance was sensed by applying a constant bias voltage across the bridge and carefully balancing it. Radiation incident on the sensitive strip changed its resistance, unbalanced the bridge, and caused a deflection of the galvanometer used to indicate bridge balance. The minimum detectable change in the temperature of the strip was 10^{-5} to 10^{-6}°C, even though the bias current flowing through the strip raised its temperature to about 25°C above the ambient. The first strips were about 1mm wide and 10 mm long and the bolometers incorporating them were about 15 times more sensitive than the best contemporary thermopiles, which had been little improved since their original development by Melloni [5, 10]. Using one of these bolometers with a grating spectrometer, Langley mapped the solar spectrum, as modified by its passage through the earth's atmosphere, out to a wavelength of 5.3 μm. He correctly identified the atmospheric constituents responsible for the observed absorption bands and recognized the importance of the major transparent regions (transmission windows) that he found at 2 to 2.5 μm and at 3 to 5 μm.

Langley continued to develop his bolometer for the next 20 years. He learned to make sensitive strips that were as narrow as 0.05 mm and he devised more efficient means of blackening them. By 1900 his bolometers were 400 times more sensitive than his first efforts. In 1901, Langley reported that the minimum detectable temperature rise in his sensitive strips was 10^{-8}°C and that his latest bolometer could detect the heat from a cow at a distance of $\frac{1}{4}$ mile [9].

It is interesting to note that one other detector from this period continues to be sold to this day. Its total sales, based on the number of units sold, probably exceed that of all other detectors combined. This detector, the radiometer, is based on the discovery by Crookes in 1876 that a set of very light vanes mounted in a partially evacuated enclosure could be set in motion by an incident beam of radiation. In 1897 Nichols refined the idea into a reliable detector, but its fragile nature limited its usefulness. In Nichols' version the vanes carried a small mirror that reflected a spot of light onto a distant scale so as to indicate the angular rotation of the vanes. In the "radiation detector" seen today, there is no limitation on the rotation of the vanes and it is a common sight to see them spinning feverishly in the display windows of novelty shops.

Not all workers needed, or could use, the long, narrow shape of the sensitive element of the bolometer. But the dramatic increase in sensitivity that it offered served to stimulate many new developments in thermopiles and thermocouples. These developments included new materials with more favorable thermoelectric properties and clever techniques for fashioning junctions with greatly reduced thermal capacities. It was recognized that the observed superiority of the thermopile over the thermocouple had no theoretical basis but was, instead, caused by shortcomings of the available galvanometers. As a result, some of the developmental efforts were directed at improving the sensitivity of galvanometers while others sought ways to increase the resistance of thermocouples so that they would offer a better electrical match to the galvanometer. The radiomicrometer, a detector that enjoyed some popularity around the turn of the century, featured an ingenious combination of a galvanometer and a thermocouple built into the same instrument. The result of all of these efforts was that, by 1910, thermopiles had been largely replaced by thermocouples and there was little difference in either sensitivity or speed of response between thermocouples and bolometers. The ultimate performance of both detectors was limited by the galvanometers used to indicate their responses. Finally, in 1925, Moll and Burger developed a way to amplify the response of a galvanometer by a factor of 100. It used an arrangement of two galvanometers in which the spot of light reflected from the first galvanometer mirror fell on a split thermopile whose output was fed to the second galvanometer. This arrangement showed, for the first time, that there were fluctuations in the output of detectors that could not be eliminated by further improvements. We now know that this was the first observation of the noise inherent in all detectors. It is this noise that places an ultimate limit on the performance of detectors.

Development of the Photon Detector

The detectors discussed thus far depend on the heating affect of the incident radiation. In 1917, Theodore W. Case introduced the first photon detector that was sensitive in the infrared. This detector, thallous sulfide, was a fundamentally new type in which the incident photons interacted directly with the electronic structure of the detector. Heating played no part in its operation. Although its response extended only to about 1.4 μm, it was the forerunner of major developments that were to occur during and after World War II which were to result in photon detectors usable throughout the entire infrared region, with peak sensitivities and response times superior to those of thermal detectors.

The modern concept of the solid divides it into two thermodynamic systems, the lattice and the electronic. The way in which the incident energy interacts with these systems gives rise to two fundamentally different detection systems, the thermal and the photon. Radiation incident on a thermal detector is absorbed by an appropriate blackening and heats the lattice. This, in turn, affects the electronic system, and, for example, results in a change in the electrical resistance of a bolometer or the generation of a thermal emf in a thermocouple. By contrast, when radiation is incident on a photon detector, the photons interact directly with the electronic system to produce a photoeffect. Thus, the response of a thermal detector is proportional to the energy absorbed while that of a photon detector is proportional to the number of photons absorbed. Because of the direct interaction between the photons and the electronic structure there is no need to blacken photon detectors and their response times are orders of magnitude shorter than those of thermal detectors.

A photoeffect is the result of an interaction between photons and matter. Many photoeffects are known, but we will be concerned only with those that have proved useful in the infrared. The reader who wishes a more detailed coverage of the subject should consult Reference 11. As early as 1890 Elster and Geitel demonstrated a photoemissive detector that consisted of an alkali-metal cathode and a collecting plate, both placed in an evacuated enclosure. When an incident photon was absorbed by an electron in the cathode, the electron might gain enough energy to escape from the surface and thence be drawn over to the plate. Since the electron actually left the cathode this is classed as an external photoeffect. Because the energy of a photon varies inversely with wavelength (or directly with frequency) there is a long-wavelength limit beyond which the energy of the photon is less than that needed by the electron to escape from the surface. The existence of a long-wavelength cutoff is characteristic of all photon detectors and provides another distinctive difference between them and thermal detectors. This long-wavelength cutoff occurs at about $1.25 \ \mu$m for known photoemissive materials, so photoemissive detectors can be used only in the very near infrared region.

Fortunately, there are numerous internal photoeffects that require less energy so that detectors utilizing them have their cutoffs at much longer wavelengths. In these effects, no electrons leave the surface of the detector. Instead, the energy transferred from the photon raises an electron from a nonconducting to a conducting state and, in so doing, produces a free charge carrier. Nearly all photon detectors are made from semiconductor materials. In an intrinsic, or pure, semiconductor the incident photon produces an electron-hole pair, consisting of the electron raised to the conducting state and the hole at the site vacated by the electron. In an extrinsic, or impurity-activated, semiconductor the photons produce charge carriers of a single sign, that is, either electrons or holes but not both. In either case, biasing the detector by placing a voltage across it creates an electric field and causes the free charge carriers to be drawn to one or the other electrode. Variations in the number of incident photons modulate the production of charge carriers, and the resulting changes in bias current can be amplified and used to provide a measure of the response of the detector to the incident radiation. This is called the photoconductive effect and it is, as can be seen, a photon-induced variation of the conductivity of the detector.

Another photoeffect can occur in a semiconductor that contains a *p-n* junction. Here the incident photon produces an electron-hole pair in the vicinity of the junction. The strong electric field across the junction separates the two carriers to yield a photovoltage. This is called the photovoltaic effect. Detectors making use of the photovoltaic effect have the advantage that they require no external bias supply since it is, in effect, already built into the *p-n* junction.

The hot electron effect is useful at very long wavelengths. Here, very long wavelength photons are absorbed directly by the free carriers in a semiconductor. At room temperature no change will be observed. But, if the semiconductor is cooled sufficiently, this absorption causes a change in the energy distribution of the electrons within the conduction band which, in turn, changes the mobility of the electrons. The result is a change in the conductivity produced by a change in carrier mobility rather than by the change in carrier concentration that is observed at shorter wavelengths.

Since considerably less energy is required to produce the various internal photo-effects than is required for photoemissivity, detectors using internal photoeffects respond much farther into the infrared before reaching their long-wavelength cutoff. Intrinsic photoconductive detectors respond out to about 7 μm while extrinsic types respond to beyond 100 μm and the hot electron types respond to beyond 1000 μm. But such performance brings with it a new complication—the requirement for cooling the detector to cryogenic temperatures. In detectors able to respond to relatively small energies associated with the long-wavelength photons, there is a high probability that free charge carriers will also be created by thermal excitation from vibrations of the crystalline lattice. Such thermally induced charge carriers degrade the performance of a detector since they provide an undesired noise in the output of the detector that makes it more difficult, or even impossible, to detect the photon-induced carriers. By cooling the detector, the number of charge carriers produced by lattice vibrations can be made negligible. As a rough rule of thumb, no cooling is required for photon detectors that do not respond beyond 3 μm (although their performance can be improved by cooling); those responding from 3 to 8 μm require moderate cooling (to 77°K, the temperature of liquid nitrogen); those responding beyond 8 μm require even lower temperatures, which may approach within a few degrees of absolute zero. One of the exciting recent developments is the tri-metal detector, such as mercury cadmium telluride or lead tin telluride, which has much less stringent cooling requirements.

In 1917, Case demonstrated to the U.S. military a transmitter and receiver for blinker signaling that used a thallous sulfide detector [12]. The response time of the thallous sulfide detector was much shorter than those of the existing thermal de-tectors and it made feasible the blinker system that is generally considered to be the first military application of infrared [13]. The response of the thallous sulfide detec-tor extended to about 1.4 μm. Unfortunately, these detectors were neither stable nor reproducible and little progress was made in improving them until World War II.

In 1939, on the eve of World War II, Robert J. Cashman, a professor of physics at Northwestern University, was the only individual in the United States working on photoconductivity [14], an astounding observation in view of the present-day importance of photoconductive detectors. In November 1941, Cashman received a contract from the Office of Scientific Research and Development under which he was to develop techniques that would lead to the successful production of thallous sulfide detectors. By 1943 he had produced nearly 800 detectors. Thus verified, his processes were released for commercial production and almost 7000 detectors were produced by the end of 1944. Cashman has noted that his thallous sulfide detectors showed a peak response at 0.95 μm, a long-wavelength cutoff at 1.45 μm, and a peak sensitivity that was 1000 times more than that of a thermocouple detector [15]. Wartime efforts in Germany also led to the successful production of thallous sulfide detectors, but the details were not generally known until after the end of the war.

In 1933, Edgar W. Kutzscher at the University of Berlin, working under the sponsorship of the German Army, discovered that lead sulfide was photoconductive and had a useful response to about 3 μm [16, 17]. This work was, of course, done under great secrecy and the results were not generally known until after 1945. Cashman, after his success with thallous sulfide, began to investigate other materials and by 1944 was concentrating his efforts on lead sulfide. In early 1945, Cashman's lead sulfide detectors were equal to, or better than, captured German ones. Cashman, in the first declassified report on his lead sulfide detectors, reported that they had a peak response at 2.5 μm, a long-wavelength cutoff at about 3.6 μm, and a sensitivity at the peak that was 1000 times greater than that of the best thermocouple detectors [15]. During this period, the Germans also investigated the effect of cooling lead sulfide detectors. They developed the first coolable packages for detectors and cooled them to a temperature of 195°K with a slug of solidified carbon dioxide or "dry ice." Their lead sulfide detectors were 100 times more sensitive when cooled than they were at room temperature [18].

During the early postwar years Cashman continued his work with lead sulfide and also found that other members of the lead-salt family showed promise as infrared detectors. Papers 11 through 13 pick up the story at this point and trace further developments. Other photoconductive and photovoltaic detectors of the intrinsic type are discussed in Papers 14 through 19, while the extrinsic types are covered in Papers 20 through 22.

Efforts to improve thermal detectors continued during the wartime years, and postwar revelations indicated several significant developments. In particular, there was a great deal of progress made in understanding the properties of thermoelectric materials and the conditions for their optimum utilization. Much of this work is summarized in Paper 24, which also describes thermocouple detectors having response times of 50 msec. For their first hundred years, thermocouples were made from bulk materials, usually in the form of wires or ribbons that were drawn or rolled very thin. In 1934, Harris and Johnson developed techniques for making thermocouples by sputtering thin metallic films [19]. Paper 25 describes a modern-day version of this technique that uses vacuum evaporation to produce the films. Langley's wire bolometer was replaced by the thermistor bolometer that was developed at Bell Telephone Laboratories [20]. In the thermistor bolometer the sensitive metal strips were replaced by thin flakes of thermistor (*therm*ally sensitive res*istor*) material made by sintering oxides of manganese, nickel, and cobalt. Their properties are described in Paper 26. To the thermistor bolometer goes the honor of being the first infrared detector to be used in space [21].

Thermal detectors have a long history as uncooled devices but, strangely enough, the most sensitive thermal detectors available today are cooled to cryogenic temperatures. The first efforts in this direction were made in 1946 with a bolometer that used superconductivity in columbium nitride cooled to about 15°K. This and further work on other cooled bolometers is described in Papers 30 through 33. Thermocouples remain the only major class of detectors that do not benefit from cooling. Unfortunately, the thermoelectric properties of usable materials decrease at low temperatures and this more than offsets any other possible gains from cooling.

Although infrared devices had only limited deployment on the battlefields of World War II, they showed sufficient merit to justify a strong postwar development program that was supported principally by military funding [22]. A by-product of this effort was the appearance of papers stressing information on engineering applications for the detector user. Some of the more outstanding appear in Papers 3 through 9. Here will be found careful descriptions of detector terminology, the physics of the detection process, methods of measuring and interpreting detector performance parameters, and the very important recent application of heterodyne techniques to the infrared.

What are the factors that ultimately limit the performance of a detector? Scattered throughout our historical summary we have noted a number of comparative statements: Melloni's thermopile was 40 times more sensitive than the best thermometer and it could detect the heat from a man at a distance of 25 or 30 feet. Langley's bolometer could detect the heat from a cow at $\frac{1}{4}$ mile. Cashman's lead sulfide detector had a peak sensitivity 1000 times greater than that of thermocouple detectors, etc. These are interesting comparisons but they provide very little information about what is ultimately achievable. In three classic papers (35 through 37) R. Clark Jones provided clear answers and, at the same time, brought badly needed order to the methods used to specify the performance of a detector.

Most modern detectors need cooling to very low temperatures. The cryogenic community, spurred by the demands of system designers in the early 1960s and supported by military funding, developed a broad range of small, reliable (but not necessarily inexpensive) detector cooling systems, some of which are described in Papers 40 through 43. Thanks to these efforts, today's system designers need not hesitate to specify the use of a cooled detector in any earth-bound, ship-, or space-borne application [23–26].

For a definitive survey of the characteristics of the infrared detectors that are available today, see Papers 1 and 2. In addition, they offer a strong assist to anyone faced with the problem of selecting the best detector for a particular application. The reader should be warned that the performance of modern-day infrared detectors is excellent and there is relatively little room left for improvement in their performance. A glance at Figure 6 in Paper 1 or Figure 1 in Paper 2 shows how close today's detectors are to the ideal photon limit—many are within a factor of 2 of the limit and virtually all are within a factor of 10. This is a happy situation for all detector users and a tribute to those who develop and produce them (many of whom are represented by papers in this volume). Remember this happy situation because from time to time you may see or hear widespread publicity announcing supersensitive detectors that are said to be thousands of times more sensitive than those previously available [27].

Prospects for the Future

What trends do we see for future developments in infrared detectors? It seems to us that there are at least four trends already evident that should have profound influence on the direction of detector developments for, perhaps, the remainder of this century. These trends are:

1. The use of heterodyne and other sophisticated signal processing techniques made possible by the development of the infrared laser.

2. The expanding system use of the background-limited (Blip) detector.

3. The growing demand for multielement arrays of detectors.

4. The continuing development of tri-metal detectors.

Lasers provide high radiance sources, but of perhaps greater import is the fact that they provide the first sources of coherent radiation available in the infrared. Coherent sources make it possible, for the first time, for the infrareder to use the same heterodyne and homodyne techniques that have long been used by the microwaver. Golay in Paper 34 points out that heterodyne techniques and coherent sources offer an increase of eight orders of magnitude in the minimum detectable signal. In Paper 7, Teich describes the theory of infrared heterodyne detection and details experimental measurements verifying this gain. The possibilities for wideband communication systems that are promised by these techniques seem almost limitless at this time.

It is remarkable that the noise in the output of a Blip detector, as shown in Paper 4, is due to fluctuations in the rate at which charge carriers are generated by photons from the detector's external environment rather than by internally generated noises. For system use Blip detectors are equipped with a cooled shield that has an aperture matching the field of view of the detector to the cone of rays coming from the system's optics. Under these conditions, the maximum detection range of the system depends upon the diameter of the optics but, strangely enough, not on their speed. Blip detectors have already improved the performance of many terrestrial-based systems, but their real potential will be realized in space-borne systems. Engineering details of system design with Blip detectors are given in References 28 and 29.

One of the most important changes evident in the current infrared system scene is the trend toward the use of multielement arrays of detectors. The theory of infrared systems shows that the maximum detection range is proportional to the fourth root of the number of individual elements in the detector [30]. Since most of the components comprising modern infrared systems are in a relatively advanced state of development, the least expensive way to gain additional detection range is to use a multielement detector array. Manufacturers of detectors have made remarkable advances in the techniques for fabricating detector arrays and in their reliability. Arrays containing 50 to 100 elements are within the capability of most manufacturers and arrays containing from 1000 to 10,000 elements appear to be feasible. For additional information on developments in array technology see Papers 13, 16, 23, and 25 as well as References 31 through 35.

The tri-metal detectors, such as mercury–cadmium–telluride or lead–tin–telluride (see Papers 1, 2, 18, and 19), provide excellent performance in the important 8 to 13 μm atmospheric window and they offer the further advantage that they have much less stringent cooling requirements. Other photon detectors sensitive to these wavelengths generally require cooling to below 20°K, whereas the tri-metal detectors operate quite satisfactorily at 77°K. In a system application it may cost twice as much to cool a detector to 20°K as it does to cool it to only 77°K. As a result, tri-metal detectors are gaining rapid acceptance by cost-conscious system designers. The theory of the Blip detector shows that it is advantageous to limit its long-wavelength cutoff. This can be done very simply with tri-metal detectors by varying the relative proportions of their constituents. It is not yet clear whether mercury–cadmium–telluride or lead–tin–telluride will emerge as the favored detector or whether some other material will be found that is better than either. Regardless of the outcome, it seems certain that tri-metal detectors will play an increasingly important role in infrared systems of the future.

References

1. Herschel, W., "Experiments on the Refrangibility of the invisible Rays of the Sun," *Phil. Trans. Roy. Soc. London,* **90,** 284 (1800).
2. Herschel, W., "Investigation of the Powers of the prismatic Colours to heat and illuminate Objects; with Remarks, that prove the different Refrangibility of radiant Heat. To which is added, an Inquiry into the Method of viewing the Sun advantageously, with Telescopes of large Apertures and high magnifying Powers," *Phil. Trans. Roy. Soc. London,* **90,** 255 (1800).
3. Herschel, W., "Experiments on the solar, and on the terrestrial Rays that occasion Heat; with a comparative View of the Laws to which Light and Heat, or rather the Rays which occasion them, are subject, in order to determine whether they are the same, or different," *Phil. Trans. Roy. Soc. London,* **90,** 293, 437 (1800).
4. Barr, E. S., "The Infrared Pioneers—I. Sir William Herschel," *Infrared Phys.,* **1,** 1 (1961).
5. Barr, E. S., "Historical Survey of the Early Development of the Infrared Spectral Region," *Amer. J. Phys.,* **28,** 42 (1960).
6. Leslie, J., "A Further Experimental Inquiry into the Nature and Propagation of Heat," *Trans. Roy. Soc. Edinburgh,* **8,** 465 (1818).
7. Barr, E. S., "The Infrared Pioneers—II. Macedonio Melloni," *Infrared Phys.,* **2,** 67 (1962).

8. Melloni, M., *La Thermochrose ou la coloration calorifique*, Baron, Naples (1850). Reprinted in a facsimile edition in 1954 by Nicola Zanichelli Editore, Bologna.
9. Barr, E. S., "The Infrared Pioneers—III. Samuel Pierpont Langley," *Infrared Phys.*, **3**, 195 (1963).
10. Langley, S. P., "Researches on Solar Heat," *Proc. Amer. Acad. Arts Sci.*, **16**, 342 (1881).
11. Hudson, R. D., Jr., *Infrared System Engineering*, Wiley, New York (1969), Chap. 7.
12. Case, T. W., "Thalofide Cell—a New Photoelectric Substance," *Phys. Rev.*, **15**, 289 (1920).
13. Arnquist, W. N., "Survey of Early Infrared Developments," *Proc. Inst. Radio Engrs.*, **47**, 1420 (1959).
14. Lovell, D. J., "Cashman Thallous Sulfide Cell," *Appl. Optics*, **10**, 1003 (1971).
15. Cashman, R. J., "New Photo-Conductive Cells," *J. Opt. Soc. Amer.*, **36**, 356 (1946).
16. Lovell, D. J., "The Development of Lead Salt Detectors," *Amer. J. Phys.*, **37**, 467 (1969).
17. Kutzscher, E. W., "Letter to the Editor," *Electro-Opt, Syst. Design*, **5**, 62 (June, 1973).
18. Oxley, C. L., "Characteristics of Cooled Lead Sulfide Photo-Conductive Cells," *J. Opt. Soc. Amer.*, **36**, 356 (1946).
19. Harris, L., and E. A. Johnson, "The Technique of Sputtering Sensitive Thermocouples," *Rev. Sci. Instr.*, **5**, 153 (1934).
20. Brattain, W. H., and J. A. Becker, "Thermistor Bolometers," *J. Opt. Soc. Amer.*, **36**, 354 (1946).
21. Hudson, *op. cit.* (ref. 11), pp. 558–561.
22. Hudson, *op. cit.* (ref. 11), Chap. 1.
23. "Detector Cooling Systems," Chap. 12 in Wolfe, W. L. (ed.), *Handbook of Military Infrared Technology*, Office of Naval Research, Department of the Navy, Washington, D.C. (1965).
24. Antonov, Ye. I., et al., *Ustroystva dlya Okhlazhdeniya Priyemnikov Izlucheniya (Radiation Sensor Cooling Devices)*, Mashinostroyeniye Press, Leningrad (1969).
25. Hudson, *op. cit.* (ref. 11), Chap. 11.
26. Hogan, W. H., and T. S. Moss, *Cryogenics and Infrared Detection*, Boston Technical Publishers, Inc., Cambridge, Mass. (1970).
27. Hudson, *op. cit.* (ref. 11), p. 289.
28. Swift, H., Performance of Background-Limited Systems for Space Use," *Infrared Phys.*, **2**, 19 (1962).
29. Hudson, *op. cit.* (ref. 11), pp. 421ff.
30. Hudson, *op. cit.* (ref. 11) p. 425.
31. Lauriente, M., et al. "Sophisticated Detector Design for Increased Infrared Sensitivity," *Infrared Phys.*, **2**, 103 (1962).
32. Johnson, T. H., "Detector Array Technology," *Infrared Phys.*, **5**, 1 (1965).
33. Turon, P. J., and D. Stefanovitch, "High Resolution Solar Imaging at 1.65 μ with a 64-Element Array," *Appl. Optics*, **11**, 2177 (1972).
34. Corsi, C., and R. Tappa, "A New Technique for Infrared Single Crystal Detector Arrays," *Infrared Phys.*, **11**, 119 (1971).
35. Corsi, C., I. Alfieri, and G. Petrocco, "Infrared Detector Arrays by r-f Sputtering," *Infrared Phys.*, **12**, 271 (1972).

A Selected Reading List of Books on Infrared Detectors

This list is arranged in chronological order.

Forsythe, W. E., *Measurement of Radiant Energy*, McGraw-Hill, New York (1937).
Margolin, I. A., and N. P. Rumyanstev, *Fundamentals of Infrared Technology*, 2nd ed. Voenizdat, Moscow (1957).

Conn, G. K. T., and D. G. Avery, *Infrared Methods,* Academic Press, New York (1960).

Hackforth, H. L., *Infrared Radiation,* McGraw-Hill, New York (1960).

Holter, M. R., et al., *Fundamentals of Infrared Technology,* Macmillan, New York (1962).

Kruse, P. W., L. D. McGlauchlin, and R. B. McQuistan, *Elements of Infrared Technology,* Wiley, New York (1962).

Jamieson, J. A., et al., *Infrared Physics and Engineering,* McGraw-Hill, New York (1963).

Kriksunov, L. Z., and I. F. Usoltsev, *Infrakrasnyye Ustroystva Samonavedeniya Upravlyayemykh Snaryadov* (Infrared Equipment for Missile Homing), Voenizdat, Moscow (1963).

Limperis, T., "Detectors," Chap. 11 in Wolfe, W. L. (ed.), *Handbook of Military Infrared Technology,* Office of Naval Research, Department of the Navy, Washington, D.C. (1965).

Chol, G., et al., *Les Detecteurs de rayonnement infra-rouge,* Dunod, Paris (1966).

Houghton, J., and S. D. Smith, *Infra-Red Physics,* Oxford University Press, London (1966).

Hadni, A., *Essentials of Modern Physics Applied to the Study of the Infrared,* Pergamon Press, New York (1967).

Kosnitser, D. M., and Yu. S. Semendyaev (eds.), *Thermal Infrared Detectors,* Naukova Dumka, Kiev (1968).

Smith, R. A., F. E. Jones, and R. P. Chasmar, *The Detection and Measurement of Infra-Red Radiation,* 2nd ed., Oxford University Press (1968).

Vasko, A., *Infra-Red Radiation,* original Czech edition of 1963 translated into English in 1968, Chemical Rubber Co., Cleveland, Ohio (1968).

Hudson, R. D., Jr., *Infrared System Engineering,* Wiley, New York (1969).

Willardson, R. K., and A. C. Beer (eds.), *Semiconductors and Semimetals,* Vol. 5, *Infrared Detectors,* Academic Press, New York (1970).

Arams, F. R., *Infrared-to-Millimeter Wavelength Detectors,* Artech House, Dedham, Mass. (1973).

Moss, T. S., G. J. Burrell, and B. Ellis, *Semiconductor Opto-Electronics,* Halsted Press, New York (1973).

Robinson, L. C., *Physical Principles of Far-Infrared Radiation,* Academic Press, New York (1973).

Seyrafi, K., *Electro-Optical Systems Analysis,* Electro-Optical Research Co., Los Angeles (1973).

Wright, H. C., *Infrared Techniques,* Oxford University Press, London (1973).

Willardson, R. K., and A. C. Beer (eds.), *Semiconductors and Semimetals,* Vol. 10, *Infrared Detectors,* Part 2, Academic Press, New York (1975).

I
Characteristics of Currently Available Infrared Detectors

Editors' Comments on Papers 1 and 2

1 Putley: *Solid State Devices for Infra-Red Detection*

2 Putley: *Modern Infrared Detectors*

Anyone contemplating the selection of a specific infrared detector faces a bewildering choice from among the many types that are available. Which detectors cover the spectral range of interest? Which of them must be cooled? What is the penalty for less than optimum cooling? Are the detector response times short enough for the intended application? Are there any peculiarities of the detector that will affect the design of its associated circuitry? These questions, and many more, must be answered before a final choice can be made. When faced with a choice such as this, one looks for an authoritative review that summarizes and compares the important characteristics of the detectors that are available. Papers 1 and 2 were selected to fill this need and they were, in turn, placed at the front of the book so that it will be easy to refer to them when they are needed.

In Paper 1, Putley describes the state-of-the-art of infrared detectors in 1966. Paper 2, written as a companion to the first one, summarizes additional developments up to 1973. Taken together, they provide the most comprehensive and authoritative summary available. Putley outlines the basic physics underlying the various detection mechanisms, develops the theoretical limits on detector performance and indicates how closely real-life detectors approach these limits, provides tables and graphs summarizing the detector parameters that are of importance to the user, and speculates on the course of future developments. Paper 2 contains some very fine guidance on the factors to be considered in the choice of the best detector for a specific application, a topic that has also been treated in detail by Hudson [1].

The reader preparing to purchase detectors would do well to heed Putley's comments in the closing paragraph of Paper 2. "It must be remembered that not only do inventors and manufacturers tend to take a generally optimistic view of their products but also the spread in performance between different samples is often an order or more." Caveat emptor!

E. H. Putley is with the Royal Radar Establishment, Malvern, Worcestershire, England.

Summary papers, such as these two by Putley, have been an important part of the infrared scene since the end of World War II. One of the earliest attempts at such a summary was the session on the Detection of Infra-Red Radiation that was held on March 7, 1946, at the Cleveland meeting of the Optical Society of America. Most of the papers reported recently declassified information on wartime detector developments. Unfortunately, the papers were not subsequently published and only abstracts exist today [2]. Among the detectors that were described at this session are superconducting, metal, and thermistor bolometers, thallous sulfide and lead sulfide photoconductors, and, perhaps, the shadow of things to come in a report of a German lead sulfide detector cooled to the temperature of dry ice. From this modest list of just two photon detectors has sprung the more than 25 that are described by Putley.

16

Summary papers provide an important chronicle of progress in the development of infrared detectors. For this reason, the following bibliography contains an unusually complete list of references to summary papers [2–23]. The reader wishing to trace the development of any detector since World War II will find that these summaries contain all the information he needs. He should, in addition, consult one or more of the books in the reading list at the end of the Introduction.

References

1. Hudson, R. D., Jr., *Infrared System Engineering,* Wiley, New York (1969), pp. 358ff.
2. "Session on Detection of Infra-Red Radiation," *J. Opt. Soc. Amer.,* **36,** 353 (1946).
3. Cashman, R. J., "Photodetectors for Ultraviolet, Visible, and Infrared Radiation," *Proc. Natl. Electronics Conf.,* **2,** 171 (1946).
4. Williams, V. Z., "Infra-Red Instrumentation and Techniques," *Rev. Sci. Instr.,* **19,** 135 (1948).
5. Jones, R. C., "Performance of Detectors for Visible and Infrared Radiation," in Marton, L. (ed.), *Advances in Electronics,* Vol. 5, Academic Press, New York (1953).
6. Beyen, W., et al., "Cooled Photoconductive Infrared Detectors," *J. Opt. Soc. Amer.,* **49,** 686 (1959).
7. Jones, R. C., "Quantum Efficiency of Detectors for Visible and Infrared Radiation," in Marton, L. (ed.), *Advances in Electronics and Electron Physics,* Vol. 11, Academic Press, New York (1959).
8. Smith, R. A., "Semiconductors and Infra-Red Spectroscopy," *Optica Acta,* **7,** 137 (1960).
9. Jamieson, J. A., "Detectors for Infrared Systems," *Electronics,* **33,** 82 (Dec. 9, 1960).
10. Bratt, P., et al., "A Status Report on Infrared Detectors," *Infrared Phys.,* **1,** 27 (1961).
11. Kovit, B., "Infrared Detectors," *Space/Aeronautics,* **36,** 104 (Nov. 1961).
12. Potter, R. F., and W. L. Eisenman, "Infrared Photodetectors: A Review of Operational Detectors," *Appl. Optics,* **1,** 567 (1962).
13. Putley, E. H., "Far Infra-Red Photoconductivity," *Phys. Stat. Sol.,* **6,** 571 (1964).
14. Levinstein, H., "Infrared Detectors," Chap. 8 in Kingslake, R. (ed.), *Applied Optics and Optical Engineering,* Vol. 2, Academic Press, New York (1965).
15. Morton, G. A., "Infrared Detectors," *RCA Rev.,* **26,** 3 (Mar. 1965).
16. Smith, R. A., "Detectors for Ultraviolet, Visible, and Infrared Radiation," *Appl. Optics,* **4,** 631 (1965).
17. Smollett, M., "The Properties and Performance of Some Modern Infra-Red Radiation Detectors," *Infrared Phys.,* **8,** 3 (1968).
18. Shapiro, P., "Infrared Detector Chart Outlines Materials and Characteristics," *Electronics,* **42,** 91 (Jan. 20, 1969).
19. Putley, E. H., "New Infrared Detectors," Chap. 5 in Drummond, A. J. (ed.), *Advances in Geophysics,* Vol. 14, Precision Radiometry, Academic Press, New York (1970).
20. "Infrared Detection—Seeing the Invisible," *Electro-Opt. Syst. Design,* **2,** 34 (June 1970).
21. Potter, R. F., "Performance Characteristics of Modern Infrared Detectors," *J. Opt. Soc. Amer.,* **62,** 1352 (1972).
22. Kutzscher, E. W., "Review on Detectors of Infrared Radiation," *Electro-Opt. Syst. Design,* **6,** 30 (Jan. 1974).
23. Levinstein, H., and J. Mudar, "Infrared Detectors in Remote Sensing," *Proc. IEEE,* **63,** 6, (Jan. 1975).

$\mathbb{1}$

Reprinted from *J. Sci. Instr.*, **43**, 857–868 (Dec. 1966)

Solid state devices for infra-red detection

E. H. PUTLEY
Royal Radar Establishment, Malvern, Worcs.
MS. received 19*th May* 1966

Abstract. Infra-red detectors can be divided into two types: those which depend upon heating effects produced by the absorbed radiation (thermal detectors) and those which make use of photoconductive effects. The characteristics of the more important members of both types are discussed. The factors which determine their performance and which affect the choice of the most suitable type for a particular application are considered. Future trends in the development of infra-red detectors are briefly mentioned.

1. Introduction

The infra-red spectrum covers the range of wavelengths which are longer than the visible wavelengths (say $\lambda \geqslant 0 \cdot 75$) but which are shorter than microwave wavelengths. The long wave limit is therefore somewhat arbitrary, but it is often set at 1000 μm or 1 mm and this limit will be used here.

With the exception of the pneumatic detector, or Golay cell, all the important infra-red detectors are solid state devices. It is true of course that the principal detectors of visible light such as the photomultiplier and the photographic plate do respond to somewhat longer wavelengths than the eye itself, but even in the near infra-red the performance of these devices can usually be bettered by the types of detector discussed in this article. It is possible to classify the types of detectors in various ways. In this article the approach will be largely based on the mode of operation. Before we can choose the most suitable detector for a given application we must specify a number of things, such as the wavelength range of interest, the detectivity (or sensitivity) which is required, the response time and whether the application imposes any restriction on the operating conditions for the detector (such as operation at room temperature or ability to withstand industrial conditions). It is also found that the mode of operation falls into one of two main types. Firstly the 'thermal detectors'. The first observations of infra-red radiation were made by instruments which registered an increase in temperature following the absorption of the radiation. By utilizing some temperature-dependent property, such as resistivity or thermoelectric effect, thermal detectors have been developed which produce an electrical output when exposed to radiation. These detectors have several advantages and are widely used over the whole infra-red spectrum. One disadvantage of the thermal detectors is that their response time is long. This is overcome in the second main type of detector—the photoconductive detector. Here the absorbed radiation causes changes in the electronic distribution and hence a change of conductivity. Since the relaxation time for electronic processes is usually much shorter than that for thermal processes, these detectors are usually much faster than thermal ones. There are several photoconductive effects which can be employed, and between them they cover the whole of the infra-red spectrum.

The characteristics of examples of both types of detector will be described in the two main sections of this article, but before doing so, the factors limiting the detectivity of an ideal infra-red detector will be discussed.

1.1. *The Ideal infra-red detector*

Suppose the detector has a sensitive area A and it forms part of the wall of a large enclosure containing black-body radiation in thermal equilibrium at a temperature T. The detector will receive black-body radiation at a rate which will fluctuate about a mean value in a way which is a characteristic of the radiation and is independent of the detector itself (Lewis 1947). If at the same time one is trying to receive a signal produced by modulating a source at a frequency f, then the smallest signal that can be observed will be determined by that part of the frequency spectrum of the black-body radiation centred at f and extending over the bandwidth Δf of the detector's amplifier. In a real detector there will be other sources of noise, but consideration of the fluctuation in the incident black-body radiation will enable the performance of a perfect detector to be determined.

Lewis (1947) has shown that the mean square fluctuation in the power of the background radiation absorbed by an area A is

$$\overline{\Delta W^2} = \frac{4(kT)^5}{c^2h^3} \alpha \cos\theta A\Delta f \int_0^\infty \frac{x^4 e^x}{(e^x - 1)^2} f(x)dx \quad (1)$$

where we are considering radiation from an element of the background subtending a solid angle α at the detector in a direction making an angle θ with the normal to the detector; $x = hc/kT\lambda$; $f(x)$ is a function to take account of either any deviation of the radiation from black-body radiation or any departure from perfect blackness of the detector itself: for black-body radiation and a black detector $f(x) = 1$.

Putting $f(x) = \epsilon$, a constant independent of wavelength, and integrating (1) gives (for a field of view of 2π steradians)

$$\overline{\Delta W^2} = \frac{16(\pi kT)^5}{15c^2h^3} \epsilon A\Delta f$$

$$= 8\epsilon\sigma kT^5 A\Delta f \quad (2)$$

where σ is Stefan's constant. This expression gives the mean square fluctuation in the power absorbed by an area A of emissivity ϵ. If A is the sensitive area of a detector which is in thermal equilibrium with the radiation field, there will be an equal fluctuation in the power radiated by the detector. If we define the noise equivalent power P_N as the signal power producing an output from the detector equal to that produced by all the sources of fluctuation affecting the detector, then in this ideal case,

$$\epsilon P_N = (2\overline{\Delta W^2})^{1/2}$$

or $P_N = 3 \cdot 56 \times 10^{-17} T^{5/2} \epsilon^{-1/2} A^{1/2} \Delta f^{1/2}$ w. $\quad (3)$

18

For a perfect detector $\epsilon = 1$ and equation (3) gives the noise equivalent power for an ideal detector responding over the whole of the spectrum and thus represents the performance of an ideal thermal detector.

The performance is frequently expressed in terms of a detectivity

$$D = P_N^{-1} \text{ w}^{-1}. \tag{4}$$

In many cases (and equation (3) is one example) we can write

$$D = D^* A^{-1/2} \Delta f^{-1/2}. \tag{5}$$

D^* is called the specific detectivity and is useful for comparing the performance of different detectors since it equals the detectivity of a detector of unit area operating into an amplifier of unit bandwidth. Unfortunately the performance of all detectors does not depend upon area in this way, so that the concept of specific detectivity must be used with care.

If the temperature of the black-body radiation is 300°K, equations (3), (4) and (5) give (for $\epsilon = 1$)

$$D^* = 1 \cdot 8 \times 10^{10} \text{ w}^{-1} \text{ cm sec}^{-1/2}. \tag{6}$$

This quantity serves as a yardstick to assess the performance of actual thermal detectors.

Unlike thermal detectors, the performance of photoconductive detectors cannot be regarded as independent of wavelength. An ideal photoconductive detector responds only to photons of energy $h\nu$ greater than the minimum energy E required to excite the electronic transition responsible for the photoconductivity. It will also be assumed that all photons of energy greater than or equal to E excite this transition, so that for $h\nu \geqslant E$, $\epsilon = 1$ while for $h\nu < E$, $\epsilon = 0$. Hence to evaluate the mean square fluctuation in the background radiation to which the detector responds, equation (1) must be integrated between the limits $x_0 = E/kT$ and ∞. In fact, for photoconductive detectors it is more convenient to calculate the mean square fluctuation in the rate of arrival of photons at the detector. Corresponding to equation (1), the expression for this is

$$\overline{\Delta n^2} = \frac{4(kT)^3}{c^2 h^3} \alpha \cos\theta A\Delta f \int_0^\infty \frac{x^2 e^x}{(e^x - 1)^2} \, dx. \tag{7}$$

When the limits of integration are determined by the spectral response of the detector, (7) must be evaluated either by approximations valid for $x > 1$ or by means of tabulated functions.

The value for the noise equivalent power for a photoconductor will depend on the wavelength. The number of quanta arriving from the source of wavelength λ must equal the r.m.s. fluctuation in the number from the background. If the number of signal quanta is n_N

$$P_N = \left(\frac{hc}{\lambda}\right) n_N = \left(\frac{hc}{\lambda}\right) (\overline{\Delta n^2})^{1/2}. \tag{8}$$

Clearly P_N will be smallest for the longest value of λ to which the detector responds, which is $\lambda_0 = hc/E$. Thus the noise equivalent power for an ideal photoconductive detector operating near its threshold wavelength is (again assuming a hemispherical field of view)

$$P_N = \frac{2\pi^{1/2}(kT)^{3/2}}{\lambda_0 h^{1/2}} (\epsilon A\Delta f)^{1/2} \left\{ \int_{x_0}^\infty \frac{x^2 e^x}{(e^x - 1)^2} \, dx \right\}^{1/2}$$

$$= 7 \cdot 08 \times 10^{-18} \frac{T^{3/2}}{\lambda_0} (\epsilon A\Delta f)^{1/2} \left\{ \int_{x_0}^\infty \frac{x^2 e^x}{(e^x - 1)^2} \, dx \right\}^{1/2}. \tag{9}$$

Using (4) or (5) the corresponding values of D or D^* can be calculated. D^* now depends upon λ_0. This is shown in figure 1 which shows D^* calculated from equations (4), (5)

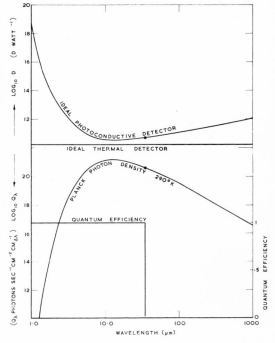

Figure 1.   Detectivity of ideal detectors.

The Planck photon distribution for 290°K black-body radiation and the detectivities for ideal thermal and photoconductive detectors exposed to this background are shown. The dot on the curve for photoconductive detectors indicates the detectivity of a detector of quantum efficiency (or spectral response) given by the lowest curve.

and (9) for 300°K background. D^* for an ideal thermal detector is also shown together with the spectral response of an ideal photoconductive detector.

Equation (9) does not include a contribution to the fluctuations resulting from the recombination of the photo-excited electrons. Application of the principle of detailed balancing (Alkemade 1959) shows that even if the detector is at a different temperature from the background the recombination will, for a photoconductive detector, be responsible for a fluctuation equal to that given by (7). To take this into account, the right-hand side of (9) must be multiplied by $\sqrt{2}$. On the other hand, this process does not contribute to the noise in devices using the photovoltaic or the photoelectromagnetic effects, so that (9) still applies to these devices. The curve shown in figure 1 is for a photoconductive device.

The evaluation of equations (1), (7) and (9) and their application to specific cases has been discussed by a number of authors, including Kruse et al. (1962) and Putley (1964a, b). If $x \gg 1$, which for $T = 300°\text{K}$ applies if $\lambda \leqslant 10 \ \mu\text{m}$,

$$\int_x^\infty \frac{x^2 \, e^x}{(e^x - 1)^2} \, dx \simeq (x^2 + 2x + 2) \, e^{-x} \qquad (10)$$

but for $\lambda > 10 \ \mu\text{m}$ this approximation is no longer accurate. It is now possible to evaluate the integral by using the transport integrals tabulated by Rogers and Powell (1958):

$$J_n(x) = \int_0^x \frac{x^n \, e^x}{(e^x - 1)^2} \, dx. \qquad (11)$$

Hence
$$\int_x^\infty \frac{x^2 \, e^x}{(e^x - 1)^2} \, dx = J_2(\infty) - J_2(x)$$
$$= 3 \cdot 290 - J_2(x). \qquad (12)$$

Figure 1 shows that D^* passes through a minimum near $10 \ \mu\text{m}$. The Planck black-body distribution function for $300°\text{K}$ is also plotted and this peaks near $10 \ \mu\text{m}$. As the threshold wavelength is increased the amount of black-body radiation exciting the detector will increase, so that the corresponding mean square fluctuation will rise. Thus P_N will increase or D will decrease. Beyond the peak, there is only a small increase in the amount of background radiation, so that at long wavelengths P_N will vary as λ_0^{-1}, or D will rise linearly.

2. Thermal detectors

2.1. *General remarks: thermal characteristics*

The radiation absorbed by a thermal detector causes its temperature to rise and this in turn causes some other property to change. By measuring this change the radiation is detected. To discuss the mechanism of a detector of this type we must first consider the thermal circuit, which relates the temperature rise to the amount of heat supplied and which determines the thermal response time of the device. Secondly, the particular temperature dependent property must be then discussed.

The thermal circuit can be represented by regarding the detector as consisting of a thermal capacity C_t coupled to an infinite heat sink at temperature T by means of a thermal conductance G_t. Then, if the detector is receiving energy at a steady rate of Q w its temperature will be $T + \Delta T$ where

$$\Delta T = Q/G_t. \qquad (13)$$

If the flow of energy is suddenly stopped, to a first approximation ΔT will decay as $e^{-t/\tau}$ where

$$\tau = C_t/G_t. \qquad (14)$$

Even in a perfect system, ΔT and Q will not be constant, but will be subject to fluctuations. In fact (Clark Jones 1953) the mean square fluctuation of Q will be

$$\overline{\Delta Q^2} = 4kT^2 G_t \Delta f. \qquad (15)$$

ΔQ takes into account power entering and leaving the detector by all processes (radiation, convection, thermal conduction, etc.) that contribute to the total thermal conductance. The noise equivalent power is now obtained by writing

$$\epsilon P_N = (\overline{\Delta Q^2})^{1/2} \qquad (16)$$

where, as before, ϵ is the emissivity of the detector.

It is interesting to compare (15) with (2). The radiation interchanges between the detector and its surroundings are responsible for a component of the thermal conductance

$$G_{tR} = 4\epsilon\sigma T^3 A \qquad (17)$$

so that if we assume that all other contribution to the thermal conductance are negligible, (15) becomes

$$\overline{\Delta Q^2} = 4kT^2 4\epsilon\sigma T^3 A \Delta f$$
$$= 16\epsilon\sigma kT^5 A \Delta f \qquad (18)$$

which is a factor of 2 greater than (2). This factor of 2 appears because (18) includes the fluctuations associated with both emission and absorption while (2) only includes the absorption. This discussion shows that the fluctuation of the background radiation falling on the detector is the limiting case of the more general temperature fluctuation. For a detector to approach the ideal performance given by (3), the coupling to its surroundings must be by radiative exchanges only.

While it is possible to design detectors in which the radiative term makes the main contribution to G_t, this may lead to other disadvantages. Consider the order-of-magnitude of the thermal capacity:

$$C_t = S\rho At \qquad (19)$$

where S is the specific heat, ρ the density and t the thickness of the detector element. Putting $S = 0 \cdot 4 \ \text{J g}^{-1} \ \text{degK}^{-1}$, $\rho = 5 \ \text{g cm}^{-3}$, $A = 4 \ \text{mm}^2$ and $t = 10 \ \mu\text{m}$, gives $C_t = 10^{-4} \ \text{J degK}^{-1}$. From equation (17), with $T = 300°\text{K}$, $G_{tR} = 6 \cdot 22 \times 10^{-4} \ \epsilon A \ \text{w degK}^{-1}$, so that assuming $\epsilon = 1$ and with $A = 4 \ \text{mm}^2$, $G_{tR} = 2 \cdot 5 \times 10^{-5} \ \text{w degK}^{-1}$. With these values of C and G_t, equation (14) gives

$$\tau = 10^{-4}/2 \cdot 5 \times 10^{-5} = 4S.$$

The numerical values used are typical for the materials used in bolometers. Thus if one attempts to approach the ideal performance the result would be a device with an inconveniently long response time. In many cases there will be some upper limit to the acceptable response time and then, having produced as small an element as is practicable to reduce the thermal capacity, it is necessary to increase the thermal conductance. This can be done by sticking the element to the heat sink by means of adhesive which is electrically insulating but thermally conducting or by attaching heavy electrical leads to the element. In this way response times of 1 msec or shorter may be achieved.

When a detector has been designed for a given response time, the best attainable performance will be given by (15) and (16) rather than by (2). With detectors that operate at room temperature, as will be seen later, the performance falls even lower, being limited by electrical noise.

The most sensitive thermal detectors do not operate at room temperature but require cooling with liquid helium. There are several reasons for this. In the first place the background radiation can be reduced and, as both equations (2) and (15) show, reducing the temperature makes possible a higher ultimate performance. Reducing the temperature also reduces considerably the thermal capacity, so improving the response time. Finally there are available at helium temperature properties (such as superconductivity) which have a much greater temperature dependence than any available at room temperature. Cooled detectors are especially important for use in the far infra-red ($\lambda > 10 \ \mu\text{m}$) where the energy available from the best sources is small compared with the room temperature background radiation at $10 \ \mu\text{m}$.

The discussion so far applies in general to all thermal detectors. It is now necessary to consider specific types, and we will discuss thermocouples, the Golay cell, resistive bolometers and finally ferroelectric bolometers.

20

2.2. *Radiation thermocouples*

Thermojunctions were first used to detect infra-red radiation a few years after the discovery of the Seebeck effect in 1826. While the radiation detector employs the same principal as the thermocouple, the construction is considerably different. Two requirements are that the thermal capacity of the junction must be kept as small as possible and it must be blackened to ensure efficient absorption of the radiation. It is also clear that the thermojunction should be constructed from materials with as large a thermoelectric figure of merit (Goldsmith 1960) as possible.

The first thermopiles used metals formed as thin wires or strips. The most widely used example was the Moll thermopile, manufactured by Kipp and Zonen, originally utilizing a copper–constantan element, but later this was replaced by a manganin–constantan one. Study of semiconductors showed that materials could be prepared which had much better thermoelectric figures of merit than the metals. The mechanical properties of these materials do not allow them to be fabricated into wires or foils, so that different construction techniques must be used. In the method developed by Schwartz and used in the thermopiles manufactured by Hilger and Watts the positive and negative semiconducting alloys are fused on to the tips of a pair of gold pins. The circuit is completed by melting a gold foil between the pins. The foil is blackened to form the receiving area. The positive material is an alloy 33% Te–2% Ag–27% Cu–7% Se–1% S; the negative material consists of 50% Ag$_2$ Se–50% Ag$_2$S. The elements may be mounted *in vacuo* or in air and they may be mounted singly or in groups. Frequently a pair of elements is placed in the same capsule so that one may be exposed to the radiation while the other is shielded from it and placed in a compensating circuit to reduce the effects of changes in ambient conditions. When they are mounted *in vacuo* the sensitivity is highest, being within about 5 of that for an ideal thermal detector, but the response time (30 msec) is rather long. When the elements are mounted in air the sensitivity falls by about one order of magnitude but the response time is also reduced. Typical performance figures are given in table 1. The spectral response of the evacuated type will be limited in the far

infra-red by the transmission of the window but at wavelengths longer than 10 μm the performance will start to fall off because the gold foil will cease to be effectively black. This is a problem common to all thermal detectors since to obtain a good performance the thermal capacity must be kept small but it is not possible to do this and at the same time coat the element with sufficient absorbing material to make it black at long wavelengths.

Similar thermocouples are produced by a number of manufacturers: Perkin–Elmer and Reeder and the Farrant Optical Co. produce couples of fine wires of Bi and Bi–Se alloys, similar to that described by Hornig and O'Keefe (1947).

Thermopiles have been designed which are almost perfectly black ($\epsilon > 0.99$) from the visible out to beyond 40 μm. These are not convenient for general use but are used in standards laboratories for measuring the characteristics of other detectors. The absorbing element is made in the form of a cone of thin electroformed copper (Eisenmann and Bates 1964) or of fine gold foil (Stair *et al.* 1965). The inside is carefully coated with specially selected black pigment while fine wire thermocouples are attached to the outside. By making the mass of the detector as small as mechanical stability will permit, the response time can be kept less than 0.2 sec, enabling the radiation to be modulated at 1 Hz.†

Comparison of the spectral response of various commercial detectors with one of these devices (figure (2)) shows that the spectral response of nominally black detectors is by no means independent of wavelength, the performance of commercial thermocouples at 35 μm falling to only one-third of the value in the near infra-red.

2.3. *The Golay cell*

Although the Golay (1947) cell is not a solid state device it must be mentioned briefly. It is one of the most important infra-red detectors that operate at room temperature and is often used as a standard of comparison for other detectors in the far infra-red.

The device consists of a small cavity containing gas (usually xenon, sometimes helium or other gases) at low pressure.

† 1 Hz = 1 c/s.

Table 1. Examples of thermal detectors

Type	Operating temp. (°K)	Useful wavelength range (μm)	Detectivity D^* or D(w^{-1})	Response time (msec)	Resistance	Remarks: availability
Semiconductor pin thermocouple	295	Vis.→30	3×10^9	30	10–100 Ω	Mounted in vacuum. Spectral response determined by window. Also available mounted in air. Manufactured by Hilger and Watts and by Reeder (U.S.A.)
Black-body thermocouple	295	Vis.→45		170		Not available commercially
Golay cell	295	Vis.→several mm	3×10^9 (D)	15		Manufactured by Unicam and by Eppley (U.S.A.)
Thermistor bolometer	295	Vis.→40	1.6×10^8 $\tau^{1/2}$	1–10	2.5 MΩ	Manufactured by Barnes Engineering (U.S.A.)
Pt foil bolometer	295	Vis.→25	1.4×10^8	16	40 Ω	Manufactured by Baird-Atomic (U.S.A.)
Superconducting Sn bolometer	3.7	>10	3×10^{11} (D)	1250	100 Ω	Not available commercially
Carbon bolometer	<2.1	>10	3×10^{10} (D)	10	100 kΩ	Not available commercially
Ge bolometer	<2.1	>10	3×10^{11} (D)	10	100 kΩ	Manufactured by Texas Instruments (U.S.A.)
TGS pyroelectric detector	295	> 2	6×10^8	1	>10 MΩ	Becoming commercially available
BaT$_i$O$_3$ dielectric bolometer	295	Vis.→20	1.1×10^8	1	125 kΩ	Becoming commercially available

21

Radiation entering the cavity is absorbed by a specially designed absorbing membrane and heats the gas. Part of the wall is made of a flexible membrane silvered on the

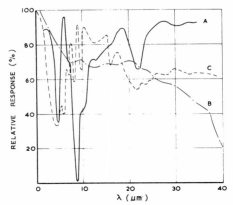

Figure 2. Response of thermal detectors relative to black-body thermocouple. A, Golay cell with diamond window; B, Reeder thermocouple with KBr window; C, Barnes thermistor bolometer with KRS-5 window.

outside, which is distorted when the gas is heated and deflects a beam of light shining on to a photoemissive cell. Thus a small amount of incident radiation causes a large change in output from the photocell. The detectivity is limited by temperature fluctuation noise and is within an order of magnitude of that for an ideal thermal detector. The spectral response is determined mainly by the window fitted to the cell and by the aperture of the cavity, which determines the long wave limit. This long wave limit is in fact several millimetres, so that the device can be used over the whole of the infra-red, overlapping into the visible and microwave regions. The response time is rather long, about 15 msec, and the device is rather sensitive to draughts, vibration, microphony and other hazards of a non-ideal environment. Figure 2 shows the spectral performance of one cell to about 40 μm relative to a black-body thermojunction. Most of the structure is produced by the diamond window fitted to that cell. At wavelengths longer than 100 μm there is some evidence that, apart from effects due to the windows, the performance is not completely independent of wavelength and probably varies from cell to cell, but precise quantitative data are not available.

2.4. *Resistance bolometers*

Since the resistance of most conductors varies with temperature a radiation detector can be made by making a resistor with a very small thermal capacity and blackening its surface or in other ways ensuring that it can absorb infra-red radiation efficiently. Devices of this type are usually termed 'bolometers'. They were first made in 1880 by Langley, who used thin platinum foils for solar observations. Platinum foil bolometers are still used, but other materials are more frequently employed now. Oxide mixtures are used in the thermistor bolometer, but when a very high detectivity is required, detectors cooled with liquid helium must be used. Suitably doped germanium or carbon composition material of the type used in the carbon resistance

thermometer can be used at helium temperatures. Another very sensitive bolometer is obtained by operating a superconductor near its transistion temperature.

An approximate expression for the responsivity (the output voltage per unit input power, v w^{-1}) of a resistance bolometer is

$$R = \frac{\epsilon i \alpha r}{G_t (1 + \omega^2 \tau^2)^{1/2}} \qquad (20)$$

where α is the temperature coefficient of resistance, r the resistance, i the current passing through the device, ϵ the emissivity, G_t the thermal conductance coupling it to the heat sink, τ the thermal time constant and ω the angular frequency at which the incident radiation is modulated.

Equation (20) gives the voltage developed across the open circuit detector per watt of incident power. Hence the voltage output produced by the noise equivalent power P_N (equation (16)) is

$$V_N = RP_N = \frac{R(\overline{\Delta Q^2})^{1/2}}{\epsilon}. \qquad (21)$$

This expression will only represent the equivalent noise voltage of an actual detector if the contributions from all other noise sources associated with the detector are small compared with V_N. These other sources include noise from the amplifier and various types of noise that may be found in actual detectors, such as noise at contacts or internal grain boundaries or noise in other ways related to the bias current i. With a good detector these may all be small but there will always be Johnson noise associated with the resistance r. Now the Johnson noise voltage V_J is

$$V_J = (4kT\Delta f r)^{1/2}. \qquad (22)$$

Comparison of equations (21) and (22) shows that for the ratio V_N/V_J to be as large as possible the ratio $R/r^{1/2}$ must be as large as possible, so that as far as the electrical characteristics are concerned, the quantity $\alpha r^{1/2}$ must be large.

In addition to having suitable electrical properties, the emissivity ϵ should be as close to unity as possible. This is not immediately apparent from consideration of equations (21) and (22), but equations (16) and (18) show that this condition is required for an ideal detector, while if the performance is limited by amplifier (or other extraneous) noise, (20) shows also that ϵ must be large.

Equation (20) shows that G_t should be small but also, if we wish to operate at an angular frequency ω, $\omega \tau < 1$. But $\tau = C_t / G_t$ so that reduction of G_t will increase τ. Hence C_t must be made as small as possible. There will be a practical limit below which C_t cannot be reduced. This will be set partly by the difficulty of fabricating very thin pieces of material and partly by the fact that ϵ will fall if the element is too thin or is not adequately blackened. As pointed out in § 1.1 it is possible to reduce G_t to the value G_{tR} given by equation (17), but with the smallest practicable value of C_t the time constant τ is likely to be of the order of 1 sec. This will be too long for many requirements. It then becomes necessary to increase G_t by mounting the device on a thermally conducting backing or by attaching heavy leads to it. In this way the response time may be reduced to 1 msec or even less. The majority of bolometers are designed for operation with a specified time constant rather than with the highest detectivity irrespective of speed.

2.4.1. *Metal bolometers*. Metal bolometers have been constructed either from thin foils or using layers evaporated on to a thin nitro-cellulose or collodion film. Nickel, bismuth and antimony have been used for the evaporated layers. Thicknesses between 0·05 and 0·1 μm have been used, and platinum ribbons down to about 0·1 μm thickness can be produced by rolling platinum foil clad with silver. The foils are mounted between a pair of electrodes and are blackened by applying platinum black electrolytically (Langton 1946, Billings et al. 1947, Chasmar et al. 1956).

These devices are used at room temperature. In the best of them, the principal source of noise is the Johnson noise of the element, although current noise may also be found. The resistance of these detectors is small (1–40 Ω) and they are rather fragile. Response times ranging from about 3 to 16 msec have been reported. Specific detectivities of about $1·4 \times 10^8$ cm Hz$^{1/2}$ w^{-1} are obtained. As this value is very similar to that of the more rugged thermistor bolometers, metal bolometers are not commonly used.

2.4.2. *Thermistor bolometers.* Thermistor materials consist of sintered mixtures of various oxides (Brattain and Becker 1946), one that is commonly used for infra-red detectors containing nickel, manganese and cobalt. This material has a negative temperature coefficient of about 4%, about ten times that for a metal, and the resistivity is also higher, ranging from 250 to 2500 Ω cm.

The thermistor material is usually prepared in the form of flakes 10 μm thick, and these are mounted on a sapphire base to obtain efficient coupling to the heat sink. The response time is determined by adjusting the backing. It is then found that the noise equivalent power is related to the time constant and the area. For one commercially available type the relation is

$$P_\mathrm{N} = 6·3 \times 10^{-10}(A/\tau)^{1/2} \ \mathrm{w} \qquad (23)$$

where A is in mm^2 and τ in msec, and the amplifier bandwidth is 1 Hz. A may vary between 0·1 and 10 mm^2. The thermistor material absorbs fairly well in the near infra-red so that although the elements are blacked, this is not so critical as in metal bolometers or radiation thermocouples. Figure 2 shows the spectral response of a bolometer relative to that of the black-body thermocouple, showing that the performance of the bolometer is maintained to at least 40 μm. The spectral response will be influenced by the window material used in the capsule, but the performance will fall off beyond 50 μm owing to inadequacies of the blacking.

For the devices whose performance is given by (23), τ may range from 1·0 to 15 msec. By using selected elements and reducing the thickness below 10 μm, both shorter response times and lower noise equivalent powers have been obtained. For this device

$$P_\mathrm{N} = 3·8 \times 10^{-10}(A/\tau)^{1/2} \qquad (24)$$

and τ may be as short as 250 μsec.

The elements are usually mounted in pairs, one being screened from the radiation. By placing the elements in the opposite arms of a bridge circuit, the effects of changes in ambient conditions are compensated. A further improvement in detectivity can be obtained in some cases by the use of immersion optics. The element is mounted in contact with a hemispherical germanium lens. Germanium is used because it transmits satisfactorily from 2 μm onwards. This optical arrangement reduces the area of the image by a factor of 16. Hence a smaller detector element may be used and as equations (23) or (24) show this should lead to an improve-

ment of 4 times in the noise equivalent power. The improvement achieved in practice is about 3½ times.

The characteristics of typical thermistor elements are included in table 1.

2.4.3. *Cooled bolometers.* There are several advantages to be gained by using a bolometer cooled with liquid helium. The element can be constructed from semiconductors or superconductors which can have much higher values of $\alpha r^{1/2}$ than any known materials at room temperature. Efficient absorption in the far infra-red can be achieved by using thicker elements without the thermal capacity becoming too large since specific heats at 4°K are much smaller than they are at room temperature. The detectivity of an ideal cooled detector will be much higher than that of one operating at room temperature. The detectivity in this case is not given simply by re-calculating equation (6) with 4°K substituted for 300°K because in most applications the detector must be exposed to some room temperature radiation. Now the bulk of the room temperature radiation occurs near $\lambda = 10$ μm and if the detector is for use at longer wavelengths than this, a cooled filter may be used to cut out this unwanted background radiation. When this is done, the noise contribution from the background radiation must be calculated by evaluating equation (1) over the pass-band of the filter and over the field of view of the detector. This calculation has been discussed by Putley (1964a) who showed that even using a filter with a narrow pass band at $\lambda > 100$ μm, the noise contribution from the 300°K background will be several orders of magnitude greater than that of the 4°K surroundings of the detector. These calculations showed that the detectivity of an ideal thermal detector responding to all wavelengths greater than 100 μm will be about two orders of magnitude greater than that given by (6). This potential improvement in performance is very important in far infra-red spectroscopy where one of the principal difficulties is the low intensity of available wide band sources.

Bolometers have been constructed from several superconductors, the most successful being that of Martin and Bloor (1961) which uses tin. The principal disadvantages are that the element is fragile and its emissivity is poor. Accurate temperature control (to better than 10^{-4} degK) is required. The performance originally obtained was $P_\mathrm{N} = 3 \times 10^{-12}$ w (1 Hz bandwidth) and $\tau = 1·25$ sec. The amplifier was the main source of noise, as indeed it still is with helium cooled detectors. Somewhat better amplifiers are available now, so that better performance could now be obtained, but other detectors with comparable performance and which are easier to use are now available. These include cooled semiconducting bolometers.

Bolometers were constructed from the Allen and Bradley resistors of the same type as used for thermometers (Richards and Tinkham 1960). These gave $P_\mathrm{N} = 10^{-11}$ w and $\tau \sim 1$ msec. This detector has the advantage that it is fairly simple to construct from readily obtainable materials but its performance is at least an order worse than the best available detectors. A more recently developed bolometer (Low 1961, 1966, Richards 1964) uses Ge doped with about 10^{17} cm^{-3} of Ga or In and fairly heavily compensated. In this material the free hole concentration is still large below 4°K. Radiation is absorbed by the free carriers and the energy absorbed raises the temperature of the whole crystal. The noise equivalent power obtainable appears to be about the same as that quoted for the superconducting bolometer (Richards, private communication) while the response time is about 5 msec. Other semiconductors such as GaAs

(Wheeler and Hill 1966) can also be used as cooled bolometers.

The properties of these cooled detectors have been described in more detail by Putley (1963, 1966a) and are summarized in table 1 and illustrated in figure 6.

2.5. Ferroelectric detectors

The properties of ferroelectric materials are very temperature-dependent in the neighbourhood of the ferroelectric Curie point. Utilization of changes in the dielectric constant (Hanel 1961), the pyroelectric effect (Cooper 1962) and in certain special cases the resistivity (Jonker 1964) can form the basis of thermal detectors.

Employment of the change in dielectric constant leads to a device analoguous to a resistance bolometer in which the reactance $1/\omega_c C$ replaces the resistance: in fact the responsivity is now

$$R = \frac{\epsilon i_c \alpha}{\omega_c C G_t (1 + \omega^2 \tau^2)^{1/2}} \tag{25}$$

which should be compared with equation (20). In (25) the bias current i_c alternates at the angular frequency ω_c and α is now the temperature coefficient of the dielectric constant. Estimates of the performance attainable with $BaTiO_3$ and related materials indicated that the performance should be superior to that of the resistive bolometers, but although some work has been done to develop this type of device, the actual performance reported is not vastly better than that of the best thermistors. Thus Burke, Hoetler and Williams (1964 private communication) obtained $P_N = 1 \cdot 02 \times 10^{-10}$ w for a device of $\frac{1}{2}$ mm diameter. For a thermistor of the same size with $\tau = 1$ msec, equation (24) gives $P_N = 1 \cdot 7 \times 10^{-10}$ w. One factor tending to degrade the performance of the $BaTiO_3$ devices is the fact that they are subject to a large Barkhausen-type noise near the Curie point.

It is a characteristic of ferroelectric materials that they show a spontaneous polarization which can be reversed by a large enough electric field and which is strongly temperature dependent (the pyroelectric effect). Cooper (1962) showed that this could be used in a thermal detector and that the responsivity is

$$R = \frac{\epsilon \omega r a q}{G_t (1 + \omega^2 r^2 C^2)^{1/2} (1 + \omega^2 \tau^2)^{1/2}} \tag{26}$$

where q is the pyroelectric coefficient, a the area of the electrodes and r the resistance shunting the capacity C of the detector. Note that in this case the reactive nature of the process tends to improve the performance at high frequencies, compensating for the effect of the thermal time constant. At high enough frequencies, the resistance–capacity time constant will attenuate the output.

Devices based on the pyroelectric effect might be expected to show more promise than those based on the dielectric effect since the polarization associated with the pyroelectric effect is much greater than that which could be produced in a dielectric material by an applied voltage too small to break it down. In addition, comparison of equations (25) and (26) (and (20)) shows that the pyroelectric device should perform better at higher frequencies.

The performance obtained with pyroelectric devices using $BaTiO_3$ has not, in fact, exceeded that of the $BaTiO_3$ dielectric bolometer. More recent experiments with triglycine sulphate and selenate pyroelectric detectors (Stanford 1965, Hadni et al. 1965, Astheimer and Beerman 1966 private communication) have produced promising results. Thus Stanford has

found $P_N = 5 \times 10^{-8}$ w with a rise time of 50 μsec while Hadni et al. have shown that triglycine sulphate absorbs strongly throughout most of the infra-red; they have been able to produce large areas far infra-red detectors which have a smaller detectivity than a Golay cell but, because of their larger area, give a performance in suitable instruments comparable to that obtainable with a Golay cell. These devices are still under development and it is too early to say whether they will replace any of the existing detectors.

3. Photoconductive detectors

There are three types of electronic transition which are utilized in photoconductive devices (figure 3). In the first to be exploited, the incident photons excite electrons from the valence band of a semiconductor into the conduction band, so producing a free hole and a free electron. These will eventually recombine, but until they do so the conductivity is increased. The simple increase in conductivity is used in many detectors but it is also possible to use other related effects such as changes in the characteristics of p–n junctions or the photoelectromagnetic effect (Pincherle 1956).

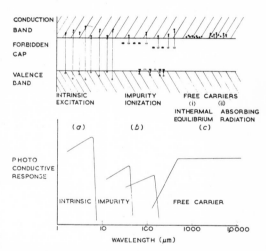

Figure 3. Photoconductive processes: (a) intrinsic photoconductivity; (b) extrinsic photoionization; (c) free electron photoconductivity.

It is clear that the energy of the photons must be greater than the intrinsic energy gap, so that this energy gap will determine the long-wave threshold of the device. In the semiconductors developed for transistors and solid state circuits, $E_G \simeq 1$ ev, giving $\lambda_c = 0 \cdot 81$ μm, barely outside the visible. Thus semiconductors with much smaller intrinsic energy gaps are required to make useful infra-red detectors. Materials are now known with $E_G < 0 \cdot 1$ ev so that intrinsic photoconductivity can occur at wavelengths greater than 10 μm.

While in fact materials with even smaller energy gaps are known, it has not been possible to use the intrinsic effect at wavelengths much longer than 10 μm. For wavelengths between 10 and 100 μm the second process shown in figure 3 is employed. Since semiconductors contain impurity states located in the band gap, the energy required to ionize these must be smaller than the intrinsic energy. Thus Ge has an intrinsic energy gap of about $0 \cdot 75$ ev but the ionization energy of the shallower impurity states is about $0 \cdot 01$ ev. By making use of appropriately doped Ge, photoconductive devices have been developed for use to beyond $100 \mu m$. The development of these devices requires material of much higher purity than is required for other types of devices, and for this reason Ge is the only satisfactory host material that has been employed so far.

Although semiconductors are known which have shallower impurity levels than Ge, difficulties of purification have prevented the use of impurity photoionization beyond $100 \mu m$. At longer wavelengths the third process must be used. In this the incident radiation is absorbed by electrons in the conduction band of a high mobility semiconductor such as InSb or Ge. Free carrier absorption occurs at room temperature but this does not lead to any change in conductivity. At helium temperature, the absorbed radiation produces a significant change in the electron distribution within the conduction band. Since the electron mobility is related to this, the result is a change in mobility (usually an increase) and hence of conductivity. Since the free carrier absorption is largest at the longer wavelengths, this effect is useful at very long wave lengths. For example, with InSb optimum results are obtained for $\lambda \gtrsim 1000 \mu m$.

Photoconductive devices have two advantages over the thermal detectors. Firstly, their response time is much shorter, typically of the order of $1 \mu sec$ rather than 1 msec. Secondly, since they do not respond to such a wide band of the spectrum the limiting background noise will be smaller, as illustrated in figure 1. One disadvantage is that the cooling requirements are more stringent. Thermal detectors operating at room temperature can be used over the whole spectrum but photoconductive devices must be operated at a temperature so low that the number of thermally excited free carriers is negligible compared with the number excited by the radiation.

3.1. Intrinsic detectors

Although photoconductivity in the near infra-red had been observed in compound semiconductors at least 50 years ago, intensive development of infra-red detectors did not start until the Second World War. One of the first materials to be studied was Tl$_2$S (Cashman 1959). Evaporated layers of this material have at room temperature a very high sensitivity in the near infra-red, but as the threshold wavelength occurs at $\lambda_c = 1 \cdot 1 \mu m$ they were soon replaced by detectors responding to longer wavelengths. There are applications where a high detectivity near $1 \mu m$ is required, but for these applications Si or Ge photodiodes would now be preferred.

The next compounds to be studied were PbS, PbSe and PbTe. Evaporated layers of the first two can be used at room temperature with $\lambda_c = 2 \cdot 5$ and $3 \cdot 4 \mu m$ respectively. On cooling λ_c moves to longer wavelengths (to about $6 \mu m$ for PbSe) and the detectivity improves. PbTe detectors do not operate without cooling. Figure 4 shows the spectral performance of this group of detectors. A second method of preparation of PbS and PbSe detectors is by controlled precipitation from solution. This process requires very stringent control but it is now possible to manufacture detectors with uniform characteristics over very large areas. At $3 \mu m$ the cooled PbS detector was, until very recently at least, the most sensitive detector available. Its performance has now been equalled by the InAs detector.

The evaporated or chemically deposited film detector suffers from two disadvantages. Firstly, the manufacturing process tends to be more of an art than a science and secondly the recombination time is rather long, as much as $100 \mu sec$ in some cases. With the development of semiconductor

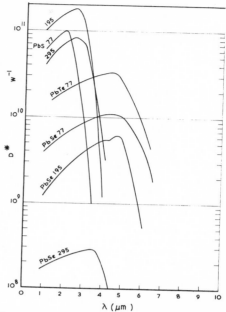

Figure 4. Spectral detectivity of PbS, PbSe and PbTe detectors.

technology, it was therefore natural to study photoconductivity in single crystals with a view to overcoming these disadvantages and also in the hope of developing materials useful at longer wavelengths. It is possible to make single crystal photoconductive devices with PbSe (Kimmitt and Prior 1961) but these have not been developed because of the difficulty of preparing single crystals of the lead salts of sufficient purity.

Study of new compounds has lead to the development of single crystal detectors using the III–V compounds InAs and InSb. Photoconductive InSb detectors are prepared from single crystals by grinding and etching plates to a thickness of a few micrometers. The operating temperature determines the optimum purity for the material. For operation at 77°K p-type material with a resistivity (at 77°K) of about 2Ω cm is required, but optimum results at higher temperatures are obtained with less pure material (Morten and King 1965). Figure 5 shows the spectral performance of detectors operating at 77, 233 and 300°K. In this detector reduction of the temperature reduces the threshold wavelength. When operated at room temperature the detectivity is limited by Johnson noise in the detector's resistance, but at 77°K the best detectors approach closely the limit set by background radiation (figure 1).

In addition to photoconductive detectors, InSb p–n junction detectors (Rieke et al. 1959) and photoelectromagnetic detectors are also produced. The performance of the junction detector is very similar to that of the photo-conductive detector. For many applications the choice will be determined by manufacturing considerations rather than performance. One advantage that p–n junctions have

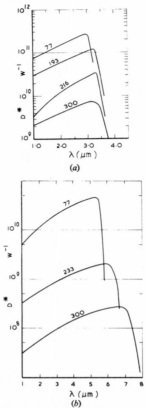

(a)

(b)

Figure 5. Spectral detectivity of (*a*) InAs detectors, (*b*) InSb detectors.

is that the recombination of the excited carriers occurs not in the high resistivity barrier region but in the adjoining p^+ or n^+ regions where the recombination times are much shorter. Thus the response times of p–n junctions can be significantly shorter than that of photoconductive devices and this is particularly useful for use with infra-red lasers. The response time of the fastest junction detectors is only a few nanoseconds or less (Lucovsky and Emmons 1965a) while that of the photoconductive detector at 77°K is about 10 μsec. The photoelectromagnetic detector (Kruse 1959) also has the advantage of a shorter response time (Zitter 1964) but has the disadvantage that its resistance is often very low ($<1\ \Omega$) so that it is difficult to use. The factors influencing the choice between these different modes of operation have been discussed by Hilsum and Simpson (1959). Characteristics of various types of InSb detector are given in table 2.

InAs has many similarities with InSb, but its energy gap is

larger and λ_c is about 3–4 μm. Both photoconductive and p–n junction devices have been produced and the latter are suitable when nanosecond response times are required. Typical characteristics are given in figure 5 and table 2.

Intrinsic photoconductive detectors usable beyond 7 or 8 μm are not commercially available. Several materials with energy gaps small enough for use in the 8–13 μm atmospheric transmission window are known (Putley 1964b) but practical difficulties with the purification or stabilization of them have held up the development of devices. At the moment, the most promising appears to be the mixed compound $Hg_x\,Cd_{1-x}Te$ where x is about 0·9 (Kruse 1965, Cohen-Solal et al. 1963). It is true that extrinsic Ge detectors are available (see § 3.2) but if the manufacturing problems can be overcome, $Hg_x\,Cd_{1-x}$ Te detectors should operate at 77°K compared with, say, 35°K required for the Ge detector; they may well have shorter response times and will lend themselves more easily to the fabrication of large area detectors or arrays.

3.2. Impurity photoionization detectors

The properties of over twenty impurities in Ge have been studied in detail, but of these the principal ones which have been used for the development of infra-red detectors are Au, Hg, Cu, Zn, B, Ga and Sb (Putley 1964b, Levinstein 1965). The Au doped detector can be used to about 10 μm and operates in liquid nitrogen. Hg doping extends the cut-off to about 14 μm but requires cooling to 35°K. Cu extends the cut-off wavelengths to nearly 30 μm, Zn to about 40 μm and B, Ga (Moore and Shenker 1965) and Sb to about 120 μm. These last four are most conveniently operated at 4°K. The spectral performance of some of these detectors is shown in figure 6. The response time of these detectors is

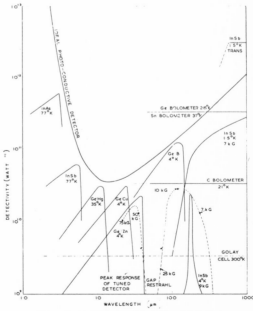

Figure 6. Performance of far infra-red detectors.

Table 2. Examples of photoconductive detectors

Type	Operating temp. (°K)	Useful wavelength range (μm)	Wavelength of max. detectivity	Detectivity D^* (or D) at optimum wavelength	Response time (μsec)	Resistance	Remarks: availability
Si–PV	295	<1·2	0·9	$2\cdot5 \times 10^{12}$	1000	450 kΩ	Manufactured by Texas Instruments (U.S.A.)
Tl$_2$S–PC layer	295	<1·2	0·9	$2\cdot2 \times 10^{12}$	500	5 MΩ	Not available commercially
Ge–PV	295	<1·8	1·6	4×10^{11}			Available from several manufacturers
PbS–PC layer	295	<3	2·5	8×10^{10}	300	500 kΩ	Available Mullards (U.K.) Santa Barbara Research Center (U.S.A.)
PbS–PC layer	193	<3·4	2·6	4×10^{11}	3500		
PbSe–PC layer	295	<4·5	3·5	10^{9}	2		
PbSe–PC layer	77	<7	5	10^{10}	40		
PbTe–PC layer	77	<5·2	4	$2\cdot7 \times 10^{9}$	25	30 MΩ	Available S.B.R.C. (U.S.A.)
InAs–PV	295	<3·7	3·4	7×10^{9}	1	20 Ω	Available from S.A.T. (France)
InAs–PV	196	<3·5	3·2	7×10^{10}	1	50 Ω	Available from several manufacturers including Texas Instruments, S.B.R.C. and Philco (U.S.A.)
InAs–PV	77	<3·4	3·0	3×10^{10}	0·05	≤10 MΩ	
InSb–PC	295	<7·8	6	3×10^{11}	5–10	5 Ω	Available from several manufacturers including Mullards (U.K.), S.A.T. (France) Texas Instruments, S.B.R.C. and Philco (U.S.A.)
InSb–PC	77	<5·5	5	3×10^{8}	<1	2000 Ω	
InSb–PV	77	<5·5	5·4	5×10^{10}	0·2	20 Ω	Available from several manufacturers including Mullards (U.K.), S.A.T. (France), Texas Instruments, S.B.R.C. Philco and Raytheon (U.S.A.)
InSb–PEM	295	<7·8	6·2	7×10^{10}	<1		
Ge–Au–IPI	60	<9	5	3×10^{8}	<1	~1 MΩ	Not available commercially
Ge–Hg–IPI	30	<14	12	3×10^{10}	0·1	>1 MΩ	
Ge–Cu–IPI	15	<30	25	3×10^{10}	~1	>3000 Ω	
Ge–Zn–IPI	4	<40	35	4×10^{10}	0·04	~1 MΩ	
Ge–Ge–IPI	4	<125	104	3×10^{11}	0·1–0·01	~10 kΩ	
Ge–B–IPI	4	<140	108	2×10^{11}	100	~10 kΩ	
InSb†	1·8	>200	1000	$2 \times 10^{12}\,(D)$	0·2		Available from Mullards (U.K.) and Texas Instruments Ltd. (U.K.)
InSb‡	1·8	>200	1000	$1\cdot2 \times 10^{11}\,(D)$	<1	~5 kΩ	
InSb§	4	determined by magnetic field	150 μm (12 kG) 26 μm (73 kG)	$2 \times 10^{10}\,(D)$		~10^{4}–10^{5} Ω	

PV, photovoltaic; PC, photoconductive; PEM, photoelectromagnetic effect; IPI impurity photoionization.
† Hot electron, no magnetic field. ‡ Hot electron in weak magnetic field. § Magnetooptical resonance.

usually less than 1 μsec but in some cases the resistance rises to such a high value that the effective response time is determined by the RC time constant of the input circuit. To obtain the best results, extreme care has to be taken to dope the Ge with the optimum amount of the desired impurity and to reduce the contamination by unwanted impurities to a minimum.

3.3. *Detectors using free electron photoconductivity*

The present available n-type InSb contains about 5×10^{13} free electrons per cm³. The impurity states (probably $\sim 10^{14}$ cm⁻³) are merged with the conduction band so that even with $T < 1°$K the material behaves in a quasi-metallic fashion. The electron mobility μ is high, being greater than 5×10^5 at 77°K and about 10^5 cm² v⁻¹ sec⁻¹ at 4°K. At 4°K this material shows marked hot electron effects, deviations from Ohm's law being seen for applied fields of less than 1 v cm⁻¹. It is therefore a very suitable material for free carrier photoconductivity. The spectral response will be determined by the free carrier absorption spectrum. The absorption will be large at low frequencies but when the frequency of the radiation $\nu > 1/\tau_c$ the absorption will vary as ν^{-2} or λ^2. Here τ_c is the conductivity relaxation time ($\tau_c = \mu(m^*/e)$) and for InSb the transition from low frequency to high frequency behaviour occurs when $\lambda \sim 1$ mm. The absorption remains sufficiently large down to $\lambda \sim 200$ μm to enable efficient detectors to operate down to this wavelength. Figure 6 gives the spectral response of an InSb submillimetre detector.

The resistivity of the present n-type InSb below 4°K is rather low ($\sim 10 \,\Omega$ cm) and since the principal source of noise is the amplifier at room temperature, the performance will be poor. By applying a moderate magnetic induction (6–7 kG), produced by a small superconducting solenoid, and reducing the temperature below 2°K, the resistivity increases by about three orders of magnitude. Under these conditions, although the detectivity is still limited by amplifier noise, it compares very favourably with that of a germanium bolometer (Richards, private communication) but the response time is now about 0·2 μsec. Thus this detector can be used for time-resolved spectroscopy of plasma discharges or pulsed lasers in the submillimetre region.

Attempts are still being made to improve the performance of this and other far infra-red detectors by developing better amplifiers. With the superconducting bolometer use of a helium cooled step-up transformer is very advantageous (Martin 1966). Kinch and Rollin (1963) have proposed the use of this technique with the InSb detector as an alternative to using a magnetic field. This should lead to a further increase in detectivity (see figure 6) and make the InSb detector the most sensitive available at 1000 μm wavelength apart from the superheterodyne receiver (Putley 1966b). The use of a cooled transformer increases the effective response time of the system to about 1 msec, so that it cannot be used when a short response time is required.

In an attempt to improve the performance at wavelengths less than 200 μm, Kimmitt applied a magnetic induction greater than 10 kG to an InSb detector. He did, indeed, find that the performance was enhanced, but it became strongly wavelength dependent, being confined to a narrow band near the cyclotron resonance wavelength (figure 6) and thus tunable by varying the magnetic field. Brown and Kimmitt (1965) have studied this effect in more detail, finding that an induction of 14 kG is required to tune it to 100 μm while 76 kG tunes it to 26 μm. So far, the main application

of this effect has been to experiments requiring a very high rejection of the higher orders from a grating spectrometer.

4. Performance of available detectors

Tables 1 and 2 summarize the characteristics of thermal and photoconductive detectors, respectively. The wavelength range is quoted on the assumption that the detector is fitted with a window of suitable transmission. The detectivity quoted is scaled to an area of 1 cm² (D^*) except in those cases in which the detectivity does not depend on the area. Here the actual detectivity D is given and this is noted in the tables. Where possible the performance figures are given for commercially available devices and source of supply are indicated. Not all detectors are available commercially. For these, the results quoted are based on the references cited in the text.

5. Future trends

The performance of the best detectors approaches closely that for an ideal wide-band detector limited by radiation fluctuation. However, the performance attainable by an ideal narrow band receiver is many orders of magnitude better. Until recently this was only of importance in the radio-frequency region, but the development of visible and infra-red lasers calls for the development of corresponding narrow band receivers. Detectors with very short response times are required for some laser applications and may be used in superheterodyne systems (Lucovsky *et al.* 1964), and while it is possible to produce fast p–n junction in Si or Ge for use in the near infra-red, the detectivity is not always adequate. A possible way to improve this is to design a device in which the signal is amplified by controlled avalanche ionization. So far, promising results have been obtained with Si and Ge (Anderson *et al.* 1965, Keil and Bernt 1966) but attempts (Lucovsky and Emmons 1965b) to apply this to InAs or InSb diodes for use at longer wavelengths have so far failed. Since fast p–n junction devices cannot be used at wavelengths longer than about 10 μm, other types of device would be required for the far infra-red. The response time of the extrinsic photoconductors can be made as short as 1 nsec or less but the reduction in sensitivity is likely to be so severe that these detectors would have to be used either with a laser amplifier or in a superheterodyne system.

Not all laser applications call for very sophisticated receivers. When advantage can be taken of the high power level available the main requirements for the detector are simplicity, ruggedness and cheapness. Further development of thermal detectors could meet this requirement as well as that for simpler instrumentation in the far infra-red.

References

ALKEMADE, C. T. J., 1959, *Physica, 's Grav.*, **25**, 1145–58.

ANDERSON, L. K., McMULLIN, P. G., D'ASARO, L. A., and GOETZBERGER, A., 1965, *Appl. Phys. Letters*, **6**, 62–4.

BILLINGS, B. H., HYDE, W. L., and BARR, E. E., 1947, *J. Opt. Soc. Amer.*, **37**, 123–321.

BRATTAIN, W. H., and BECKER, J. A., 1946, *J. Opt. Soc. Amer.*, **36**, 354.

BROWN, M. A. C. S., and KIMMITT, M. F., 1965, *Infrared Phys.*, **5**, 93–7.

CASHMAN, R. J., 1959, *Proc. Inst. Radio Engrs, N.Y.*, **47**, 1471–5.

CHASMAR, R. P., MITCHELL, W. H., and RENNIE, A., 1956, *J. Opt. Soc. Amer.*, **46**, 469–77.

CLARK JONES, R., 1953, *Adv. Electronics*, **5**, 1–96.

COHEN-SOLAL, G., Bailey, F., VERIE, C., and MARFAING, Y., 1963, *C.R. Acad. Sci., Paris*, **257**, 863–6.

COOPER, J., 1962, *Rev. Sci. Instrum.*, **33**, 92–5.

EISENMANN, W. L., and BATES, R. L., 1964, *J. Opt. Soc. Amer.*, **54**, 1280–1.

GOLAY, M. J. E., 1947, *Rev. Sci. Instrum.*, **18**, 357–62.

GOLDSMID, H. J., 1960, *Applications of Thermoelectricity* (London: Methuen).

HADNI, A., HENNINGER, Y., THOMAS, R., VERGNAT, P., and WYNCKE, B., 1965, *J. Phys.*, **26**, 345–60.

HANEL, R. A., 1961, *J. Opt. Soc. Amer.*, **51**, 220–4.

HORNIG, D. F., and O'KEEFE, B. J., 1947, *Rev. Sci. Instrum.*, **18**, 474–82.

HILSUM, C., and SIMPSON, O., 1959, *Proc. Instn Elect. Engrs*, **106B**, 398–401.

JONKER, G. H., 1964, *Solid St. Electron.*, **7**, 895–903.

KEIL, G., and BERNT, H., 1966, *Solid St. Electron.*, **9**, 321–5.

KIMMITT, M. F., and PRIOR, A. C., 1961, *J. Electrochem. Soc.*, **108**, 1034–8.

KINCH, M. A., and ROLLIN, B. V., 1963, *Brit. J. Appl. Phys.*, **14**, 672–6.

KRUSE, P. W., 1959, *J. Appl. Phys.*, **30**, 770–8.

—— 1965, *Appl. Opt.*, **4**, 687–92.

KRUSE, P. W., McGLAUCHLIN, L. D., and McQUISTAN, R. B., 1962, *Elements of Infrared Technology* (New York: Wiley).

LANGTON, W. G., 1946, *J. Opt. Soc. Amer.*, **36**, 355.

LEWIS, W. B., 1947, *Proc. Phys. Soc.*, **59**, 34–40.

LEVINSTEIN, H., 1965, *Appl. Opt.*, **4**, 639–47.

LOW, F. J., 1961, *J. Opt. Soc. Amer.*, **51**, 1300–4.

—— 1966, *Proc. Inst. Elect. Electron. Engrs*, **54**, 477–84.

LUCOVSKY, G., and EMMONS, R. B., 1965a, *Appl. Opt.*, **4**, 697–702.

—— 1965b, *Proc. Inst. Elect. Electron. Engrs*, **53**, 180.

LUCOVSKY, G., EMMONS, R. B., and ALTMOSE, H., 1964, *Infrared Phys.*, **4**, 193–7.

MARTIN, D. H., and BLOOR, D., 1961, *Cryogenics*, **1**, 159–65.

MARTIN, D. H., 1966, *Millimetre and Submillimetre Waves*, ed. D. H. Martin (Amsterdam: North-Holland).

MOORE, W. J., and SHENKER, H., 1965, *Infrared Phys.*, **5**, 99–106.

MORTEN, F. D., and KING, R. E. J., 1965, *App. Opt.*, **4**, 659–63.

PINCHERLE, L., 1965, *Photoconductivity Conference*, ed. R. G. Breckenridge, B. R. Russell and E. E. Hahn, (New York: Wiley), pp. 307–20.

PUTLEY, E. H., 1963, *Proc. Inst. Elect. Electron. Engrs*, **51**, 1412–23.

—— 1964a, *Infrared Phys.*, **4**, 1–8.

—— 1964b, *Phys. Stat. Sol.*, **6**, 571–614.

—— 1966a, *Millimetre and Submillimetre Waves*, ed. D. H. Martin (Amsterdam: North-Holland).

—— 1966b, *Proc. Inst. Elect. Electron. Engrs*, **54**, 1096–8.

RICHARDS, P. L., and TINKHAM, M., 1960, *Phys. Rev.*, **49**, 66–9.

RICHARDS, P. L., 1964, *J. Opt. Soc. Amer.*, **54**, 1474–84.

RIEKE, F. F., DeVAUX, L. H., and TUZZOLINI, A. J., 1959, *Proc. Inst. Radio Engrs*, **47**, 1475–8.

ROGERS, W. M., and POWELL, R. L., 1958, *National Bureau of Standards Circular* 595.

STAIR, R., SCHNEIDER, E. W., WATERS, W. R., and JACKSON, J. K., 1965, *App. Opt.*, **4**, 703–10.

STANFORD, A. L., 1965, *Solid-St. Electron.*, **8**, 747–55.

WHEELER, R. G., and HILL, J. C., 1966, *J. Opt. Soc. Amer.*, **56**, 657–65.

ZITTER, R. N., 1964, *Rev. Sci. Instrum.*, **35**, 594–6.

Reprinted from *Physics in Technology*, **4**, 202–222 (Dec. 1973)

Modern infrared detectors

E. H. PUTLEY

Abstract. This review describes recent progress on infrared detectors. After a brief introduction discussing the general principles of infrared detection, there follow detailed accounts of new types of photon detectors and of modern thermal detectors. These sections concentrate on work carried out since an earlier review in *Journal of Scientific Instruments*. The present review concludes with a short section and table summarizing the characteristics of the detectors described.

1. Introduction

There have been a number of books and review articles published within the last few years discussing both the fundamental principles of infrared detection and particular detectors in detail. (See for example Kruse *et al* 1962, Hudson 1969, Willardson and Beer 1970, 1973, Chol *et al* 1966, Hogan and Moss 1970, *IERE Conference Proceedings No 22—Infrared Techniques* 1971, Moss 1965, 1973, Pell 1971, Kosnitser and Semendyaev 1968, Putley 1964, 1966.) This article attempts to review the more significant advances which have occurred recently and to discuss the factors governing the choice of detectors for specific applications.

Before going further, however, it is worthwhile recalling some of the general considerations which are discussed in more detail in the references quoted in the first paragraph.

The infrared region extends from the visible to the microwave region. There is no clear-cut boundary between infrared and microwaves, but usually it is placed at 1000 μm or 1 mm.

Some types of detectors used in the visible region such as photoemissive cathodes and photographic plates can be used in the near infrared, while microwave receivers can operate at wavelengths shorter than 1000 μm; this article will only discuss detectors which are of principal use in the infrared. The feature which distinguishes the infrared from other regions of the electromagnetic spectrum is that the bulk of the thermal radiation from objects at normally accessible temperatures occurs within it. Thus 50 % of the radiation from the sun (6000 K) occurs at wavelengths longer than 0·7 μm, while the proportion of infrared radiation from bodies at lower temperatures is even greater. Taking red heat to correspond to 700 °C, then less than 0·0002 % of the emission occurs in the visible whilst 50 % occurs at wavelengths longer than 4·2 μm. Not only does much of the interest in the infrared arise from this fact, but also it has a significant bearing on the operation of infrared detectors.

The first group of infrared detectors to be developed was the thermal detector. This employs materials possessing some strongly temperature-dependent property. The incident radiation raises the temperature of the detecting element producing a change in the property being used to detect the infrared radiation. The thermopile and the bolometer are typical examples of detectors of this sort.

30

In the second main group of detectors the infrared radiation induces an electronic transition which leads to a change in electrical conductivity (photoconductivity) or to an output voltage appearing across the terminals of the device, as in photovoltaic devices and some others.

The majority of infrared detectors can be classified in one or other of these two groups. A large number of physical effects are used. In the space available in this article it will only be possible to summarize recent developments.

1.1. Ideal thermal and photon detectors

The infrared radiation emitted by the objects surrounding the detector not only provides in many cases the signal to be detected, but also is responsible for a background signal which determines the ultimate sensitivity attainable in a perfect detector. When experiments are carried out under normal ambient conditions the detector will receive radiation from surroundings at approximately 300 K. This radiation is of maximum intensity near 10 μm wavelength. Insofar as it constitutes a steady background signal its presence can usually be ignored but there will always be a fluctuation in its intensity. This will set the value of the minimum detectable signal for the ideal case in which the detector is completely free of all internal fluctuations or noise. No real detector ever satisfies this condition, but the best modern detectors can detect signals within a factor of two or three of this ideal limit.

The procedure for calculating the performance of an ideal detector is described elsewhere at some length (Smith *et al* 1968, Kruse *et al* 1962, Jones 1953, 1959, Putley 1970). It is usually expressed as a noise equivalent power P_N, that is the incident signal power required to give an output equal to that produced by the background fluctuations, or as a detectivity D which is the reciprocal of P_N. With a real detector P_N or D will depend on the wavelength, the modulation frequency of the signal and on the area of the detector and will of course have a value inferior to that of an ideal detector. Since the magnitude of the background fluctuation affecting a particular detector will depend on the spectral range to which the detector responds, its area, and the field of view over which background radiation is incident upon it, there is no single value for the performance of an ideal detector. The simplest case is that of a thermal detector operating at room temperature. The perfect detector is supposed to respond to the whole EM spectrum (ie be perfectly black) and to receive background radiation emanating from a complete hemispherical field of view. It is not usually supposed that it is subject to radiation from the complete sphere because in principle one surface of the detector can be covered with a perfect reflector so that it only absorbs radiation (signal or background) on one face. The value obtained for the noise equivalent power of this ideal detector is (Putley 1970)

$$P_N = (16\sigma k T^5 A \Delta f / \eta)^{1/2} \tag{1}$$

where σ is Stefan's constant, k Boltzmann's constant, T the absolute temperature, Δf the amplifier bandwidth and η the emissivity of the detector. For the ideal detector $\eta = 1$. The NEP varies as $A^{1/2}$.

Putting $A = 1$ cm^2 and $\Delta f = 1$ Hz, we define the specific detectivity D^* by writing

$$D^* = 1/P_N = (16\sigma k T^5)^{-1/2} = 2 \cdot 83 T^{-5/2} \quad \text{W}^{-1} \text{ cm Hz}^{1/2}. \tag{2}$$

Whilst thermal detectors have a broad band response, the response of photon detectors is usually set by some characteristic energy transition ϵ. Thus they can only respond to quanta satisfying the condition

$$h\nu \geqslant \epsilon. \tag{3}$$

When calculating the detectivity of an ideal photon detector, it is usual to assume that the incident signal consists of photons of energy

$$h\nu = \epsilon \tag{4}$$

and that the detector responds only to those photons in the background satisfying (3). The response to background photons such that

$$h\nu < \epsilon \tag{5}$$

is assumed zero. Thus the performance of the ideal photon detector is a function of ϵ. This means that to obtain best results for a given frequency ν a detector must be chosen to satisfy (4). It is now no longer adequate to assume that the detector is exposed to the complete hemispherical field of view. If it were operating at the same temperature as the background this would be fair enough, but many photon detectors require low temperatures for satisfactory operation. It then becomes possible to use cooled radiation shields to restrict the field of view, reducing the incident fluctuations and increasing the detectivity.

The expression derived for the noise equivalent power of this idealized photon detector is given by Putley (1970)

$$P_N = (8\pi A \Delta f/\eta)^{1/2}\{(kT)^{3/2}/\lambda_c h^{1/2}\} \left(\int \frac{\alpha \cos\theta}{\pi}\right)^{1/2} (J_2(\infty) - J_2(x_c))^{1/2}. \tag{6}$$

Here A is the detector area and Δf the amplifier bandwidth, k is Boltzmann's constant and h Planck's constant. λ_c is the cut-off wavelength of the detector and η its quantum efficiency, so that for the ideal detector $\eta = 1$. $x = h/\lambda_c T$ and

$$J_2(x) = \int_0^x \frac{x^2 \exp x \, dx}{(\exp x - 1)^2}$$

which has been tabulated by Rogers and Powell (1958) who give $J_2(\infty) = 3\cdot2899$.

The term

$$\int \frac{\alpha \cos\theta}{\pi}$$

in equation (6) introduces the field of view. It is assumed that the surface of the detector obeys Lambert's law. α is an element of solid angle in a direction making an angle θ with the normal to the detector's surface. If as commonly occurs with cooled photoconductive detectors the field of view is restricted by a cooled shield the amount of incident background radiation will be reduced by a factor determined by this integral. When the field of view is a complete hemisphere, $\int \alpha \cos\theta = \pi$ so that the factor reduces to unity. At short wavelengths and normal temperatures such that

$$x = h/\lambda_c T \gg 1$$

equation (6) can be simplified to give

$$P_N = (8\pi A \Delta f / \eta)^{1/2} \{(kT)^{3/2} / \lambda_c h^{1/2}\} (x_c^2 + 2x_c + 2)^{1/2} \exp\left(-\frac{x_c}{2}\right) \left(\int \frac{\alpha \cos \theta}{\pi}\right)^{1/2}. \qquad (7)$$

This expression is valid for $\lambda \sim 10 \, \mu$m and $T \sim 300$ K and is discussed in detail by Kruse *et al* (1962).

Using equations (6) or (7), putting $A = 1$ cm², $\Delta f = 1$ Hz and $\eta = 1$, P_N can be calculated for the ideal detector as a function of λ_c and as before the specific detectivity D^* is defined by

$$D^* = 1/P_N. \qquad (8)$$

Curves of D^* are usually plotted for a hemispherical field of view, but when considering the performance of an actual detector it should of course be compared with the specific detectivity of an ideal detector having the same field of view. This can be found from the value for a hemispherical field of view by using the appropriate value of $\int \alpha \cos \theta / \pi$. Equations (2) and (8) are illustrated in figures (4) and (1) respectively.

1.2. Choice of a suitable infrared detector

There are now so many different types of infrared detector available that the intending user may have some difficulty in choosing the most suitable. The only consideration which arises in defining the ideal detector is the noise behaviour and it is easy to measure how closely any actual detector approaches the ideal in this respect. However other factors must be considered, and some of them may be of greater importance than the noise characteristics or detectivity.

One obvious factor is the spectral response. Although in principle the thermal detectors can be used over the whole electromagnetic spectrum, their actual performance will be limited to fairly broad spectral bands by factors such as the absorption characteristics of the detector material or the transmission of the window when, as is almost always the case, it is necessary to completely encapsulate the detector. Photon detectors have an even more restricted spectral response, limited in the more straightforward types by a long wave threshold. It is best to choose a photon detector whose spectral response matches as closely as possible the spectral band being used. It would be a mistake to choose a detector covering a much wider band since firstly the unused portion of the spectral response band merely adds noise so degrading the performance, and secondly the broader the response the lower the detector operating temperature is likely to be, introducing penalties of inconvenience and extra cost.

A second important factor is the response time. One can specify for any application the time available to acquire the required data or the information bandwidth required. For many applications the time available may be relatively long, so that the comparatively slow thermal detectors will be suitable. These have response times typically of the order of 1 ms whilst the response time of photon detectors is typically $< 1 \, \mu$s. Some electro-optic systems require bandwidths of 1 GHz or greater. Detectors satisfying this requirement can be designed, but usually at the price of some sacrifice in detectivity. This in fact may not be very serious since this requirement implies a laser system where the use of heterodyne detection is practicable. The performance of a heterodyne receiver does not depend directly on the detectivity but on the quantum efficiency of the detector. Thus by ensuring that a high proportion of signal quanta are absorbed by the detector a good signal-to-noise ratio can be achieved by applying sufficient local oscillator power to the detector. The optimum design for a laser detector may therefore be significantly different from one intended for use with incoherent radiation. The laser detector may require a much shorter response time and be able to operate when irradiated by a relatively high level of laser power but it may not need the ultimate in detectivity.

In addition to factors set by the application, considerations of robustness, convenience, reliability and cost will also influence the choice. Where possible detectors operating at room temperature will normally be preferred, but when the best attainable performance is required in the 5–15 μm spectral region operation at temperatures down to liquid nitrogen is still necessary. At longer wavelengths (15 μm–1000 μm) operation at even lower temperatures down to liquid helium will be required. Despite this, however, the number of applications at the longer wavelengths for which adequate performance can be obtained from uncooled detectors is steadily increasing.

2. New photon detectors

In the majority of photon detectors the incident photons excite electrons from a bound to a mobile state, thus enhancing the conductivity, as in a photoconductor or allowing the electrons to drift from the bound centres so creating an electric field as in the photovoltaic or photoelectromagnetic detectors. New detectors of these types have been produced using new materials but in addition other processes have been utilized of a rather different type. The first of these, which has now been exploited for some time, was the free electron photoconductive or electron bolometer effect which is used in the so-called Putley or Rollin detectors (Putley 1960, 1973). The use of two-photon excitation processes (Gibson *et al* 1968) to enable low energy photons to excite a higher energy electronic transition is perhaps a special case of the normal photoconductive effect, but the photon drag effect (Gibson *et al* 1970, Kimmitt 1971) in which momentum is transferred from a photon stream to the free carriers in a semiconductor bears a relation to free electron photoconductivity similar to that between the normal Seebeck effect and the phonon drag thermoelectric effect.

The recent developments in photoconductive materials have enabled intrinsic 10 μm detectors to be produced so that several types of detector using mercury cadmium telluride (Chiari and Jervis 1971, Long and Schmidt 1970, Cohen-Solal and Riant 1971) or lead tin telluride (Melngailis and Harman 1970, Rolls *et al* 1972) are now available commercially. These detectors operate at temperatures of 77 K or higher but their performance is comparable with that of the earlier copper or mercury doped germanium detectors which can only operate at much lower temperatures. Because of their greater convenience, the new intrinsic detectors are replacing the older extrinsic germanium detectors for applications requiring high performance 10 μm detectors. However at longer wavelengths, say 15–160 μm, extrinsic germanium detectors are still required when the highest performance is essential.

Two other types of detector which do not fit into the simple pattern describing infrared detectors are both junction devices. The first is the metal–oxide–metal point contact detector (Hocker *et al* 1968, Bradley *et al* 1972) which operates in a very similar way to the more familiar microwave catswhisker diode but which has been used at wavelengths as short as 5 μm. The second is the Josephson superconducting junction detector (Grimes *et al* 1968, Blaney 1971) which has a number of applications in the sub-mm region. Both of these detectors are useful for research applications when a very wide frequency response is required.

2.1. *Intrinsic photon detectors*

The search for semiconductors with intrinsic energy gaps small enough for excitation by 10 μm radiation started after the discovery of the PbS, PbSe, PbTe group of photoconductors. This search was soon rewarded in the sense that narrow gap semiconductors were discovered relatively early (Lawson *et al* 1959). However it is only within the last few years that it has been possible to utilize some of these materials to produce practicable devices. At present two materials are being used—mixed compounds of mercury telluride/cadmium telluride and of lead telluride/tin telluride. By varying respectively the Hg/Cd or the Pb/Sn ratios the energy gap can be adjusted to match wavelengths within the range 5–15 μm. Both photoconductive and photovoltaic HgCdTe devices are being manufactured whilst PbSnTe devices are usually photovoltaic. The development of these detectors is not yet completed, but at present photoconductive HgCdTe detectors are available which operate at 77 K and have a detectivity comparable with that of Hg or Cu doped Ge intrinsic detectors, which of course require much lower operating temperatures. The response time of the most sensitive HgCdTe detector is about 1 μs, which is relatively slow for photon devices. However the PbSnTe photovoltaic detectors (also operating at 77 K) (Rolls *et al* 1972, Rolls and Eddolls 1973) have response times of less than 50 ns, and the most sensitive of these devices have a detectivity approaching that of the best HgCdTe detectors. Photovoltaic HgCdTe detectors are being produced with even shorter response times ($\sim$1 ns) (Vérié and Sirieix 1972). The most sensitive have a detectivity comparable with that of the PbSnTe devices (Cohen-Solal and Riant 1971). It is unlikely that PbSnTe devices can be produced with such short response times because compounds of this type have extremely large dielectric constants so that the response time will be limited by the dielectric relaxation time. In addition to 10 μm detectors, HgCdTe detectors have been designed for operation in the 5 μm region. These detectors operate at higher temperatures than 77 K, at $-80\,°C$ or at room temperature (Chiari and Jervis 1971, Elliott *et al* 1974). They have comparable or greater detectivity than InSb detectors so that they could well replace the InSb detector for some applications. At present the HgCdTe and PbSnTe offer comparable performance for many applications, although as the development of these devices is completed it is probable that one will largely replace the other (although I am not prepared to guess which one!) except for 10 μm laser applications requiring the shortest response time. Here the HgCdTe detector is the more suitable, although this in its simplest form may not be as fast as the fastest extrinsic Ge detector. If it proves feasible to develop more complex graded gap structures (Marfaing *et al* 1964), then these could have response times as short as 10^{-10} s which would be comparable with that of the fastest Ge devices.

The performance of some of these detectors is shown in figures 1 and 4.

2.2. *Extrinsic photon detectors*

The first photon detectors useful at 10 μm or longer wavelengths employed the photoexcitation of carriers from impurity centres in semiconductors such as B in Si or Zn, Cu or more recently Hg in Ge. Their characteristics have been discussed by Putley (1966). These detectors usually operate at temperatures in the range 4–30 K which, whilst not inconvenient for research applications, do restrict their industrial use.   For the detection of thermal radiation in the 8–13 μm band these detectors are being replaced by the new intrinsic photoconductors, but there are several areas where the extrinsic photon detectors are still required.   For wavelengths in the band 15–120 μm they are still the only convenient photon detectors and are especially useful for research into lasers, plasma discharges and infrared astronomy in this part of the spectrum.   The long wavelength limit is set by the ionization energy of the shallowest impurity centres in Ge, such as B, Ga or Sb.   By using even shallower centres in other semiconductors such as GaAs (Stillman *et al* 1968, 1971) this limit can be extended to about 400 μm (see figure 1) and free carrier photoconductivity in InSb is widely used for wavelengths of the order of 100 μm to beyond 1000 μm or 1 mm.

At 10 μm the extrinsic detectors have not yet been completely replaced by the intrinsic ones.   The extrinsic detectors are used where very short response times are required.   The response time is determined by the concentration of compensating

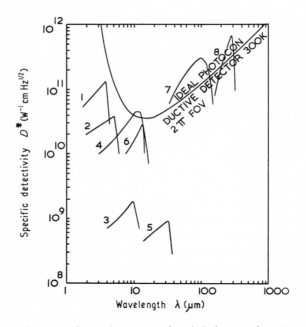

Figure 1.   Spectral response of cooled photoconductors.

(1) $Hg_{1-x}Cd_xTe$	× 33 %	⎫	
(2) $Hg_{1-x}Cd_xTe$	× 27 %	Temperature −80 °C	180° field of view
(3) $Hg_{1-x}Cd_xTe$	× 18 %	⎭	
(4) $Hg_{1-x}Cd_xTe$	× 20 %	Temperature 77 K	60° field of view
(5) $Hg_{1-x}Cd_xTe$	× 18 %	Temperature 24 K	24° field of view
(6) Pb_xSn_xTe photovoltaic	× 80 %	Temperature 77 K	180° field of view
(7) Ga doped Ge		Temperature 4 K	25° field of view
(8) Extrinsic GaAs		Temperature 4 K	15° field of view

Curves (1), (2) and (3) are based on Chiari and Jervis (1971), curve (4) on the Mullard Infrared Detector Handbook, curve (5) on Saur (1968), curve (6) on Rolls *et al* (1972), curve (7) on Moore and Shenker (1965) and Jeffers and Johnson (1968) and curve (8) on Stillman *et al* (1971).

impurity centres so that, by deliberately making this concentration large, response times of the order of 0·1 ns can be achieved (Arams *et al* 1970). These are required for laser systems encountering large Doppler frequencies.

In some space and infrared astronomical applications, the thermal background radiation is considerably less than in normal terrestrial applications. Under these circumstances the detectivity of a detector will ultimately be limited by internal noise sources such as Johnson noise. It turns out that in this case the extrinsic detectors have a significantly higher detectivity than the intrinsic detectors, although the penalty to be paid is that their resistance becomes very high and their response times rather long.

2.3. Nonlinear photon effects

With the high photon densities encountered using intense laser sources effects occur which can be employed for detecting high power sources. For applications where sensitivity is not important it is thus possible to make simple, rugged, fast detectors operating at room temperature.

One of the first effects to be discovered was two-photon excitation of an intrinsic semiconductor. A photon of approximately 5 μm wavelength is required to raise an electron from the valence to the conduction band in InSb. However when InSb is irradiated at room temperature with a sufficiently intense beam of 10 μm CO_2 laser radiation a small increase in intrinsic carrier concentration is observed. The energy of a single photon is insufficient to raise an electron across the gap, but the magnitude of the effect varies as the square of the photon intensity, implying a two-photon process which has been confirmed by detailed comparison of theory and experimental results. By utilizing this effect standard room temperature InSb detectors can be used to monitor high intensity CO_2 lasers. Because the process is a fast one, it is suitable for pulsed or Q-switched lasers (Gibson *et al* 1968).

A second effect which has been utilized for monitoring high power CO_2 lasers is the interaction between photons and conduction electrons in a semiconductor. A sufficiently intense beam of photons will impart in its direction of propagation a momentum component to the electrons. The effect is thus similar to that employed in the Crookes rotating vane radiometer. Since the net electron current must be zero, an electric field will be set up in the direction of propagation of the incident radiation of magnitude to balance the photon drag force. Devices using this effect operate at room temperature, are robust, and have a very short response time so that they are useful for studying the pulse shape of high power Q-switched CO_2 lasers. The responsivity is rather low, although proportional to incident intensity, so that this detector is not suitable when a high sensitivity is required (Kimmitt 1971).

2.4. Metal–oxide–metal point contact detector

This detector evolved from the microwave point contact diode. Its response time is so short that it has been used as a mixer for different laser wavelengths down to about 5 μm, enabling their frequencies to be compared directly with microwave sources (Bradley *et al* 1972) (figure 2). It is probably the most sensitive detector available for this application but it is an extremely fragile device so that its use will probably remain restricted to such very specialized purposes as absolute measurement of the frequency of laser radiation. This, combined with absolute measurement of the wavelength, provides a new method for the absolute determination of the velocity of light.

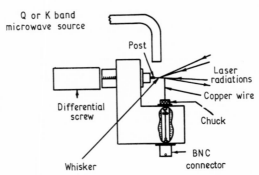

Figure 2. Schematic diagram of Metal–Oxide–Metal junction detector. With acknowledgments to Bradley *et al* (1972).

In a microwave detector a finely pointed tungsten wire is pressed against a semi-conductor forming a rectifying contact. For efficient operation the capacity across the contact must be kept as small as possible. This means that the tungsten catswhisker must be etched to a very fine point and that only very light contact pressure can be used. As the frequency is increased these requirements become more stringent. During experiments to determine the optimum configuration for a sub-mm detector it was found that the optimum contact pressure was so low that when the tungsten wire was pressed directly against the brass stud on which the semiconductor wafer was normally mounted, the pressure was too small to drive the point through the thin oxide layer normally present on the metal surface and this metal–oxide–metal contact performed as efficiently as the usual metal–semiconductor contact. More recently it has been found that the M–O–M detector can be used down to wavelengths as short as $3.39\,\mu$m (Evenson *et al* 1973). For efficient operation a detector of this sort must be small compared with the wavelength. Even in the microwave region this condition can only be satisfied approximately. It is virtually impossible to make the length of the catswhisker small compared with $10\,\mu$m, but by applying antenna theory to the whole detector configuration (Kwok *et al* 1971) it is possible to arrive at a relatively efficient design. Because the contact region itself must be small the device is extremely fragile and even when used with the greatest care the average lifetime of a usable contact has been reported to be only about one hour. Whether by the application of modern microcircuit construction techniques it will be possible to evolve a more practical design is rather dubious, but if it were possible this type of device would provide a most useful laser detector since it operates at room temperature and has the highest reported detectivity of any $10.6\,\mu$m room temperature heterodyne detector (Abrams and Gandrud 1970).

2.5. *The Josephson detector*

The Josephson effect occurs at the junction between two superconductors. It is responsible for non-ohmic contact effects as revealed by the direct current–voltage characteristics. These show discontinuities which are associated with the emission and absorption of photons of energy comparable with that of the superconducting energy gap. Thus the effect can be used to detect sub-mm radiation. It has been used as a sensitive fast detector for radiation of wavelength about 1 mm and it is especially useful as a mixing element for combining sub-mm and microwave radiation, thus enabling the frequency of sub-mm sources to be compared directly with the primary frequency standard (figure 3). Since the Josephson detector consists essentially of a small area metal–metal contact in liquid helium, it tends to be a rather fragile device (although not as fragile as the M–O–M detector!). Recently improved designs (Blaney 1972) have been produced so that although it will probably remain a laboratory tool it should prove a reasonably reliable one, not only for use as a detector but also as a harmonic generator used in infrared frequency synthesis (Halford *et al* 1972).

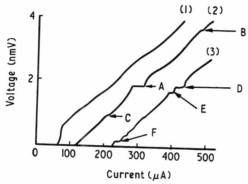

Figure 3. Current–voltage characteristics of a niobium–niobium point contact Josephson junction showing detection and mixing of HCN laser lines. With acknowledgments to Blaney (1972).

Curve (1), no monochromatic radiation incident on junction.

Curve (2), with radiation from an HCN laser operating at 891 GHz (337 μm); current zero for this curve at 100 μA on scale. A, fundamental laser frequency step $2eV_0 = \hbar\omega$; B, harmonic step $2eV_0 = 2\hbar\omega$; C, subharmonic step $2 \times 2eV_0 = \hbar\omega$.

Curve (3), with radiation from an HCN laser operating at two frequencies 891 and 805 GHz; current zero at 200 μA on scale. D, fundamental laser frequency step for 891 GHz $2eV_0 = \hbar\omega_1$; E, fundamental laser frequency step for 805 GHz $2eV_0 = \hbar\omega_2$; F, step corresponding to mixture of 891 and 805 GHz lines $2eV_0 = \hbar(\omega_1 - \omega_2)$.

3. Modern thermal detectors

In principle, any thermometric system can be used as a thermal infrared detector, and in fact the first infrared detectors were simple gas or liquid-in-glass thermometers. However a simple thermometer does not make an efficient infrared detector. Unless suitably designed, it will not absorb the radiation efficiently and its thermal mass will be so large that its response time will be extremely long. The first true infrared detector was the thermopile designed by Nobilli and used by Melloni in his studies of thermal radiation (see Scott Bar 1962). This was followed some years later by the resistive metal bolometer invented by Langley. Both these types of detector are still widely used and in fact have been the subject of recent developments. A newer thermal detector which has tended to replace them for spectroscopic and other applications requiring high sensitivity is the Golay cell. One limitation of the Golay cell has been its reliability (or lack of ruggedness?), but this has been improved significantly by the use of modern components in its construction. A detector which has been developed since the Golay cell and which now appears to be equalling or bettering it in performance while being more convenient in use is the pyroelectric detector. Recent work on several other types of thermal detector has shown that there are others which could compare favourably with those already in use (Benton and Mytton 1971, Walser *et al* 1971).

One of the principal advantages of the thermal detectors is that they can operate at room temperature. Nevertheless, an improvement in detectivity can be obtained by designing thermal detectors to operate at liquid helium temperatures. This was first achieved by the superconducting bolometer, but more recently cooled semiconductor bolometers have been used in infrared and sub-mm astronomy where the highest sensitivity is required. Such detectors do not in principle offer any advantage over photon detectors operating at low temperatures, but one practical advantage is that they offer a broader spectral response. Where only a narrow band response is needed, this can be selected by placing a cooled filter in front of the detector. Since the limiting noise is that associated with the warm background radiation transmitted by the filter, the ideal detectivity of a detector of this type is no different from that of a photon detector used with the same filter. Cooled semiconducting bolometers may be slightly easier to construct than photon detectors, especially for use at the longer wavelengths, because the semiconductors required may be easier to produce than those for photon detectors, but this advantage is rather marginal.

Figure 4 shows the performance of several detectors at room temperature.

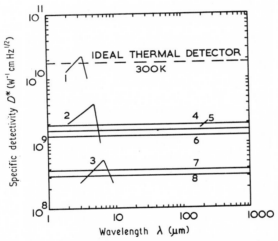

Figure 4. Performance of some room temperature detectors.

    (1) $Hg_{1-x}Cd_xTe$ $\times 35$ %
    (2) $Hg_{1-x}Cd_xTe$ $\times 26$ %
    (3) $Hg_{1-x}Cd_xTe$ $\times 20$ %
    (4) TGS pyroelectric detector
    (5) Golay Cell
    (6) Spectroscopic Thermopile
    (7) Thermistor Bolometer
    (8) Thin Film Thermopile.

    Curves (1), (2) and (3) are based on Chiari and Jervis (1971). Curves (7) and (8) are based on the Barnes Engineering Detector Brochure. Curves (4) to (8) are idealized in that they assume perfect transmission through the detector window, and uniform absorption by the detector itself. In practice the spectral transmission of the window will determine the spectral response obtainable and in addition the performance of thermopiles and bolometers tends to fall off beyond about 20 μm due to incomplete absorption by the blackening of the detector.

3.1. Thermopiles

A recent review by Stevens (1970) discusses the characteristics of modern thermopiles. The first thermopile used metal–metal junctions and ones similar to those designed originally by Moll employing copper–constantan and manganin–constantan junctions are still widely used. They are useful where a robust, stable, highly reproducible detector is required but where the highest sensitivity is not needed. A somewhat better sensitivity can be obtained using silver–bismuth junctions at a small sacrifice in robustness. Applications for these detectors include absolute radiometric measurements (Gillham 1970), determination of geophysical properties such as the solar constant and in radiation pyrometers for high temperature measurements in industrial installations such as steel plants.

Thermopiles using semiconductor elements have a greater sensitivity than those using metal couples, but are much more fragile (Schwarz 1952). They are still widely used in spectrometers and similar instruments but their design has remained unchanged for some years (figure 4). Since there have been no recent improvements in semi-conducting thermoelements the performance attainable from these detectors is not likely to improve and it is likely that these detectors will be replaced by the new pyro-electric detectors (see below) which can offer both a better performance and greater reliability.

By making use of modern techniques for vacuum deposition and photolithography thin film thermopiles of small area and complex form are being produced for use mainly in space applications but they also are used in some radiation pyrometers and as laser detectors. The method of construction is to evaporate patterns of two different metals on a substrate, with the patterns overlapping to form the junctions. The choice of substrate depends on the application. If a sensitive but relatively slow device is required a thin plastic membrane is used as the substrate (figure 4) since this will make a negligible contribution to the thermal characteristic of the device, but if a fast detector is required an insulating substrate with a good thermal conductivity is needed to form a heat sink. Sapphire or beryllium oxide are suitable for this application (Day *et al* 1968). The metals usually used are antimony and bismuth since these have good thermoelectric properties and evaporate easily, but silver–bismuth is sometimes used. Thermopiles constructed this way can have response times less than 30 ns and have been used as heterodyne detectors of CO_2 laser radiation (Contreras and Gaddy 1971). However the responsivity is rather low ($\sim 5 \times 10^{-6}$ V W^{-1}) which means that they are useful as simple monitors working at high power level but would not be used where a sensitive laser detector is required.

3.2. Bolometers

Both metal film and thermistor bolometers operating at room temperature have remained in use. Recently attempts to develop improved bolometers by using newer materials with larger temperature coefficients have been made. A promising material is semi-conducting lanthanum-doped barium strontium titanate (Mendelsohn *et al* 1966). This has a positive temperature coefficient as large as 10 %, compared with values of about 4 % in thermistor elements. Use of positive temperature coefficient material gives more stable operating conditions. Attempts to utilize this new material (Benton and Mytton 1971) have not so far succeeded in exploiting it. The performance obtained has been limited to a D^* of about 4×10^5 which is set by the large current noise found in experimental devices. It should be possible to reduce this by developing improved techniques for fabricating the elements.   Cheap bolometers using thin films of chalco-genide glasses are being developed (Bishop and Moore 1973).   These can be produced very easily and could be employed as robust but simple heat sensors.

Liquid helium cooled bolometers are still employed in far infrared spectroscopy and astronomy where the performance of the uncooled detectors is inadequate.   In addition to the carbon and germanium bolometers described earlier, useful bolometers employing silicon and gallium arsenide have been introduced recently.   It is not clear whether from the point of view of performance these newer bolometers offer any signifi-cant advantage over the Ge bolometer.   The selection of the optimum material for constructing this type of bolometer is very critical.   Any advantage the newer ones will have over the Ge bolometer will be in ease of manufacture, but these devices are used in such small numbers that it is not yet clear whether there is much to choose between them.

A detailed study of the Ge bolometer has been published by Zwerdling *et al* (1968). The performance of these bolometers improves as the temperature is reduced below 4 K. The simplest procedure is to pump the helium below the λ point.   In this way an opera-ting temperature between 1 K and 2 K can be obtained, but Drew and Sievers (1969) have described a He^3 cryostat operating at about 0·37 K.   An In-doped Ge bolometer was used in this cryostat, and a noise equivalent power of 3×10^{-14} W Hz$^{-1/2}$ was claimed. A comparable performance has been claimed by Kinch (1971) for a Si bolometer.

3.3. The Golay cell

This is one of the most widely used thermal detectors.   The incident radiation is absorbed by a metal film designed to match into free space.   The heat absorbed by the film is transferred to the gas surrounding it (usually xenon) which expands, deflecting a mirror forming part of the wall of the enclosure.   The movement of the mirror deflects a beam of light directed at a photocell which produces the output signal.   Because the main source of noise in the Golay cell is temperature noise associated with the thermal exchange between the absorbing film and the gas, it is possible to achieve a good detectivity, within about one order of that of an ideal thermal detector (Golay 1947) (figure 4). The principal disadvantages of the Golay cell are that it is rather slow, is sensitive to vibration and does not tolerate careless handling.   Recently its reliability has been improved by using a GaAs diode for the lamp and replacing the photocell by a Si photodiode (Hickey and Daniels 1969, Firth and Davies 1971).   The use of these components reduces the heat dissipation inside the Golay cell's case and this may be the reason why the reliability has improved.   The detectivity of the improved Golay detector is essentially the same as that obtainable from the earlier design.

3.4. Pyroelectric detectors

Considerable progress has been made in developing pyroelectric detectors (Putley 1971, Hadni 1971).   It now appears that for most purposes triglycine sulphate (TGS) or its derivatives are the most suitable materials, although other pyroelectrics such as strontium barium niobate (SBN) (Glass and Abrams 1971), lithium sulphate or members of the lead zirconate titanate family (Mahler *et al* 1972, Yamaka *et al* 1972) may be more suitable for some specific applications.

The optimum choice of material properties has been discussed in detail by Putley (1970).   It is not possible to define a single figure of merit applicable to all situations, but the relevant material properties are the pyroelectric coefficient, the complex dielectric constant and the thermal capacity. In a desirable material the pyroelectric coefficient should be large, both components of the dielectric constant small and the thermal capacity small. TGS does not possess the largest known pyroelectric coefficient at room temperature, higher values being found in SBN and in some doped lead zirconate titanate ceramics, but the materials with larger pyroelectric coefficients have much larger dielectric constants with large loss factors (high equivalent conductivity). For this reason their overall performance is inferior. In the best pyroelectric detectors the dominant source of noise is the Johnson noise associated with the dielectric loss, followed by the amplifier noise. The amplifier noise becomes dominant at high frequencies ($\geqslant 10$ kHz) and in small area detectors at lower frequencies ($\leqslant 10$ Hz). Where a detector of area less than 10^{-3} cm² is required for use at low frequencies then a better performance may be obtained using one of the materials with a higher dielectric constant. The capacity of a small TGS detector could fall so low that it could become less than the amplifier input capacity. This would lead to a deterioration in performance offsetting the advantages of TGS relative to SBN or one of the ceramics. There is therefore an application for these other materials despite the general superiority of TGS. Detectors made from these materials are somewhat more robust than TGS, although this advantage has been exaggerated, and they are being produced to meet requirements for simple cheap detectors for applications not requiring the highest performance (Doyle 1970, 1971).

The improvement in the performance of TGS detectors has been brought about by improvements in the quality of TGS and of the performance of the associated high input impedance low noise amplifier. One of the problems with the earlier TGS detectors was a tendency for multi-domain samples to depole if the temperature rose above the Curie point (49 °C). This cannot happen if single domain elements are used, but their production from TGS of suitable quality is very difficult (Furuhata 1970). The discovery that the introduction of about 1 % of the optically active amino acid alanine introduces an internal bias, which prevents the TGS domain reversing, removed the problem of the detectors depoling spontaneously (Lock 1971). It also reduced the dielectric loss associated with domain switching, which has contributed to the improvement in sensitivity.

The performance of a recent pyroelectric detector is shown in figure 4. The NEP obtained compares very favourably with that of a Golay cell, and in addition the frequency response and the reliability are much superior (Baker *et al* 1972). At low frequencies the detectivity is within an order of that of an ideal thermal detector, but some further improvement resulting from both further improvements in the material and the amplifier is still possible. The results put the pyroelectric detector amongst the best uncooled thermal detectors and indicate that although at present pyroelectric detectors are not very widely used, their importance in future is likely to be considerably greater.

The possibility of using a pyroelectric detector as a heterodyne laser detector has been considered in some detail, and heterodyne operation with both the 337 μm HCN laser (Gebbie *et al* 1967) and the 10·6 μm CO_2 (Abrams and Glass 1969) laser has been demonstrated.

The attractions of using a pyroelectric detector in this way are that it operates at room temperature and it can be of robust construction. The main objection is that, although as thermal detectors go it is a fast device, it cannot compete with the fastest photon detectors. Since many laser applications require a very wide bandwidth, the pyroelectric detector is at a disadvantage. In principle by applying enough local oscillator drive the sensitivity can be made to approach that of an ideal detector, as has been achieved with photon detectors at $10.6 \, \mu$m, but so far this has not been demonstrated with the pyroelectric detector.

For those applications, such as monitoring high power pulsed lasers, not requiring a high sensitivity, the output of a pyroelectric detector can be corrected to give an un-distorted reproduction of μs or shorter pulses. The sensitivity attained is then limited by amplifier noise but the ultimate frequency of operation is set by piezoelectric resonances in the pyroelectric element. With properly designed elements these will occur at frequencies greater than 100 MHz so that the effective upper frequency limit is probably set by the amplifier noise.

3.5. *Other types of thermal detector*

Since practically every material property has some temperature dependence, the number of possible types of thermal detector is very large. Several effects have been studied in some detail but have not been fully exploited. The utilization of temperature-dependent dielectric properties of various types of dielectric including ferroelectrics has been considered several times (Moon and Steinhardt 1938, Hanel 1960, Maserjian 1970). The pyromagnetic effect is the magnetic analogue of the pyroelectric effect and leads to a corresponding family of devices (Walser *et al* 1971). The Nernst effect in mixed crystals of InSb–NiSb (Paul and Weiss 1968), Cd_3As_2–InAs (Goldsmid *et al* 1972) and bismuth and bismuth–antimony alloys (Washwell *et al* 1970, 1971) has also been considered. In addition to the simple superconducting bolometer, superconducting detectors employing a non-uniform temperature distribution arranged so that only a part of the element is superconducting depending upon the signal level (Franzen 1963) and employing the change in the reactive (inductive) component of the superconductor's impedance (Naugle and Porter 1971) have both been proposed.

Whilst these detectors could be developed further the considerable effort required hardly seems justified since none of them is likely to offer a significant advantage over the ones already available. Since the best available detectors approach closely the fundamental performance limits emphasis of new development is passing to such aims as cheapness of production, reliability and ease of use. These objectives are more likely to be attained by further refinement of the already established designs.

Finally, one elegant detecting system whose performance approaches very closely the fundamental limit for an uncooled detector, but because of its relative complexity is not likely to be widely used, is Jones' optical lever which uses the linear expansion of a strip of constantan (Jones and Richards 1959). With a strip of area 5×0.2 mm^2 a NEP of 10^{-11} W Hz$^{-1/2}$ was achieved, corresponding to a D^* of 10^{10}, within a factor of two of the fundamental limit, and better than the values reported for any other thermal detector operating at room temperature.

4. Conclusion. Summary of detector characteristics

The characteristics of the principal types of detector discussed are summarized in table 1. This is intended to give a general guide. It must be remembered that not only do inventors and manufacturers tend to take a generally optimistic view of their products but also the spread in performance between different samples is often an order or more. In table 1 we have tried to select above average but realistic values. Table 1 only refers to detectors discussed in this paper. The tables in the earlier review (Putley 1966) should be consulted for the other principal types of detector.

Table 1.

Type	Operating temperature	Typical area (cm²)	Responsivity (V W⁻¹)	Response time	Spectral response peak wavelength (λ_p)	Specific peak wavelength detectivity (D^*) (cm W⁻¹ Hz$^{1/2}$) or Noise-equivalent power (NEP) (W Hz$^{-1/2}$)	Notes	Availability
Photoconductive cadmium-mercury telluride	77 K	$3 \cdot 8 \times 10^{-3}$ / $6 \cdot 4 \times 10^{-5}$	100 / 1000	< 1 μs	$9 \cdot 5$–15 μm / ~10 μm for Cd 20 %	$D^*(\lambda_p)$ 4×10^{10}	D^* measured with 60° field of view. Below 1 kHz performance degraded by $1/f$ noise. Element resistance 20–1000 ohms	Available from Mullard (UK), Honeywell (USA), SBRC (USA)
	−80 °C			200 ns / < 50 ns	λ_p $3 \cdot 6$ μm / λ_p 5 μm / λ_p $9 \cdot 5$ μm	$D^*(\lambda_p)$ $1 \cdot 5 \times 10^{11}$ / $D^*(\lambda_p)$ 4×10^{10} / $D^*(\lambda_p)$ $1 \cdot 8 \times 10^9$	33 % Cd / 27 % Cd / 18 % Cd — D^* measured with 180° field of view at 20 kHz modulation frequency	experimental samples from Mullard
	+20 °C			300 ns	λ_p $3 \cdot 4$ μm / λ_p $4 \cdot 9$ μm / λ_p $6 \cdot 5$ μm	$D^*(\lambda_p)$ $2 \cdot 5 \times 10^{10}$ / $D^*(\lambda_p)$ 4×10^9 / $D^*(\lambda_p)$ 5×10^8	35 % Cd / 26 % Cd / 20 % Cd — D^* measured with 180° field of view at 20 kHz modulation frequency	
Photovoltaic cadmium-mercury telluride	77 K			5–0·5 ns	λ_p as for pc HgCdTe	$D^*(\lambda_p)$ ~ 10^{10}	Even shorter response times may be obtained from graded gap structures	Available from SAT (France)
Photovoltaic lead-tin telluride	77 K	4×10^{-4} to 4×10^{-3}	up to 500	20–100 ns	λ_p 12 μm	$D^*(\lambda_p)$ 3×10^{10}	~80 % Pb. D^* measured 180° field of view at 800 Hz. Surface anti reflection coated with ZnSe quantum efficiency ≮ 80 %. Slope resistance at operating bias 20–300 ohms	Available from Plessey (UK)

45

Table 1. (*Continued*)

Type	Operating temperature	Typical area (cm²)	Responsivity (V W⁻¹)	Response time	Spectral response peak wavelength (λ_p)	Specific detectivity (D^*) (cm W⁻¹ Hz$^{1/2}$) or Noise-equivalent power (NEP) (W Hz$^{-1/2}$)	Notes	Availability
Extrinsic photoconductive Ge doped Hg	$\ngtr$ 35 K		up to 10^4	$\sim$ 100 ns; down to 0·3 ns	λ_p 11 μm	$D^*(\lambda_p)$ 2×10^{10} (60° FOV)	Lightly compensated material for high detectivity. Heavily compensated for wide band application. Quantum efficiency > 50%. Heterodyne NEP $1·9 \times 10^{-19}$ W Hz⁻¹ with 50 MHz bandwidth	Available from Mullard (UK), Texas Instruments (USA), SBRC (USA), Raytheon (USA)
Ge doped Cu	4 K			$\sim$ 100 ns; down to 0·1 ns	λ_p 25 μm	$D^*(\lambda_p)$ $1·5 \times 10^{10}$	Lightly compensated high detectivity material. With reduced background D^* can rise to 10^{15}. Heavily compensated for wide band applications NEP $\sim 10^{-20}$ W Hz⁻¹ measured in narrow band heterodyne system. Frequency response > 1 GHz can be obtained	Available commercially
GaAs	4 K	0·2	$> 10^5$	< 10 ns	λ_p 282 μm	$D^*(\lambda_p)$ 5×10^{11}	Use n type epitaxial GaAs. D^* measured using cooled long pass filter cut off 150 μm. Modulation frequency 260 Hz	Not available commercially

Table 1. (*Continued*)

Type	Operating temperature	Typical area (cm²)	Responsivity (V W⁻¹)	Response time	Spectral response peak wavelength (λ_p)	Specific detectivity (D^*) (cm W⁻¹ Hz$^{1/2}$) or Noise-equivalent power (NEP) (W Hz$^{-1/2}$)	Notes	Availability
Two-photon InSb	room		output ∝ square of input intensity	≯ 200 ns	λ_p double that for normal photoconductivity	very low	InSb photoconductive detectors as used at 5 μm are used (fitted with suitable windows)	Commercially available InSb detectors can be used
Photon drag Ge	room	~1 or larger	≯ 10⁻⁶	~0·1 ns	determined by free carrier absorption	~10³	Fast, robust monitors for high power lasers	Manufactured by Rofin (UK)
Metal-oxide-metal junction	room	very small	~2×10⁻²	< 3×10⁻¹⁴ s	extend from microwaves to a few μm	NEP ~10⁻⁷	Useful as heterodyne detector, frequency mixer and harmonic generator. Heterodyne performance at 10·6 μm and 30 MHz IF 9×10⁻¹⁴ W Hz NEP	Not commercially available
Josephson junction	4 K	very small		≯ 0·1 ns	0·3–1 mm	NEP 10⁻¹³–10⁻¹⁴		Not commercially available
Thermopiles	room	matched to spectrometers slits	~5	~10 ms	visible–medium IR depending on blacking	~10⁹	For spectrometers and similar instruments	Available commercially
	room	10⁻²–10⁻⁴	~100	~100 μs		~3×10⁸	Thin film evaporated onto film support.	
	room		10⁻⁶	~30 ns		~10⁶	For space instrumentation Thin film on high thermal conductivity substrate. Laser detector	

47

Table 1. (*Continued*)

Type	Operating temperature	Typical area (cm²)	Responsivity (V W⁻¹)	Response time	Spectral response peak wavelength (λ_p)	Specific detectivity (D^*) (cm W⁻¹ Hz$^{1/2}$) or Noise-equivalent power (NEP) (W Hz$^{-1/2}$)	Notes	Availability
Bolometer (Ba, Sr, La)TiO₃	room	10^{-3}	~ 10	~ 1 ms	broad—determined by blacking	$3 \cdot 5 \times 10^{5}$	Performance limited by excessive noise in material	Further material studies required
Tl₂SeAs₂Te₃ bolometer	room	10^{-3}–10^{-4}	~ 100	~ 1 s		$\sim 1 \cdot 5 \times 10^{7}$	Of moderate performance but easy to manufacture	Still in experimental stage
Compensated Ge bolometer	4–0·4 K	$\sim 3 \times 10^{-1}$	2×10^{4} -2×10^{6}	$100\ \mu s$ -10 ms	Broad band extending to sub mm and mm regions	NEP 10^{-12} -3×10^{-14}	NEP measured at $f < 500$ Hz. Limited by amplifier noise but improved by reducing temperature which increases responsivity and time constant	Can be obtained commercially
Compensated Si bolometer	1·5 K	10^{-2}–10^{-3}	$\sim 10^{6}$	10 ms	extending to mm region	NEP $2 \cdot 5 \times 10^{-14}$	Measured at 13 Hz	Not commercially available
Golay cell	room	6×10^{-2}		15 ms	visible to mm waves depending on window	NEP 2×10^{-10}		Available from Cathodeon (UK), Eppley (USA)
Pyroelectric	room	10^{-3} to >1	100–1000 V W⁻¹ (depends on area and operation frequency)	With suitable circuits effective response time $<1\ \mu s$	visible to mm waves mainly depending on windows	$D^* > 10^{9}$ at 10 Hz	TGS detectors best for high performance but ceramic or polymer film detectors useful where maximum robustness and simplicity required	Available from several manufacturers, principally Mullard (UK), Barnes (USA) and Laser Precision (USA)

References

Abrams R L and Gandrud W B 1970 *Appl. Phys. Lett.* **17** 150–2
Abrams R L and Glass A M 1969 *Appl. Phys. Lett.* **15** 251–3
Arams F R, Sard E W, Peyton B J and Pace F P 1970 *Semiconductors and Semimetals* (New York: Academic Press) **5** ch 10 409–34
Baker G, Charlton D E and Lock P J 1972 *Radio Electron. Engr* **42** 260–4
Benton R K and Mytton R J 1971 *Conf. Infrared Techniques. IERE Conf. Proc. No* 22 pp41–54
Bishop S G and Moore W J 1973 *Appl. Opt.* **12** 80–3
Blaney T G 1971 *Conf. Infrared Techniques, IERE Conf. Proc. No* 22 61–8
—— 1972 *Radio Electron. Engr* **42** 303–8
Bradley C C, Edwards G and Knight D J E 1972 *Radio Electron. Engr* **42** 321–7
Chiari J A and Jervis M H 1971 *Conf. Infrared Techniques* Unpublished paper
Chol G, Marfaing Y, Munsch M, Thorel P and Combette P 1966 *Les Détécteurs de Rayonnement Infra-Rouge* (Paris: Dunod)
Cohen-Solal G and Riant Y 1971 *Appl. Phys. Lett.* **19** 436–8
Contreras B and Gaddy O L 1971 *Appl. Phys. Lett.* **18** 277–8
Day G W, Gaddy O L and Iversen R J 1968 *Appl. Phys. Lett.* **13** 289–90
Doyle W M 1970 *Laser Focus* July 34–7
—— 1971 *Electro Optics* Sept. 7–8
Drew H D and Sievers A J 1969 *Appl. Opt.* **8** 2067–71
Elliott C T, Jervis M H and Phillips J B 1974 *Proc. Int. Conf. Physics of Semimetals and Narrow Gap Semiconductors, Nice–Cardiff 1973* to be published
Evenson K M, Wells J S, Petersen F R, Danielson B L and Day G W 1973 *Appl. Phys. Lett.* **22** 192–5
Firth J R and Davies L B 1971 *Conf. Infrared Techniques, IERE Conf. Proc. No* 22 pp227–30
Franzen W 1963 *J. Opt. Soc. Am.* **53** 596–603
Furuhata Y 1970 *J. Phys. Soc., Japan* **28** suppl. 425–7
Gebbie H A, Stone N W B, Putley E H and Shaw N 1967 *Nature* **214** 165–6
Gibson A F, Kent M J and Kimmitt M F 1968 *J. Phys. D: Appl. Phys.* **1** 149–54
Gibson A F, Kimmitt M F and Walker A C 1970 *Appl. Phys. Lett.* **17** 75–7
Gillham E J 1970 *Adv. Geophys.* (New York: Academic Press) **14** 53–81
Glass A M and Abrams R L 1971 *Proc. Symp. Sub mm Waves* (New York: Polytechnic Institute of Brooklyn) 281–94
Golay M J E 1947 *Rev. sci. Instrum.* **18** 357–62
Goldsmid H J, Savvides N and Uher C 1972 *J. Phys. D: Appl. Phys.* **5** 1352–7
Grimes C C, Richards P L and Shapiro S 1968 *J. Appl. Phys.* **39** 3905–12
Hadni A 1971 *Proc. Symp. Sub mm Waves* (New York: Polytechnic Institute of Brooklyn) 251–66
Halford D, Hellwig H and Wells J S 1972 *Proc. IEEE* **60** 623–5
Hanel R A 1960 *J. Opt. Soc. Am.* **51** 220–4
Hickey J R and Daniels D B 1969 *Rev. sci. Instrum.* 732–3
Hocker L O, Sokoloff D R, Daneu V, Szoke A and Javan A 1968 *Appl. Phys. Lett.* **12** 401–2
Hogan H and Moss T S (eds) 1970 *Cryogenics and Infrared Detection* (Cambridge, Mass: Boston Technical Publishers)
Hudson R D Jr 1969 *Infrared System Engineering* (New York: John Wiley)
Jeffers W Q and Johnson C J 1968 *Appl. Opt.* **7** 1859
Jones R C 1953 *Adv. Electron.* (New York: Academic Press) **5** 1–96
—— 1959 *Adv. Electron.* (New York: Academic Press) **11** 87–183
Jones R V and Richards J C S 1959 *J. sci. Instrum.* **36** 90–4
Kimmitt M F 1971 *Conf. Infrared Techniques, IERE Conf. Proc. No* 22 69–74
Kinch M A 1971 *J. appl. Phys.* **42** 5861–3
Kosnitser D M and Semendyaev Yu S (eds) 1968 *Thermal Infrared Detectors* (Kiev: Naukova Dumka)
Kruse P W, McGlauchlin L D and McQuistan R B 1962 *Elements of Infrared Technology* (New York: John Wiley)
Kwok S P, Haddad G I and Lobov G 1971 *J. appl. Phys.* **42** 554–63
Lawson W D, Nielsen S, Putley E H and Young A S 1959 *J. Phys. Chem. Solids* **9** 325–9
Lock P J 1971 *Appl. Phys. Lett.* **19** 390–1
Long D and Schmit J L 1970 *Semiconductors and Semimetals* (New York: Academic Press) **5** ch 5 175–255

Mahler R J, Phelan R J Jr and Cock A R 1972 *Infrared Phys.* **12** 57–9

Marfaing Y, Cohen-Solal G and Bailey F 1964 *Physics of Semiconductors, Proc. 7th Int. Conf.* ed M Hulin (Paris: Dunod) 1245–50

Maserjian J 1970 *Appl. Opt.* **9** 307–15

Melngailis I and Harman T C 1970 *Semiconductors and Semimetals* (New York: Academic Press) **5** ch 4 111–74

Mendelsohn L I, Orth E D, Curran R E and Robie E D 1966 *Am. Cer. Soc. Bull.* **45** 771–6

Moon P and Steinhardt R L 1938 *J. Opt. Soc. Am.* **28** 148–62

Moore W J and Shenker H 1965 *Infrared Phys.* **5** 99–106

Moss T S 1965 *Rep. Prog. Phys.* **28** 15–60

—— 1973 *J. Luminesc.* **7** 359–89

Naugle D G and Porter W A 1971 *Proc. Special Meeting on Unconventional Infrared Detectors* (Ann Arbor: University of Michigan, Infrared Information and Analysis Center) 43–8

Paul B and Weiss H 1968 *Solid St. Electron.* **11** 979–81

Pell E M (ed) 1971 *Proc. 3rd Int. Conf. Photoconductivity, Stanford* 1969 (Oxford: Pergamon)

E H Putley 1960 *Proc. Phys. Soc.* **76** 802–5

—— 1964 *Phys. Stat. Solidi* **6** 571–614

—— 1966 *J. sci. Instrum.* **43** 857–68

—— 1970 *Semiconductors and Semimetals* (New York: Academic Press) **5** 259–85

—— 1971 *Proc. Symp. Sub mm Waves* (New York: Polytechnic Institute of Brooklyn) 267–80

—— 1973 *Semiconductors and Semimetals* (New York: Academic Press) **11** to be published

Rogers W M and Powell R L 1958 *Tables of Transport Integrals* National Bureau of Standards Circular 595 July 5

Rolls W H and Eddolls D V 1973 *Infrared Phys.* **13** 143–7

Rolls W H, Waterfield T J, Simkins R S, Sherring C W and Rogers C J 1972 *Radio Electron. Engr* **42** 317–20

Saur W D 1968 *Infrared Phys.* **8** 255–8

Scott Barr E 1962 *Infrared Phys.* **2** 67–73

Schwarz E 1952 *Research* **5** 407–11

Smith R A, Jones F E and Chasmar R P 1968 *Detection and Measurement of Infrared Radiation* Second Edition (London: Oxford University Press)

Stevens N B 1970 *Semiconductors and Semimetals* (New York: Academic Press) **5** ch 7 287–318

Stillman G E, Wolfe C M, Melngailis I, Parker C D, Tannenwald P E and Dimmock J O 1968 *Appl. Phys. Lett.* **13** 83–4

Stillman G E, Wolfe C M and Dimmock J O 1971 *Proc. Symp. Sub mm Waves* (New York: Polytechnic Institute of Brooklyn) 345–59

Vérié C and Sirieix M 1972 *IEEE J. Quantum Electron.* **QE–18** 180–4

Walser R M, Bené R W and Carruthers R E 1971 *IEEE Trans. Electron Devices* **ED–18** 309–15

Washwell E R, Hawkins S R and Cuff K F 1970 *Appl. Phys. Lett.* **17** 164–6

—— 1971 *Proc. Spec. Meeting on Unconventional Infrared Detectors* (Ann Arbor: University of Michigan, Infrared Information and Analysis Center) 89–111

Willardson R K and Beer A C (eds) 1970 *Semiconductors and Semimetals* **5** *Infrared Detectors Part I*

—— 1973 *Semiconductors and Semimetals* **10** *Infrared Detectors Part II*

Yamaka E, Kayashi T, Matsumoto M, Nakamura K, Shigiyama K, Akiyama M and Kitahori K 1972 *Nat. Tech. Rep. Japan* **18** 141–52

Zwerdling S, Smith R A and Theriault J P 1968 *Infrared Phys.* **8** 271–336

II
Fundamentals of Infrared Detection

Editors' Comments on Papers 3 Through 9

These seven papers provide a tutorial introduction to the technical topics that are fundamental to a thorough understanding of infrared detectors. Those who are new to the subject, or those preparing to take on added responsibility in the field, will find these papers of particular value. Four of the papers appeared originally in the Infrared Physics and Technology issue of the *Proceedings of the Institute of Radio Engineers* [**47**(9), 1959]. Professor S. S. Ballard, the Editor of this series of Benchmark Books, was the guest editor of that memorable issue, which stands as a true benchmark in the literature of the infrared.

All who work with infrared detectors owe a particular debt of gratitude to R. Clark Jones, for it is he, more than any other individual, who brought order out of chaos in the methods used to describe the performance of a detector. Jones has recalled that a detector was once described to him as having "a peak-signal-to-rms-noise-ratio of 23 db when there was 45 volts across the cell and 45 volts across the load resistor; when the source was at a temperature of 500°K, was 2.5 cm in diameter, and was 40 cm from the cell; the cell had a sensitive area of 9 mm²; the radiation was square wave chopped at 450 cps, and the noise had a bandwidth of 9 cps" [1]. Compare this mouthful with the present-day description of a detector by its D^* (pronounced dee-star). Jones introduced D^* in 1957 and it remains as sound a description of a detector's performance today as it was then.

Paper 3 is an outstanding example of Jones' precise and lucid style. Detectors are viewed as black boxes with an input and an output, and the terms used to describe them are explained in detail. The paper ends with a discussion of the figures of merit M_1 and M_2 and their use to compare such diverse detectors as phototubes, ideal heat detectors, lead sulfide detectors, radio antennas, the human eye, and photographic film. Further discussion of M_1 and M_2 will be found in Papers 35 through 37. The newcomer, and many old-timers as well, should note Jones' warning: "sensitivity . . . has been used to mean so many different concepts that its use in scientific and technical discourse is to be deplored." Other papers of interest are References 2 and 3.

R. Clark Jones, Research Fellow in Physics at the Polaroid Corporation, Cambridge, Massachusetts, was the 1972 recipient of the Frederic Ives Medal of the

Optical Society of America for his distinguished contributions in optics. His scientific promise had been recognized earlier, when in 1944 he was awarded the Adolph Lomb Medal of the Optical Society.

Contrasting nicely with Jones' "black box" treatment, Paper 4, by Petritz, contains a careful analysis of the basic physics of detection processes. Photoconductive, photo-voltaic, and photoelectromagnetic effects in photon detectors are described as are the effects utilized in thermal detectors. Both signal and noise properties are described, and this leads to the theory of the ideal radiation detector operating in the back-ground limited (Blip) condition. Many modern detectors are capable of Blip opera-tion and Petritz gives precise guidelines for achieving this condition. Other papers of interest are References 4 through 9.

Richard Petritz is President of New Business Resources, Inc., Dallas, Texas.

In Paper 5, Jones picks up and enlarges on Petritz' discussion of the electrical noises found in the output of detectors. Since they place the ultimate limit on detector performance, it is essential for the detector user to have a clear understanding of the characteristics of these noises. Jones distinguishes the various kinds of noise by their physical source and finds that there are a total of eight types that may occur in the output of detectors. Other discussions of noise in detectors will be found in Ref-erences 10 through 13.

In 1962 the U.S. Department of Defense, in a key move, established the Joint Services Infrared Sensitive Element Testing Program (JSIRSETP) at the Naval Ordnance Laboratory, Corona, California (now located at Naval Electronics Labora-tory Center, San Diego, California). The JSIRSETP has played an important role in support of U.S. detector development. It has provided much needed leadership in the development of techniques that have become the standard methods for the measurement of detector performance in the United States. The JSIRSETP, early in their existence, established a program in which the performance of detectors from all manufacturers was measured and the results made readily available to those who needed them. In addition, it has acted as an impartial referee in disputes over proper measurement techniques and in the interpretation thereof. In Paper 6, Potter, Per-nett, and Naugle give a hands-on description of JSIRSETP measurement techniques and the rationale underlying them. Additional information on the measurement of detector performance is contained in References 14 through 18. Hudson gives a particularly detailed treatment complete with specific recommendations for commer-cially available test equipment that can be combined to form an integrated detector test set [19].

A detector-testing laboratory may be called upon to make two types of meas-urements of detector performance. First is the measurement of various detector parameters such as D^*, time constant, resistance, and noise spectra. Second is the calibration of the response of the detector so that it can be used to make absolute measurements of radiation. Essentially the same test equipment can be used for both determinations. As is evident from the description given by Potter, Pernett, and

Naugle in Paper 6, it is customary to use a carefully designed and constructed blackbody as a standard source of radiation. Current understanding of the theory of such sources limits calculations of their radiation output to an accuracy of about ±1 percent [20]. One approach to greater accuracy is to forego the idea of a standard source in favor of an absolute or self-calibrating detector. An example of such a detector is a thermopile with fine heater wires attached. Radiation incident on the thermopile is absorbed, its temperature rises, and a thermal voltage is generated that is proportional to the temperature rise. The incident energy can be measured by shielding the thermopile and passing a current through its heater wires so as to duplicate the heating produced by the incident energy. Thus the response of the detector can be determined in terms of measurable electrical quantities without recourse to any sort of standard source. The primary reference on self-calibrating detectors is that by Gillham [21], while other important material will be found in References 22 through 25.

R. E. Potter is Head, Electronic Materials Science Division, Naval Electronics Laboratory Center, San Diego, California.

The laser, providing, as it does, a high-radiance source of coherent radiation, now makes it possible to use heterodyne and homodyne techniques in the infrared just as they have long been used in the rf and microwave portions of the spectrum. Teich, in Paper 7, develops the theory of heterodyne detection in the infrared. He shows how this theory was applied to the design of a CO_2 laser system operating at 10.6 μm and presents measurements of the minimum power detectable with the system. This work has great significance for infrared communication and radar systems as well as for heterodyne spectroscopy. For other work in this area, see References 26 through 35.

Malvin C. Teich is an associate professor in the Department of Electrical Engineering and Computer Science, Columbia University, New York. Teich received the Browder J. Thompson Memorial Prize Award of the IEEE for Paper 7.

Paper 8, by Melchior, Fisher, and Arams, contains a survey of detectors suitable for use in optical and infrared communication systems. The requirements that must be met for communication use are not the same as those normally considered for detectors to be used in conventional search and detection applications. For communication use a detector must have a high responsivity and an instantaneous bandwidth that is wide enough to accommodate the information bandwidth of the incoming signal. Low-noise demodulation techniques are also crucially important. Additional information on these subjects will be found in References 36 and 37.

Hans Melchior is a Member of the Technical Staff, Bell Telephone Laboratories, Inc., Murray Hill, New Jersey.

A background limited (Blip) detector is customarily equipped with a cooled radiation shield that matches the field of view of the detector to the cone of rays coming from the optics. The purpose of the shield is to prevent photons from the surround from reaching the detector and producing noise in the output of the detector. The value of D^* for a shielded detector is proportional to the amount of shielding provided (this is why the field of view is stated for some of the detectors shown in Papers 1 and 2). In Paper 9, Jones introduces a modified detectivity, called D^{**} (pronounced dee-double-star), that is particularly useful for the description of shielded detectors. D^{**}, like D^*, denotes the detectivity normalized to an area of 1 cm^2 and a bandwidth of 1 Hz, and additionally to an (effective weighted) solid angle of π steradians (which, in most cases, is equivalent to a hemispherical surround). As a result, D^{**}, unlike D^*, is independent of the details of the shielding.

One way to increase the D^* of a detector is by optical immersion, i.e., by placing it in optical contact with a lens made from a material that has a high index of refraction. The theory of immersed detectors shows that when the detectivity is limited by internally generated noise, the value of D^* can be increased by a factor equal to the index of refraction of the immersion element [38]. The use of a germanium immersion lens having an index of refraction of 4 is a relatively inexpensive way of increasing the D^* of a detector by a factor of 4. Techniques and materials for producing immersed detectors are discussed in References 39 through 44.

References

1. Jones, R. C., "A Method of Describing the Detectivity of Photoconductive Cells," *Rev. Sci. Instr.,* **24,** 1035 (1953).
2. Levinstein, H., "Characterization of Infrared Detectors," Chap. 1, in Willardson, R. K., and A. C. Beer (eds.), *Semiconductors and Semimetals,* Vol. 5, *Infrared Detectors,* Academic Press, New York (1970).
3. Wolfe, W. L., "Photon Number D^* Figure of Merit," *Appl. Optics,* **12,** 619 (1973).
4. Rose, A., "Performance of Photoconductors," *Proc. Inst. Radio Engrs.,* **43,** 1850 (1955).
5. Long, D., "Properties of Semiconductors Useful for Sensors," *IEEE Trans. Electron Devices,* **ED-16,** 836 (1969).
6. Keyes, R. J., and R. H. Kingston, "A Look at Photon Detectors," *Phys. Today,* **25,** 48 (Mar. 1972).
7. Sharma, B. L., R. K. Purohit, and S. N. Mukerjee, "Detectivity Calculations for Photovoltaic Heterojunction Detectors," *Infrared Phys.,* **10,** 225 (1970).
8. Sharma, B. L., S. N. Mukerjee, and J. K. Modi, "Detectivity Calculations for n-p Heterojunction Detectors," *Infrared Phys.,* **11,** 207 (1971).
9. Cruse, P. M., and J. S. Lee, "Low Level Switching of Infrared Detectors with Solid State Devices," *Infrared Phys.,* **5,** 65 (1965).
10. Oliver, B. M., "Thermal and Quantum Noise," *Proc. IEEE,* **53,** 436 (1965).
11. Long, D., "On Generation–Recombination Noise in Infrared Detector Materials," *Infrared Phys.,* **7,** 169 (1967).
12. van Vliet, K. M., "Noise Limitations in Solid State Photodetectors," *Appl. Optics,* **6,** 1145 (1967).
13. van der Ziel, A., "Noise in Solid-State Devices and Lasers," *Proc. IEEE,* **58,** 1178 (1970).
14. Jones, R. C., D. Goodwin, and G. Pullan, "Standard Procedure for Testing Infrared Detectors and for Describing Their Performance," Office of Director of Defense Research and Engineering, Washington, D.C., Sept. 12, 1960.

15. Bradshaw, P. R., "Improved Checkout for IR Detectors," *Electronic Industries*, **22**, 82 (Oct. 1963).
16. Brown, M. A. C. S., P. Porteous, and D. J. Solley, "Time Constants of Some Fast Photodetectors Measured Using an Indium Arsenide Laser," *J. Sci. Instr.*, **44**, 419 (1967).
17. Baryshev, N. S., "Measurement of the Time Constant of Radiation Detectors with the Aid of InAs Injection Lasers," *Sov. J. Opt. Tech.*, **36**, 733 (1969).
18. Corcoran, V. J., and W. T. Smith, "Bandwidth Measurement of Wideband Infrared Detectors," *Appl. Optics*, **11**, 881 (1972).
19. Hudson, R. D., Jr., *Infrared System Engineering*, Wiley, New York (1969), Chap. 9.
20. *Ibid.*, pp. 67–82.
21. Gillham, E. J., "Recent Investigations in Absolute Radiometry," *Proc. Roy. Soc. London*, **269A**, 249 (1962).
22. Eppley, M., and A. R. Karoli, "Absolute Radiometry Based on a Change in Electrical Resistance," *J. Opt. Soc. Amer.*, **47**, 748 (1957).
23. Kendall, J. M., Sr., and C. M. Berdahl, "Two Blackbody Radiometers of High Accuracy," *Appl. Optics.*, **9**, 1082 (1970).
24. Geist, J., and W. R. Blevin, "Chopper-Stabilized Null Radiometer Based upon an Electrically Calibrated Pyroelectric Detector," *Appl. Optics*, **12**, 2532 (1973).
25. Phelan, R. J., Jr., and A. R. Cook, "Electrically Calibrated Pyroelectric Optical-Radiation Detector," *Appl. Optics*, **12**, 2494 (1973).
26. Somers, H. S., Jr., W. B. Teutsch, and E. K. Gatchell, "Demodulation of Low-Level Broad-Band Optical Signals with Semiconductors—I," *Proc. IEEE*, **51**, 140 (1963); "Analysis of the Photoconductive Detector—II," **52**, 144 (1964); "Experimental Study of the Photoconductive Detector," **54**, 1553 (1966).
27. DeLange, O. E., "Optical Heterodyne Detection," *IEEE Spectrum*, **5**, 77 (Oct. 1968).
28. Schemel, R. E., "The Signal/Noise Ratio of Quantum Detectors in Coherent Light Systems," *Radio Electronic Engr.*, **35**, 89 (Feb. 1968).
29. Vlasov, V. G., and E. V. Lazneva, "Heterodyne Method in Photodiode Detection of Radiation Amplitude-Modulated in the 5–50 MHz Range," *Sov. J. Opt. Tech.*, **35**, 608 (1968).
30. Teich, M. C., "Homodyne Detection of Infrared Radiation from a Moving Diffuse Target," *Proc. IEEE*, **57**, 786 (1969).
31. Teich, M. C., "Field-Theoretical Treatment of Photomixing," *Appl. Phys. Letters*, **14**, 201 (1969).
32. Teich, M. C., "Three-Frequency Heterodyne for Acquisition and Tracking of Radar and Communication Signals," *Appl. Phys. Letters*, **15**, 420 (1969).
33. Arams, F. R., et al., "Infrared Heterodyne Detection with Gigahertz IF Response," Chap. 10 in Willardson, R. K., and A. C. Beer (eds.), *Semiconductors and Semimetals*, Vol. 5, *Infrared Detectors*, Academic Press, New York (1970).
34. Keyes, R. J., and T. M. Quist, "Low-Level Coherent and Incoherent Detection in the Infrared," Chap. 8 in Willardson, R. K., and A. C. Beer (eds.), *Semiconductors and Semimetals*, Vol. 5, *Infrared Detectors*, Academic Press, New York (1970).
35. Teich, M. C., "Coherent Detection in the Infrared," Chap. 9 in Willardson, R. K., and A. C. Beer (eds.), *Semiconductors and Semimetals*, Vol. 5, *Infrared Detectors*, Academic Press, New York (1970).
36. Anderson, L. K., M. D. Domenico, Jr., and M. B. Fisher, "High Speed Photodetectors for Microwave Demodulation of Light," in Young, L. (ed.), *Advances in Microwaves*, Vol. 5, Academic Press, New York (1970).
37. Sommers, H. S., Jr., "Microwave-Biased Photoconductive Detector," Chap. 11 in Willardson, R. K., and A. C. Beer (eds.), *Semiconductors and Semimetals*, Vol. 5, *Infrared Detectors*, Academic Press, New York (1970).
38. Jones, R. C., "Immersed Radiation Detectors," *Appl. Optics*, **1**, 607 (1962).

39. Dreyfus, M. G., "Wedge-Immersed Thermistor Bolometer," *Appl. Optics,* **1,** 615 (1962).
40. DeWaard, R., and S. Weiner, "Miniature Optically Immersed Thermistor Bolometer Arrays," *Appl. Optics,* **6,** 1327 (1967).
41. Weiner, S. I., "Infrared Imaging System Comprising an Array of Immersed Detector Elements," U.S. Patent No. 3,397,314, Aug. 13, 1968.
42. DeWaard, R. D., D. H. Fisher, and A. Hvizdak, "Immersed Thermistor Bolometers with Radiation Impervious Mask on Back of Active Area," U.S. Patent No. 3,109,097, Oct. 29, 1963.
43. Norton, B., "Method of Producing Immersed Bolometers," U.S. Patent No. 3,420,688, Jan. 7, 1969.
44. Packard, R. D., "A Bonding Material Useful in the 2–14 μm Spectral Range," *Appl. Optics,* **8,** 1901 (1969).

3

Reprinted from *Proc. IRE*, 47(9), 1495–1502 (1959)

Phenomenological Description of the Response and Detecting Ability of Radiation Detectors[*]

R. CLARK JONES[†]

I. Introduction

PREVIOUS articles discuss the nature of the physical processes that go on inside the active elements of radiation detectors. In this article, however, the radiation detector will be viewed as a black box with an input and an output.

Input and Output

The input of a radiation detector is radiant power, and the output is an electrical signal. Various applications will have appropriately different ways of describing both the input and the output. The input may be expressed in lumens, lumen-seconds, watts, ergs, or any of these quantities per unit area. The output is usually expressed as a voltage or a current. For the sake of specificity, we shall suppose that the input is expressed in watts of radiant power, and output is expressed in volts.

In the testing of radiation detectors, the input radiation is nearly always chopped at a uniform frequency. It is then desirable that the same measure be used for the input and the output. Usually the output is measured by the root-mean-square amplitude of its component of fundamental frequency; then the same measure is used for the input.

Responsivity

Most detectors have a range over which the output is proportional to the input, and over this range the ratio of the output to the input is called a responsivity. Even when the relation is not linear, one may still define an incremental responsivity as the ratio of a change in the output to a change in the input. The units of the responsivity R are volts per watt, usually rms volts per rms watt.

The responsivity, however, tells us nothing about the detecting ability of a radiation detector; that is to say, it tells us nothing about how small a radiation input is detectable.

Noise Equivalent Power

To evaluate the detecting ability, we must know the characteristics of the electrical noise in the output of the detector. If one divides the rms voltage of the noise by the responsivity (in volts per watt), one obtains the noise equivalent power P_N of the detector.

Thus it is clear that responsivity and noise equivalent power are important characteristics of radiation detectors. A detector that produces a greater output (from a given radiation signal) has a greater responsivity. But a detector that has a greater detecting ability has a smaller noise equivalent power. Thus the noise equivalent power is an upside-down measure of the detecting ability of a detector.

Detectivity

The reciprocal of the noise equivalent input is called the detectivity.[1] The detectivity D is a right-side-up measure of the detecting ability. The detectivity D is defined in general as the responsivity divided by the rms noise voltage.

If we use P to denote the radiation input power, S the voltage output, and N the rms noise voltage, then the following relations hold.

$$S = RP, \qquad (1)$$

$$S/N = DP. \qquad (2)$$

That is to say, to determine the electrical signal output, multiply the radiation signal P by R, and to obtain the signal-to-noise ratio, multiply P by D.

The Term Sensitivity

There is a widespread tendency to use the term sensitivity to mean either the responsivity or detectivity, or their reciprocals. Just because the word sensitivity is part of everyday speech, there is a natural human tendency to use this word to mean whatever the writer happens to have in mind. In fact, it has been used to mean so many different concepts that its use in scientific and technical discourse is to be deplored.[1]

Individual Detectors and Types of Detectors

An individual detector is a single sample of a given type of detector. A type of detector is a class of individual detectors with some common properties. Heat detectors, bolometers, thermistor bolometers, photoconductive cells, and lead sulfide photoconductive cells with gold electrodes. are all examples of types of detectors.

[*] Original manuscript received by the IRE, June 29, 1959.
[†] Research Laboratories, Polaroid Corp., Cambridge, Mass.

[1] R. C. Jones, "Detectivity, the reciprocal of noise equivalent input of radiation," *Nature*, vol. 170, pp. 937–938; 1952. Also *J. Opt. Soc. Am.*, vol. 42, p. 286; 1952.

II. Parameters of the Radiation Signal

In testing the properties of detectors, one usually uses a radiation signal that is more or less monochromatic, or a radiation signal from a blackbody source. The radiation signal is thus described by its wavelength λ or by its blackbody temperature T_{bb}.

In testing detectors, it is also customary to modulate the signal at a given frequency f. It is advisable to state the amplitude of the modulated radiation power by its root-mean-square value. Most measuring systems respond only to the fundamental component of the detector output, and therefore if the source is not modulated in a sinusoidal manner, one should state the root-mean-square amplitude of the fundamental component.

In addition to the signal radiation that is to be detected or measured, there is usually additional steady radiation that is incident on the detector. Blackbody radiation from the environment, or daylight, are examples of steady ambient radiation. The power of this steady ambient radiation is denoted P_a.

III. Detector Parameters

Detector Temperature T

The detector temperature T is the temperature of the responsive element of the detector. Usually the temperature of the responsive element is substantially uniform over its volume, and there is no difficulty in defining the temperature of the responsive element.

Responsive Area A

Roughly speaking, we may say that the responsive area A of the detector is the area over which the detector is more or less equally responsive. In some types of detectors, such as wire-type thermocouples, there is no question of how the responsive area should be specified. In some other types of detectors, such as evaporated photoconductive films, there may be a real question how the area A is to be specified.

To define the area A more precisely, it is necessary to introduce the concept of the local responsivity of a detector. If the radiant power is concentrated on a very small area centered at the point x, y on the surface of the responsive element, the local responsivity $R(x, y)$ is the ratio of the electrical output to the incident power.

If the local responsivity has a flat maximum in the central region of the responsive area and decreases smoothly away from the center, then a useful measure of the responsive area A is

$$A = \iint R(x, y) dx\, dy / R_{\max} \tag{3}$$

where $R_{\max}$ is the maximum value of $R(x, y)$.

If, however, the local responsivity is an irregular function (corresponding to the presence of small regions of particularly high responsivity), then the observer may elect to replace $R_{\max}$ in the above expression by a suita-

bly chosen average over the central region of the detector.

Specifying the value of the responsive area A is particularly important because the original measurement of the responsivity usually involves measuring the ratio R_H of the electrical output to the irradiance of the cell. To obtain the responsivity R in the form electrical output divided by the incident power, R_H must be divided by A.

$$R = R_H / A. \tag{4}$$

In addition to having an area specified in square centimeters, the detector also has a size and shape. In many cases it will be sufficient to describe the two principal dimensions of a rectangular detector, but if the local responsivity is an irregular function, the only adequate way to express the shape of the local responsivity is to show a contour diagram or several profiles of the relative local responsivity.

The Gain g

Most kinds of individual detectors have externally adjustable parameters that permit variation of the responsivity (and of the detectivity). Examples of these adjustable parameters are the biasing current in bolometers and photoconductive cells, the applied potentials in multiplier phototubes and simple phototubes, the biasing voltage in back-biased junctions, the emitter current in phototransistors, the several adjustable parameters of the Golay detector, the pre-exposure of photographic films, and the magnetic field in photoelectromagnetic detectors. Indeed, only the thermocouple and photovoltaic detectors seem to be free of such adjustable parameters.

All of these parameters have the effect of varying the "internal gain" of the detectors, and as a generic way of referring to the variation of any of these parameters, we shall in the following refer to varying the gain g of the detectors. The gain g may be considered to mean any of the adjustable parameters. Since the gain g is used in this report only in a qualitative sense, we do not need to define g more precisely.

The Bandwidth Δf

The noise bandwidth Δf is the bandwidth in cycles per second of the noise that is effective in the measurement of the noise equivalent power P_N. If $N(f)$ denotes the mean-square-noise-voltage (or current) per-unit bandwidth [$N(f)$ is often called the power spectrum of the noise] at the point in the circuit where the noise equivalent power is measured, the bandwidth Δf is defined by

$$\Delta f = \int_0^\infty N(f) df / N_{\max} \tag{5}$$

where $N_{\max}$ is the maximum value of $N(f)$ with respect to variation of f.

59

Time Constants

The time constant of a detector is usually defined in terms of the way that the responsivity R depends on the frequency f. If, for example, the responsivity depends on f in accordance with the simple relation

$$R(f) = R_0/(1 + (2\pi f \tau_r)^2)^{1/2} \qquad (6)$$

then τ_r is called the responsive time constant of the detector. More generally, the responsive time constant may be defined by

$$\tau_r = \frac{R_m^2}{4 \int_0^\infty [R(f)]^2 df} \qquad (7)$$

where R_m is the maximum value of $R(f)$ with respect to frequency.

Much more fundamental is the detective time constant, which is defined in general by

$$\tau_d = \frac{D^*_m{}^2}{4 \int_0^\infty [D^*(f)]^2 df} \qquad (8)$$

where D^*_m is the maximum value of $D^*(f)$ with respect to f, and where $D^*(f)$ is defined by (20) below. The detective time constant τ_d cannot be changed by changing the gain vs frequency characteristic of the amplifier, and is thus an invariant measure of the speed of response of the detector.

IV. The Parameters on Which R, N and D Depend

The Responsivity R

The responsivity of an individual detector operated at a given temperature T and at a given gain g depends on these parameters and on the parameters of the radiation, giving

$$R = R(\lambda, f, P_a, T, g). \qquad (9)$$

But the responsivity of a type of radiation detector, in which one is free to vary the detective time constant τ_d and the area A, will depend on these parameters also. Thus, one has

$$R = R(\lambda, f, P_a, T, g, \tau_d, A). \qquad (10)$$

The Noise Power Spectrum N

The power spectrum N of the noise at the output of an individual radiation detector depends only on the modulation frequency f, the detector temperature T, and the gain g, provided that the noise is not affected by the amount of the ambient power P_a. But if the noise power does depend on the ambient power P_a, it also de-

pends on the wavelength λ of this power. Thus, one has in general

$$N = N(f, T, g; \lambda, P_a). \qquad (11)$$

The noise spectrum of a type of radiation detector, in which one is free to vary the detective time constant τ_d and the responsive area A, depends also on these variables.

$$N = N(f, T, g, \tau_d, A; \lambda, P_a). \qquad (12)$$

The Detectivity D

The detectivity D of a radiation detector depends on the same parameters as the responsivity, and in addition it depends on the bandwidth Δf of the noise.

Thus, for an individual detector, one has

$$D = D(\lambda, f, P_a, T, g, \Delta f). \qquad (13)$$

But if one is free to vary the detective time constant τ_d, and the area A, the detectivity of a type of radiation detector will depend also on these parameters

$$D = D(\lambda, f, P_a, T, g, \Delta f, \tau_d, A). \qquad (14)$$

In terms of the responsivity R and the power spectrum N of the electrical noise, the detectivity D is defined in general by

$$D(\lambda, f, P_a, T, g, \Delta f, \tau_d, A)$$
$$= \frac{R(\lambda, f, P_a, T, g, \tau_d, A)}{(\Delta f N(T, g, \tau_d, A; \lambda, P_a))^{1/2}}. \qquad (15)$$

V. The Detectivity D^*, and the Figures of Merit M_1 and M_2

The Detectivity D^* in the Reference Condition E

Most of the material in this section was condensed from the first three sections of a chapter[3] written in 1952. The material relating to D^* was first presented to an IRIS meeting on November 13, 1956, and later appeared in the *Proceedings of the IRIS*.[4]

In this section, it is supposed that the detectivity D does not depend appreciably on the ambient power P_a. The detectivity thus has the following functional dependence.

$$D = D(\lambda, f, T, g, \Delta f, \tau_d, A). \qquad (16)$$

No one has yet been able to perceive any useful systematic trends in the dependence of the detectivity on the wavelength λ and the detector temperature T. But it is possible to perceive systematic relations between the detectivity and the other five parameters. For the present, we suppress the symbols λ and T, and consider that the detectivity has the following functional dependence.

$$D = D(f, g, \Delta f, \tau_d, A). \qquad (17)$$

---

[3] R. C. Jones, "Performance of detectors for visible and infrared radiation," in "Advances in Electronics," L. Marton, ed., Academic Press, Inc., New York, N. Y., vol. 5, pp. 27–30; 1953. The responsive time constant τ_r was there called the physical time constant τ_p, and the detective time constant τ_d was there called the reference time constant τ.

[3] *Ibid.*, pp. 1–96.
[4] R. C. Jones, "Method of rating the performance of photoconductive cells," *Proc. IRIS*, vol. 2, pp. 9–12; June, 1957.

It is customary to adjust the gain g by varying the internal parameters of the detector so that the detectivity is maximized. To be sure, this maximum will usually depend on the frequency f, but some compromise adjustment is usually selected. Once this choice is made, the detectivity no longer depends on the gain g, so that one has

$$D = D(f, \Delta f, \tau_d, A). \qquad (18)$$

We consider next the dependence of the detectivity D on the responsive area A. On the basis of certain idealized assumptions, one may prove that, other conditions being equal, D varies inversely as the square root of the area A.

$$DA^{1/2} = \text{independent of } A. \qquad (19)$$

This relation is valid when the dominant noise of the detector is radiation noise. The relation also follows if the area A is varied by connecting together several identical adjacent detectors. In practice, one sometimes finds deviation from the above relation, and in all cases known to the writer, these deviations occur because the properties per unit area depend on the total area. 1) In thermistor bolometers, more power per unit area can be dissipated in very small detectors, because the lines of heat flow are then radial instead of normal to the surface. 2) In evaporated photoconductive cells, the surface often contains a small area of exceptionally high detectivity. With cells of very small area, one can select the cells that contain a "hot spot," and reject the others; in cells of large area that contain many hot spots, this is not possible. 3) In indium antimonide photovoltaic cells, there is often $1/f$ noise that arises from the boundary. This noise varies as the circumference of the cell and causes the detectivity to vary as $A^{-1/4}$, but the $1/f$ noise can be eliminated by suitable back bias or by use of a dc short circuit, in which case (19) is found to hold. In the first two examples, the exponent is greater than one-half; in the third, it is less than one-half. For comparing the performance of a wide variety of detectors, the exponent one-half in (10) is clearly a better choice than any other number.

Departures from (19) may also occur in a different kind of situation. Some detectors have a directivity pattern that is not Lambertian. With an antenna array, the quotient of the area A and the directive gain is a constant. The same relation holds when the effective area is changed by immersing the responsive element in a high-index medium. In both of these situations (19) fails, but may be made to hold by replacing A by the ratio of A to the directive gain.

We consider next the dependence of the detectivity on the noise bandwidth Δf. There are no perfectly general statements that can be made in this respect, because we have introduced so far no restrictions on the way that the power spectrum $N(f)$ of the noise depends on the frequency f. We now introduce the assumption that over the band of frequencies we wish to discuss,

the power spectrum $N(f)$ is independent of f. (The amplifier can always be equalized so that this assumption is justified.) Then it follows that the detectivity D, other parameters being constant, varies as the inverse square root of the bandwidth Δf.

We now define $D^*(f, \tau_d)$ as the detectivity measured with a bandwidth Δf of one cycle per second and reduced to a responsive area of one cm^2. By the three preceding paragraphs, one has

$$D^*(f, \tau_d) = (A \Delta f)^{1/2} D(f, \tau_d, A, \Delta f). \qquad (20)$$

D^* is pronounced "D-star." (An author may desire a formal name for D^*; the name "detectivity in the reference condition E" is recommended.) The dimensions of D^* are length-(frequency)$^{1/2}$/power. If A is measured in cm^2, and f in cycles/second, the units of D^* are cm-(cps)$^{1/2}$/watt. The writer recommends that these particular units be used for D^* unless there is specific indication to the contrary.

The Figures of Merit M_1 and M_2

For most detectors, including lead sulfide photoconductive cells, there is a systematic relation between the detectivity D^* and the parameters f and τ_d. In fact, detectors may be placed in two mutually-exclusive classes[5] on the basis of the way that D^* depends on the detective time constant τ_d. In order to define these two classes, we first introduce the quantity $D^*_m(\tau_d)$, which is defined as the maximum value of $D^*(f, \tau_d)$ with respect to the frequency f.

A Class I detector is then defined as a type of detector for which D^*_m is independent of the value of the detective time constant, and a Class II detector is a type of detector for which D^*_m is proportional to the square root of τ_d. It is an experimental fact that all detectors fall into one of these two classes, except for those types of detectors for which insufficient information is available to specify the Class membership.

Class I detectors include photoemissive tubes, ideal heat detectors, the Golay heat detector, radio antennas, and some photoconductive cells. Class II detectors include thermocouples and bolometers, photographic films, the human eye, and most lead sulfide cells. These statements are subject to a number of conditions and assumptions. A full discussion will be found in Jones.[3]

In 1949, a figure of merit was introduced[5] for each of the two Classes. The figure of merit M_1 for Class I detectors was defined as the ratio of the measured value of D^*_m to the value of D^*_m for an ideal heat detector at the temperature 300°K. M_1 may be written

$$M_1 = (5.52 \times 10^{-11} \text{ watt/cm-(cps)}^{1/2}) D^*_m \qquad (21)$$

where D^*_m must be expressed in cm-(cps)$^{1/2}$/watt. M_1 is dimensionless. Similarly, the figure of merit for Class II detectors was defined[5] as the ratio of D^*_m to the value

[5] R. C. Jones, "Radiation detectors," *J. Opt. Soc. Am.*, vol. 39, pp. 327–356; 1949

of D^*_m for a detector that accords with Havens' limit.[6,7] Havens' limit is an engineering estimate of the maximum possible detectivity of a thermocouple or bolometer. The value of M_2 may be written

$$M_2 = (6 \times 10^{-11}\ \text{watt-sec/cm}) D^*_m / \tau_d^{1/2} \qquad (22)$$

where D^*_m must be expressed in cm-(cps)$^{1/2}$/watt and τ_d in seconds. M_2 is dimensionless.

Since the detective time constant is defined by

$$\tau_d = \tfrac{1}{4}(D^*_m)^2 \Big/ \int_0^\infty [D^*(f)]^2 df \qquad (23)$$

M_2 may also be written

$$M_2 = (12 \times 10^{-11}\ \text{watt/cm-cps}) \left[\int_0^\infty [D^*(f)^2 df \right]^{1/2} \qquad (24)$$

where D^* must be expressed in cm-(cps)$^{1/2}$/watt, and f in cps.

In these expressions, the figures of merit M_1 and M_2 are dimensionless, and depend only on the wavelength λ (or the temperature T_{bb} of the blackbody source) and on the temperature T of the detector. (D^* depends on λ and T, and also on f and τ_d.)

M_1 and M_2 are plotted vs the wavelength for a number of different types of detectors in Figs 1 and 2. These figures are reproduced from Jones.[3]

VI. The Detective Quantum Efficiency Q_D

Most of the material in this section is condensed from the first four sections of a review written in 1958, to be published[8] in 1959. Some of the material has already been published.[9-12]

There are two kinds of quantum efficiency. The more familiar kind is always defined as the ratio of the number of output events to the number of input events. This kind of quantum efficiency is thus a ratio, and is moreover a ratio of the numbers of two kinds of countable events. A familiar example is the quantum efficiency of a photoemissive tube, which is usually defined as the ratio of the number of emitted photoelectrons to the number of incident photons. This kind of quantum efficiency is clearly related to the responsivity of a detector, and for this reason it has been named *responsive* quantum efficiency.

[6] R. J. Havens, "Theoretical comparison of heat detectors," *J. Opt. Soc. Am.*, vol. 36, p. 355; 1946.

[7] R. J. Havens, "Theoretical limit for the sensitivity of heat detectors," *Proc. IRIS*, vol. 2, pp. 5–8; June, 1957.

[8] R. C. Jones, "Quantum Efficiency of Detectors for Visible and Infrared Radiation," in "Advances in Electronics and Electron Processes," Academic Press, Inc., New York, N. Y., vol. 12; 1959.

[9] R. C. Jones, "Quantum efficiency of photoconductors," *Proc. IRIS*, vol. 2, pp. 13–17; June, 1957.

[10] R. C. Jones, "On the quantum efficiency of photographic negatives," *Phot. Sci. Engrg.*, vol. 2, pp. 57–65; 1958.

[11] R. C. Jones, "Quantum efficiency of human vision," *J. Opt. Soc. Am.*, vol. 49, pp. 645–653; July, 1959.

[12] R. C. Jones, "On the detective quantum efficiency of television camera tubes," *J. Soc. Mot. Pict. Telev. Engrs.*, to be published.

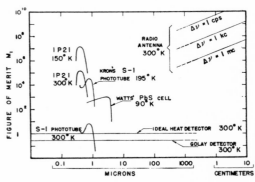

Fig. 1—The figure of merit M_1 as a function of the wavelength for several types of Class I detectors; both coordinate scales are logarithmic, but the ordinate scale is compressed relative to the abscissa scale. Of particular interest is the extent to which the detectivity of S-1 phototubes can be increased by cooling. The improvement of the 1P21 multiplier phototube by cooling is also marked.

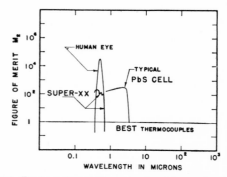

Fig. 2—The figure of merit M_2 as a function of the wavelength for several types of Class II detectors; the long-wavelength portions of the curves for human vision and for Super-XX film are coincident. References to the detectors described in Figs. 1 and 2 will be found in the literature.[8]

There is another kind of quantum efficiency that is related to the detectivity of a detector, and for this reason, it has been named *detective* quantum efficiency. The concept of detective quantum efficiency was first formulated by Rose[13] in 1946.

The detective quantum efficiency may be defined without regard to the amount of the ambient power P_a, but it acquires its chief interest only when the ambient power P_a is sufficient to affect appreciably the detectivity D. Thus the situation discussed in this section is complementary to that in Section V of this paper.

In the presence of a given amount of ambient power of the same wavelength as the radiation signal to be de-

[13] A. Rose, "A unified approach to the performance of photographic film, television pickup tubes, and the human eye," *J. Soc. Mot. Pict. Telev. Engrs.*, vol. 47, pp. 273–294; 1947.

tected, there is a very fundamental limit to the detectivity. This limit is due to the quantum fluctuations in the "steady" ambient radiation.

To bring out the nature of this limit in a simple way, we consider the number of photons in the radiation incident on a detector of area A in the period of duration T. The average number of signal photons is denoted M_s, and the average number of photons in the ambient radiation is denoted M_a. All of the photons are assumed to have the same wavelength. It is important to note that M_a is the average number, where the average is over a large number of successive periods, each of length T. In any given period, however, the number of actual photons $\mathfrak{M}_a$ will usually be different from the average number M_a. The number varies from period to period even in the most steady ambient radiation fields, because the individual photons arrive at random, like drops of rain on a roof. As shown by Fry,[14] the distribution of the number will be approximate to a Poisson distribution, from which it follows that the root-mean-square fluctuation of the number is equal to the square root of M_a.

$$N \equiv \Delta M = \langle (\mathfrak{M}_a - M_a)^2 \rangle_{\text{Av}}^{1/2} = M_a^{1/2}. \quad (25)$$

Since this fluctuation is a noise, it has been denoted by N in the above equation.

Similarly, the number S of signal photons is the number M_s,

$$S = M_s \quad (26)$$

from which one finds that the signal-to-noise ratio is given by

$$\frac{S}{N} = \frac{M_s}{M_a^{1/2}}. \quad (27)$$

This is the signal-to-noise ratio present in the radiation incident on the detector. Under the given conditions, no detector may have a higher signal-to-noise ratio in its output. In fact, no detector can have quite as high a signal-to-noise ratio.

The squared ratio of the actual signal-to-noise ratio $(S/N)_{\text{meas}}$ to the maximum possible signal-to-noise ratio as given by (26) is the detective quantum efficiency,

$$Q_D = \frac{(S/N)_{\text{meas}}^2}{M_s^2/M_a}. \quad (28)$$

Thus, one sees by its very definition that the maximum possible value of the detective quantum efficiency is unity.

The detective quantum efficiency must be defined as the *square* of the ratio of signal-to-noise ratios, in order to make the detective quantum efficiency equal to the responsive quantum efficiency of an ideal photoemissive tube. The details are here omitted.

[14] T. C. Fry, "Probability and its Engineering Uses," D. Van Nostrand Co., Inc., New York, N. Y., pp. 216–227; 1929.

The detective quantum efficiency is equivalently defined as the squared ratio of the noise equivalent power in the incident radiation to the noise equivalent power of the actual detector measured under the same conditions.

$$Q_D = (P_N)_{\text{rad}}^2 / (P_N)_{\text{meas}}^2. \quad (29)$$

One may show that the mean-square fluctuation in the power of the ambient radiation is given by

$$(P_N)_{\text{rad}}^2 = 2EP_a\Delta f \quad (30)$$

where E is the energy of a single photon. Combining the last two relations then yields

$$Q_D = 2EP_a\Delta f / (P_N)_{\text{meas}}^2. \quad (31)$$

As indicated by (29) above, the reciprocal of the detective quantum efficiency Q_D^{-1} is a kind of noise figure. It is a noise figure in which the reference noise is the noise in the ambient radiation, whereas the noise figure ordinarily used in electronic equipment has Johnson noise as the reference noise. These two reference noise levels are in general different in amount and in concept. In a radio antenna, however, the two become identical.

The expressions for the radiation noise used above are based on Maxwell-Boltzmann statistics. Actually, ensembles of photons must be described by Bose statistics. So long, however, as hc/λ is large compared with kT_{rad}, where T_{rad} is the temperature of the source of the ambient radiation, the approximation used above is valid. The approximation is valid for the infrared spectrum (although it begins to break down for a high temperature source at wavelengths longer than about ten microns), but it is not correct in the radio spectrum. See Jones[8] for a full discussion.

Fig. 3 shows a plot with logarithmic scales of the detective quantum efficiency plotted vs the amount of ambient radiation M_a. The curve shown is for an imaginary detector. This plot has a number of interesting properties. On the plot, any detector that has a detectivity D independent of the ambient power is presented by a straight line with a slope of plus one. Thus the point on the curve where the detectivity has its maximum value is the point where the slope of the curve is plus one, as at the point A.

There is another kind of detectivity, named the contrast detectivity D_c, which is defined as the reciprocal of the noise equivalent value of the contrast M_s/M_a. D_c may be written

$$D_c = P_aD. \quad (32)$$

D_c is dimensionless. Any detector that has a contrast detectivity that is independent of the ambient power is represented in Fig. 3 by a straight line with a slope of minus one. Thus, the point on the curve where the contrast detectivity of the detector has its maximum value is the point where the slope is minus one, as at the point C.

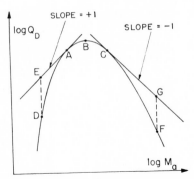

Fig. 3—A schematic plot showing the detective quantum efficiency Q_D plotted against the irradiation of the detecting surface; if the irradiation is less than at the point A (where the curve has a slope of plus one), the detector is underloaded, and Q_D can be increased by adding additional steady irradiation. If the irradiation is more than that at the point C (where the curve has a slope of minus one), the detector is overloaded, and Q_D can be increased by placing a neutral filter over the detector. All detectors can be overloaded, but not all detectors can be underloaded.

At any point on the curve where the slope is more than plus one, the detector is underloaded, which means that the detectivity D can actually be increased, as from the point D to E, by adding a local source of ambient radiation that raises the total ambient radiation to the value at A. Similarly, at any point on the curve where the slope is less than minus one, the detector is overloaded, by which is meant that the contrast detectivity can actually be increased, as from F to G, by placing an attenuating filter over the detector that reduces the ambient radiation to the value at C.

Thus, the range from A to C is the working range of the detector. Nothing is ever gained by operating outside of this range. All detectors can be overloaded, but most detectors cannot be underloaded. The classic example of a detector that is easily underloaded is the photographic film, where it is well known that it is necessary to pre- or post-expose the film to record the weakest images.

Figs. 4 and 5 show the detective quantum efficiency of human vision, two photographic films, and three television camera tubes. In Fig. 4, Q_D is plotted vs the ambient exposure with all of the parameters adjusted to their optimum values. In Fig. 5, Q_D is plotted against wavelength with all of the other parameters adjusted to their optimum values.

This section is concluded with the relation between the detective quantum efficiency Q_D and the detectivity D^*,

$$AQ_D = 2EP_a(D^*)^2. \tag{33}$$

This relation is easily derived from (31) and above, and holds for an individual detector in which the same conditions of measurements are used for Q_D and D^* (same values of λ, f, P_a, g and T).

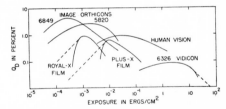

Fig. 4—The detective quantum efficiency Q_D plotted against the ambient exposure in ergs per square centimeter, for three varieties of image-forming detectors: television camera tubes, photographic negatives, and human foveal vision. The dashed lines have slopes of plus one and minus one, and correspond to the lines in Fig. 3. The results shown are for optimum choice of all of the other parameters that affect the value of Q_D, such as wavelength, size of signal area, signal duration, etc. The details of the method by which the curves are calculated are given in the literature.[8]

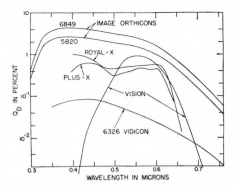

Fig. 5—The detective quantum efficiency Q_D plotted vs the wavelength for the same detectors as in Fig. 4. The results are for optimum choice of all of the other parameters that affect the value of Q_D, such as the amount of steady irradiation, size of signal area, signal duration, etc.

VII. CONCEPTS RESTRICTED TO PHOTOCONDUCTIVE CELLS

The great importance of photoconductive cells in infrared detection has led to a number of concepts and notations whose application is restricted to photoconductive cells.

"Jones' S"

The first concept proposed[15] specifically for photoconductive cells was "Jones' S." This concept is now obsolete, but it is still important because all of the detector reports issued by the Naval Ordnance Laboratory at Corona, Calif., use this concept. Jones' S is related to D^* by

[15] R. C. Jones, "A method of describing the detectivity of photoconductive cells," *Rev. Sci. Instr.*, vol. 24, pp. 1035–1040; 1953. Jones' S was proposed in a privately circulated report, "Proposal of a Figure of Merit for Photoconductive Cells," dated February 1, 1950, and the proposal was presented at Prof. Nottingham's Photoconductivity Conf. at Massachusetts Institute of Technology, Cambridge, Mass., March 28, 1950.

Jones' $S = f^{1/2}/D^*$.   (34)

Jones' S is defined as the noise equivalent power reduced to unit area and to a noise bandwidth in which the upper edge of the band has a frequency $e = 2.718$ times the frequency of the lower edge. Jones' S was intended to be used for low frequencies where S is constant if the responsivity is constant and the power spectrum $N(f)$ of the noise varies as $1/f$. According to (34), Jones' S has the dimensions of power/length, and is usually expressed in watt/cm.

Specific Noise and Specific Responsivity

Shortly after Jones' S was proposed, Mundie and Kirk[16] proposed a specific noise N_1 and a specific responsivity S_1 defined by

$$N_1 = \frac{A^{1/2}V_{n,e}}{V} \cdot \frac{(R_c + R_L)^2}{4R_cR_L} \qquad (35)$$

and

$$S_1 = \frac{V_s}{FV} \cdot \frac{(R_c + R_L)^2}{4R_cR_L} \qquad (36)$$

where V is the bias voltage across the cell, V_s is the signal voltage produced by a signal irradiation of F watts per square centimeter, and $V_{n,e}$ is the rms noise voltage measured in a bandwidth with the frequency ratio e. R_c is the (variable) resistance of the cell, and R_L is the (variable) resistance of the load. They have the property that N_1 divided by S_1 is equal to Jones' S.

Jones' $S = N_1/S_1$.   (37)

Since Jones' S and N_1 were introduced, it has become increasingly evident that the power spectrum of the noise in many cases deviates quite importantly from a $1/f$ shape[17] and that, therefore, the appropriateness of the definitions of Jones' S and N_1 is reduced. Jones' S, S_1 and N_1 continue to be used in the Corona reports[17] up to the time of writing.

Petritz' Specific Noise

Petritz[18] has introduced a modified specific noise

$$N_s = \frac{A^{1/2}V_n}{(\Delta f)^{1/2}V} \cdot \frac{(R_c + R_L)^2}{4R_cR_L} \qquad (38)$$

where V_n is the rms noise voltage measured within the bandwidth Δf. D^* is related to S_1 and N_s by

$$D^* = S_1/N_s. \qquad (39)$$

N_1 is usually expressed in cm, S_1 in cm²/watt, and N_s in cm/(cps)$^{1/2}$.

[16] L. G. Mundie and D. D. Kirk, "Notes for Users of Photoconductive Detectors," NBS-Corona Rept. No. 30-E-109; August 15, 1952.

[17] See the papers by van Vliet and his co-workers in *Physica* and PROC. IRE, in 1956 and following years. Also see the NBS-Corona and NOL-Corona series of reports, beginning in 1952 and continuing through the present.

[18] R. L. Petritz, "Theory of photoconductivity in semiconductor films," *Phys. Rev.*, vol. 104, pp. 1508–1516; 1956.

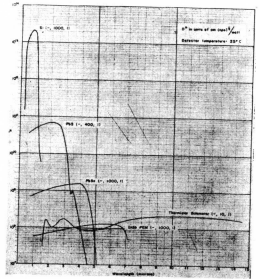

Fig. 6—Comparative spectral curves of several representative photodetectors at room temperature.

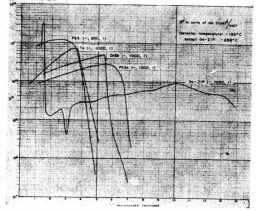

Fig. 7—Comparative spectral curves of several representative cooled photodetectors.

In conclusion, through the courtesy of R. F. Potter of the Infrared Division, U. S. Naval Ordnance Laboratory, Corona, Calif., it is possible to show, in Figs. 6 and 7, curves summarizing the current state-of-the-art of photodetectors. Detectivity in the reference condition E is plotted against wavelength. The hyphen in the parentheses following the label on each curve indicates that the data are for spectral measurements; the next number is the chopping frequency in cycles per second, and the final number is the frequency bandwidth, in each case unity.

Reprinted from *Proc. IRE*, **47**(9), 1458–1467 (1959)

Fundamentals of Infrared Detectors*

RICHARD L. PETRITZ†, MEMBER, IRE

INTRODUCTION

TWO fundamental types of infrared detectors are thermal detectors and photodetectors. Incident radiation changes the electrical properties in each of the detectors. Both thermal and photodetectors are quantum detectors, since radiation is quantized; however, a distinction can be made if the nature of the solid is considered.

The modern description of the solid divides it into two thermodynamic systems—lattice and electronic. The electronic system is characterized by a system of energy levels: the conduction band, forbidden band impurity levels, and the valence band (see Fig. 1). The lattice is composed of atoms or molecules which constitute the solid and can be characterized by lattice vibrations which have mathematical properties analogous to radiation (lattice *phonons* are analogous to *photons*).

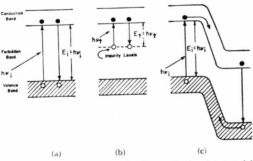

Fig. 1—Generation of carriers in photoconductive detectors (a) Intrinsic absorption. (b) Impurity absorption. (c) *P-n* junction.

The fundamental difference between photo- and thermal-type detectors can now be stated precisely. In thermal detectors, the radiation is absorbed by the lattice, causing heating of the lattice. The change in the temperature of the lattice causes a change in the electronic system. Two of the most common thermal detectors are thermocouples which develop a thermal voltage, and bolometers which change electrical resistance as a result of lattice heating. In photodetectors, radiation is absorbed directly by the electronic system to cause changes in the electrical properties. Photodetectors include photoconductive, photovoltaic, and photoelectromagnetic detectors.

* Original manuscript received by the IRE, June 26, 1959.
† Texas Instruments Inc., Dallas, Texas.

PHOTODETECTORS

Contemporary infrared photodetectors [1] with response beyond 1.5 microns are semiconductors with energy level systems similar to those shown in Fig. 1. For the visible and near infrared, photoemissive tubes are ordinarily used. These tubes are discussed by Morton and Forque [1a].

After the radiation flux is converted to an electrical signal, the electrical signal can be amplified to the level desired. Detection systems generally have sufficient amplification to cause the system to be limited by noise; therefore, the signal-to-noise ratio is of basic interest in the physics of detectors.

Signal Properties

In a photodetector, the infrared radiation which is usefully absorbed excites electrons (and/or holes) to the conducting state. The generation of carriers per unit volume is given by

$$\text{generation rate} = \alpha J_s, \tag{1}$$

where J_s is the number of incident photons/cm² sec and α is that part of the absorption coefficient relating to carrier excitation. The excess electrons, Δn, return to the valence band at a rate related to the carrier lifetime, τ, by the equation

$$\text{recombination rate} = \Delta n/\tau. \tag{2}$$

In the steady state

$$\Delta n = \alpha J_s \tau. \tag{3}$$

Since the signal response is proportional to Δn, (3) shows the importance of the absorption coefficient, determined by the generation process, and the lifetime, determined by the recombination process. These are discussed in turn.

Generation Process: Generation of carriers can occur by main band transitions or by transitions involving impurity levels. In a main-band transition [Fig. 1(a)], a photon is absorbed, exciting an electron from the valence band to the conduction band, creating a hole-electron pair. Detectors of this type include the lead salts and indium antimonide. The long wavelength limit of spectral response, λ_i, is determined by the intrinsic energy gap, E_i, of the material where

$$\lambda_i = hc/E_i. \tag{4}$$

The second mechanism of generation involves impurity levels which are located in the forbidden energy gap as shown in Fig. 1(b). In this type of detector, the long wavelength cutoff is determined by the impurity level relative to the conduction or the valence band. Gold-doped germanium is the best known infrared detector of this type.

The radiation intensity decreases with the distance, x, into the semiconductor as

$$J_s(x) = J_{s0}e^{-\alpha x}. \tag{5}$$

If most of the radiation is to be absorbed, (5) shows that the thickness of the crystal should be of the order of $1/\alpha$.

In impurity-type detectors, the absorption coefficient is given by

$$\alpha = \sigma N_t \tag{6}$$

where σ is the cross section of the impurity level and N_t is the density of impurity level atoms. The cross section for simple impurity levels can be approximated by hydrogen atom formulas [2]. The density of impurity levels is usually determined by the solubility limits of the atoms in the semiconductors. In germanium, the maximum solubility of gold is about 2×10^{16} atoms/cm³, giving an $\alpha \cong 0.14$ cm⁻¹; therefore, a germanium crystal would have to be approximately 10 cm thick to absorb most of the radiation. This rather small absorption coefficient is characteristic of most impurity semiconductors [3].

In the intrinsic absorption process, the absorption coefficient near the main-band region is usually of the order of 10^4 cm⁻¹, and the radiation is effectively absorbed within about 1 micron of the surface. In the case of the lead-salt photoconductive detectors [4], the material is deposited in films approximately 1 micron thick. This assures that the radiation affects the total volume of the detector.

In broad-area diffused or alloyed p-n junction-type detectors [5], [Fig. 1(c)], such as indium antimonide, the main generation occurs within the first micron of the surface; however it is not necessary to place the junction this close to the surface. The critical distance is set by the diffusion length, L, of the minority carriers and by the surface recombination velocity, S. For best performance, the distance, w, of the junction from the surface must satisfy

$$w \ll L, \tag{7a}$$

and

$$S \ll D/w, \tag{7b}$$

where $L = \sqrt{D\tau}$ and D is the diffusion coefficient for the minority carriers. The crystal can be much larger than either w or L.

Recombination Process: Some of the more important mechanisms by which carriers can recombine are discussed below and shown in Fig. 2.

The inverse of the main-band absorption process is the direct radiative recombination process where the electron falls from the conduction band to the valence band and gives up its energy by radiation [Fig. 2(a)]. It is desirable that the carrier lifetime be limited by direct

radiative recombination; however, in most detectors the lifetime is limited by another mechanism.

In the normal recombination center (bulk and surface) shown in Fig. 2(b), the minority carrier is captured by an impurity center and then falls to the valence band in a two-step process described by Shockley and Read [6]. When the cross sections for the two steps are nearly equal, the rate limiting step is set by capture of the minority carrier by the impurity center, and the electron and hole lifetimes are equal.

Another type of impurity center is called a non-recombining trap [7] (bulk and surface) and is shown in Fig. 2(c). This method of recombination differs from the normal recombination in that the cross section for the second step, the majority carrier, is relatively small. Under these conditions, the majority carrier is free for an extended period of time after the minority carrier is captured. The increase in majority carrier lifetime will be proportional to the time the minority carrier spends in the trap. Thus, minority carrier trapping is a mechanism for increasing the lifetime of majority carriers, which increases the signal response as shown in (3).

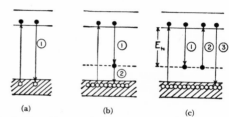

Fig. 2—Recombination processes in photoconductors. (a) Direct recombination. (b) Normal recombination center. (c) Non-recombining trap.

The recombination process can be used to classify the mechanism of detection. The impurity level detector deals with the majority carrier, and the time constant is the lifetime of the majority carrier. In the case of single crystal minority carrier detectors such as photovoltaic detectors, the important lifetime is that of the minority carrier.

Various researchers disagree on the basic recombination model in the thin-film lead-salt detectors. In one model [8], [9], the majority carrier lifetime is increased by the oxygen sensitization process which incorporates minority carrier traps. Another model [10] suggests that an array of p-n junctions is generated as a result of the oxygen sensitization process. The interested reader is referred to other sources, [4], [8]–[10], for further discussion of this subject.

Noise Properties

Noise mechanisms [11] that limit detector performance are considered in the following sections.

Background Radiation: An infrared detector must, if it is to be used in a physically interesting situation, have background radiation of some sort fall upon it. For detectors used in laboratories, the background radiation is that of a blackbody at the ambient temperature. The background radiation has a random characteristic due to random arrival of photons which sets an ultimate limit to the smallest radiation signals that can be detected. The statistics of this noise have been analyzed by a number of authors, and a summary is given by Jones [12]. Background radiation is one source of generation-recombination noise; the other main source is the lattice (see below). An over-all goal of infrared detector physics and technology is to achieve detectors that are background radiation-noise limited.

Signal Fluctuations: In contemporary infrared detectors, the inherent fluctuations in the signal can be neglected, since background radiation fluctuations are of a much greater intensity. However, in future detectors which may operate in outer space, where effective background temperatures can be very low, conditions where fluctuations in the signal are of interest may be encountered. These signal fluctuations can be handled by the same methods used for background fluctuations.

Generation-Recombination Noise by Lattice Excitation and Recombination: Detector noise caused by fluctuations in the rate of generation of carriers by the lattice and the corresponding recombination process [13] can and usually does exceed background radiation noise. The magnitude of the lattice-induced noise is related to the lattice temperature, and cooling of the lattice reduces this noise.

Nyquist-Johnson Noise of Conduction Electrons: The electrons in the conduction band have velocity fluctuations which are generally the source of the Nyquist-Johnson noise, characteristic of all resistors [14]. This is the only noise present in the detector at equilibrium. In most photodetectors, the generation-recombination noise can be made to exceed the Nyquist-Johnson noise by proper biasing.

1/f Noise: Contemporary semiconductors have a component of noise power characterized by a $1/f$ spectrum which is proportional to the square of the biasing current. Considerable work is being directed toward determining the basic foundation of this noise and reducing its magnitude [15]. Experimental evidence indicates that $1/f$ noise is associated with surfaces, contact phenomena, and bulk phenomena such as dislocations. The subject is too complex to be discussed here, but $1/f$ noise must be recognized as an important source of noise in infrared detectors.

Amplifier Noise: The aim of amplifier design for use with infrared detectors is to assure that detector noise, not amplifier noise, limits the signal-to-noise ratio. This condition can usually be achieved by careful design [16].

Signal-to-Noise Ratio

The signal-to-noise ratio for an infrared detector is the basis for the absolute detectivity or sensitivity of a detector [17]. It can be calculated and/or measured with knowledge of the signal properties and the noise properties. For the purpose of characterizing detectors the concept of signal-to-noise ratio per unit of signal radiation, the detectivity, is useful. The concept of noise-equivalent-power, which is the radiation power required to give a signal equal to noise, is also useful.

BACKGROUND RADIATION LIMITED CONDITION OF OPERATION (BLIP)

As stated above, an important goal of infrared-detector physics and technology is the achievement of the true radiation limit of detectivity. In order to show the physics behind this condition, a discussion of the intrinsic photoconductor will be presented. This discussion is based on [13] where the background limited photoconductor was first analyzed with explicit consideration given to both the lattice-induced and the background radiation-induced noise. Recently, Burstein and Picus [18] have discussed the background limited condition for impurity level photoconductors and have coined the term *Blip* (background limited infrared photoconductor to describe it.

Intrinsic Photoconductor [19]

The instantaneous electrical conductivity of an intrinsic semiconductor is

$$\sigma(t) = a(\mu_e n(t) + \mu_h p(t)) = q\mu N(t)/V, \quad (8)$$

where

$$V = Ad, \mu = \mu_e + \mu_h, \text{ and } N = \mu V = pV$$

Here, q is the electronic charge, $N(t)$ is the total number of conduction electrons, V is the volume, A is the surface area, d is the thickness of the detecting filament, and μ_e and μ_h are the mobilities of electrons and holes, respectively. The radiation signal is measured by the change in conductance caused by a change in N; therefore, the smallest signal that can be detected will be limited by the random fluctuation in N.

The two fundamental mechanisms by which N can fluctuate are phonon and photon absorption and emission, due to the lattice and background radiation. Other noise mechanisms present in the semiconductor are neglected for this discussion.

Since incident radiation will penetrate into the semiconductor to a depth of about $x = 1/\alpha$, material at a depth greater than $1/\alpha$ sees little incident radiation and shunts the radiation effects. The analysis can be simplified if the filament thickness is assumed to be $d \leq 1/\alpha$ and the radiation is considered to be uniform over this region. Small densities of impurity levels, both bulk and surface, may be present for lattice excitation and recombination processes [Fig. 2(b)], but the electron and hole distribution is assumed to be that of the intrinsic semiconductor.

The small signal properties are governed by

$$d\Delta N_s/dt = A\eta_s J_s - \Delta N_s/\tau, \quad (9)$$
$$\Delta N_s = N(t) - N,$$

where N is the equilibrium number of electrons in the absence of the signal radiation J_s, η_s is the responsive quantum efficiency, and τ is the electron-hole lifetime.

These quantities can be expressed as

$$J_s = \frac{c}{4} \int_{E_i/h}^{\infty} n_s(\nu, T_s) d\nu, \tag{10}$$

$$\eta_s = \frac{c}{J_s 4} \int_{E_i/h}^{\infty} \eta(\nu) n_s(\nu, T_s) d\nu/, \tag{11}$$

$$\eta(\nu) = (1 - R)[1 - e^{-\alpha(\nu)d}]/[1 - Re^{-\alpha d}], \tag{12}$$

and

$$1/\tau = 1/\tau_r + 1/\tau_l, \tag{13}$$

where $n_s(\nu, T_s)$ is the density of photons in the radiation signal, T_s is the temperature of the signal source, $\eta(\nu)$ is the responsive quantum efficiency at the spectral frequency ν, R is the reflectance of the detector surface, τ_r is the radiative lifetime, and τ_l is the lattice lifetime which includes the effects of both bulk and surface recombination centers [Fig. 2(b)].

The solution of (9) for a sinusoidal signal of modulation frequency, $f = \omega/2\pi$, is

$$|\Delta N_s(f)| = A\eta_s J_s \tau/[1 + (\omega\tau)^2]^{1/2}. \tag{14}$$

The signal response can be increased by increasing the lifetime and by increasing the quantum efficiency. It can be seen that the carrier lifetime is the responsive time constant of the detector. The fractional change in conductivity is

$$\frac{|\Delta N_s(f)|}{N} = \frac{\eta_s J_s \tau}{nd[1 + (\omega\tau)^2]^{1/2}}, \tag{15}$$

showing the importance of making the equilibrium density of carriers as small as possible.

The fluctuation in N, which is the fundamental noise, is analyzed in Appendix A of [13]. The mean-square fluctuation of N in a small frequency interval, Δf, is

$$\langle \Delta N^2(f) \rangle = \frac{\Delta f 2\tau^2(F_r + R_r + F_l + R_l)}{1 + (\omega\tau)^2}. \tag{16}$$

F_r is the mean rate of generation of carriers by background radiation, R_r is the mean rate of recombination with emission of radiation, F_l is the mean rate of generation of carriers by absorption of lattice phonons, and R_l is the mean rate of recombination with emission of lattice phonons. F_r can be expressed as

$$F_r = n_r J_r A = \frac{Ac}{4} \int_{E_i/h}^{\infty} \eta(\nu) n_r(\nu, T_r) d\nu, \tag{17}$$

$$n_r(\nu, T_r) = \frac{8\pi\nu^2}{c^3(e^{h\nu/kT_r} - 1)}, \tag{18}$$

where J_r is the background radiation flux incident on the detector, and T_r is the temperature of the background radiation. A hemispherical field of view is assumed in (17).

When $E_i \gg kT_r$, the -1 in (18) can be neglected and

$$n_r J_r = B_r(T_r) e^{-E_i/kT_r}, \tag{19}$$

where

$$B_r(T_r) = 2\pi c \left(\frac{kT_r}{hc}\right)^3 \left[\left(\frac{E_i}{kT_r}\right)^2 + 2\left(\frac{E_i}{kT_r} + 1\right)\right]. \tag{20}$$

In the absence of an exact distribution function for lattice phonons, an analogy with the radiation field yields,

$$F_l = \eta_l J_l A, \tag{21}$$

$$\eta_l J_l = B_l(T_l) e^{-E_i/kT_l}, \qquad E_i \gg kT_l, \tag{22}$$

where J_l is the lattice phonon flux and T_l is the lattice temperature. The lattice excitation involves the same activation energy, E_i, as the radiative process.

When a dynamic equilibrium has been established,

$$F_r + F_l = R_r + R_l \tag{23}$$

represents a state in which the total rate of generation must be equal to the total rate of recombination. Substituting (23) into (16),

$$\langle \Delta V^2(f) \rangle = \frac{\Delta f 4\tau^2 A(\eta_r J_r + \eta_l J_l)}{1 + (\omega\tau)^2}, \tag{24}$$

which expresses the noise in terms of the rates of generation of carriers by the radiative and lattice processes. The recombination rates can be expressed [13] in terms of the lifetimes as

$$R_r = N/2\tau_r, \qquad R_l = N/2\tau_l, \tag{25}$$

and

$$R_r + R_l = N/2\tau. \tag{26}$$

The detailed balance (23) can be expressed in terms of the semiconductor parameters by combining (23) and (25),

$$n_r J_r + \eta_l J_l = \frac{nd}{2}\left(\frac{1}{\tau_r} + \frac{1}{\tau_l}\right) = \frac{nd}{2\tau}. \tag{27}$$

When (27) is substituted into (24), an expression for the noise in terms of semiconductor parameters is obtained,

$$\langle \Delta V^2(f) \rangle = \frac{2\tau\Delta f n A d}{1 + (\omega\tau)^2}. \tag{28}$$

Combining (14), (24), and (28) two equivalent expressions for the signal-to-noise ratio are obtained:

$$\frac{|\Delta N_s(f)|}{\sqrt{\langle \Delta N^2(f) \rangle}} = \frac{\eta_s J_s \sqrt{A}}{[4\Delta f(\eta_r J_r + \eta_l J_l]^{1/2}}, \tag{29}$$

and

$$\frac{|\Delta N_s(f)|}{\sqrt{\langle \Delta N^2(f) \rangle}} = \frac{\eta_s J_s \sqrt{A}}{(2nd\Delta f/\tau)^{1/2}}. \tag{30}$$

Eqs. (29) and (30) allow for a discussion of the signal-to-noise ratio of an intrinsic photodetector under dynamic as well as thermal equilibrium. Eq. (29) is especially useful for analyzing the effects of cooling the background radiation and/or the lattice since J_r and J_l are functions of T_r and T_l, respectively, as shown in (19) and (22). Conversely, (30) is useful for discussing the effects of the semiconductor parameters on the signal-to-noise ratio. It is important to emphasize that (29) and (30) are equivalent because of the detailed balance condition (27). These equations will now be used to discuss actual and ideal detectors, and methods of obtaining an ideal detector.

Ideal Radiation Detector (Blip)

An ideal radiation detector must satisfy two criteria:
1) It must detect with unit quantum efficiency all of the radiation in the signal for which the detector is designed,

$$\eta_s = 1. \tag{31}$$

2) The noise should be that of the background radiation that comes from the field of view of the detector. In terms of (29) this means that

$$\eta_r J_r \gg \eta_l J_l; \tag{32}$$

that is, the carriers should be generated by background radiation, not by the lattice.

$\eta_r J_r$ is given by (17) for detectors which have a hemispherical field of view, but if apertures are used,

$$\eta_r J_r = (\eta_r J_r)_{FV} + (\eta_r J_r)_{EXT}. \tag{33}$$

In an ideal detector only $(\eta_r J_r)_{FV}$ should be present.

While (32) assures that lattice noise is dominated by radiation noise, it does not assure that the radiation noise dominates other noises such as $1/f$ noise. For this reason (32) is a necessary, but not a sufficient condition, for ideal operation. Background radiation must dominate all other sources of noise if ideal conditions are to be achieved. The term *Blip* [18] is quite suitable since it stresses the importance of the background radiation.

The signal-to-noise ratio for the ideal *Blip* detector for intrinsic photoconductors can be obtained by substituting (31) and (32) into (29);

$$(S/N)_{Blip} = \sqrt{A} J_s/(4\Delta f - J_r)^{1/2}, \tag{34}$$

where $\eta_r = 1$ is required over the spectral region of interest by definition of the *Blip* condition. For certain detectors the $\sqrt{4}$ in the denominator of (34) is replaced by $\sqrt{2}$. An example of this is the broad area p-n junction detector, where only excitation noise is present and the recombination term is absent. This point is further discussed by Pruett and Petritz [5].

Methods for Achieving Blip Operation

An optimization of the responsive quantum efficiency can be understood from an analysis of (12). For thick crystals, $\alpha d \gg 1$,

$$\eta(\nu) = 1 - R, \tag{12a}$$

while for thin crystals, $\alpha d \ll 1$,

$$\eta(\nu) = \frac{(1 - R)\alpha d}{[1 - R(1 - \alpha d)]} \tag{12b}$$

A loss in signal due to a low responsive quantum efficiency will result from a crystal that is too thin, while some of the signal will be shunted by a crystal that is too thick. A good compromise discussed after (5) is $\alpha d \cong 1$, where

$$\eta(\nu) = \frac{(1 - R)(1 - e^{-1})}{(1 - Re^{-1})}. \tag{12c}$$

A reduction of reflection by optical coating techniques will yield a further improvement in $\eta(\nu)$.

The reduction of noise to the condition of being generated only by background radiation will now be discussed. At room temperature, contemporary photoconductors have $\eta_l J_l \gg \eta_r J_r$; that is, the generation of carriers occurs primarily through lattice excitation. It can be seen from (22) that the generation of carriers through lattice excitation can be reduced by cooling the lattice. It has been shown experimentally that cooling the lattice, while keeping the background radiation constant, does improve the detectivity of the detectors toward the *Blip* condition.

It is of considerable interest to determine whether or not the *Blip* condition of operation can be achieved without cooling the lattice. At thermal equilibrium, there exist additional equations of detailed balance, other than (27), for radiative and lattice processes separately. These are

$$\eta_r J_r = nd/2\tau_r \tag{35}$$

and

$$\eta_l J_l = nd/2\tau_l. \tag{36}$$

These equations require a balanced condition of absorption and emission of photons and of lattice phonons separately. The over-all equation of detailed balance (27) is obtained by adding (35) and (36).

From (35) and (36),

$$\eta_l J_l/\eta_r J_r = \tau_r/\tau_l. \tag{37}$$

It is required that

$$\tau_r/\tau_l \ll 1, \tag{38}$$

and

$$\tau \rightarrow \tau_r, \tag{39}$$

to achieve the condition of (32). This means that the electronic system must be decoupled from the lattice to allow the lifetime to approach the radiative lifetime required in (39). If this can be done, the *Blip* condition can be achieved without cooling the lattice.

The next question is how the electronic system can be decoupled from the lattice. It may be possible to do this in single crystal detectors. In the case of germanium, improved crystal technology has increased the lifetime from microseconds to the order of milliseconds. Improved crystal technology may lead to increases in carrier lifetime in single crystal detectors such as indium antimonide. It may be possible to achieve its calculated radiative lifetime of 3×10^{-7} second [20] at room temperature. However, it remains to be done and this is pointed out merely as an approach.

When the *Blip* condition is approached, (35) and (39) yield

$$\eta_r J_r = nd/2\tau. \qquad (40)$$

Since $\eta_r J_r$ is fixed by the background radiation level, n and τ are no longer independent variables. Comparison of (40) with (27) indicates that in the *Blip* condition, the ratio n/τ has a minimum value. The necessary reduction from the value in (27) can be achieved by a reduction of n, an increase of τ, or both. When the lattice is strongly coupled to the electronic system and the lattice is cooled, n will decrease while τ remains essentially unchanged. If the lattice is decoupled from the electronic system and no cooling is employed, τ will increase and approach τ_r, the radiative lifetime. At the same time, n will remain constant because at thermal equilibrium n is a thermodynamic quantity in an intrinsic semiconductor.

Whether or not the complete *Blip* operation is achieved at room temperature is not the only point of this discussion. The degree of cooling necessary to achieve *Blip* operation is related to the strength of coupling to the lattice. If the carrier lifetime is increased by decreasing the lattice coupling, a corresponding lessening in the cooling requirement is obtained; thus, by improved crystal technology, the cooling requirement can be steadily reduced. One suggested goal would be an attempt to move the cooling requirement from the liquid nitrogen temperature to the dry ice temperature. The next few years of detector research and development may make this possible.

Blip Conditions for Other Types of Infrared Detectors

While the preceding discussion has been in terms of the intrinsic photoconductor, the requirements for *Blip* operation are basically the same for all types of infrared detectors. Analysis of impurity level detectors by Burstein and Picus [18] and the lead salts by Petritz [8] shows that (32) is a necessary condition for *Blip* operation. The over-all signal-to-noise ratio is reduced, due to increased noise, if the carriers are generated by a mechanism other than by background radiation.

While the intrinsic detector may reach *Blip* operation without extensive cooling, certain types of detectors require cooling for additional reasons.

Impurity Level Photoconductors [3], [18]: Impurity-type photoconductors are of interest for detecting infrared radiation beyond 3 microns. Since such impurities have very low energy levels in the semiconductor [see Fig. 1(b)], without cooling, the centers would be thermally ionized and very little absorption of radiation

would take place at the impurity levels. Cooling is necessary in the case of impurity-type photoconductors, first, in order to have the carriers in the impurity levels where they may absorb radiation, and second, to reduce lattice noise (32).

Minority Carrier P-N Junction Infrared Detectors [5]: The p-n junction [Fig. 1(c)] infrared detectors are also of considerable interest for detection beyond 3 microns. In these detectors, notably indium antimonide, cooling is required to obtain good p-n junctions; the intrinsic numbers of carriers is so large at room temperature that the minority carrier effects are completely masked. Long-wavelength p-n junctions require cooling to obtain good p-n junctions and to reduce lattice noise. It is conceivable that shorter wavelength p-n junctions could reach the *Blip* condition without cooling.

Lead-Salt Film Detectors [4]: Lead sulfide is a very good photoconductor at room temperature although it does not reach *Blip* operation [21]. Cooling improves these detectors by reducing the lattice contribution to the noise, and the *Blip* condition is approached at solid carbon dioxide temperatures except for a component of $1/f$ noise which is usually present.

The longer wavelength lead-salt detectors such as lead selenide and lead telluride are far from the *Blip* condition at room temperature, and cooling improves these detectors in a more complex manner than in lead sulfide. Cooling appears to improve the signal response by an increase in the lifetime of the carriers [see (3)]. At temperatures of the order of 77°K, in both lead selenide and lead telluride, the carriers are generated principally by background radiation. However, both of these detectors still have $1/f$ noise at these temperatures, and the main improvement that can be obtained at low temperatures will be from a reduction of $1/f$ noise.

Photoelectromagnetic Detectors (PEM) [5], [22]: PEM detectors have been reported to operate quite well without cooling, and offer promise for still higher performance without cooling. This detector is a single crystal device. A magnetic field is used to separate photo-generated holes from electrons, causing a transverse voltage in the presence of radiation. The minority carriers are the important carriers, similar to the p-n junction detectors, but no p-n junction is required to separate the carriers. It is conceivable that this detector may approach *Blip* operation at or near room temperature, but further study is required to understand the competition between the Nyquist-Johnson noise of conduction electrons and the generation-recombination noise.

Information Capacity and the Blip Condition

When a detector is limited by generation-recombination noise, whether of lattice- or background-radiation origin, the frequency response of the signal and the spectrum of the noise are identical, as shown explicitly in (14) and (24). Under these conditions, electronic techniques can be used to boost the high-frequency response of both the signal and the noise without impairing the signal-to-noise ratio; therefore, the information capacity of the detector is not limited by the carrier lifetime. Now, when the detector is in the *Blip* condition these arguments are still valid, and moreover the signal-to-noise ratio is a maximum. Thus *Blip* should be achieved even at the expense of an increased time constant. Electronic techniques can then be used to increase the information capacity of the system.

While this discussion has been in terms of intrinsic photodetectors, it is basic for all photodetectors and has been subjected to experimental proof in recent work on lead sulfide [23]. A lead-sulfide detector was studied at room temperature and at −50°C. The time constant changed from 500 microseconds ($f_c = 1/2\pi\tau = 310$ cps) at room temperature to 5 milliseconds ($f_c = 31$ cps) at −50°C. The 1-cycle NEP was improved more than a factor of 10 by cooling. This improvement was not gained at the expense of information capacity, since a bandwidth for the cooled detector greater than the original 310 cps was obtained by electronic techniques.

This experiment shows that a most important objective is to approach as nearly as possible the *Blip* condition of operation by whatever techniques must be employed. After this condition is achieved, one can then use electronic techniques to increase the bandwidth of the detector. Such an approach will achieve the maximum signal-to-noise ratio for a given bandwidth requirement.

These statements are based on the assumption that the noise is generation-recombination noise. Usually the noise will have a 1/f component at low frequencies, and at sufficiently high frequencies, must go into white noise of the resistor and amplifier. This limits the extent to which frequency compensation can be employed and requires a more detailed study. Such a study has been made in two papers on the information capacity of detectors [23], [24].

Noise Figure—A Figure of Merit for Detectors

The degree to which a detector approaches an ideal detector can be expressed by the noise figure, F, defined as

$$F = (S/N)^2_{Blip}/(S/N)^2_{actual}$$
$$= (NEP)^2_{actual}/(NEP)^2_{Blip} = (D^*)^2_{Blip}/(D^*)^2_{actual}. \quad (41)$$

$(S/N)^2_{Blip}$ is the signal-to-noise ratio expected from an ideal detector, and $(S/N)^2_{actual}$ is the actual value of the ratio for the detector. D^* is the signal-to-noise ratio referred to unit radiation power, unit bandwidth and unit area [25]. F is called the noise figure since it corresponds to the noise figure as defined for microwave and radio-frequency receivers [26]. F approaches unity for a *Blip* detector and is greater than unity for a nonideal detector.

This can be seen in more detail by considering the theoretical expressions for the intrinsic detector operating under generation-recombination noise. Substituting (29) and (34) into (41) yields,

$$F = (\eta_r/\eta_s)^2[1 + (\eta_l J_l/\eta_r J_r)]. \quad (42)$$

F is greater than unity because of the two effects defined by (31) and (32).

When the detector is being limited by lattice noise,

$$F = \eta_l J_l/\eta_s^2 J_r, \quad (43)$$

where

$$\eta_l J_l \gg \eta_r J_r,$$

and increasing the quantum efficiency improves F as $1/\eta_s^2$. When the detector is background radiation noise limited,

$$F = \eta_r/\eta_s^2 \cong 1/\eta_s, \quad (44)$$

where

$$\eta_l J_l \ll \eta_r J_r,$$

and the improvement of F is with $1/\eta_s$ rather than $1/\eta_s^2$ because an improvement of quantum efficiency now affects the noise.

To facilitate the use of the noise figure, it is useful to have a plot of D^*_{Blip} as a function of the long wavelength threshold, λ_i. The definition of D^*_{Blip} is

$$D^*_{Blip}(\lambda_i) = (S/N)_{Blip}\sqrt{\Delta f}/(E_s J_s\sqrt{A}), \quad (45)$$

where $E_s = E_i = hc/\lambda_i$. Substituting (34) into (45)

$$D^*_{Blip}(\lambda_i) = \frac{1}{E_i\sqrt{4J_r}}, \quad (46)$$

where J_r is calculated from (17) with $\eta(\nu) = 1$ for $\nu \geq \nu_i$, and $\eta(\nu) = 0$ for $\nu \leq \nu_i$. A hemispherical blackbody radiation at $T_r = 300°K$ is assumed. A discontinuity exists as $E_i \rightarrow 0$, and

$$D^*_{Blip}(E_i \rightarrow 0) = (16\sigma kT_r^5)^{-1/2} = 1.81 \times 10^{10}, \quad (47)$$

has been calculated for an ideal blackbody thermal detector [12]; σ is the Stefan-Boltzmann constant. $D^*_{Blip}(\lambda_i)$ is plotted vs λ_i and E_i in Fig. 3.

While the question of the characterization of detectors requires more discussion than can be given here, it is suggested that the noise figure be considered as a figure of merit for detectors. The reasons are: F is a quantitative measure of the departure from an ideal detector, and the methods for determining F are consistent with methods used for other radiation receivers. Because of its generality and related uses in the microwave and radio-frequency bands, it is suggested that the noise figure, rather than its reciprocal [27], be employed in the infrared region in order to avoid new terms. For an example of the use of the noise figure for an infrared detector see Pruett and Petritz [5].

THERMAL DETECTORS

The theory of thermal detectors is given in [12] and [28], and an up-to-date summary of their characteristics appears in this volume [29]. The discussion here will be confined to the basic characteristics of thermal detectors and to the possibilities for this type of detector to reach the *Blip* condition.

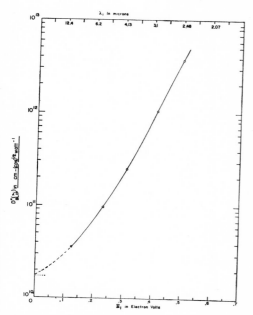

Fig. 3—Detectivity D*$_{Blip}$ of background-limited infrared photoconductors plotted against their long wavelength cutoff and corresponding energy gap. This curve assumes both generation and recombination processes contribute to the noise, if only one does, the values of D*$_{Blip}$ are increased by a factor $\sqrt{2}$.

Signal Properties

In general, thermal detectors can be considered as blackbody detectors; that is, they will absorb radiation rather uniformly over the infrared spectrum. Currently, they are the only detectors that will operate without cooling out to wavelengths as far as 14 microns. This is one of the inherent advantages of thermal detectors over photodetectors.

The photon absorption in thermal detectors causes a temperature rise in the lattice, which then causes a change in an electronic property. The rise in temperature will depend upon the thermal capacity of the system; therefore, great care is taken to make these detectors as small as possible.

Consider next how temperature equilibrium is established in the detector. If the detector is coupled to its environment only through the background-radiation field, temperature equilibrium is established by reradiation from the detector element. Such a condition in principle, could be achieved by suspending the detector in a high vacuum, with very low thermal conductivity leads attached to it. Such a detector has a very-long-time constant because the rate of reradiation is quite slow.

In actual practice, higher-speed detectors are desirable. These are made by mounting the sensitive wafer on a body which rapidly conducts heat away from the detector. Therefore, the total thermal conductivity is the sum of that due to reradiation plus that through the backing material. Usually the thermal conduction through the backing material completely dominates the reradiation effect. Much of today's technology on thermal detectors is to decrease the time constant by increasing the thermal conductance and by decreasing the thermal mass of the detector

Noise Properties

The ultimate noise of the thermal detector, as in all infrared detectors, is established by fluctuations in the background radiation. Such noise would manifest itself as a temperature fluctuation in the detector element. However, thermal detectors as presently manufactured do not approach this noise mechanism, but are limited by the Nyquist-Johnson noise of the conduction electrons.

Present thermal detector technology, which aims towards increasing the speed of response by thermal conduction to the backing plate, is going in the wrong direction to achieve the *Blip* operation. The temperature fluctuations under this condition will result from heat flow to the backing plate and will not be due to background radiation. The heat conduction to the backing plate plays an analogous role to the $\eta_i J_i$ term in the photodetector, [see (29)]; therefore, increasing this term leads away from the *Blip* condition. However, so long as these detectors are limited by Nyquist-Johnson noise, the speed of response improvements are achieved without raising the noise level; thus, they are entirely in order. This is pointed out to indicate that such approaches will not achieve the *Blip* condition.

The use of cooling to improve the performance of thermal detectors is discussed by Smith, Jones, and Chasmer [28], but the reported experimental results to date have fallen well below the theoretical expectations. One particular cooled thermal detector, the superconducting bolometer, has received considerable study, but for a number of reasons [28] has yet to be developed into a practical device.

An approach to the *Blip* condition in thermal detectors, which might not require cooling, is to suspend the detector element in a vacuum such that it is coupled predominantly to the radiation field. The resulting very-long-time constant could be tolerated and electronic means could be used to speed up the response of the detector if the true *Blip* condition were achieved. However, in addition, it would require the overcoming of the Nyquist-Johnson noise, which seems difficult in terms of present technology.

In summary, thermal detectors are of great importance for long wavelength detection at room temperatures, but rather difficult problems must be overcome before they will approach the *Blip* condition.

Future Trends

An important goal will be to develop *Blip* detectors to efficiently cover the infrared spectrum. It appears that cooling will be allowed for high-quality military and commercial detection systems; therefore his approach will be discussed first. Nearly all contemporary photodetectors are approaching the condition that the generation of carriers is mainly by background radiation if the detector is cooled sufficiently. However, none of these detectors has quite reached the total *Blip* condition because of 1/f noise. Therefore, more emphasis will be placed on improving the noise characteristics of detectors. This means more research on contact phenomena, surfaces and other sources of 1/f noise.

A second improvement in contemporary detectors will be to utilize fully the signal radiation; that is, to improve the responsive quantum efficiency, η_s. Antireflection coatings, optimizing detector thickness relative to absorption coefficient, and other techniques will lead to over-all improvements in η_s.

A third and important area will be that of increased emphasis on the physics and chemistry of detector preparation to optimize production yields. Better uniformity of response and methods of assuring no deterioration under adverse storage conditions will be realized.

Another goal will be to provide the user with a broader range of detectors that do not require cooling. Here the physics of hole-electron recombination processes becomes especially important and research will be done to decouple the electronic system from the lattice.

Concerning thermal detectors, further work aimed at improving their speed of response, uniformity of response, and general methods of preparation, can be expected. The present approach to achieve higher speeds of response does not appear to be compatible with reaching *Blip* conditions. However, good performance of thermal detectors at room temperature will make them valuable detectors for the indefinite future. More work will be done in the area of cooling thermal detectors and improvements in performance should result.

Finally, a completely new device, the infrared maser [30], can be expected to receive attention and it may have a large impact on the over-all infrared picture.

Acknowledgment

The author wishes to acknowledge the many stimulating discussions he has had with his colleagues and friends in the field of infrared. In particular, when at the Naval Ordnance Laboratory, his discussions with James Humphrey, Frances Lummis, Wayne Scanlon, and Robert Talley were invaluable. Discussions with Eli Burstein on the subject of *Blip* detectors have been very helpful.

Since joining Texas Instruments, the author's discussions with Werner Beyen and George Pruett have been especially helpful, both in regard to the subject of single-crystal detectors and to the preparation of this paper.

References

[1] A good general reference on the subject of photodetectors is "Photoconductivity Conference," eds., R. G. Breckenridge, B. R. Russell, and E. E. Hahn, John Wiley & Sons, Inc., New York, N. Y.; 1956.
[1a] G. A. Morton and S. V. Forque, "The photoconductive beam-scanning pickup tube," this issue.
[2] E. Burstein, G. Picus, and N. Sclar, "Optical and Photoconductive Properties of Silicon and Germanium," [1], pp 353–413.
[3] For discussion and references to impurity-type photodetectors, see H. Levinstein, "Impurity photoconductivity in germanium," paper 3.3.6, this issue, p. 1478.
[4] For discussion and references to the lead-salt photoconductors, see R. J. Cashman, paper 3.3.4, this issue, p. 1471. Another recent review is, R. L. Petritz and J. N. Humphrey, "Research on photoconductive films of the lead salts," *Proc. IRIS*, vol. 3, pp. 65–79; March, 1958.
[5] For discussion and references to the indium antimonide photovoltaic and PEM detector, see F. F. Reike, L. H. DeVant, and A. J. Tuzzolino, "Single Crystal infrared detectors based upon intrinsic absorption," paper 3.3.5, this issue, p. 1475 and G. R. Pruett and R. L. Petritz, "Detectivity and pre-amplifier considerations for indium antimode photo-voltaic detectors," paper 4.1.9, this issue, p. 1524.
[6] W. Shockley, W. T. Read, Jr., "Statistics of the recombination of electrons and holes," *Phys. Rev.*, vol. 87, pp. 835–842; September, 1952. W. Shockley, "Electrons, holes, and traps," *Proc. IRE*, vol. 46, pp. 973–990; July, 1958. G. Bemski, "Recombination in semiconductors," *Proc. IRE*, vol. 46, pp. 990–1004; July, 1958.
[7] Nonrecombining traps have been discussed by A. Rose, *Phys. Rev.*, vol. 97, pp. 322–333; January, 1955, for insulator photodetectors; by J. R. Haynes and J. A. Hornbeck, *Phys. Rev.*, vol. 97, pp. 311–321; January, 1955, for silicon; by H. Y. Fan, *Phys. Rev.*, vol. 92, pp. 1424–1428; December, 1953, and vol. 93, p. 1434; March, 1954, for germanium; and by J. N. Humphrey and R. L. Petritz, *Phys. Rev.*, vol. 105, pp. 1736–1740; March, 1957, and R. H. Harada and H. T. Minden, *Phys. Rev.*, vol. 102, pp. 1258–1262; June, 1956, for the lead salts. Also Shockley [6] and Bemski [6].
[8] R. L. Petritz, "Theory of photoconductivity in semiconductor films," *Phys. Rev.*, vol. 104, pp. 1508–1515; December, 1956; Harada-Minden and Humphrey-Petritz [7].
[9] J. F. Woods, "Investigation of the photoconductive effect in lead sulfide films using Hall and resistivity measurements," *Phys. Rev.*, vol. 106, pp. 235–240; April, 1957.
[10] J. C. Slater, "Barrier theory of photoconductivity in lead sulfide," *Phys. Rev.*, vol. 103, pp. 1631–1644; September, 1956; G. W. Mahlman, "Photoconductivity of lead sulfide films," *Phys. Rev.*, vol. 103, pp. 1619–1630; September, 1956; E. S. Rittner, "Electron processes in photoconductors," in [1], p. 215.
[11] A recent review of noise in photodetectors, K. M. van Vliet, "Noise in semiconductors and photoconductors," *Proc. IRE*, vol. 46, pp. 1004–1018; June, 1958.
[12] R. C. Jones, "Performance of detectors for visible and infrared radiation," in "Advances in Electronics," ed. L. Marton, Academic Press Inc., New York, N. Y., vol. 5, pp. 1–96; 1953.
[13] R. L. Petritz, "The Relation Between Lifetime, Limit of Sensitivity, and Information Rate in Photoconductors," [1], pp 49–77.
[14] For a discussion of Nyquist-Johnson noise, see J. L. Lawson and G. E. Uhlenbeck, "Threshold Signals," McGraw-Hill Book Co., Inc., New York, N. Y., ch. 4; 1948.
[15] For reviews of 1/f noise, see: A. van der Ziel, "Noise in junction transistors," *Proc. IRE*, vol. 46, pp. 1019–1038; June, 1958. "1/f Noise and Germanium Surface Properties," "Semiconductor Physics," Ed. R. H. Kingston University of Pennsylvania Press, Philadelphia, Pa., pp. 207–225; 1957; R. L. Petritz, "Recent semiconductor noise studies," 5ᵃ Rassegna Internazl. Elett.-Nuclear, Rome, Italy, pp. 217–226; 1958.
[16] For a discussion of preamplifiers for infrared-detectors, see G. R. Pruett and R. L. Petritz, [5].
[17] A good discussion and extensive bibliography of the statistical theory of detection is, D. Middleton and D. Van Meter, "Detection and extraction of signals in noise from the point of view of statistical decision theory," *J. Soc. Indust. Appl. Math.*, vol. 3, pp. 192–253; December, 1955, and vol. 4, pp. 86–119; June, 1956.
[18] E. Burstein and G. S. Picus, "Background Limited Infrared Detection," paper presented at IRIS; February 3, 1958.

[19] The analysis outlined here is based on [13], in particular, Sections 1, 3–7, and Appendix A. These sections can be read independently of the information theory aspects of the article; the latter has been superseded by [24].

[20] G. K. Wertheim, "Carrier lifetime in indium antimonide," *Phys. Rev.*, vol. 104, pp. 662–664; November, 1956. Also Bemski [6].

[21] F. L. Lummis and R. L. Petritz, "Noise-time constant, and Hall studies on lead sulfide photoconductive films," *Phys. Rev.*, vol. 105, pp. 502–508; January, 1957.

[22] O. Simpson and C. Hilsum, "The design of single-crystal infrared photocells," *Proc. IRIS*, vol. 3, pp. 115–120; September, 1958. P. W. Kruse, "Indium antimonide photoelectromagnetic infrared detector," *J. App. Phys.*, vol. 30, pp. 770–778; May, 1959.

[23] R. L. Petritz and I. Stiglitz, "Improved Use of Infrared Detectors by Methods Suggested by Information Theory," paper presented at IRIS; January 6, 1959. ("Statistics of Signal Detection," to be published.)

[24] R. L. Petritz, "Information theory of the performance of radiation detectors," *Proc. IRIS*, vol. 2, pp. 18–34; June, 1957.

[25] For the definition of D* see R. C. Jones, "Phenomenenological description of the response and detecting ability of radiation detectors," paper 4.1.1, this issue, p. 1481. Also Jones' paper, "Methods of rating the performance of photoconductive cells," *Proc. IRIS*, vol. 2, pp. 9–12; June, 1957.

[26] A. Rose, "A unified approach to the performance of photographic film, television pickup tubes, and the human eye," *J. Soc. Mot. Pict. Engrs.*, vol. 47, pp. 273–294; October, 1946. H. C. Torrey and C. A. Whitmer, "Crystal Rectifiers," McGraw-Hill Book Co., Inc., New York, N. Y., ch. 2; 1948.

[27] R. C. Jones, "Quantum efficiency of photoconductors," *Proc. IRIS*, vol. 2, pp. 13–17; June, 1957; "Quantum efficiency of detectors for visible and infrared radiation," in "Advances in Electronics and Electron Processes," vol. 11; 1959 (in press). Also [25], this issue.

[28] R. A. Smith, F. E. Jones, and R. P. Chasmar, "The Detection and Measurement of Infrared Radiation," Oxford University Press, Oxford, Eng., ch. 3; 1957.

[29] R. DeWaard and E. M. Wormser, "Description and properties of various thermal detectors," paper 4.1.3, this issue, p. 1508.

[30] A. L. Schawlow and C. H. Townes, "Infrared and optical masers," *Phys. Rev.*, vol. 112, pp. 1940–1949; December, 1958.

5

Reprinted from *Proc. IRE,* **47**(9), 1481–1486 (1959)

Noise in Radiation Detectors*

R. CLARK JONES†

INTRODUCTION

THE kinds of noise we discuss in this paper are the kinds of electrical noise that one may find in the output of radiation detectors. The kinds of noise are distinguished by the physical source of the noise, not by any particular characteristic of the noise as measured at the output of the detector.

The discussion will be restricted to descriptions of the noise that relate to its mean-square amplitude. Other more complex measures may be used, but so far these have found little use in the application of radiation detectors.

The tool that will be used to describe the electrical noise is the power spectrum, defined as the mean-square fluctuation of the voltage (or current) about its mean value per unit frequency interval. A more general kind of spectrum is the mean-square fluctuation of Y (about the mean value $\overline{Y}$) per unit frequency interval measured in cycles per unit of X. Only when Y is a voltage or current is it appropriate to call this more general kind of spectrum a power spectrum. When Y is not a voltage or current, the term power spectrum is clearly inappropriate, and is obviously *wrong* when the variable Y is itself a power. The author recommends (and uses in this report) the name "Wiener spectrum" for the more general kind of spectrum. The name "variance spectrum" may also be used.

* Original manuscript received by the IRE, June 26, 1959.
† Res. Lab., Polaroid Corp., Cambridge, Mass.

THE POWER SPECTRUM

We use e to denote the instantaneous amplitude of a noise voltage. (The current i may be substituted for the voltage e throughout the following discussion.) The voltage is considered to be a function of the time t, where the epoch of the origin of t is supposed to be chosen arbitrarily. The mean value of $e(t)$ is supposed to be zero.

Consider that part of the noise voltage that occupies the time interval from $-T$ to T. The Fourier transform of this voltage is defined by

$$F(f) = \int_{-T}^{T} e(t)e^{2\pi i f t}dt \qquad (1)$$

where f is the frequency in cycles per second. Next we form the square of the absolute value of $F(f)$, divide by $2T$; and multiply by 2:

$$|F(f)|^2/T. \qquad (2)$$

This expression is very close to what is meant by a power spectrum. To form what a mathematician means by a power spectrum, however, we must take an appropriate limit of this expression.

There are two different but equivalent ways to define the power spectrum. The first, and more simple, involves the average over an ensemble of independently generated noise voltages of the same kind. If an ensemble average is denoted by angular brackets, the power spectrum $P(f)$ is defined by

$$P(f) = \langle T^{-1} | F(f) |^2 \rangle. \qquad (3)$$

The second equivalent definition of the power spectrum is formulated by first averaging the expression (2) over the bandwidth $2B$, and then taking the limit of the resulting expression as the time interval $2T$ becomes infinite and the bandwidth $2B$ approaches zero in such a way that BT becomes infinite:

$$P(f) = \lim_{\substack{BT \to \infty \\ B \to 0}} \frac{1}{2BT} \int_{f-B}^{f+B} | F(f') |^2 df'. \qquad (4)$$

If the power spectrum determined by these operations turns out to be independent of the epoch of the origin of the time t, then the noise is said to be stationary. All of the noises discussed in this paper are supposed to be stationary.

The power spectrum defined above is defined for positive frequencies only. The total mean-square fluctuation of the voltage $\overline{e^2}$ is obtained by integrating $P(f)$ over positive frequencies:

$$\overline{e^2} = \int_0^\infty P(f) df. \qquad (5)$$

Anyone who contemplates measuring a power spectrum or a Wiener spectrum would be well advised to read first the excellent pair of papers by Blackman and Tukey.[1]

Attention is called to an important paper by Callen and Welton.[2] These authors show, by use of quantum statistical mechanics, that wherever there is a dissipation process there is an associated fluctuation, and that the Wiener spectrum of the fluctuation can be calculated from the magnitude of the dissipation. Callen and Welton give three examples of the application of their theory. They show that the dissipation involved in the viscous drag on a moving airborne particle leads to a fluctuation in the position and velocity of the particle that is in exact accord with the accepted formulas for Brownian motion. They show that the radiation re-

sistance offered to the acceleration of an electrical charge leads to a fluctuation of the electrical field that is equivalent to blackbody radiation; this is a particularly interesting and significant result. Finally, they show that the acoustic radiation resistance of an infinitesimal pulsating sphere leads to the accepted formulas for the pressure and density fluctuation of a gas. The author[3] has further exemplified their theory by using it to derive the Einstein formula for the fluctuation of the temperature of a small body that is in equilibrium with a radiation field or in contact with a heat reservoir; here, the dissipation process is the loss of negentropy in the thermal conductance between the body and its surroundings.

Eight Kinds of Noise

Table I names eight different kinds of noise that may be present in the output of a radiation detector. The table also indicates the kind of detector in which the noise may be important, and indicates briefly the mechanism.

Some of the kinds of noise listed in Table I have an ambiguous meaning. The term "temperature noise," for example, could mean either the fluctuation in the temperature of a bolometer element, or it could mean the consequent voltage fluctuation in the output of the detector. In this review, we avoid the ambiguity by using the eight names in Table I to mean always the *electrical* noise in the output of the detector.

TABLE I

The Eight Kinds of Noise, the Type of Detector in which the Type of Noise May Be Important, and a Brief Indication of the Physical Mechanism That Generates the Noise

Kind of Noise	Detector Concerned	Physical Mechanism
Shot Noise	Thermionic Detector	Random emission of electrons by thermionic emission
Johnson Noise	All Detectors	Thermal agitation of current carriers
Temperature Noise	Thermal Detectors	Temperature fluctuation
Radiation Noise	All Detectors	Bose-Einstein fluctuation of radiation photons
Generation-Recombination Noise		Fermi-Dirac fluctuation of the current carriers
Modulation Noise	Photoconductive cells, Bolometers	Resistance fluctuation in semiconductors
Contact Noise		Resistance fluctuation at contacts
Flicker Noise	Thermionic Detector	Fluctuation of the work function

The power spectra of the various kinds of noise must be added in order to obtain the power spectrum of the total electrical noise at the output of the detector. Thus, for example, the power spectrum of the total noise of a bolometer may be regarded as the sum of the power spectra of the Johnson noise, the temperature noise, the modulation noise, and perhaps also the contact noise.

Shot Noise

The usual expression for the power spectrum of the shot noise in a temperature-limited diode is

$$P_i(f) = 2\epsilon \overline{I} \qquad (6)$$

[1] R. B. Blackman and J. W. Tukey, "The measurement of power spectra from the point of view of communication engineering," *Bell Syst. Tech. J.*, vol. 37, pts. 1 and 2, pp. 185–282 and 485–569; 1958.

[2] H. B. Callen and T. A. Welton, "Irreversibility and generalized noise," *Phys. Rev.*, vol. 83, pp. 34–40; 1951.

[3] R. C. Jones, "Performance of detectors for visible and infrared radiation," in "Advances in Electronics," Academic Press, New York, N. Y., vol. 5, pp. 1–96; 1953.

77

where ϵ is the absolute value of the charge of the electron, and $\bar{I}$ is the mean current. The subscript i on P indicates that the independent variable is the current —that is to say, the power spectrum $P_i(f)$ is the mean-square fluctuation of the current per unit bandwidth.

The above formula is derived easily if one supposes that the charge is transported discontinuously in quanta of size ϵ and that the transportations occur randomly in time. In a temperature-limited thermionic diode these conditions are quite well satisfied for frequencies above the range where flicker noise is significant.

The presence of space charge reduces the magnitude of the noise below that given by (6) without changing the flatness of the spectrum. The effect of finite transit time[4] within the diode and of shunt capacity causes the spectrum to fall off at high frequencies.

The idea behind shot noise—that the current carriers have a random element—has caused the term to be used to describe almost every noise except Johnson noise. Such a broad use of a term weakens its meaning. In this review, the term shot noise will be confined to the noise of a thermionic diode. More explicit terms are available for all of the other purposes for which the term is sometimes used.

In temperature-limited diodes, the existence of shot noise was predicted by Schottky[5] in 1918. The prediction was confirmed by Johnson[6] and by Hull and Williams,[7] and was confirmed later with high precision by Williams and Vincent[8] in 1926.

Johnson Noise

Johnson noise is the thermal agitation noise that appears at the output of every radiation detector. If the detector is in thermal equilibrium with its surroundings, both with respect to radiation and conduction, and if the steady electric current through the detector is zero, then the only noise at the output of the detector is Johnson noise.

The power spectrum of Johnson noise is given by

$$P_e(t) = 4kTR \cdot \frac{hf/kT}{e^{hf/kT} - 1} \tag{7}$$

where R is the real part of the electrical impedance. Except at very high frequencies, where hf is not very small compared with kT, the last factor may be omitted.

In 1906, Einstein[9] applied his recently developed

theory of Brownian motion to the motion of the electrons inside of a wire. He found that the mean-square charge transported across any given cross section in the period τ was equal to

$$\overline{q_\tau^2} = \frac{2kT\tau}{R} \tag{8}$$

where the wire is supposed to be part of a loop of resistance R. This result was extended by de Haas-Lorentz.[10] She supposed that the transport of charge indicated by (8) was due to a fluctuating electromotive force in series with the loop. Let E be the instantaneous value of this electromotive force and let e_τ be the average value of E over a given period of duration τ. She showed that $\overline{e_\tau^2}$ is given by

$$\overline{e_\tau^2} = 2RkT/\tau. \tag{9}$$

The power spectrum of Johnson noise was not worked out until 1928. In that year, Nyquist[11] published a derivation of the expression (7) by a procedure that essentially involved counting the independent modes of vibration in a transmission line.

During the years between 1928 and 1941, the use of the equation for Johnson noise was pushed to higher and higher frequencies. According to Southworth,[12] it gradually became apparent to those working in the field that Johnson noise was the one-dimensional form of blackbody radiation. This was established in 1941 by Burgess.[13] He showed that if one connected an antenna through a transmission line to a resistor, the energy radiated from the antenna due to the Johnson noise fluctuation in the resistor was just balanced by the blackbody radiation received by the antenna, provided only that the temperature of the radiation field was equal to the temperature of the resistor.

Johnson noise is also called Nyquist noise, thermal noise, and sometimes thermal fluctuation noise.

Temperature Noise

Temperature noise is the electrical noise in the output of a detector due to the temperature fluctuation of the active element. Temperature noise is important primarily in connection with thermal detectors, which are defined as detectors whose conversion of incident radiation into an electrical output is mediated by a change in the temperature of the active element. Thermal detectors include the bolometer, the thermocouple, the Golay pneumatic detector, and the thermionic detector.

The Wiener spectrum of the temperature fluctuation is given by

[4] S. S. Solomon, "Thermal and shot fluctuations in electrical conductors and vacuum tubes," *J. Appl. Phys.*, vol. 23, pp. 109–112; 1952.

[5] W. Schottky, "Über spontane Stromschwankungen in verschiedenen Elektrizitätsleitern," *Ann. Physik.*, vol. 57, pp. 541–567, 1918.

[6] J. B. Johnson, "The Schottky effect in low frequency circuits," *Phys. Rev.*, vol. 26, pp. 71–78; 1925.

[7] A. W. Hull and N. H. Williams, "Determination of elementary charge from measurements of shot effect," *Phys. Rev.*, vol. 25, pp. 147–173; 1925.

[8] N. H. Williams and H. B. Vincent, "Determination of elementary charge e from measurements of shot effect in aperiodic circuits," *Phys. Rev.*, vol. 28, pp. 1250–1264; 1926.

[9] A. Einstein, "Zur Theorie der Brownschen Bewegung," *Ann. Physik.*, vol. 19, pp. 371–381; 1906.

[10] G. L. de Haase-Lorentz, "Die Brownsche Bewegung und einige verwandte Erscheinungen," Braunschweig, Ger.; 1913.

[11] H. Nyquist, "Thermal agitation of electric charge in conductors," *Phys. Rev.*, vol. 32, pp. 110–113; 1928.

[12] G. C. Southworth, "Microwave radiation from the sun," *Phys. Rev.*, vol. 239, pp. 285–297; 1945.

[13] R. E. Burgess, "Noise in receiving-aerial systems," *Proc. Phys. Soc. (London)*, vol. 53, pp. 293–304; 1941.

$$W_t(f) = \frac{4kT^2 g(f)}{g^2 + (h + 2\pi fC)^2} \cdot \frac{hf/kT}{e^{hf/kT} - 1} \qquad (10)$$

where C is the thermal capacity of the active element, and $g(f) + ih(f)$ is the complex thermal conductance of the active element to the environment, including the conductance due to radiation exchange. $W_t(f)$ is the mean-square fluctuation of the temperature about its mean value per unit frequency bandwidth.

It follows directly from a result given by Einstein[14] in 1904 that if two bodies, A and B, are in thermal contact with one another, and if they have been in contact long enough so that their mean temperatures are equal, the mean-square difference in their temperature is given by

$$\overline{t^2} = kT^2/C \qquad (11)$$

where C is given by

$$C = \frac{C_A C_B}{C_A + C_B} \qquad (12)$$

and C_A and C_B are the thermal capacities of the two bodies. Einstein's result may be derived from (10).

The Wiener spectrum of the fluctuation of the temperature was first determined by Milatz and Van der Velden.[15] The term temperature noise was introduced by the author[16] in 1949.

In order to obtain the power spectrum $P_e(f)$ of the temperature noise, the Wiener spectrum $W_t(f)$ must be multiplied by the square of factor $Q(f)$ that converts a change of temperature into a change of output voltage:

$$P_e(f) = Q^2(f)W_t(f). \qquad (13)$$

Radiation Noise

Radiation noise is the noise in the output of a radiation detector that is due to the fluctuation in the radiant power that acts on the detector, and is often called photon noise.

The radiation noise may be formulated in a variety of ways, and several different methods are used in the discussion by Jones.[3] In a narrow band of radiation frequencies, the Wiener spectrum of the incident power is given by

$$W_p(f) = 2h\nu\overline{P}/(1 - e^{-h\nu/kT}) \qquad (14)$$

where $\overline{P}$ is the mean incident power, and ν is the radiation frequency. $W_p(f)$ is the mean-square fluctuation of incident power per unit frequency bandwidth.

When the bandwidth of the radiation frequencies is not narrow, the right-hand side of the above expression must be integrated over the radiation frequency. If the radiation spectrum is that of blackbody radiation of temperature T, the result of the integration is

$$W_p(f) = 8kT\overline{P} = 8Ak\sigma T^5 \qquad (15)$$

where A is the active area of the detector, σ is the Stefan-Boltman radiation constant, and where it is assumed that the detector is immersed in the blackbody radiation field.

In order to obtain the power spectrum $P_e(f)$ of the radiation noise, the Wiener spectrum $W_p(f)$ must be multiplied by the square of the responsivity $R(f)$:

$$P_e(f) = R^2(f)W_p(f). \qquad (16)$$

The theory of the fluctuation of the power in a beam of thermal radiation was first worked out by Lewis,[17] although, as he points out, the theory is based on the older theory of the fluctuation phenomena in Bose-Einstein ensembles. Radiation noise is considered from a number of points of view by Jones.[3]

Generation-Recombination Noise

Generation-recombination noise is the noise in the output of a photoconductive cell or a semiconductor bolometer that is due to the fluctuation in the number and lifetime of the thermally-generated carriers.

Simple and general expressions for G-R noise do not exist. As Bernamont[18] showed, if the carriers are generated randomly in time and have an exponential distribution of lifetimes, the power spectrum of the noise can be expressed by

$$P_i(f) = 4\overline{I}^2\tau/n_0(1 + (2\pi f\tau)^2) \qquad (17)$$

where τ is the mean lifetime of the carriers, and n_0 is the mean number of carriers in the semiconductor. This expression may also be written

$$P_i(f) = 4\epsilon\overline{I}G/(1 + (2\pi f\tau)^2) \qquad (18)$$

where ϵ is the absolute charge of the electron, and G is the semiconductor gain[19] defined by

$$G = \tau/T_r = \tau\mu V/L^2 \qquad (19)$$

where μ is the mobility, V is the voltage across the electrodes, L is the distance between electrodes, and T_r is the interelectrode transit time.

The factor 4 in (18) instead of the factor 2 in the shot noise formula (6) is brought about by the fluctuation in the lifetime of the carriers as well as in their number.

The above expressions for the power spectrum, based on Maxwell-Boltzmann statistics, do not have general

[14] A. Einstein, "Zur allgemeinen molekularen Theorie der Wärme," *Ann. Physik.*, vol. 14, pp. 354–362; 1904.
[15] J. M. W. Milatz and H. A. Van der Velden, "Natural limit of measuring radiation with a bolometer," *Physica.*, vol. 10, pp. 369–380; 1943.
[16] R. C. Jones, "Radiation detectors," *J. Opt. Soc. Am.*, vol. 39, pp. 327–356; 1949.

[17] W. B. Lewis, "Fluctuations in streams of thermal radiation," *Proc. Phys. Soc. (London)*, vol. 59, pp. 34–40; 1947.
[18] J. Bernamont, "Fluctuations de potentiel aux bornes d'un conducteur metallique de faible volume parcouru par un courant," *Ann. Phys.*, vol. 7, pp. 71–140; 1937.
[19] A. Rose, "Performance of photoconductors," Proc. IRE, vol. 43, pp. 1850–1869; December, 1955.

validity. In fact, the only special case where they are usually considered to be valid is an extrinsic semiconductor that is only slightly ionized. If the degree of ionization is γ, a simple argument due to van der Ziel[20] suggests that the factor $1 - \gamma$ should be inserted in (17) and (18).

The analysis by Fermi-Dirac statistics of even a three-level system leads to complex relations.[21,22] These complications are beyond the scope of this short review.

Modulation Noise

The kinds of noises that have been discussed so far in this review have been rather fundamental in character. The three remaining types of noise, however, do not appear to be fundamental, and from some points of view can be considered as due to defects that, in principle, are avoidable. On the other hand, they.occur so commonly in radiation detectors, and are of such great practical importance, that they must be included in any realistic discussion of noise in radiation detectors.

The term modulation noise introduced by Petritz[23] refers to noise in semiconductors that is due to modulation of the conductivity due to other causes than Fermi-Dirac fluctuations in the number of carriers. The power spectrum of modulation noise varies approximately as $1/f$:

$$P_i(f) = CI^2/f^\alpha \tag{20}$$

where α is an exponent close to unity.

The frequency range over which modulation noise has been observed is remarkably large. In germanium filaments, it has been followed[24] down to 2×10^{-4} cps, and in point contact diodes it has been found[25,26] as low as 6×10^{-6} cps. Hyde[27] has observed $1/f$ noise in germanium crystals up to 4×10^6 cps.

Unlike the first five kinds of noise discussed in this review, the physical mechanism of the last three is obscure.

The physical location of the source of at least part of the modulation noise is at the surface of the crystals; the level of modulation noise in single-crystal detectors can often be reduced by suitable treatment of the surface. The physical mechanism may involve shallow sur-

face trapping states for carriers, or diffusion of impurity atoms, perhaps along dislocations.

Contact Noise

Contact noise is similar to modulation noise and is often confused with it.

Contact noise differs from modulation noise in that it has its source at the contacts and can sometimes be distinguished from modulation noise by the use of four-terminal methods. Like modulation noise, contact noise has a spectrum that often varies approximately as the inverse first power of the frequency:

$$P_i(f) = CI^\beta/f^\alpha \tag{21}$$

where β is close to two, and α is near unity.

The noise in granular resistors and in carbon microphones is usually ascribed to contact noise. The first mention of contact noise appears to be that of Hull and Williams.[28] The first comprehensive study of contact noise was made by Christensen and Pearson.[29] They studied single carbon contacts, and carbon resistors. They found the average value of α and β to be about 1.0 and 1.85 at audio frequencies.

Flicker Noise

Flicker noise is the low-frequency noise that is in excess of shot noise in vacuum tubes. More specifically, when one measures the shot noise in temperature-limited thermionic emission, one finds that at low frequencies the power spectrum rises above the value predicted by the shot-noise formula (6).

Flicker noise was discovered by Johnson[6] in 1925, and was given its name by Shottky[30] in 1926. Johnson's measurements indicated that the power spectrum of flicker noise varied with frequency roughly as $1/f$ for oxide-coated cathodes, and roughly as $1/f^2$ for tungsten cathodes. He further found that the power spectrum of the fluctuation in the current was proportional to the square of the current, $\bar{I}$, whereas in shot noise, the spectrum is proportional to the first power of the current. Johnson suggested that the cause of flicker noise was a fluctuation in the state of the cathode.

Schottky pointed out that if the fluctuation in the state of the cathode was due to a diffusion process that was independent of the amount of the cathode current $\bar{I}$, then one would predict that the spectrum of the fluctuation should be proportional to $\bar{I}^2$. The most extensive investigation of flicker noise in the temperature-limited condition is that of Graffunder[31] in 1939. Graffunder studied the level of the flicker noise in a number of dif-

[20] A. van der Ziel, "Shot noise in semiconductors," *J. Appl. Phys.*, vol. 24, pp. 222–223; 1953.
[21] K. M. van Vliet, "Noise in semiconductors and photoconductors," Proc. IRE, vol. 46, pp. 1004–1018; June, 1958.
[22] S. Teitler, "Generation-recombination noise in a two-level impurity semiconductor," *J. Appl. Phys.*, vol. 29, pp. 1585–1587; 1958.
[23] R. L. Petritz, "On the theory of noise in *p-n* junctions and related devices," Proc. IRE, vol. 40, pp. 1440–1456; November, 1952.
[24] R. V. Rollin and I. M. Templeton, "Noise in semiconductors at very low frequencies," *Proc. Roy. Phys. Soc. B*, vol. 66, pp. 259–261; 1953.
[25] D. Baker, "Flicker noise in germanium rectifiers at very low frequencies," *J. Appl. Phys.*, vol. 25, pp. 922–924; 1954.
[26] T. Firle and H. Winston, "Noise measurements in semiconductors at very low frequencies," *J. Appl. Phys.*, vol. 26, p. 716; 1955.
[27] F. J. Hyde, "Excess noise spectra in germanium," *Proc. Roy. Phys. Soc. B*, vol. 69, pp. 242–245; 1956.

[28] A. W. Hull and N. H. Williams, "Determination of elementary charge *e* from measurements of shot effect," *Phys. Rev.*, vol. 25, pp. 147–173; 1925.
[29] C. J. Christensen and G. L. Pearson, "Spontaneous resistance fluctuations in carbon microphones and other granular resistances," *Bell Sys. Tech. J.*, vol. 15, pp. 197–223; 1936.
[30] W. Schottky, "Small shot effect and flicker effect," *Phys. Rev.*, vol. 28, pp. 74–103; 1926.
[31] W. Graffunder, "Das Röhrenrauschen bei Niederfrequenz," *Telefunken Röhre*, no. 15, pp. 41–63; 1939.

ferent types of commercial tubes, including tubes with pure tungsten, thoriated tungsten, bariated tungsten, and oxide-coated cathodes. His study was limited to the range of frequencies from 40 to 10,000 cps. He found that the measured power spectrum of the flicker noise accords fairly well with the 1/f law in this range of frequencies.

Many more investigators have studied the spectrum of flicker noise, but not in the temperature-limited condition. Among these investigators are Graffunder[31] Bogle,[32] Orsini,[33] Kronenberger and Seitz,[34] Brynes,[35] and van der Ziel.[36] Kronenberger and Seitz[34] have found the 1/f law to extend down to 0.1 cps for oxide-coated cathodes.

All investigators find that the flicker noise of pure tungsten and thoriated-tungsten cathodes is much less than that of most oxide-coated cathodes.

The power spectrum of the fluctuation in the current in temperature limited emission due to both shot noise and flicker noise may be written

$$W_i(f) = 2e\bar{I}\left(1 + \frac{F}{f}\cdot\frac{\bar{I}}{A}\right) \qquad (22)$$

where F is a constant that characterizes the nature of the cathode surface, and A is the surface area of the cathode. The appearance of the cathode area A in the formula follows from the assumption that the noises contributed by separate regions of the cathode are statistically independent.

The value of the constant F has been evaluated here for the various surfaces chiefly from the results of Graffunder.[31] The relatively small amount of data published by Johnson[6] and by Williams[37] has also been employed.

A summary of the results is shown in Table II. The first column shows the cathode surface, the second column shows the range of values of F, and the third column shows the number of tubes on which the evaluation is based.

RELATIONS AMONG THE VARIOUS KINDS OF NOISE

Perhaps the most fundamental relation is that between radiation noise and Johnson noise. Radiation noise is due to the fluctuation in the power incident from a three-dimensional radiation field. Johnson noise is a

---

TABLE II
VALUES OF THE PARAMETER F INTRODUCED IN (22)

Cathode	F in (cps-cm²/amp)	Number of tubes
W	250–2500	4
Th-W	1000–4000	4
Ba-O-W	2500–15,000	4
Oxide	$10^4 - 10^8$	7

---

one-dimensional form of blackbody radiation. Thus there is a sense in which radiation noise is the fluctuation in the power of the radiation field, whereas Johnson noise *is* the power of the radiation field.

Many efforts have been made to find a relation between shot noise and Johnson noise, but all of them have been unsuccessful.[38]

In the realm of validity of Wien's law, where $h\nu$ is is large compared with kT, the radiation photons are statistically independent. Thus in any detector that produces one current carrier for one effective photon (examples: photoemissive tubes, backbiased p-n junctions), the radiation will produce a noise current that accords with the relation given for shot noise. In this restricted sense, radiation noise and shot noise are identical. Indeed, the term shot noise is often used where the term radiation noise or photon noise would be more specific.

In the special case of the thermocouple radiation detector, the Peltier effect produces a coupling between the electrical and thermal circuits, whose result is that the effective resistance of the thermocouple is higher than it would be in the absence of the Peltier effect. The extra resistance has been called a dynamic resistance by Fellgett.[39] The noise generated by the dynamic resistance is the same whether it be viewed as the Johnson noise of the extra resistance or as the temperature noise of the detector. There is a similar dynamic resistance in a bolometer,[3] but in this case the extra noise due to the temperature fluctuation is *not* equal to the (calculated) Johnson noise of the dynamic resistance. (In general, the theory of Johnson noise is applicable only in situations where the detector is in complete thermodynamic equilibrium.)

In the special case where the active element of a thermal detector is in thermal contact with its surroundings only by radiation exchange, radiation noise and temperature noise are identical.

The term current noise, not used so far in this review is a catch-all term that denotes the noise in excess of Johnson noise when current is passed through the element. The concept thus includes the temperature noise of bolometers, generation-recombination noise, modulation noise, and contact noise.

[32] R. W. Bogle, "Low Frequency Fluctuation Voltages in Vacuum Tubes," Applied Physics Lab., The Johns Hopkins University, Baltimore, Md., Tech. Monograph 137; 1948.

[33] L. de Queiroz Orsini, "L'amplification sélective en basse fréquence," *Onde Elec.*, vol. 29, pp. 408–413; 1949; vol. 29, pp. 449–456; 1949; vol. 30, pp. 91–102; 1950.

[34] K. Kronenberger and F. Seitz, "Experimentelle Untersuchung der Schwankungserscheinungen, die die Verstärkung von Gleichspannungs- und Tiefstfrequenzverstärkern begrenzen," *Z. Angew. Phys.*, vol. 3, pp. 1–5; Month, 1951.

[35] F. X. Byrnes, "The effect of tube noise on the equivalent noise pressure of transducer systems," *J. Acoust. Soc. Am.*, vol. 24, 452(A); 1952.

[36] A. van der Ziel, "Study of the Cause and Effect of Flicker Noise in Vacuum Tubes," Electron Tube Res. Lab., University of Minnesota, Minneapolis, Minn., Quart. Repts. from October 15, 1951.

[37] F. C. Williams, "Fluctuation voltage in diodes and in multielectrode valves," *J. IEE*, vol. 79, pp. 349–360; 1936.

[38] An account of these efforts is given in Jones, footnote 3, pp. 14–15.

[39] P. B. Fellgett, "Dynamic impedance and sensitivity of radiation thermocouples," *Proc. Roy. Phys. Soc. B*, vol. 62, pp. 351–359; 1949.

6

Reprinted from *Proc. IRE*, **47**(9), 1503–1507 (1959)

The Measurement and Interpretation of Photodetector Parameters*

R. F. POTTER†, J. M. PERNETT†, AND A. B. NAUGLE†

DEFINITIONS OF SYMBOLS

J = rms value of the fundamental component of the energy flux measured in watts/cm².

V = rms value of the fundamental component of the signal voltage, as measured with the entire detector exposed.

N = rms noise voltage.

A = area of the cell, as defined by the electrodes, in cm².

P_n = noise equivalent power.

f = modulating frequency in cps.

Δf = frequency bandwidth in cps.

E = bias voltage applied across the detector and the load resistor.

R_e = detector dark resistance.

R_L = load resistance.

INTRODUCTION

INFRARED technology is including an increasing number of challenging problems within its domain. In order to meet and solve these challenges engineers and scientists must design and develop novel systems built on the physical principles and techniques of infrared technology. In many instances the key to the solution of an unexplored problem may be in understanding and realizing the potential utility of the photodetector.

The Infrared Sensitive Element Evaluation Program[1] which has been carried on for several years as a joint services effort at the U. S. Naval Ordnance Laboratory, Corona (previously known as the National Bureau of Standards, Corona Laboratories) is designed to provide up-to-date quantitative measurements on all types and kinds of photodetectors. For technical application, there are several important parameters which determine the detector's utility, and it is these parameters and the measuring techniques used at Corona which are described below.

In understanding and appreciating how these parameters are measured, the designer may find that particular detectors have more application potential than that utilized at present. The physical picture of photodetector operation has been presented in the preceding section on physics. In order to compare and evaluate

detectors, several figures of merit[2] have been defined, and those in common use today, which are listed in the Corona reports, are given in Table I along with definitions of terms.

The following will be discussed: determination of optimum operating conditions, response to standard radiation sources, relative spectral response, determina-

TABLE I
FIGURES OF MERIT

Figure of Merit	Definition	Units
NEI	$\text{NEI} = \dfrac{JN}{V}$	Noise Equivalent Input in watts/cm².
NEP = P_n	$\text{NEP} = \dfrac{JNA}{V}$	Noise Equivalent Power in watts.
Jones' S	$S = \dfrac{P_n}{A^{1/2}}\left(\dfrac{f}{\Delta f}\right)^{1/2}$	Jones' S in watts/cm.
D^*	$D^* = \dfrac{A^{1/2}(\Delta f)^{1/2}}{P_n}$	The detectivity normalized to unit area and unit bandwidth in cm/watt.
S_1	$S_1 = \dfrac{V}{JE}\dfrac{(R_e + R_L)^2}{4R_eR_L}$	Specific Sensitivity in cm²/watt.

tion of absolute response levels, frequency response, noise spectrum, and the sensitivity contours, as well as the manner of presentation of these data in the NOLC photodetector series of reports. In most instances, examples characteristic of a cooled PbSe photodetector will be given. This detector is one reported in NOLC detector series Report No. 36. There is a large variety of types of detectors with differing capabilities and operating conditions. This particular cell required cooling to −196°C.

OPTIMUM BIAS AND STANDARD RADIATION SOURCES

The standard sources in use at Corona include a 500°K blackbody with a flux density of 7.7 μw/cm² at the detector plane, and a 1.1 μ He line with a flux density of 8.5 μw/cm². The flux density of the sources is obtained by comparison of the response of a radiation

* Original manuscript received by the IRE, June 26, 1959.
† Naval Ordnance Lab., Corona, Calif.
[1] NOLC Reports, "Properties of Photoconductive Detectors," a series of reports of measurements performed under the Joint Services Infrared Sensitive Element Testing Program, NE 120713-5; 1952–1959.

[2] For further discussion of these figures of merit, see R. C. Jones, "Phenomenological description of the response and detecting ability of radiation detectors," paper 4.1.1, this issue, p. 1495.

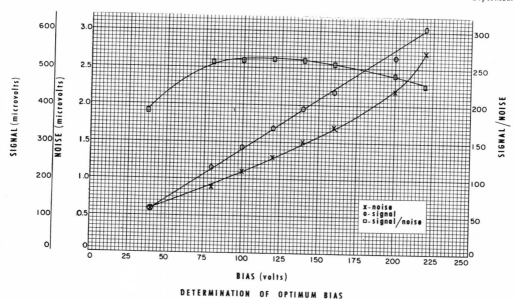

DETERMINATION OF OPTIMUM BIAS

Fig. 1—The signal V and noise voltage N and the ratio V/N determined with a 1.1 μ He standard radiation source as a function of bias voltage. The dark cell resistance is 2.4 megohms and the load resistance is 2.4 megohms.

thermopile to the radiation from the sources, with the response to a National Bureau of Standards standard radiant energy lamp. The sources are mechanically modulated at 90 cps.

The signal and noise for the detector with load resistance is determined as a function of bias voltage with the 1.1 μ He radiation being used for the signal measurements. The amplifier consists of a cathode follower, a Tektronix model 122 preamplifier, and a General Radio model 736A wave analyzer tuned to 90 cps. The bias current found to give the highest signal-to-noise ratio is subsequently used for all other measurements of the detector. For most detectors the optimum bias does not appreciably change with the modulating frequency. Fig. 1 shows the bias graph for a cooled PbSe detector, with optimum bias at 24.5 μamps.

Relative Spectral Response

The relative spectral response is a very significant detector characteristic since it indicates over which portion of the infrared spectrum the detector will be useful. Two monochromators are in normal use at Corona; one, a Leiss double monochromator has CaF$_2$ prisms giving a range from 0.6 μ to 8 μ, the other a Perkin-Elmer Model 98 has NaCl prisms, giving a range from 2 μ to 15 μ.

The energy flux from the exit slit is kept at a constant value over that portion of the spectrum used, by com-

paring the flux at each wavelength with a thermopile or thermocouple. This is done with the internal thermocouple in the Model 98 while an external arrangement is used for the Leiss monochromator. For the latter, the light from the glower source is modulated at both 10 cps and 90 cps. The energy from the exit slit falls on a "black" thermopile and the signal is amplified with a 10-cps amplifier. The energy is raised or lowered to an arbitrary value by opening or closing the entrance slit, with the middle and exit slits usually remaining fixed. Once this level is set the energy flux is allowed to fall onto the detector and the response is rated with the same arrangement as for the standard source measurement.

The relative response curve in the example of Fig. 2 is normalized to unity at 1.2 μ. The interesting thing about detectors of this general type is their relatively high sensitivity out to 5 μ.

Using the Planck distribution function for a 500°K blackbody,[3] the detector response to such a source, and the relative spectral response, the absolute spectral response is determined in the following manner. The product of the fraction of blackbody energy in each 0.1 μ interval and the spectral relative response is numerically integrated over all wavelengths to give a factor γ which indicates how effective the detector is, compared to a "black" or "gray" detector (*i.e.*, a de-

[3] R. A. Smith, F. E. Jones, and R. P. Chasmar, "The Detection and Measurement of Infrared Radiation," Oxford Press, New York, N. Y., ch. 2; 1957.

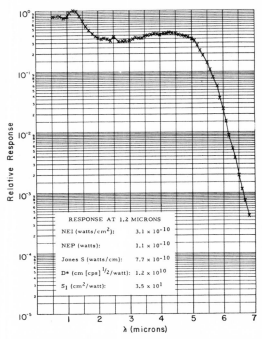

RESPONSE AT 1.2 MICRONS

NEI (watts/cm^2):	3.1×10^{-10}
NEP (watts):	1.1×10^{-10}
Jones S (watts/cm):	7.7×10^{-10}
D* (cm [cps]$^{1/2}$/watt):	1.2×10^{10}
S$_1$ (cm^2/watt):	3.5×10^{1}

Fig. 2—The relative spectral response of the PbSe detector measured with the Leiss model monochromator. The curve was normalized to unity at the peak response at 1.2 μ. The detector has 40 per cent of its peak response at 4.5 μ.

tector with constant response at all wavelengths). The measured response of the detector to the 500°K blackbody is divided by this factor to determine the absolute response at the wavelength at which the relative response was normalized. The absolute spectral response is easily determined at any other wavelength from the relative spectral response. In practice, such figures of merit as S_1 and D^* are evaluated at peak response by taking the quotient of the figure determined with the blackbody and the factor of effectiveness γ. Merit figures such as NEI and NEP are set by taking the product of the 500°K blackbody value and the effectiveness factor.

FREQUENCY RESPONSE AND NOISE SPECTRUM

A knowledge of a detector's characteristic frequency response and noise spectrum are of great importance to the systems designer. They tell him at what frequencies he may operate the detector and still retain sufficient signal for his purposes; also at what frequencies the detector will operate to maximize the signal-to-noise ratio, V/N.

The frequency response is determined at Corona by means of a variable speed chopper giving a frequency range of 100 to 40,000 cps. Radiation from a Nernst

glower is sinusoidally modulated by the chopper and is usually filtered by a selenium coated germanium window. The signal from the detector is measured by putting the output of a cathode follower and a Tektronix model 121 preamplifier into the y-axis input of a DuMont model 304-A oscilloscope. An incandescent tungsten source is simultaneously modulated by the chopper and activates a 931 photomultiplier tube whose signal is fed into a Tektronix model 121 preamplifier and a Hewlett-Packard model 500-A tachometer; the latter's output is proportional to frequency and is put on the Y axis of the oscilloscope. The screen display is photographed as the chopper slows down from its maximum speed.

When the photon excited carriers in the semiconductor have a simple decay mechanism[4] the response to a sinusoidal varying signal can be written

$$\frac{R(\omega)}{R_0} = [1 + \omega^2 T^2]^{-1/2}, \tag{1}$$

where $R(\omega)$ is the response as a function of $\omega = 2\pi f$, f is the frequency of the exciting signal, and T is the time constant for the decay mechanism. An effective time constant is reported in the NOLC reports based on (1) and the frequency response curves. The effective time constant is a very useful parameter for comparing frequency responses of different detectors. For example, the effective time constant from the curve of Fig. 3 is 28 μsec. A note of caution here: because there are distributed capacities present the load resistance must be kept as low as possible in order to approach the detector time constant. The practice at Corona is to reduce the load resistor to 1/10 the dark cell resistance, and this is noted in the report.

The noise spectrum, measured from 10 cps to 10,000 cps with a 5 cps bandwidth is shown in Fig. 4. As the noise theory has been treated in detail elsewhere in this issue, it will suffice to discuss the utility of the noise spectrum for evaluating detectors. In general, the noise has a negative frequency characteristic, but approaches a constant value at relatively high frequencies. For the systems designer, who is using a detector in a system which is cell noise limited, a great deal of flexibility in realizing greater detectivity is available by varying the chopping rate; most detectors may be operated at higher modulation frequencies with greatly improved detectivities. Of course, when system noise is the limitation, the signal response, S_1 for example, is the important parameter. Complete information is presented in the Corona reports to permit the designer to determine V/N values for a wide frequency range; thus one can determine a value for D^* at any frequency, f, from the value at a given frequency (*e.g.*, 90 cps):

$$D^*_f = D^*_{90} \frac{(V/N)_f}{(V/N)_{90}} \tag{2}$$

[4] *Ibid.*, p. 141.

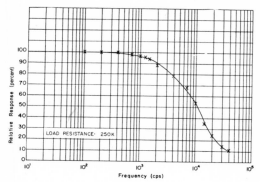

Fig. 3—The frequency response of the PbSe detector. The response is down 3 db at 58 kc.

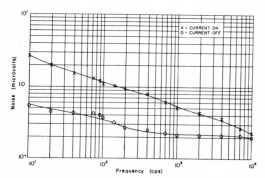

Fig. 4—The noise spectrum for the PbSe detector is shown with and without the bias current.

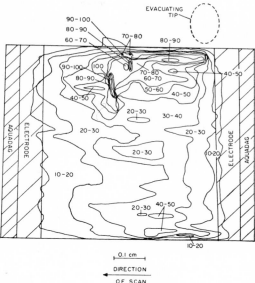

Fig. 5—The sensitivity contour for the PbSe detector. It appears that the upper portion of the detector is more sensitive than the lower.

A similar relation applies for such merit figures as NEI and NEP.

$$(NEI)_f = (NEI)_{90}(V/N_{90}/V/N_f). \qquad (3)$$

By way of example, D^* for the PbSe detector for 500°K blackbody radiation is 1.0×10^9 cm/watt at 90 cps, and 2.0×10^9 cm/watt at 810 cps, with only a slight decrease in signal.

Sensitivity Contours

Contours of the sensitive areas of a detector are of principal concern when the optical system associated with the detector does not utilize the full area. The arrangement at Corona uses a microtable which allows the cell to be moved a small measured amount. The table is linked through a system of gears to a plotting table which gives up to a 36/1 increase in the scale. The exciting radiation is from an incandescent tungsten bulb chopped at 90 cps, and is passed in reverse through a microscope such that a spot 0.066 mm in diameter is focused onto the detector. As the detector is moved beneath this radiation, the relative response at 10 per cent

intervals is noted on the plotting table. Lines connecting points of equal sensitivity are drawn. Such a plot for the PbSe detector is shown in Fig. 5.

Conclusion

This brief discussion outlines the detector parameters and the methods used in measuring them as they appear in the NOLC photodetector series of reports. It has also indicated how the detector user can handle this information in order to determine which detector is best suited for his particular purposes. The completed data summary sheet as it appears in the Corona report is shown in Table II for the sample PbSe detector. Values of the figures of merit for these detectors are shown in Table III.

Test procedures, of course, must be modified, improved, and even replaced as the detectors themselves are modified and improved, and as new types become available. An example is the measurement of time constants which are generally getting shorter. Although effective time constants as short as 1.0 μsec can be determined with existing equipment at NOLC, designs are under study to modify the frequency response apparatus to enable measurements of time constants less than 0.5 μsec. A "flying shutter" device is now being tested which gives rise and decay time constants in this range and is expected to be of use for the range $10 > T > 0.1$ μsec.

TABLE II

Data Summary Sheet for PbSe Detector as it Will Appear in Photoconductive Series Report

Cell Sensitivity		Conditions of Measurement	
Effective time constant (μsec):	2.8×10^1	Chopping frequency (cps):	90
500°K blackbody response:		Bandwidth (cps):	5
Noise equivalent input (watts/cm²):	3.8×10^{-9}	Humidity (per cent):	44
Noise equivalent power (watts):	1.3×10^{-9}	Cell temperature (°C):	-195
Jones' S (watts/cm):	9.4×10^{-9}	Dark resistance (megohms):	2.4
D^* (cm/watt):	1.0×10^9	Load resistance (megohms):	2.5
Position of spectral peak (microns):	1.2	Cell current (μamps):	24.5
Response at spectral peak:		Cell noise (μvolts):	1.3
Noise equivalent input (watts/cm²):	3.1×10^{-10}	Blackbody flux density (μwatts/cm², rms):	7.7
Noise equivalent power (watts):	1.1×10^{-10}	Spot diameter (mm):	0.066
Jones' S(watts/cm):	7.7×10^{-10}	Spot energy (μwatts, rms):	0.0295
D^* (cm/watt/sec$^{1/2}$):	1.2×10^{10}		
"S_1" (cm²/watt) at 1.2 microns:	3.5×10^1	Cell Description	
Sensitivity contour:			
Maximum sensitivity (volts/watt):	3.25×10^4	Type: PbSe	
Masking factor:		Electrode material: Aquadag	
		Window: Sapphire	
		Method of preparation: Evaporated	
		Area (mm): 5.3×6.5	
		Age:	

TABLE III

Figures of Merit for a PbSe Photoconductive Cell Measured at 90 CPS

Figure of Merit	500°K Black-body Response	Response at Spectral Peak (1.2 microns)	Response at 4.5 μ
NEI $\left(\dfrac{\text{watts}}{\text{cm}^2}\right)$	3.8×10^{-9}	3.1×10^{-10}	7.8×10^{-10}
NEP (watts)	1.3×10^{-9}	1.1×10^{-10}	2.8×10^{-10}
Jones' $S\left(\dfrac{\text{watts}}{\text{cm}}\right)$	9.4×10^{-9}	7.7×10^{-10}	1.9×10^{-9}
$D^*\left(\dfrac{\text{cm}}{\text{watt}}\right)$	1.0×10^9	1.2×10^{10}	4.8×10^9
$S_1\left(\dfrac{\text{cm}^2}{\text{watt}}\right)$	2.9	3.5×10^1	1.4×10^1
T (μsec)	2.8×10^1		

ACKNOWLEDGMENT

The joint services program for detector evaluation has been carried on at the National Bureau of Standards and the U. S. Naval Ordnance Laboratory, Corona, with coordination and welcome assistance by O. H. Hunt and C. S. Woodside of the Bureau of Ships, U. S. Navy. Many of the basic procedures used in the program were developed by L. G. Mundie while at the Naval Ordnance Laboratories, Whiteoak, Md., and Corona, Calif. A. J. Cussen made significant contributions during his association with the program from its inception in 1952, until December, 1957, and provided leadership during much of that period. Finally the authors wish to acknowledge the aid and encouragement received from C. J. Humphreys, Head of the Research Department, U. S. Naval Ordnance Laboratory, Corona.

86

Reprinted from *Proc. IEEE*, **56**(1), 37–46 (1968)

Infrared Heterodyne Detection

M. C. TEICH, MEMBER, IEEE

Abstract—Heterodyne experiments have been performed in the middle infrared region of the electromagnetic spectrum using the CO_2 laser as a radiation source. Theoretically optimum operation has been achieved at kHz heterodyne frequencies using photoconductive Ge:Cu detectors operated at 4°K, and at kHz and MHz frequencies using $Pb_{1-x}Sn_xSe$ photovoltaic detectors at 77°K. In accordance with the theory, the minimum detectable power observed is a factor of $2/\eta$ greater than the theoretically perfect quantum counter, $h\nu\Delta f$. The coefficient $2/\eta$ varies from 5 to 25 for the detectors investigated in this study. A comparison is made between photoconductive and photodiode detectors for heterodyne use in the infrared, and it is concluded that both are useful.

Heterodyne detection at 10.6 μm is expected to be useful for communications applications, infrared radar, and heterodyne spectroscopy. It has particular significance because of the high radiation power available from the CO_2 laser, and because of the 8 to 14 μm atmospheric window.

I. INTRODUCTION

THE USE of heterodyning as a coherent detection method in the radiowave, microwave, and optical[1]–[6] regions of the electromagnetic spectrum is well known. It is the purpose of this paper to show that optimum heterodyne performance may be attained well into the infrared. Experiments performed with several different detector configurations and materials, at the CO_2 laser wavelength (10.6 μm), will be discussed.

Coherent detection differs in several significant respects from direct detection, or simple photon counting. The con-

Manuscript received August 10, 1967; revised October 2, 1967.
The author was formerly with the M.I.T. Lincoln Laboratory, Lexington, Mass. (Operated with support from the U. S. Air Force.) He is presently with the Department of Electrical Engineering, Columbia University, New York, N. Y.

figuration for a generalized heterodyne receiver is shown in Fig. 1. Its operation is based on the square-law response of the photodetector to the incident radiation electric field. As a result, two electromagnetic waves of different frequencies (ω_1 and ω_2) mix at the photodevice to produce a signal at the difference frequency $\omega_1 - \omega_2$. When one of these beams is strong (it may be locally produced and is then called the local oscillator or LO beam), the sensitivity for the process is considerably greater than in the straight detection or video case because of the high conversion gain between power at the input and at the difference frequency.[2] In addition to this high conversion gain, the heterodyne detector exhibits both strong directivity and frequency selectivity. The frequency selectivity of the coherent detection process, in turn, permits the noise bandwidth to be reduced to a very small value. It is also observed that the heterodyne detector is linear in that the detector output power is proportional to the input signal radiation power.

At optical and infrared frequencies, the heterodyne detector acts, in effect, as both an antenna and a receiver,[7] and requires careful alignment between the LO and signal beams in order to maintain a constant phase over the surface of the photodetector. As a result, the use of coherent detection in a communication system is limited by the atmospheric distortion of the wavefront, which imposes a restriction on the maximum achievable signal-to-noise ratio.[8] Heterodyne detection is therefore most useful for detecting weak signals which are coherent with a locally produced source. It is a relatively insensitive detector for thermal radiation,[7] although it should be pointed out that it is

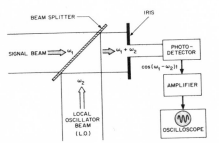

Fig. 1. The generalized infrared or optical heterodyne receiver.

capable of furnishing information about the frequency spectrum.[6]

In the case where both the signal and the LO derive from the same source (such as in the experiments described in this paper), the heterodyne signal can provide information about the velocity of a target through the Doppler shift.[1] Heterodyne detection is also useful for heterodyne spectroscopy[11],[9] and in the study of physical processes occurring in materials. Use of the technique has already been made in the design of a laser Doppler velocimeter, which measures localized flow velocities in gases and liquids.[10],[11]

Coherent detection experiments have been previously reported in the visible and the near infrared using photoemissive devices,[5],[12] photodiodes,[2],[13] and photoconductors.[5],[14],[15] The use of an InAs diode has permitted heterodyne measurements to be extended to 3.5 μm.[16] In the submillimeter region, an improvement in sensitivity with heterodyne operation[17],[18] has been demonstrated for InSb, pyroelectric, and Golay cell detectors.

The measurements reported here have been performed at 10.6 μm in the middle infrared region. It is the availability of the high radiation power from the CO_2 laser coupled with the 8 to 14 μm atmospheric window which makes sensitive detection at 10.6 μm important for systems use. Furthermore, it is at these longer wavelengths that the higher sensitivity available from coherent detection is particularly valuable, since it enables the user to discriminate against various noise sources including the blackbody radiation from objects at room temperature, which is appreciable at 10.6 μm. In the experiments reported below, we have observed a minimum detectable radiation power which is within a factor of 5 of the theoretical quantum limit, $h\nu\Delta f$. (Here, $h\nu$ is the photon energy and Δf is the receiver bandwidth.)

Because the experimental setup employed in these experiments detects the scattered radiation from a diffusely reflecting moving surface, it is, in effect, a miniature prototype CO_2 laser radar. A full-scale CO_2 laser radar based on a similar experimental configuration has been recently set up and successfully operated on targets as far as two miles from the transmitter by Bostick.[19]

---

[1] This is still possible if the LO and signal beams arise from different, but frequency locked, lasers.[3],[6]

II. Theory

A parameter which is of interest in evaluating the usefulness of a receiving technique is the signal-to-noise ratio. In this section, we discuss the operation of an infrared (optical) heterodyne receiver and calculate the expected signal-to-noise ratio at the output of the detector.

Consider two parallel electromagnetic waves of frequencies ω_1 and ω_2 impinging normally on the photodetector of an infrared heterodyne receiver[3]-[5] (see Fig. 1). The total electric field vector E_t is given by

$$E_t = E_1 \cos \omega_1 t + E_2 \cos \omega_2 t \qquad (1)$$

where E_1 and E_2 are the amplitudes of the individual incident waves. Assuming that E_1 and E_2 have the same polarization, the response r from the detector is proportional to the intensity of the radiation or to the square of the electric field

$$r \propto E_t^2 = E_1^2 \cos^2 \omega_1 t + E_2^2 \cos^2 \omega_2 t$$
$$+ E_1 E_2 \cos(\omega_1 - \omega_2)t + E_1 E_2 \cos(\omega_1 + \omega_2)t. \quad (2)$$

Because the detector cannot follow the instantaneous intensity at infrared frequencies, it will respond to the average value of the first, second, and fourth terms in (2) above. These average values are $E_1^2/2$, $E_2^2/2$, and zero, respectively. However, it is assumed that the detector has a sufficiently high frequency response to follow the signal at the difference frequency $\omega_1 - \omega_2$. Thus, the response of the detector to the two incident waves is given by

$$r = \beta \left[\frac{E_1^2}{2} + \frac{E_2^2}{2} + E_1 E_2 \cos(\omega_1 - \omega_2)t \right] = r_{dc} + r_{IF} \quad (3)$$

where β is a proportionality constant containing the detector quantum efficiency.

If we confine measurement of the signal to a bandpass about the difference or heterodyne frequency (also called the intermediate frequency or IF), then it follows that

$$r_{IF} = \beta \cdot E_1 E_2 \cos(\omega_1 - \omega_2)t. \quad (4)$$

But, since $r_{dc} = (\beta/2)(E_1^2 + E_2^2)$, the detector response may be written in terms of its dc component:

$$r = \left[1 + \frac{2 E_1 E_2 \cos(\omega_1 - \omega_2)t}{E_1^2 + E_2^2} \right] r_{dc}. \quad (5)$$

For a very strong LO beam, $E_2 \gg E_1$, and it follows that

$$r_{IF} = 2 \frac{E_1}{E_2} r_{dc} \cos \omega_{IF} t. \quad (6)$$

The mean-square photodetector response is then given by

$$\langle r_{IF}^2 \rangle = 2 \frac{P_1}{P_2} r_{dc}^2, \quad (7)$$

where P_1 and P_2 are the radiation powers in the signal beam and the LO beam, respectively.

If we now consider the noise response r_n in the detector as arising from shot noise,[20],[21] which is the case for the photoemitter and the ideal reverse-biased photodiode, then the mean-square noise response is given by the well-known shot-noise formula

$$\langle r_n^2 \rangle = 2er_{dc}\Delta f. \tag{8}$$

Hence, the signal-to-noise ratio, $(S/N)_{power}$, may be written

$$(S/N)_{power} = \frac{\langle r_{IF}^2 \rangle}{\langle r_n^2 \rangle} = \frac{P_1}{e\Delta f}\left(\frac{r_{dc}}{P_2}\right). \tag{9}$$

However, since r_{dc} arises from the comparatively large LO, it is related to the LO beam power P_2 by the quantum efficiency η

$$r_{dc} = \frac{\eta e}{h\nu}P_2. \tag{10}$$

Thus, the result for the signal-to-noise ratio becomes[20],[21]

$$(S/N)_{power} = \eta P_1/h\nu\Delta f$$

(photoemitter and reverse-biased photodiode). (11a)

From this relation, it is seen that the value of the signal beam radiation power necessary to achieve a $(S/N)_{power} = 1$ is given by

$$P_s^{min} = \frac{h\nu\Delta f}{\eta}$$

(photoemitter and reverse-biased photodiode). (11b)

This quantity is defined as the minimum detectable power, and is denoted by P_s^{min}.

If the two radiation beams impinging on the detector are not parallel to within a certain angular tolerance,[7],[22] and do not illuminate the same area, or if the radiation is not normally incident upon the photodetector,[23] then (S/N) and P_s^{min} will differ from the expressions in (11) given above. In the experiments reported in this work, however, the conditions required for the validity of (11) have been satisfied. For a sufficiently large LO power, the theory derived in the form given above has been experimentally verified both for the case of photoemitters,[5] and for back-biased photodiodes.[13],[16] In particular, Hanlon and Jacobs[24] have recently verified (11) in a bandwidth of 1 Hz, using an InAs diode detector.

For the case of a photoconductor, the noise behavior differs from simple shot noise, and the results derived above are not directly applicable. Photoconductor noise is a complicated phenomenon,[25] and depends to a great extent on the nature of the photoconductor.[2] In the limit of large LO powers, however, extrinsic Ge:Cu is expected to display simple generation-recombination $(g$-$r)$ noise.[26] Since the behavior for simple g-r noise is the same as that for shot noise except for a factor of two,[15],[26]–[28] it may be shown that the signal-to-noise ratio for Ge:Cu has a value just one-half as large as that for a photoemitter or a nonleaky reverse-biased photodiode of the same quantum efficiency.

2 van Vliet[25] has separated photoconductors into four classes, each of which behaves differently: intrinsic, minority trapping model, two-center model, and extrinsic.

The same result has also been obtained as a special case[3] of a relation derived by DiDomenico and Anderson[29] for CdSe.

In the photovoltaic cell, on the other hand, the same processes occur as in the reversed-biased photodiode. However, instead of generating a current, a voltage results from the dipole-layer charge since the cell is effectively open-circuited. The detectivity and the real noise equivalent power (RNEP) for both the reverse-biased p-n junction and the photovoltaic detector have recently been discussed by van Vliet,[25] who has shown that the RNEP for the photovoltaic cell is higher than that for the reverse-biased photodiode by a factor of $\sqrt{2}$. It follows that the (electronic) noise power, which is proportional to the square of the RNEP, is a factor of 2 greater for the photovoltaic configuration. Therefore, the signal-to-noise ratio for the photovoltaic device, as for the photoconductor, is just one-half that for the photoemitter or the reverse-biased photodiode. It should be pointed out, however, that the advantage gained in signal-to-noise ratio for reverse-biased photodiode operation can only be realized for detectors having a high reverse-bias dynamic resistance, as will be seen later.

The signal-to-noise ratio and minimum detectable power for the extrinsic photoconductor and for the photovoltaic junction are therefore given by

$$(S/N)_{power} = \eta P_1/2h\nu\Delta f \tag{12a}$$

$$P_s^{min} = \left(\frac{2}{\eta}\right)h\nu\Delta f$$

(photoconductor and photovoltaic diode). (12b)

These devices are a factor of 2 less sensitive than a photoemitter or ideal reverse-biased photojunction of the same quantum efficiency [compare (11)], and a factor of $2/\eta$ less sensitive than the perfect quantum counter. (For the photoconductor, although both the signal and the noise depend on the photoconductor gain G, the ratio may be shown to be independent of this parameter.[14])

The operation of photoconductive Ge:Cu as a heterodyne detector near the theoretical limit given by (12) has been demonstrated earlier by Teich, Keyes, and Kingston.[30] More recently, similar experiments performed on Ge:Hg by Buczek and Picus[31] have also been found to agree closely with the predictions of (12).

In later sections, we discuss in detail the experimental results of heterodyne measurements on photoconductive Ge:Cu and on photovoltaic $Pb_{1-x}Sn_xSe$. In both of these cases, the experimental agreement with the theory outlined in this section is quite good.

III. EXPERIMENT

A schematic diagram of the experimental arrangement used for the heterodyne measurements in photoconductive Ge:Cu is shown in Fig. 2. The radiation from a CO_2-N_2-He laser, emitting approximately 10 W at 10.6 μm, was incident on a modified Michelson interferometer. One mirror of the

3 In the absence of trapping.

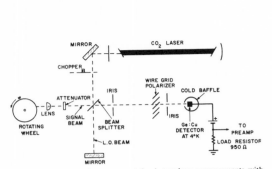

Fig. 2. Experimental arrangement for heterodyne measurements with a Ge:Cu detector. The electric field vector lies perpendicular to the plane of the paper.

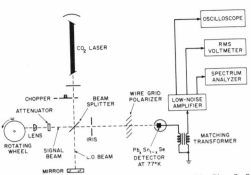

Fig. 3. Experimental arrangement for measurements with a $Pb_{1-x}Sn_xSe$ detector. The arrangement is similar to that shown in Fig. 2, with the exception of the detector output circuitry.

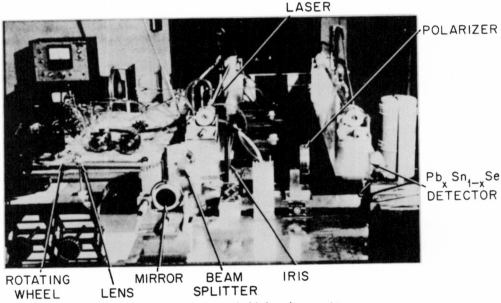

Fig. 4. Photograph of the heterodyne apparatus.

conventional interferometer was replaced by an off-center rotating aluminum wheel with a roughened surface.

The diffusely scattered radiation from the wheel provided a Doppler-shifted signal which was recombined at the beam-splitter with the unshifted LO radiation reflected from the mirror of the other interferometer leg. Heterodyne detection measurements with scattered radiation at 0.6328 μm have been made previously by Gould et al.[32] and by others.[33] Siegman has calculated the maximum radiation power to be returned by a random scatterer.[7],[34]

With the exception of the rotating wheel and the chopper, the experimental apparatus was mounted on a granite slab supported by compressed fiberglass blocks. To further minimize the effect of acoustic vibrations, the 1.25-m-long Brewster-window sealed laser tube was set on shock mounts and enclosed in a shield constructed of acoustic tile. The laser was operated well above threshold and on a single line and mode. An uncoated Irtran II flat (of thickness

0.64 cm) served as a beam splitter, and front surface mirrors were of standard aluminum-coated glass.[4]

A 2.54-cm-focal-length Irtran II lens inserted in the signal beam focused the radiation to a single spot on the rim of the rotating wheel. The purpose of the lens was twofold. It served to collect sufficient scattered radiation to permit an incoherent (nonheterodyne) measurement of the scattered signal power at the detector for calibration purposes, and it also insured spatial coherence of the scattered radiation over the receiver aperture. The procedure is analogous to that used to obtain spatially coherent thermal radiation, where the source is focused onto a pinhole aperture stop. This insures that all points on the wavefront emanating from the pinhole arise from the same source point and are therefore correlated. The coherence properties may be deduced from the van Cittert-Zernike theorem.[32],[35]

Irises were used to maintain the angular alignment of the wave fronts of the two beams to within λ/a, the required angular tolerance for optimum photomixing (a is the detector aperture).[7] It should be noted that this angular alignment restriction is twenty times less stringent than in the visible region of the spectrum. A Perkin-Elmer wire-grid polarizer insured that the recombined beams had a common linear polarization. The beams impinged normally on the photodetector. The output from the detector was fed through a controlled-bandwidth, low-noise amplifier to a thermocouple type rms voltmeter. Alternately, the signal was fed simultaneously to an oscilloscope and to a spectrum analyzer.

The experimental setup used for the heterodyne measurements with photovoltaic $Pb_{1-x}Sn_xSe$ is shown in Fig. 3. It is essentially identical to the arrangement for Ge:Cu, with the notable exception of the detector output circuitry. For the high-impedance photoconductor (dark resistance ~ 600 kΩ), a 1-kΩ load resistor is used to convert the photocurrent to a voltage suitable for amplification. For the low-impedance photovoltaic device (~ 1.5 Ω), on the other hand, the voltage is both amplified and transformed in impedance to the standard 50 Ω by the use of a matching transformer. A photograph of the actual experimental equipment used in these measurements is shown in Fig. 4.

IV. RESULTS FOR PHOTOCONDUCTIVE Ge:Cu

The copper-doped germanium detectors used in the heterodyne experiments were made by indiffusion of Cu into high-resistivity n-type germanium host material for a period of 16 hours at 760°C. The samples, which were 2 mm × 2.2 mm × 3 mm in size, were then quenched in air. The resulting copper atom concentration was 6.8×10^{15} cm^{-3}, and the compensation by the original donors was such as to produce a free hole lifetime of about 2×10^{-9} s at 4°K. With a bias voltage of 13.5 V on the detector, its (incoherent) low-power responsivity was found to be 0.2

[4] These mirrors must have high reflectivity to prevent thermal distortion and consequent deformation of the wavefront of the reflected radiation.

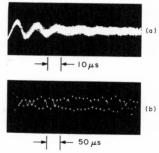

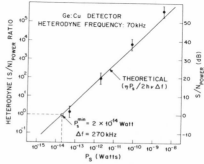

Fig. 5. (a) A multiple-sweep display of the heterodyne signal from a Ge:Cu detector. The loss of definition of the waveform in the third cycle reflects the finite bandwidth of the heterodyne signal. (b) A single-sweep of the signal shown in (a), but with a longer time scale. The modulation of the signal envelope arises from the random nature of the scattering surface.

Fig. 6. The data points, obtained from a typical run, represent the observed signal-to-noise ratio of the heterodyne signal in Ge:Cu, $(S/N)_{power}$, for a given signal-beam radiation power (P_s). The theoretical curve, given by the expression $(S/N)_{power} = \eta P_s/2h\nu\Delta f$, is in good agreement with the data. The minimum detectable power P_s^{min} (defined as that signal beam power for which the heterodyne S/N is unity) corresponds, in a 1-Hz bandwidth, to 7×10^{-20} W.

A/W by calibration with a blackbody source of known temperature. The detector was operated near liquid helium temperature.

Fig. 5(a) shows a multiple-sweep display of the heterodyne signal obtained at the detector output with a signal beam radiation power of 1×10^{-8} W. The loss of definition of the waveform in the third cycle reflects the finite bandwidth of the heterodyne signal. Fig. 5(b) shows a single trace of this signal for a longer time scale. The modulation bandwidth is caused by statistical fluctuations of the heterodyne signal arising from the moving diffuse surface of the wheel.

The results of a typical experimental measurement of the heterodyne signal-to-noise ratio for the detector are shown in Fig. 6. The filled circles represent the observed signal-to-noise power ratio data points, $(S/N)_{power}$, as a function of the signal beam radiation power (P_s or P_1). Only noise arising from the presence of the LO beam (which was the dominant contribution to the noise) is considered. Various values of P_s were obtained by inserting calibrated CaF_2 attenuators in the signal beam, while the unattenuated power was measured by chopping the signal beam in the absence of the LO. As indicated earlier, the presence of the lens facilitated this measurement.

A plot of the theoretically expected result,

$$(S/N)_{power} = \eta P_s/2h\nu\Delta f,$$

is also shown in Fig. 6. Using an estimated quantum efficiency $\eta = 1/2$, it is seen to be in good agreement with the experimental data. Had noise from sources other than the LO been taken into account in computing the S/N, the experimental values would still be within a factor of two of the theoretical curve. Measurements were made with an LO power of 1.5 mW.

With a heterodyne signal centered at about 70 kHz, and an amplifier bandwidth of 270 kHz, the experimentally observed minimum detectable power P_s^{min} (defined as that signal beam power for which the heterodyne S/N is unity) is seen to be 2×10^{-14} W. In a 1-Hz bandwidth, this corresponds to a minimum detectable power of 7×10^{-20} W, which is to be compared with the expected value $(2/\eta)h\nu\Delta f \simeq 7.6 \times 10^{-20}$ W. The experimental measurement is therefore within 6 dB of the theoretically perfect quantum counter, and is in substantial agreement with the expected result for the Ge:Cu detector used in these experiments.

Because the roughness of the wheel ($\sim 10~\mu m$) is comparable to the radiation wavelength λ, the bandwidth of the noise modulation B should be approximately[30] v/d, where v is the velocity at which the illuminated spot traverses the surface, and d is the diameter of the focused spot on the wheel ($\sim 50~\mu m$). This follows from the fact that every d/v seconds, a completely new area of the wheel is illuminated, giving rise to scattered radiation which is uncorrelated with that of the previous time interval. The coherence time is therefore $\sim d/v$, and the frequency bandwidth is given by the inverse coherence time. With $v = r\dot\theta$ and $d \sim F\lambda/D$, B is given approximately by $r\dot\theta D/F\lambda$. Here v is the tangential velocity of the wheel (157 cm/s), $\dot\theta$ is its angular velocity ($10\pi~s^{-1}$), and r its radius (5.05 cm). F represents the focal length of the lens (2.54 cm), while D is the diameter of the radiation beam at the output of the laser (~ 5 mm).

Using these values, we obtain a calculated noise modulation bandwidth $B \sim 30$ kHz, which is comparable with the value obtained from the power-spectral-density trace shown in Fig. 7. A Panoramic model SB-15a ultrasonic spectrum analyzer operated with a trace sweep speed of $\simeq 4~s^{-1}$ was used for the observations.[5] The center frequency of 70 kHz is seen to correspond to the period of 14 μs observed in Fig. 5(b). Both traces were obtained directly across the (1-kΩ) photoconductor load resistor.

A discrepancy between the observed values of signal and noise (individually, rather than the ratio) and the values calculated on the basis of the measured responsivity has

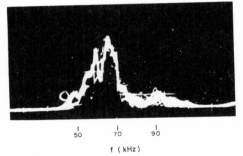

I I I
50 70 90

f (kHz)

Fig. 7.   A typical power-spectral-density trace of the heterodyne signal from Ge:Cu. The trace sweep speed was 4 s^{-1}. The center frequency of 70 kHz corresponds to the period of 14 μs observed in Fig. 5(b).

not been resolved.[30] Experiments have shown, however, that the photoconductor gain does not depend either on the chopping frequency of the incident radiation or on the heterodyne frequency, both possible causes for the disagreement. Other experiments, which have been performed by placing attenuators in various positions in the optical path, indicate that amplification of frequency-shifted (scattered) radiation[36] by the laser is not responsible for the effect.[6] Measurements of the photoconductor gain as a function of the LO power were inconclusive, and it remains possible that this effect has some bearing on the problem.

The results reported in this section differ only slightly from those reported previously for Ge:Cu by Teich, Keyes, and Kingston.[30] Buczek and Picus,[31] in their experiments with Ge:Hg, used two independent CO_2 lasers oscillating at slightly different frequencies. The minimum detectable power which they obtained (referred to a 1-Hz bandwidth) was P_s^{min} (Ge:Hg) $= 1.73 \times 10^{-19}$ W, which is in good agreement with the results obtained by us for Ge:Cu, using a completely different experimental configuration.

V. Results for Photovoltaic $Pb_{1-x}Sn_xSe$

The $Pb_{1-x}Sn_xSe$ diodes used as heterodyne detectors have been fabricated from Bridgman-grown crystals by Melngailis and by Calawa et al.[37]–[39] The bandgap of these diffused p-n junction devices varies with composition (x) so that the wavelength for peak responsivity may be adjusted by varying x. The devices which we used achieved their maximum responsivity (~ 1 V/W, 77°K) at the CO_2 laser wavelength, and had the composition $Pb_{0.936}Sn_{0.064}Se$. The nature and inversion properties of the conduction and valence bands for these alloys have been discussed in detail both for the diodes[37]–[39] and for single-crystal thin films.[40] The inversion behavior of the bands in $Pb_{1-x}Sn_xSe$ is similar to that observed for $Pb_{1-x}Sn_xTe$.[41] The detectivity of the devices (D^*) was $> 3 \times 10^9$ cm $\cdot s^{-1/2} \cdot W^{-1}$, and the carrier concentration was $\sim 10^{17}$ cm^{-3}.

---

[5] A smooth, bell-shaped curve may be obtained by integrating and then recording the power spectral density curve.

[6] The author is grateful to A. E. Siegman of Stanford University for discussing this problem with him and suggesting such a possibility.

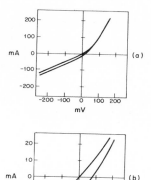

Fig. 8. (a) Current-voltage (I-V) characteristic of the $Pb_{0.936}Sn_{0.064}Se$ diode used in the heterodyne experiments. The upper trace is the dark characteristic while the lower trace is the characteristic with the (18 mW) LO applied. (b) Same characteristic on expanded I and V scales.

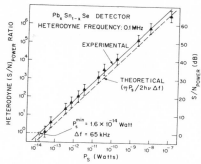

Fig. 9. The solid line is the observed signal-to-noise ratio for the heterodyne signal in $Pb_{1-x}Sn_xSe$ as a function of the signal beam radiation power. The heterodyne frequency is 110 kHz and the detection bandwidth is 65 kHz. The theoretical curve, $(S/N)_{power} = \eta P_s/2h\nu\Delta f$, lies within the limit of experimental accuracy.

The diodes had a 1-mm-diameter active area and were operated at 77°K in the photovoltaic mode. The thin n-type layer (~ 10 μm) was exposed to the LO and signal beam radiation. The I-V characteristic of diode no. 37, both in the absence and in the presence of the LO, is shown in Fig. 8. It can be seen from these curves that the zero-current impedance, as well as the reverse impedance, of the detector is $\simeq 1.5$ Ω. This value, which is very low, is essentially independent of the presence of the LO. Using a calibrated thermopile and the I-V characteristic of Fig. 8, the quantum efficiency and responsivity for the device were directly determined to be 8.5 percent and 0.9 V/W, respectively. The efficiency could presumably be further improved by depositing an antireflection coating on the diodes. The numerical values for the quantum efficiency and the responsivity are consistent with those obtained by Melngailis using a different method at much lower radiation powers.

The arrangement used in the heterodyne experiments (see Fig. 3) has been described in detail in Section III. A transformer at the output of the detector transformed its impedance to a level appropriate for matching to the low-noise amplifier. The experimental procedure was identical to that described for measurements on Ge : Cu in Section IV: various values of the signal beam radiation power P_s were obtained by inserting calibrated CaF_2 attenuators in the signal beam. The unattenuated power was determined from the known responsivity of the diode by chopping the signal beam in the absence of the LO, and then using phase-sensitive detection. In all cases, the direct response of the detector was ascertained to depend linearly on the LO radiation power. In calculating the signal-to-noise ratio, only noise arising from the presence of the LO was considered. The noise figure of the amplifier was such that with

modest LO powers ~ 15 mW, the noise associated with the LO was typically ~ 25 percent of the total noise. It appears that higher LO powers could have been used without any difficulty; however, it would have required a rearrangement of our apparatus to obtain LO powers in excess of 20 mW.

Experiments were performed in two different regions of heterodyne frequency and bandwidth: an IF of 110 kHz with a bandwidth of 65 kHz; and an IF of 2.05 MHz with a bandwidth of 10.0 MHz. They are described in the following paragraphs.

A. Heterodyne Detection at kHz Frequencies

A Princeton Applied Research Model AM-2 input transformer (frequency range 5 kHz to 150 kHz; turns ratio 1 to 100) was used to couple the detector output to the high input-impedance low-noise amplifier (PAR Model CR4-A). Measurements were made with an LO power of 9 mW.

The results of a typical experiment are shown in Fig. 9. The solid line is the observed signal-to-noise power ratio, $(S/N)_{power}$, of the heterodyne signal as a function of the signal beam radiation power P_s. With a heterodyne signal centered at 110 kHz, and a transformer-amplifier bandwidth of 65 kHz, the experimentally observed minimum detectable power P_s^{min} is seen to be 1.6×10^{-14} W. The dashed line in Fig. 9 represents the theoretical result. Using the relation $(S/N)_{power} = \eta P_s/2h\nu\Delta f$, and a quantum efficiency $\eta = 0.085$, it is seen to lie within the limit of experimental accuracy. The observed minimum detectable power corresponds, in a 1-Hz bandwidth, to 2.5×10^{-19} W. Since the experiments were performed using a scattering surface, however, it must be kept in mind that the observation bandwidth for the heterodyne signal must be greater than the noise modulation bandwidth (~ 50 kHz for an IF of 100 kHz).

B. Heterodyne Detection at MHz Frequencies

The behavior of the $Pb_{1-x}Sn_xSe$ heterodyne detectors at MHz frequencies was investigated by rotating the scattering wheel faster. This was accomplished by replacing the 300-r/min synchronous motor driving the scattering wheel

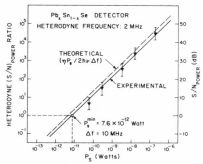

Fig. 10.   Signal-to-noise ratio as a function of signal beam radiation power for 2.05 MHz heterodyne signal from $Pb_{1-x}Sn_xSe$. The agreement of theory and experiment, as in Fig. 9, is good.

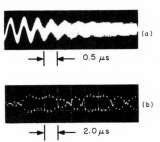

Fig. 11.   (a) A multiple-sweep display of the heterodyne signal in $Pb_{1-x}Sn_xSe$. The loss of definition of the waveform in the fifth cycle reflects the finite bandwidth of the signal. (b) A single sweep of the heterodyne signal shown in (a), but with a longer time scale. This figure is similar to Fig. 5 for Ge:Cu; note the very different time scales, however.

with a 3600-r/min motor. A small matching transformer[7] provided an impedance of approximately 50 Ω at the input of a wide-bandwidth low-noise integrated-circuit amplifier. The effective bandwidth of the transformer-amplifier combination was 10.0 MHz. The LO power was determined from Fig. 8 (and the known responsivity of the detector) to be 18 mW.

The signal-to-noise ratio for the heterodyne signal at 2.05 MHz is shown in Fig. 10. This plot is similar to that of Fig. 9, except for the IF and the bandwidth. The minimum detectable power for this experiment is 7.6×10^{-12} W, which is larger than that of Fig. 9 because of the increased bandwidth. The dashed line, representing the theoretical result, predicts a P_s^{min} of 4.8×10^{-12} W, which is within the experimental bracket. The observed minimum detectable power, extrapolated to a 1-Hz bandwidth, is 7.6×10^{-19} W, which may be compared with the expected value $(2/\eta)hv\Delta f \simeq 4.8 \times 10^{-19}$ W.

Fig. 11(a) shows a multiple sweep display at the detector output which is similar to that shown for Ge:Cu in Fig. 5. The loss of definition of the waveform in the fifth cycle reflects the finite bandwidth of the heterodyne signal. Fig. 11(b) shows a single trace of this signal for a longer time scale. Since the noise modulation bandwidth B and the heterodyne frequency are both proportional to the angular velocity of the scattering wheel $\dot\theta$, their ratio is independent of the IF and depends only on geometrical factors. Therefore, Figs. 5 and 11 appear very much alike in spite of their very different time scales.

Heterodyne detection has also been observed in $Pb_{1-x}Sn_xTe$ diodes[37],[42],[43] operated in the photovoltaic mode. The particular alloy composition used had $x = 0.17$ ($Pb_{0.83}Sn_{0.17}Te$), which has its peak response at 10.6 μm when operated at 77°K. The detector output voltage was observed to be proportional to the square root of the signal power ($\propto \sqrt{P_s}$), as is expected for heterodyne operation. The responsivity of these preliminary diodes was too

---

[7] Turns ratio 11 to 55, no. 30 wire, on a Ferroxcube Corporation 7F160 cup core. This transformer exhibited a sizeable resonance at about 200 kHz.

low, however, to observe the noise associated with the LO. This, of course, is necessary for optimum heterodyne detection.

VI. A COMPARISON BETWEEN PHOTOCONDUCTORS AND PHOTODIODES IN THE INFRARED

It has been demonstrated in the previous sections that optimum heterodyne detection has been achieved in the infrared using both photoconductive and photovoltaic detectors. The question of the advantages of each naturally arises.

The signal-to-noise ratio for heterodyne detection was given in Section II, where it was shown that, for equal quantum efficiency, the nonleaky reverse-biased photodiode has a $(S/N)_{power}$ which is superior to that of the photoconductor and the photovoltaic device by a factor of two. Therefore, from the point of view of S/N, it is preferable to operate a (sufficiently high reverse-impedance) diode in a back-biased, rather than in a photovoltaic or photoconductive, configuration. This statement is also valid for direct detection, where the detectivity D^* for reverse-biased operation is augmented by $\sqrt{2}$ over photovoltaic and photoconductive operation.[25] On the other hand, a leaky photodiode characteristic may give rise to adverse effects when operated back-biased, as discussed by Pruett and Petritz.[44]

Aside from the possible improvement in signal-to-noise ratio, another advantage in operating a photodiode in the reverse-biased configuration may be increased frequency response. DiDomenico and Svelto[45] and Lucovsky et al.[13] have shown that the frequency response for a heterodyne photodiode is either transit-time or RC limited. Reverse-biasing increases the diode depletion layer, reducing the capacity of the device and therefore increasing its frequency response. (Reducing the carrier density also will decrease the diode capacity.) However, the $Pb_{1-x}Sn_xSe$ photodiodes which we employed had RC time constants ~ 1.5 ns (with $R \simeq 1.5$ Ω and $C \simeq 1100$ pF), which was considerably less than the 20-ns response time. (The response time was measured by connecting the diode directly to a properly terminated 50-Ω line, and illuminating it with a 1-ns risetime GaAs injection-laser pulse.) We conclude that these diodes are presently limited by transit time (presumably through the ~ 10-μm n-type layer) to the junction. This transit time could be reduced by decreasing the thickness of the n-type layer without loss of responsivity.

Photovoltaic operation may be preferred in certain cases. For example, with diodes having a low reverse impedance, a reverse voltage could cause undue heating. In photovoltaic operation, the circuitry is simpler,[25] and with low reverse-resistance devices (< 50 Ω), the use of a broadband transformer might be adequate for impedance transformation and a satisfactory amplifier noise figure for frequencies up to ~ 1 GHz.[46]

For the photoconductor with ohmic contacts, the basic frequency response is similar to that of the photodiode, except that it is lifetime or RC limited[14],[45],[47] rather than transit-time limited. Using fast pulse techniques in 2-mm³ samples of uncompensated and Sb-compensated Ge:Cu ($C \sim 10$ pF), Bridges[48] has recently observed a frequency response of ~ 1 ns, which is quite close to the RC limit for the 50-Ω system which he used. Similar measurements have been made by Buczek and Picus[31] in the several-hundred-MHz region. It should be mentioned that, using proper compensation, Ge:Cu detectors with lifetimes as short as 10^{-12} s have been made at this laboratory. However, it must be kept in mind that when high-frequency response is obtained by matching into a 50-Ω system, the responsivity of the high-impedance photoconductor is considerably reduced.

For optimum heterodyne detection, it is necessary that the LO be sufficiently strong so as to provide the dominant source of noise (to overcome the amplifier noise). A high responsivity is therefore desirable so that the LO radiation power may be kept moderate. Because the photoconductor responsivity is proportional to the photoconductor gain G, which is given by τ/T where τ is the free-carrier lifetime and T is the transit time across the device,[28] it is higher for thin photoconductors. Therefore, a compromise between responsivity and RC frequency response must be made. A discussion of the trade-offs necessary for optimum photoconductor heterodyne operation at high frequencies ($\rightarrow 2$ GHz) has been given by Arams et al.,[49] who have fabricated thin Ge:Cu detectors for this purpose. On the other hand, photodiodes having high reverse impedances should have high responsivity and, since the gain is unity, should in general require less LO than the photoconductor.

Finally, perhaps the most striking characteristic of the $Pb_{1-x}Sn_xSe$ (as well as the $Pb_{1-x}Sn_xTe$ and $Cd_xHg_{1-x}Te$) photodiode detectors is their ability to operate at liquid nitrogen temperatures (77°K). By contrast, Ge:Cu requires near liquid helium temperatures (4°K) while Ge:Hg requires liquid hydrogen temperatures (18°K). The diodes are therefore more convenient to operate and more suitable for field use than are the photoconductors. Nevertheless, the quantum efficiency of the photodiode reported in this work is below that of the photoconductor by a factor of ~ 4, and the minimum detectable power is, therefore, correspondingly higher.

Both photoconductors and photodiodes are seen to be useful for infrared heterodyne detection, the choice of a particular device depending on the desired application.

VII. Conclusion

It has been shown that heterodyne techniques, which have been used extensively in the radiowave and microwave regions, and more recently in the optical (visible) portion of the electromagnetic spectrum, are equally as valuable in the infrared.

Coupled with the high power of the CO_2 laser, and the 8 to 14 μm atmospheric window, the optimum-detection heterodyne experiments reported in this work are expected to be significant for communications applications. The operation of the system as an infrared radar has also been demonstrated. The technique might prove useful for infrared heterodyne spectroscopy. It should be pointed out that coherent detection in the infrared is expected to be more sensitive than in the optical region because of the smaller photon energy (the minimum detectable power is proportional to the photon energy).

Further improvements in heterodyne sensitivity may be expected since the quantum efficiency of detectors such as $Pb_{1-x}Sn_xSe$ and $Pb_{1-x}Sn_xTe$, which are presently $\simeq 8$ to 15 percent, show promise of being greater in the future. $Pb_{1-x}Sn_xSe$ detectors have already been operated at dry-ice temperatures ($-78°C$) with a response which is down by only a factor of 20 from the response at 77°K. Furthermore, diodes such as $Cd_xHg_{1-x}Te$[50] (which peak at 10.6 μm with $x = 0.195$) and $Pb_{1-x}Sn_xTe$ have now been fabricated with reverse impedances of ~ 50 Ω so that impedance matching is less of a problem. With the availability of these higher impedances at the amplifier input, an added advantage is that the noise figure of the amplifier is improved, thus requiring less LO to overcome amplifier noise. Furthermore, if the diode reverse impedance could be raised to a level much greater than that of the load resistance, an additional factor of two could be gained in the signal-to-noise ratio with reverse-biased operation. Thin Ge:Cu photoconductive detectors with high gain as well as short lifetime have been reported by Arams et al.[49] so that heterodyne detection in the GHz range is expected to be possible soon. Therefore, a general relaxation of the few conditions which are still required for efficient infrared heterodyne detection at or near the theoretical limit may be anticipated.

Acknowledgment

It is a pleasure to thank I. Melngailis for supplying the $Pb_{1-x}Sn_xSe$ and $Pb_{1-x}Sn_xTe$ detectors used in these experiments, and to acknowledge many valuable discussions with him, with R. J. Keyes, and with R. H. Kingston. I am grateful to F. D. Carroll for capable technical assistance, to A. Ross for design of the high-frequency amplifier, and especially to Mary L. Barney for fabricating the Ge:Cu detectors. Finally, I would like to express my appreciation to the reviewers for their helpful comments.

REFERENCES

[1] A. T. Forrester, "Photoelectric mixing as a spectroscopic tool," *J. Opt. Soc. Am.*, vol. 51, pp. 253–259, March 1961.

[2] A. E. Siegman, S. E. Harris, and B. J. McMurtry, "Optical heterodyning and optical demodulation at microwave frequencies," in *Optical Masers*. J. Fox, Ed. New York: Wiley, 1963, pp. 511–527.

[3] S. Jacobs, "The optical heterodyne," *Electronics*, vol. 36, p. 29, July 12, 1963.

[4] M. E. Lasser, "Detection of coherent optical radiation," *IEEE Spectrum*, vol. 3, pp. 73–78, April 1966.

[5] S. Jacobs and P. Rabinowitz, "Optical heterodyning with a CW gaseous laser," in *Quantum Electronics III*, P. Grivet and N. Bloembergen, Eds. New York: Columbia Univ. Press, 1964, pp. 481–487.

[6] L. Mandel, "Heterodyne detection of a weak light beam," *J. Opt. Soc. Am.*, vol. 56, pp. 1200–1206, September 1966.

[7] A. E. Siegman, "The antenna properties of optical heterodyne receivers," *Proc. IEEE*, vol. 54, pp. 1350–1356, October 1966.

[8] D. L. Fried, "Atmospheric modulation noise in an optical heterodyne receiver," *IEEE J. Quantum Electronics*, vol. QE-3, pp. 213–221, June 1967; also "Optical heterodyne detection of an atmospherically distorted wave front," *Proc. IEEE*, vol. 55, pp. 57–67, January 1967.

[9] H. Z. Cummins, N. Knable, and Y. Yeh, "Observation of diffusion broadening of Rayleigh scattered light," *Phys. Rev. Lett.*, vol. 12, pp. 150–153, February 1964.

[10] Y. Yeh and H. Z. Cummins, "Localized fluid flow measurements with an He-Ne laser spectrometer," *Appl. Phys. Lett.*, vol. 4, pp. 176–178, May 1964.

[11] J. W. Foreman, Jr., W. W. George, J. L. Jetton, R. D. Lewis, J. R. Thornton, and H. J. Watson, "Fluid flow measurements with a laser Doppler velocimeter," *IEEE J. Quantum Electronics*, vol. QE-2, pp. 260–266, August 1966.

[12] A. E. Siegman, S. E. Harris, and B. J. McMurtry, "Microwave demodulation of light," in *Quantum Electronics III*, P. Grivet and N. Bloembergen, Eds. New York: Columbia Univ. Press, 1964, pp. 1651–1658.

B. J. McMurtry and A. E. Siegman, "Photomixing experiments with ruby optical maser and traveling wave microwave phototubes," *Appl. Optics*, vol. 1, pp. 51–53, January 1962.

[13] G. Lucovsky, M. E. Lasser, and R. B. Emmons, "Coherent light detection in solid state photodiodes," *Proc. IEEE*, vol. 51, pp. 166–172, January 1963.

G. Lucovsky, R. B. Emmons, B. Harned, and J. K. Powers, "Detection of coherent light by heterodyne techniques using solid state photodiodes," in *Quantum Electronics III*, P. Grivet and N. Bloembergen, Eds. New York: Columbia Univ. Press, 1964, pp. 1731–1738.

[14] M. DiDomenico, Jr., R. H. Pantell, O. Svelto, and J. N. Weaver, "Optical frequency mixing in bulk semiconductors," *Appl. Phys. Lett.*, vol. 1, pp. 77–79, December 1962.

R. H. Pantell, M. DiDomenico, Jr., O. Svelto, and J. N. Weaver, "Theory of optical mixing in semiconductors," in *Quantum Electronics III*, P. Grivet and N. Bloembergen, Eds. New York: Columbia Univ. Press, 1964, pp. 1811–1818.

G. Lucovsky, R. F. Schwartz, and R. B. Emmons, "Photoelectric mixing of coherent light in bulk photoconductors," *Proc. IEEE (Correspondence)*, vol. 51, pp. 613–614, April 1963.

[15] P. D. Coleman, R. C. Eden, and J. N. Weaver, "Mixing and detection of coherent light in a bulk photoconductor," *IEEE Trans. Electron Devices*, vol. ED-11, pp. 488–497, November 1964.

[16] F. E. Goodwin and M. E. Pedinoff, "Application of CCl4 and CCl2:CCl2 ultrasonic modulators to infrared optical heterodyne experiments," *Appl. Phys. Lett.*, vol. 8, pp. 60–61, February 1966.

[17] E. H. Putley, "The use of an InSb detector as a mixer at 1 mm," *Proc. IEEE (Correspondence)*, vol. 54, pp. 1096–1098, August 1966.

[18] H. A. Gebbie, N. W. B. Stone, E. H. Putley, and N. Shaw, "Heterodyne detection of sub-millimetre radiation," *Nature*, vol. 214, pp. 165–166, April 1967.

[19] H. A. Bostick, "A carbon dioxide laser radar system," *IEEE J. Quantum Electronics*, vol. QE-3, p. 232, June 1967.

[20] B. M. Oliver, "Signal-to-noise ratios in photoelectric mixing," *Proc. IRE (Correspondence)*, vol. 49, pp. 1960–1961, December 1961.

[21] H. A. Haus, C. H. Townes, and B. M. Oliver, "Comments on 'Noise in photoelectric mixing,'" *Proc. IRE (Correspondence)*, vol. 50, pp. 1544–1546, June 1962.

[22] V. J. Corcoran, "Directional characteristics in optical heterodyne detection processes," *J. Appl. Phys.*, vol. 36, pp. 1819–1825, June 1965.

[23] A. J. Bahr, "The effect of polarization selectivity on optical mixing in photoelectric surfaces," *Proc. IEEE (Correspondence)*, vol. 53, p. 513, May 1965.

[24] J. Hanlon and S. F. Jacobs, "Narrowband optical heterodyne detection," *IEEE J. Quantum Electronics*, vol. QE-3, p. 242, June 1967.

[25] K. M. van Vliet, "Noise limitations in solid state photodetectors," *Appl. Optics*, vol. 6, pp. 1145–1169, July 1967.

[26] H. Levinstein, "Extrinsic detectors," *Appl. Optics*, vol. 4, pp. 639–647, June 1965.

[27] R. C. Jones, "Noise in radiation detectors," *Proc. IRE*, vol. 47, pp. 1841–1846, September 1959.

[28] A. van der Ziel, *Fluctuation Phenomena in Semi-Conductors*. New York: Academic Press, 1959, pp. 22–45 and 65–82.

[29] M. DiDomenico, Jr., and L. K. Anderson, "Signal-to-noise performance of CdSe bulk photoconductive detectors," Bell Telephone Labs., Murray Hill, N. J., unpublished memo.

[30] M. C. Teich, R. J. Keyes, and R. H. Kingston, "Optimum heterodyne detection at 10.6 μm in photoconductive Ge:Cu," *Appl. Phys. Lett.*, vol. 9, pp. 357–360, November 1966.

[31] C. J. Buczek and G. S. Picus, "Heterodyne performance of mercury doped germanium," *Appl. Phys. Lett.*, vol. 11, pp. 125–126, August 1967.

G. S. Picus and C. Buczek, "Far infrared laser receiver investigation," Hughes Research Labs., Malibu, Calif., Interim Tech. Rept. 4, Contract AF33(615)-3847; 1967. See also Repts. 1–3.

[32] G. Gould, S. F. Jacobs, J. T. LaTourrette, M. Newstein, and P. Rabinowitz, "Coherent detection of light scattered from a diffusely reflecting surface," *Appl. Optics*, vol. 3, pp. 648–649, May 1964.

[33] R. D. Kroeger, "Motion sensing by optical heterodyne Doppler detection from diffuse surfaces," *Proc. IEEE (Correspondence)*, vol. 53, pp. 211–212, February 1965.

G. A. Massey, "Photomixing with diffusely reflected light," *Appl. Optics*, vol. 4, pp. 781–784, July 1965.

[34] A. E. Siegman, "A maximum-signal theorem for the spatially coherent detection of scattered radiation," *IEEE Trans. Antennas and Propagation*, vol. AP-15, pp. 192–194, January 1967.

[35] M. Born and E. Wolf, *Principles of Optics*. New York: Pergamon, 1959, pp. 505–510.

[36] W. M. Doyle, W. D. Gerber, and M. B. White, "Use of an oscillating laser as a heterodyne receiver preamplifier," *IEEE J. Quantum Electronics*, vol. QE-3, p. 241, June 1966.

[37] I. Melngailis, "Properties of Pb$_{1-x}$Sn$_x$Te and Pb$_{1-x}$Sn$_x$Se infrared detectors," presented at the IRIS Infrared Detector Specialty Group Meeting, Wright-Patterson AFB, March 1967.

[38] A. R. Calawa, I. Melngailis, T. C. Harman, and J. O. Dimmock, "Photovoltaic response of Pb$_{1-x}$Sn$_x$Se diodes," presented at the Solid State Device Research Conf., University of California at Santa Barbara, June 1967.

[39] J. F. Butler, A. R. Calawa, I. Melngailis, T. C. Harman, and J. O. Dimmock, "Laser action and photovoltaic effect in Pb$_{1-x}$Sn$_x$Se diodes," *Bull. Am. Phys. Soc.*, vol. 12, p. 384, March 1967.

[40] A. J. Strauss, "Inversion of conduction and valence bands in Pb$_{1-x}$Sn$_x$Se alloys," *Phys. Rev.*, vol. 157, pp. 608–611, May 1967.

[41] J. O. Dimmock, I. Melngailis, and A. J. Strauss, "Band structure and laser action in Pb$_x$Sn$_{1-x}$Te," *Phys. Rev. Lett.*, vol. 16, pp. 1193–1196, June 1966.

[42] I. Melngailis and A. R. Calawa, "Photovoltaic effect in Pb$_x$Sn$_{1-x}$Te diodes," *Appl. Phys. Lett.*, vol. 9, pp. 304–306, October 1966.

[43] I. Melngailis, A. R. Calawa, J.F. Butler, T. C. Harman, and J. O. Dimmock, "Photovoltaic effect in Pb$_x$Sn$_{1-x}$Te diodes," presented at the Internat'l Electron Devices Meeting, Washington D. C., October 1966.

[44] G. R. Pruett and R. L. Petritz, "Detectivity and preamplifier considerations for indium antimonide photovoltaic detectors," *Proc. IRE*, vol. 47, pp. 1524–1529, September 1959.

[45] M. DiDomenico, Jr., and O. Svelto, "Solid state photodetection: A comparison between photodiodes and photoconductors," *Proc. IEEE*, vol. 52, pp. 136–144, February 1964.

[46] C. L. Ruthroff, "Some broad-band transformers," *Proc. IRE*, vol. 47, pp. 1337–1342, August 1959.

[47] O. Svelto, P. D. Coleman, M. DiDomenico, Jr., and R. H. Pantell, "Photoconductive mixing in CdSe single crystals," *J. Appl. Phys.*, vol. 34, pp. 3182–3186, November 1963.

[48] T. Bridges: to be published.

[49] F. Arams, E. Sard, B. Peyton, and F. Pace, "10.6 micron heterodyne detection with gigahertz IF capability," *IEEE J. Quantum Electronics*, vol. QE-3, pp. 241–242, June 1967; also "Infrared 10.6-micron heterodyne detection with gigahertz IF capability," *IEEE J. Quantum Electronics*, vol. QE-3, pp. 484–492, November 1967.

[50] C. Vérié and J. Ayas, "Cd$_x$Hg$_{1-x}$Te infrared photovoltaic detectors," *Appl. Phys. Lett.*, vol. 10, pp. 241–243, May 1967.

Reprinted from *Proc. IEEE*, **58**(10), 1466–1486 (1970)

Photodetectors for Optical Communication Systems

HANS MELCHIOR, MEMBER, IEEE, MAHLON B. FISHER, MEMBER, IEEE,
AND FRANK R. ARAMS, FELLOW, IEEE

Abstract—The characteristics of high-sensitivity photodetectors suitable for wide bandwidth optical communication systems are summarized. Photodiodes, photomultipliers, and photoconductive detectors for wavelengths from 0.3 μm to 10.6 μm are covered. The use of internal current gain by means of avalanche and electron multiplication and by means of optical heterodyne detection to increase sensitivity of high speed photodetectors is discussed. The application to visible and infrared laser communication systems is reviewed.

I. INTRODUCTION

DEMODULATION of signals in optical communication systems requires photodetectors and detection systems which combine wide instantaneous bandwidth with high sensitivity to weak light signals. This paper will discuss the basic requirements of photodetection systems suitable for the demodulation of laser signals and present a review of the relevant characteristics of different types of high speed photodetectors.

Vacuum and Si photodiodes with response in the ultraviolet, visible, and near-infrared part of the spectrum up to about 1 μm will be described. Photomultipliers and avalanche photodiodes with wide instantaneous bandwidths that operate in the same wavelength region but provide internal gain for the photocurrent are discussed. It will be shown that internal current gain as well as optical heterodyne detection significantly improve the sensitivity of photodetection systems to weak light signals. A comparison of the characteristics of heterodyne and direct detection systems will be presented. The most important infrared photodetectors with response to a few micrometers including Ge, InAs, and InSb solid-state photodiodes with and without internal current gain will be mentioned. The extrinsic and intrinsic photoconductors and the mixed crystal photodetectors which may be utilized as detectors for 10.6-μm radiation will be discussed.

II. GENERAL REQUIREMENTS FOR PHOTODETECTORS IN OPTICAL COMMUNICATION SYSTEMS

Proper design of a demodulation system for optical signals requires specially designed photodetectors that are efficient and fast. The major requirements imposed on photodetectors and detection systems for optical communication applications thus include

Manuscript received June 18, 1970.
H. Melchior is with the Bell Telephone Laboratories, Inc., Murray Hill, N. J. 07922.
M. B. Fisher is with Sylvania Electronic Products, Seneca Falls, N. Y.
F. R. Arams is with AIL, a division of Cutler-Hammer, Melville, N. Y.

1) large response to the incident optical signal,
2) sufficient instantaneous bandwidth to accommodate the information bandwidth of the incoming signal
3) minimum of noise added by the demodulation process.

Involved is the optimization of the entire detection system with respect to speed of response and sensitivity to weak light signals [1]–[4]. Attention must not only be given to the choice of the optimum detector and to the design of its load circuit and associated low noise amplifier but also to other system parameters, such as desired field-of-view [5]–[7], optical bandwidth, possible relative motion between transmitter and receiver (possibly leading to Doppler shift and/or a requirement for the generation of pointing-error information [8]), and interference due to the sun or other radiation.

Photodetectors convert the absorbed optical radiation into electrical output signals. They are square law detectors that respond to the intensity of light averaged over a number of optical cycles [5], [9]. This is because the speed of response is determined by carrier transport and relaxation processes within the photodetector. These processes do not have sufficiently short time constants to reproduce field variations which occur at optical frequencies. The general expression for the conversion of an incident photon stream of average power P_{opt} and optical frequency v into a primary photocurrent I_{ph} is

$$I_{ph} = \eta \frac{q P_{opt}}{hv} \qquad (1)$$

with P_{opt}/hv = average number of incident photons per unit time, I_{ph}/q = average number per unit time of electrons emitted from the photocathode, or electron–hole pairs collected across the junction region of photodiode, or mobile electrons and holes excited within the photoconductor, and η = conversion or quantum efficiency.

Utilization of wide information bandwidths for optical communication systems—bandwidths of several hundred MHz or even GHz are under consideration—requires demodulation systems with correspondingly wide instantaneous bandwidths. The information bandwidth of the communication system sets the longest time constants that are permissible in the photodetector and associated output amplifier and load circuit. Circuit considerations play a key role in determining the bandwidth and sensitivity of a detection system. Detector shunt capacitance, series resistance, and circuit parasitics must be considered. Because of the capacitance inherent in any photodetector and output circuit, the load or amplifier input resistance required to achieve a sufficiently large bandwidth is, with the exception of certain photoconductors and cooled long wavelength photodiodes, much smaller than the internal shunt resistance of the photodetector. As a consequence, the sensitivity to weak light signals can be quite low for large bandwidth direct detection systems. In the absence of a gain mechanism for the photosignal ahead of or within the photodetector, the thermal noise of the load resistance or the amplifier noise dominates over the quantum or shot noise of the optical signal, thus limiting the sensitivity for demodulation of weak signals.

An improvement in sensitivity is possible through the use of wide-band output transformers such as the helical coupling structure of traveling wave phototubes [10]. However, the most significant improvement in sensitivity for the demodulation of weak light signals with wide bandwidths is brought about by the introduction of gain for the photosignal before it reaches the detector output. Practically useful gain mechanisms are 1) current gain within photomultipliers, avalanche photodiodes, and photoconductors, and 2) optical mixing, that is, heterodyne detection [11]–[18]. In addition, optical preamplification [19]–[22] and parametric upconversion [23]–[28] constitute potential means for obtaining high detection sensitivity.

A. Direct Detection

The signal-to-noise ratio and the minimum detectable signal are major criteria by which the sensitivity of detection systems to weak optical signals is judged. The signal-to-noise ratio at the output of a generalized direct photodetection system that comprises a photodetector with internal current gain M and a load circuit as shown in Fig. 1 is given by [1], [29]

$$\frac{S}{N} = \frac{\frac{1}{2}(mI_{ph})^2|M(\omega)|^2}{\left[2q(I_{ph} + I_B + I_D)|M(\omega)|^2 F(M) + 4k\frac{T_{eff}}{R_{eq}}\right]B} \quad (2)$$

In this equation m = modulation index of the light, $M(\omega)$ = current gain or multiplication factor within the photodetector, $2q(I_{ph}+I_B+I_D)B = \overline{i_s^2}$ = mean-square shot-noise current, I_B = background radiation induced photocurrent [3], [5], I_D = dark current component that is multiplied within photodetector, B = electrical bandwidth of detection system, $F(M)$ = factor that accounts for the increase in noise induced by the internal current gain process. The term $4k(T_{eff}/R_{eq})B$ represents the effective mean-square thermal noise current, with R_{eq} being the equivalent resistance of the photodetector and output circuit [1], and T_{eff} is an effective noise temperature that takes into account thermal noise due to the detector and load resistor, if any, and the following amplifier.

The factor $F(M)$ accounts for the increase in noise that is induced by the current gain process. While this noise factor is unity in reverse-biased solid-state, and vacuum photodiodes, it is larger than unity for photomultipliers, and

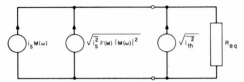

Fig. 1. Equivalent circuit of generalized photodetection system showing principal signal and noise sources of photodetector and load circuit.

especially for avalanche photodiodes. For photoconductors and unbiased photodiodes, $F = 2$ at low frequencies because, as will be shown in Section III-C, the magnitude of the radiation induced generation–recombination noise is twice as large as the shot noise.

From (2) the minimum signal that can be detected for a given bandwidth and signal-to-noise ratio can be determined.

With a direct detection system, the highest sensitivity is reached if the minimum detectable signal is limited by fluctuations in the average signal current itself. In this limit, the peak value of the minimum detectable optical power becomes

$$P_{s\,min} = mP_{opt} = 4hv\,\frac{B}{m}\left(\frac{S}{N}\right)\frac{F}{\eta} \quad (3)$$

where (3) is the quantum noise limited sensitivity of direct detection.

In practical broad-band direct photodetection systems without internal current gain, the minimum detectable signal is usually limited by the thermal noise of the detector and load resistance and by amplifier noise

$$P_{s\,min} = mP_{opt} = \frac{hv}{q}\left(B\left(\frac{S}{N}\right)\right)^{1/2}\left(\frac{8kT_{eff}}{\eta^2 R_{eq}}\right)^{1/2} \quad (4)$$

(thermal or amplifier noise limited sensitivity).

An appreciable improvement in sensitivity over the thermal or amplifier noise limited case (4) is possible through the use of photodetectors with internal current gain such as photomultipliers, avalanche photodiodes, and photoconductors. The ideal quantum noise limited sensitivity of (3) can, however, not be fully reached because the current gain in photomultipliers and especially in avalanche photodiodes provides excess noise ($F(M)$) and because practical quantum efficiencies are smaller than unity. In a practical photodetection system with internal current gain the highest sensitivity will be reached when the shot noise; which is induced by the average light signal, the background radiation, and the leakage currents; is multiplied to a level comparable to the thermal noise [1], [30]. For this optimum current gain M_{opt}, the minimum detectable signal will be lower by approximately a factor of $2/M_{opt}$ as compared to the thermal noise limited case without gain (4) [1]. If more gain than this optimum value is used in photomultipliers and especially in avalanche photodiodes, the sensitivity decreases again because of the excess noise associated with carrier multiplication.

Sufficient current gain as determined by the previously mentioned noise considerations cannot be reached in all wide bandwidth detection systems because gain-bandwidth limitations ($M \cdot B$) of the current amplification mechanism set an upper limit to the maximum achievable gain M_{max}. In this case, the sensitivity increases only by a factor M_{max}.

In detection systems with moderate bandwidths, both those with and without current gain, the sensitivity is often not limited by thermal or signal induced noise but by noise generated by background radiation (I_B) [5], [6] or by dark current (I_D). Noise due to leakage currents are of importance in photoemitters and photodiodes. Noise due to background radiation often sets the sensitivity limit in direct detection systems that operate at infrared wavelengths [4]–[6]. However, by limiting the size, spectral acceptance bandwidth, and field of view of the detector and through cooling, these currents can be usually lowered considerably.

B. Optical Heterodyne Detection

In optical mixing or heterodyne detection [2]–[4], [10], [11]–[18], a coherent optical signal P_s is mixed with a laser local oscillator P_{LO} at the input of a photodetector as shown schematically in Fig. 2. The mean value of the photocurrent generated at the intermediate or difference frequency is then [11], [16], [31] for $P_{LO} \gg P_s$

$$i_{IF} = \frac{\eta q}{h\nu} \sqrt{2 P_{LO} P_s}. \tag{5}$$

For the generalized photodetection circuit of Fig. 1, the heterodyne transducer gain G_T defined [32] as the ratio of actual IF power delivered to the output amplifier with input resistance R_A, divided by the available optical signal power P_S, is

$$G_T = \frac{P_{IF}}{P_S} = 2\left(\frac{\eta q}{h\nu}\right)^2 P_{LO} \frac{(M(\omega)R_{eq})^2}{R_A}. \tag{6}$$

This equation indicates that high conversion gain is possible using sufficient local oscillator power.

For proper mixing action, and for best response at the intermediate frequency, the polarizations of the signal and the local oscillators should be the same [33]–[35]. The signal and the local oscillator must both be normally incident and maintain parallel wavefronts over the entire sensitive area of the detector, the dimensions of which should be minimized to ease the alignment requirement. The minimum detectable signal is then as determined from (2) replacing $m I_{ph}/\sqrt{2}$ by i_{IF}, so that $m = 2(P_S/P_{LO})^{\frac{1}{2}}$

$$P_{S\,min} = \frac{S}{N} B \left\{ h\nu \frac{F}{\eta} + \left(\frac{h\nu}{g\eta}\right)^2 \frac{1}{P_{LO}M^2} \right.$$
$$\left. \cdot \left(2k \frac{T_{eff}}{R_{eq}} + q(I_B + I_D)M^2 F \right) \right\} \tag{7}$$

(sensitivity of heterodyne detection). (7)

For sufficient LO power, the second term representing thermal noise of detector, load and amplifier, and shot or

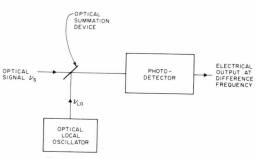

Fig. 2. Optical heterodyne detection system.

generation–recombination noise due to background and dark current, becomes small. This is the case of a well-designed heterodyne receiver in which noise is dominated by local-oscillator induced shot or generation–recombination noise.

Heterodyne detection thus allows realization of optical receivers with wide bandwidths and high sensitivity almost equal to the theoretical quantum noise limit ($h\nu B$). The availability of high conversion gain eases the design requirements for photodetectors considerably. High quantum efficiency and sufficient speed of response of the photodetector and the output circuit are necessary, but the internal current gain is not. Operation at higher temperature than direct detection may also be possible.

Practical application of heterodyne detection is most successful at infrared wavelengths because the alignment requirements are easier to maintain. For 10.6 μm, where strong and stable single line local oscillators exist, heterodyne detection systems with 1-GHz base-bandwidth and sensitivities that approach the ideal limits have been reported [17].

Heterodyne detection systems have a diffraction-limited field of view [33] and small spectral acceptance bandwidth. This makes them relatively insensitive to background radiation and interference effects. Optical mixing preserves the frequency and phase information of the input signal and can thus be used for intensity, phase, and frequency modulation.

III. REVIEW OF DETECTOR CHARACTERISTICS

A. Photoemissive Devices

The external photoelectric effect which describes the emission of electrons into a vacuum from a material which has absorbed optical radiation provides a basic mechanism for the detection of modulation on an optical carrier. Photoemission from photoelectric materials, that is photocathodes [36], [37], is used in both vacuum photodiodes and photomultipliers for the conversion of optical radiation into an electron current. The time constants involved in the photoemission process are sufficiently short [9] that the intensity modulation present on the incoming radiation is converted to a similar modulation of the electron current emitted from the photocathode for modulation frequencies

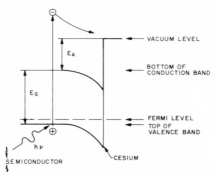

Fig. 3. Idealized energy band diagram of semiconductor photocathode. Photoexcitation of electrons from valence band over energy bandgap E_G and electron affinity E_A into vacuum is indicated.

TABLE I

BAND GAP ENERGY, ELECTRON AFFINITY, AND QUANTUM EFFICIENCY FOR SOME SEMICONDUCTORS WITH HIGH PHOTOEMISSIVE EFFICIENCY

Material	S Number	Energy-Gap E_G[eV]	Electron Affinity E_A[eV]	Quantum Efficiency η (%)	Ref.
Na$_2$KSB		1.0	1.0	30	[37]
(Cs)Na$_2$KSB	S-20	1.0	0.55	40	[37]
Cs$_3$Sb	S-17	1.6	0.45	30	[37]
GaAs-Cs$_2$O		1.4	−0.55	35 at 0.4μm	[38]–
				15 at 0.8μm	[40], [43]
In$_{0.16}$Ga$_{0.84}$As-Cs$_2$O		1.1		1 at 0.9μm	[42]
InAs$_{0.15}$P$_{0.85}$-Cs$_2$O		1.1	−0.25	0.8 at 1.06μm	[43]

extending into the microwave range. In a vacuum photodiode the modulated electron current is directly collected at an anode and is passed through an external load resistance to generate the output signal. In a photomultiplier, however, the electron current is first amplified in a chain of secondary emission dynodes.

The characteristics of vacuum photodiodes and photomultipliers which are of importance in the performance of optical receivers are 1) the efficiency with which the photocathode converts optical radiation to electron current, 2) time dispersion effects within the device which may limit bandwidth, 3) internal gain mechanisms which may be used to amplify the primary photocurrent, and 4) internal noise sources which may limit the sensitivity of the system. These characteristics will now be discussed with particular emphasis upon the factors which affect the performance of optical communication systems.

The net efficiency with which the incident photons are converted to emitted electrons, termed the external quantum efficiency of the photocathode, is dependent upon the incident photon energy hv, the effective diffusion length of electrons within the photosensitive material, and the work function of the photosurface [37]. Most high efficiency photocathodes are semiconductors [37] and an energy diagram of a typical semiconductor photocathode is shown in Fig. 3 where the parameters of interest are the bandgap energy E_G and the electron affinity E_A. The work function of the photosurface E_W, which defines the long wavelength threshold for photoemission, is then the sum of the bandgap energy and the electron affinity. Representative values of bandgap energy, electron affinity, and maximum quantum efficiency for several high efficiency photosurfaces are given in Table I. The newest photoemitters characterized by zero or negative electron affinity [38] such as GaAs[Cs$_2$O] [39]–[42] and InAs$_{0.15}$P$_{0.85}$[Cs$_2$O] [43] are of particular importance both because of high quantum efficiency and extended wavelength response. The long wavelength threshold for these surfaces is essentially determined by the bandgap of the photosensitive material rather than the work function of the surface and substantial improvements

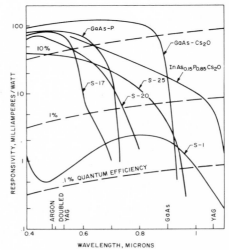

Fig. 4. Wavelength dependence of responsivity and quantum efficiency for several high efficiency photosurfaces.

in sensitivity at near infrared wavelengths, where the GaAs and the Nd:YAG lasers operate, appear possible.

The wavelength variation of the quantum efficiency is dependent upon the reflection and absorption properties of the photosurface as well as on the escape probability of the photoexcited electron in the material [37]. The typical variation of the current responsivity (A/W) and the quantum efficiency with wavelength for several photosurfaces is shown in Fig. 4 together with some of the more important laser wavelengths.

As Fig. 4 and (1) show the magnitude of the current from a photocathode is typically very small. For example, the current from an $S-20$ surface illuminated by one microwatt of argon laser power at 4880 Å is approximately 7×10^{-8} amperes. The output signal generated in a simple vacuum photodiode by passing this photocurrent directly through a load resistor is thus very small, particularly in the case of wide bandwidth devices which require small load impedances. Thus the signal-to-noise ratio of the photodiode system will be limited by noise associated with the preamplifier unless some means of amplifying the primary photocurrent prior to the load resistor is utilized. In a photomultiplier the process of secondary electron emission provides such an internal current gain with a very low noise factor. The typical photomultiplier geometry utilizes reflection dynodes in which the secondary electrons are emitted from the same side of the dynode struck by the primary beam. In most commonly used devices the electron beam is focused through the series of dynodes by a suitably shaped electrostatic field. Magnetic focusing has been used in some very wide bandwidth photomultipliers.

The gain of an electron multiplier is given by δ^N where δ is the average gain per stage and N is the number of stages. Commonly used secondary emission surfaces such as cesium antimonide, magnesium oxide, or beryllium oxide provide gains per stage in the range of 3 to 5 [36]. Recently developed negative electron affinity surfaces such as GaP[Cs] [44] or GaAs[Cs$_2$O] can provide gains per stage of 20 to 50. An overall gain of 10^6 to 10^8 can be achieved using any of these surfaces although obviously a larger number of stages will be required in the case of conventional low gain dynodes. The choice of dynode material must primarily be made on the basis of desired bandwidth and noise factor rather than overall gain as will now be described.

The factors which limit the bandwidth of photomultipliers are time dispersion effects within the secondary electron multiplier and capacitance associated with the anode output circuit [45], [46]. The time dispersion within the electron multiplier is caused by nonuniform electron transit times for 1) electrons emitted from different positions on the dynode surface because of spatially varying electric fields, and 2) electrons emitted from the same position on the dynode surface but with finite initial velocity distributions. The effect of nonuniform electric fields can be minimized by proper device geometry. The time dispersion due to the unavoidable initial velocity distribution can only be minimized by employing high interstage voltages and reducing the number of stages. For example, the upper half-power frequency limit of a rather idealized electron multiplier in which the electric field is assumed to have no spatial variation is given by [45]

$$f_{3 \text{ dB}} = 0.094 \frac{2qE}{m_0 \bar{v} N^{\frac{3}{2}}} \tag{8}$$

where E = electric field between dynodes, m_0 = electron mass, $\bar{v}$ = most probable initial velocity, and N = number of stages in the electron multiplier. Thus the bandwidth is proportional to the dynode voltage (for a given dynode spacing) and inversely proportional to the square root of the number of stages. The dynode voltage, however, also determines the secondary emission gain since most commonly used dynode materials exhibit a peak in secondary

emission gain as a function of the voltage between the stages [36]. Although this peak is rather broad it still results in a bandwidth that is not independent of the gain. Additional shot noise is introduced due to the statistics involved in the secondary emission process which amplifies both signal and primary shot noise currents [47]. The mean-square shot noise of the primary current leaving the photocathode $i_n^2 = 2q(I_{ph} + I_B + I_D)B$, with components due to signal, background, and dark current, increases through this multiplication process to

$$\overline{i^2} = 2q(I_{ph} + I_B + I_D)M_0^2 F B \tag{9}$$

where $M_0 = \delta^N$ = the average total multiplication, N = the number of stages, δ = the average gain per stage, and δ is assumed to have a Poisson distribution. The noise factor F of the photomultiplier is given by [47]

$$F = \frac{\delta^{(N+1)} - 1}{\delta^N(\delta - 1)} \tag{10}$$

or, for $\delta \gg 1$ which is the case for any practical secondary emission dynode material

$$F \approx \frac{\delta}{\delta - 1}. \tag{11}$$

Thus it may be observed that a photomultiplier has a very low noise factor, a typical example being $F \approx 1.5$ for $\delta = 3$. The new high gain dynodes result in even lower noise factors. Since the electron multiplier structure is in effect a cascaded series of amplifiers, the gain of the first stage is more important in determining excess shot noise than the gain of the remaining stages.

The optimum photomultiplier for an optical communication system operating at a specific wavelength should have the following characteristics based upon the previous discussion. The photosurface should be chosen for maximum quantum efficiency at the operating wavelength using antireflection coatings and multiple-pass techniques to enhance optical absorption whenever possible [36], [37]. High gain dynodes such as GaP should be utilized, particularly for the first stage, since excess shot noise can be minimized in this manner. These high gain dynodes typically require high dynode voltages for maximum gain and the resulting high electric field reduces transit time dispersion. In addition, their high gain reduces the number of required stages with a resultant further reduction in transit time dispersion and increase in bandwidth. Finally, the anode lead geometry or output coupling circuit must be designed to provide a bandwidth consistent with that of the communications system. Relatively narrow bandwidth photomultipliers employ simple wire connections from the anode to the output header pin, while in wide bandwidth devices the output current must be focused into a coaxial anode and output transmission system. Factors such as photocathode area and dynamic range must be considered in addition. It is desirable to minimize the cathode area to reduce the dark current noise, consistent with optical and mechanical design requirements. The dynamic range is limited at the low end

101

by noise and at the high end by saturation of the output current or by space charge effects that originate in the dynode chain.

The dynamic range of photomultipliers to be used in wide bandwidth systems must be carefully considered with respect to the desired S/N ratio. The S/N ratio measured at the output of the photomultiplier in a direct detection system with sufficient gain to achieve shot noise limited operation is, after (2), given by

$$\frac{S}{N} = \frac{\frac{1}{2}(mI_{ph}M)^2 R_{eq}}{2qI_{ph}M^2 FBR_{eq}}. \tag{12}$$

However, if the output current $mI_{ph}M$ becomes limited by saturation effects to I_{sat}, the S/N ratio has a limiting value of

$$\frac{S}{N} \propto \frac{I_{sat}}{M \cdot B} \tag{13}$$

which might not be sufficient to reach signal shot noise limited sensitivity.

Of the commercially available photomultipliers, the conventional electrostatically focused multipliers have bandwidths up to about 100 MHz. If a third high voltage electrode is placed between each pair of dynodes to reduce transit times, bandwidths of several hundred MHz result [49]. Improved designs that still maintain the conventional dynode chain, but use high efficiency photoemitters and the new high gain dynodes have been reported with subnanosecond response times [50]. Magnetically focused crossed-field photomultipliers can provide a bandwidth of 6 GHz by using a well-matched coaxial output line [46].

B. Solid-State Photodiodes and Avalanche Photodiodes

1) *Photodiodes:* Efficient high speed photodiodes, and avalanche photodiodes with internal current gain, have been fabricated by using specially designed p–n, p–i–n or metal–semiconductor junctions as shown in Fig. 5. These diodes are usually operated in the reverse bias region. In general, photon excited electrons and holes that are generated 1) within the high field region of the junction [51] and 2) in the bulk region and then diffuse to the junction [52], are collected as photocurrent across the high field region. However, diffusion processes are slow compared to the drift of carriers in the high field region. Therefore, in high speed photodiodes, the carriers have to be excited within the high field region of the junction or so close to the junction that diffusion times are shorter than, or at least comparable to, carrier drift times [15], [29]. Carriers are then collected across the junction at scattering limited velocities that are of the order of $v_{sat} = 10^6$ to 10^7 cm/s. Under these circumstances, the peak ac photocurrent is given by [51] (for photoexcitation of carriers at the junction edge)

$$i_{ph} = \frac{\eta q}{h\nu} mP_{opt} \frac{1 - \exp(-j\omega T_r)}{j\omega T_r} \tag{14}$$

where T_r = drift transit time. As space charge layer widths (w_i) can be in the order of micrometers, carrier transit times $T_r = w_i/v_{sat}$ in the subnanosecond region are possible.

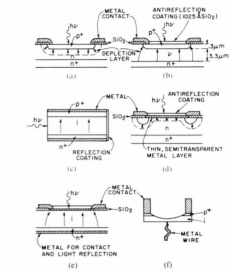

Fig. 5. Construction of different high speed photodiodes. (a) p–n diode. (b) p–i–n diode (Si optimized for 0.63 μm). (c) p–i–n diode with illumination parallel to junction. (d) Metal–semiconductor diode. (e) Metal–i–n diode. (f) Semiconductor point contact diode.

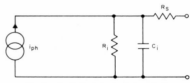

Fig. 6. Equivalent circuit of photodiode.

The ac characteristics of photodiodes can be described in terms of an equivalent circuit [15], [29], [31], [52], [54] that contains a photocurrent generator i_{ph}, diode capacitance C, series resistance R_S, and shunt resistance R_i, as shown in Fig. 6. The shunt resistance R_i is very high for diodes operating in the visible but is included to account for the relatively low leakage resistances of infrared photodiodes.

In the visible and near infrared well-designed diodes with high quantum efficiency, fast speed of response, low dark currents, and low series resistances can be obtained. High quantum efficiency requires minimization of the light reflection at the diode surface and placement of the junction in such a way that most photons are absorbed within or close to the high field region of the junction. Solid-state photodiodes are usually designed for light incidence normal to the junction plane. The quantum efficiency for a particular photodiode depends on the wavelength of operation and on the depth and width of the junction region. The spectral dependence of the quantum efficiency for a number of photodiodes is shown in Figs. 7 and 8.

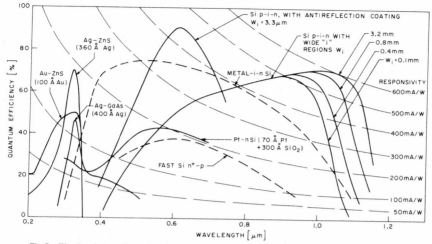

Fig. 7. Wavelength dependence of quantum efficiency and responsivity for several high speed photodiodes.

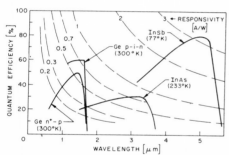

Fig. 8. Quantum efficiency and responsivity of photodiodes
that operate between 1 and 6 μm.

The speed of response of photodiodes is determined by RC time constants [29], [54], and the already mentioned carrier diffusion or drift transit times. The characteristic cutoff frequency of a photodiode is given by $f_C = 1/2\pi R_S C(1 + (R_S/R_i))$. Practical detection systems have, however, lower cutoff frequencies because of the finite load resistance and the additional capacitance (and inductances) of the load or amplifier circuit. Fast photodiodes have usually planar junctions that are small, 50 to 1000 μm in diameter, in order to keep the diode capacitance and leakage currents small [53]. Light has to be focused onto these diodes. The highest speeds of response and the smallest capacitances might be obtained from point contact diodes [Fig. 5(f)] [55] but the focusing of the light onto the small light sensitive areas is more difficult. Low series resistances in small area photodiodes are possible if the thickness of undepleted semiconducting bulk regions is kept small, as in p–i–n or metal–i–n diodes with fully de-

pleted space charge layers or in p$^+$–n diodes with an epitaxial base. For operation with baseband widths up into the GHz region, high speed photodiodes (and avalanche photodiodes) are commonly mounted into especially designed coaxial headers [54], [56].

The choice of a photodiode depends mainly on the wavelength of operation (Table II). Because of their developed technology, silicon diodes are preferentially used in the ultraviolet, visible, and near infrared part of the spectrum up to about 1 μm. With germanium diodes the response can be extended to over 1.5 μm. The geometry of both Si and Ge photodiodes can be optimized for particular applications. At short wavelengths, where light is absorbed close to the semiconductor surface it makes sense to use metal–semiconductor photodiodes with thin semitransparent metal layers [56]–[58]. Carriers are then separated in the high field region close to the surface thus yielding high quantum efficiencies and short response times. Several metal–semiconductor photodiodes [Fig. 5(d) and (e)] with response at short wavelengths have been fabricated from Si [57], GaAs [59], and ZnS [60] as indicated in Fig. 7 and Table II. For the visible range of the spectrum, where light penetrates a few micrometers into silicon, diffused p–n [61] and p–i–n junctions [53], [54], [62] are used [Fig. 5(a) and (b)]. Diffused junctions exhibit somewhat smaller reverse dark currents than metal–semiconductor junctions. For longer wavelengths close to the bandedge of the diode material, light penetrates deeply into the material. High quantum efficiency thus requires wide space charge layers. This leads to relatively long carrier transit times. For these diodes a tradeoff exists between quantum efficiency and speed of response [63]. As an example Fig. 7 shows the quantum efficiency at long wavelengths around 1 μm for Si p–i–n diodes with various space charge layer widths

103

TABLE II

PERFORMANCE CHARACTERISTICS OF PHOTODIODES

Diode	Wave-length Range (μm)	Peak Efficiency (η_o) or Respon-sivity	Sensitive Area (cm^2)	Capacitance (pF)	Series Resistance (Ω)	Response Time (seconds)	Dark Current	Operating Temperature (K)	Comments	Ref.
Silicon n$^+$–p	0.4–1	40	2×10^{-5}	0.8 at -23 V	6	130 ps with 50-Ω load	50 pA at -10 V	300	avalanche photo-diode	[61]
Silicon p–i–n	0.6328	>90	2×10^{-5}	<1	~1	100 ps with 50-Ω load	$<10^{-9}$ A at -40 V	300	optimized for 0.6328 Å	[62]
Silicon p–i–n	0.4–1.2	>90 at 0.9 μm >70 at 1.06 μm	5×10^{-2}	3 at -200 V 3 at -200 V	<1 <1	7 ns 7 ns	0.2 μA at -30 V	300		[65]
Metal–i–nSi	0.38–0.8	>70	3×10^{-2}	15 at -100 V		10 ns with 50-Ω load	2×10^{-2} A at -6 V	300		[57]
Au–nSi	0.6328	70	2			<500 ps		300	Schottky barrier. antireflection coating	[56]
PtSi–nSi	0.35–0.6	~40	2×10^{-5}	<1		120 ps		300	Schottky barrier avalanche photo-diode	[58]
Ag–GaAs	<.36	50						300		[59]
Ag–ZnS	<.35	70						300		[60]
Au–ZnS	<.35	50						300		[60]
Ge n$^+$–p	0.4–1.55	50 uncoated	2×10^{-5}	0.8 at -16 V	<10	120 ps	2×10^{-8}	300	Germainum avalanche photo-diode	[30]
Ge p–i–n	1–1.65	60	2.5×10^{-5}	3		25 ns at 500 V		77	illumination entering from side	[65] [66]
GaAs point contact	0.6328	40		0.027	30					[55]
InAs p–n	0.5–3.5	>25	3.2×10^{-4}	3 at -5 V	12		$<10^{-6}$	77		[15] [29]
InSb p–n	0.4–5.5	>25	5×10^{-4}	7.1 at -0.2 V	18		5×10^{-6}	77		[15] [29] [67]
InSb p–n	2–5.6		5×10^{-4}				1 MΩ shunt resistance	77	Reverse break down voltages 30 V	[68]
Pb$_{1-x}$Sn$_x$Te x = 0.16	9.5 μm	45 V/W $\eta = 60$	4×10^{-3}			~10^{-9}		77	shunt resistance $R_i = 10$ Ω	[72] [74]
Pb$_{1-x}$Sn$_x$Se x = 0.064	11.4 μm	3.5 V/W $\eta = 15$	7.8×10^{-3}			~10^{-9}		77	shunt resistance $R_i = 2.5$ Ω	[73] [74]
Hg$_{1-x}$Cd$_x$Te x = 0.17	15 μm	$\eta \sim 10$–30	4×10^{-4}		8		$<3 \times 10^{-9}$	77	shunt resistance $R_i > 100\Omega$	[70] [71]. [136]

[64], [65]. The corresponding carrier transit times or diode response times and bias voltages can be determined from Fig. 9 [63]. Germanium diodes [53] are used at wavelengths below 1 μm to beyond 1.5 μm. However, due to the narrower bandgap, germanium diodes exhibit larger dark currents. Reduced dark currents and thus higher sensitivity can be obtained in both Si and Ge diodes by cooling. The problems encountered in the design of Ge diodes that are fast and efficient for 1.53 (HeNe laser) and 1.54 μm (Erbium laser) are similar to those encountered with Si diodes for

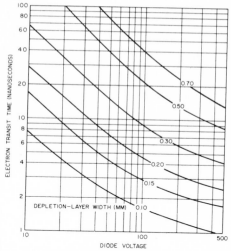

Fig. 9.   Carrier transit times for Si p-i-n diodes with different depletion layer widths (after McIntyre [63], [64]).

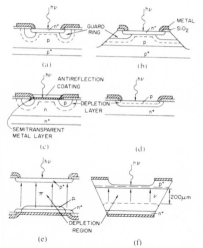

Fig. 10.   Construction of different avalanche photodiodes. (a) Si guard ring structure. (b) Ge mesa structure with guard ring. (c) Metal–semiconductor structure with guard ring. (d) Planar p–n$^+$ structure with low fields at junction edges. (e) Si–n$^+$–p–π–p$^+$ structure. (f) Si structure with wide depletion region and slant surfaces.

0.9 and 1.06 μm. Better results might be achieved if light is allowed to penetrate from the side [66], parallel to the junction as shown in Fig. 5(c). This may result in a very narrow sensitive area.

For wavelengths beyond 1.5 μm, up to 3.6 μm and 5.6 μm, respectively, InAs and InSb photodiodes [29], [32], [67], [68] are suitable as can be seen from the quantum

efficiency curves of Fig. 8. For maximum sensitivity these diodes have to be operated at temperatures between 77 and 135°K.

For longer wavelengths in the 6- to 30-μm range and especially for 10.6 μm, where the CO_2 laser operates, mixed crystal photodiodes such as $Hg_{1-x}Cd_xTe$ [69]–[71] and $Pb_{1-x}Sn_x(Te, Se)$ [72]–[74] are under development. Dynamic shunt resistances in $Hg_{1-x}Cd_xTe$ in the range 10^2 to 10^5 ohms at 77°K have been obtained [70], [71], so that photodiodes for 10.6 μm that are fast and efficient are becoming available at 77°K or even 110°K.

Recently, (Hg, Cd) Te photodiodes with wider bandgaps have been investigated for use at shorter wavelengths down to 1–2 μm [69].

2) Avalanche Photodiodes: Current gain in solid-state photodiodes is possible through avalanche carrier multiplication [75], observed at high reverse bias voltages where carriers gain sufficient energy to release new electron–hole pairs through ionization. Substantial current gains have been achieved by this process even at microwave frequencies [61], [75]. Despite the fact that excess noise is introduced by the multiplication process [30], [61], [76]–[78], significant improvements in overall sensitivity are possible using silicon or germanium avalanche photodiodes for photodetection systems with wide instantaneous bandwidths.

Similar criteria apply for avalanche photodiodes, with respect to quantum efficiency and speed of response, as for conventional nonmultiplying photodiodes. Additional attention must, however, be given to the current gain and its limitations and to the noise properties of avalanche photodiodes.

In the design of avalanche photodiodes, special precautions must be taken to assure spatial uniformity of carrier multiplication over the entire light sensitive diode area. Microplasmas, i.e., small areas with lower breakdown voltages than the remainder of the junction, and excessive leakage at the junction edges can be eliminated through the use of guard ring structures [61], [79] as indicated in Fig. 10. The selection of defect-free material and cleanliness in processing allows fabrication of microplasma-free diodes. Highly uniform carrier multiplications in excess of 10^4 and 200 have been reached at room temperatures in small area silicon [61] and germanium [30] diodes, respectively. In large area diodes that are free of microplasmas, the spatial uniformity of carrier multiplication is limited either by doping inhomogeneities of the starting material or by inhomogeneities in the diffusion profile. Typical variations can be 20 to 50 percent at an average multiplication of 10^3.

The highest current gains are observed if the diodes are biased to the breakdown voltage. This is illustrated by Fig. 11. where the voltage dependence of the dark current of a Ge avalanche photodiode as well as the response to pulses from a phase locked 6328 Å laser are shown. At low reverse bias voltages, no carrier multiplication takes place. As the reverse bias voltage is increased, carrier multiplication sets in as indicated by the increase in the current pulse.

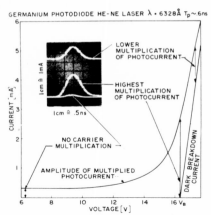

Fig. 11.   Current voltage characteristic and pulse response of germanium avalanche photodiode.

The maximum pulse amplitude and multiplication of photocarriers is obtained at the breakdown voltage. At voltages above the breakdown voltage, a self-sustained avalanche current flows that becomes less and less sensitive to photon-excited carriers.

The maximum gain of an avalanche photodiode is limited either by current induced saturation effects [30] or by a current gain bandwidth product [80]–[83], [85]. Current induced saturation of the carrier multiplication is observed because the carriers that emerge from the multiplication region reduce the electric field within the junction and cause voltage drops across the series and load resistance of the diode [30]. This leads to a current dependent reduction of the carrier multiplication factor. This saturation manifests itself for high light intensities, at which the multiplied current will only increase as the square root of the photocurrent instead of being proportional to the photocurrent as at low light intensities. For low light intensities, the dark current will set a limit to the average value for the maximum carrier multiplication that can be achieved at low frequencies. The aforementioned maximum multiplication of 200 for germanium diodes [30] is due to the high dark current. Lowering of the temperature in diodes with sufficiently high breakdown voltages (> 20 volts for Si and Ge) so that no internal field emission takes place, decreases the dark current and leads to higher values for the maximum carrier multiplication.

At high modulation frequencies or for fast optical pulses, the current gain–bandwidth product sets a limit to the maximum achievable gain [1], [80]–[85]. For silicon [61] and germanium [30] n$^+$–p diodes gain–bandwidth products of 100 and 60 GHz have been reported as indicated in Table III. This current gain–bandwidth product is due to the fact that both electrons and holes excite electron–hole pairs through ionization in silicon and germanium diodes. Secondary electrons and holes thus travel back and forth

through the multiplication region long after the primary carrier has left the junction.

The current gain–bandwidth product is inversely proportional to the average transit time of the carriers through the multiplication region [80]–[83], [85] and depends on the ratio between the electron and hole ionization rates [84]. For Ge, the current gain–bandwidth product is independent of excitation because the ionization coefficients are almost equal, but for Si, higher current gain–bandwidth products result if the avalanche is initiated with electrons because they have a higher ionization rate. If one can realize carrier multiplication in solid-state diodes in which only one type of carrier ionizes, then the current gain would not be limited by a current gain–bandwidth product. The response time at high multiplications would then be about twice as long as without multiplication. Although indications exist that only electrons cause ionizations in InSb [86] and Schottky barrier GaAs [87] photodiodes, no speed of response measurements have been reported to date.

Excess noise in avalanche photodiodes is induced by fluctuations in the carrier multiplication process [76]–[79]. For practical diodes, the shot noise of the average photo and dark current has been found [30], [61], [78], [79] to increase faster than the square of the carrier multiplication M, approximately as

$$\overline{i^2} = 2q(I_{ph} + I_D)M^{2+x}B \qquad (15)$$

in reasonably close agreement with theoretical calculations based on a spatially uniform avalanche ionization region [77]. The noise factor $F = M^x$ depends on the ratio of the ionization rates and on the type of carriers that initiate the avalanche [1], [30], [77]–[79]. For germanium diodes [30], where the ionization coefficients for holes and electrons are about equal, the excess noise factor increases as the carrier multiplication ($x = 1$). This relatively high noise is due to the fact that for equal ionization rates, a relatively low number of carriers is present within the multiplication region at any one time so that fluctuations in the ionization events have a large influence on the avalanche process. Much lower noise is observed in junctions with highly unequal ionization rates if the avalanche is initiated with carriers of the higher ionization rate. Initiation of avalanche carrier multiplication by electrons leads to low excess noise in Si ($F = M^{0.4}$) [78], [79], InSb [86], and Schottky barrier GaAs [87] diodes. If only one type of carrier can ionize, theoretical calculations show that the noise factor does not follow (15), but would be equal to two at high multiplications [77]. The results for InSb and for certain GaAs diodes indicate that only electron ionization is of importance in these materials.

The different types of avalanche photodiodes that have been constructed to date are shown schematically in Fig. 10 and their performance data are listed in Table III. As Table III shows, various Si n$^+$–p guard ring avalanche photodiodes have been developed with diameters of the light sensitive area between 40 and 200 μm [61], [79], [89]–

106

TABLE III

CHARACTERISTICS OF AVALANCHE PHOTODIODES

Diode	Construction	Wavelength Range (µm)	Sensitive Area (cm²)	Dark Current	Avalanche Breakdown Voltage (volts)	Maximum Gain	Multiplication Noise $i^2 \sim M^{2+x}$	Current Gain-Bandwidth Product (GHz)	Capacitance (pF)	Series Resistance (Ω)	Ref.
Silicon n+–p	Fig. 10(a)	0.4–1	2×10^{-5}	50 pA at −10 V	23	10^4	$x \sim 0.5$	100	0.8 at −23 V	6	[61]
Silicon n+–p π–p+	Fig. 10 (e)	0.5–1.1	2×10^{-3}		~88	200	$x \sim 0.4$	high			[91], [92]
Silicon n+–i–p+	Fig. 10(f)	0.5–1.1			200 to 2 000		low	not very high			[89]
PtSi–nSi	Fig. 10(c)	0.35–0.6	4×10^{-5}	~1 nA at −10 V	50	400	$x \sim 1$ for visible illumination	40 for UV excitation	<1		[58]
Pt–GaAs	Fig. 10(c)	0.4–0.88			~60	>100	very low	>50			[87]
Germanium n+–p	Fig. (10b)	0.4–1.55	2×10^{-5}	2×10^{-8} A −16 V and 300 K	16.8	250 at 300°K >10^4 at 80°K	$x \sim 1$	60	0.8 at −16 V	<10	[30]
Germanium n+–p	Fig. 10(a)	for 1.54			150						[91], [92]
InAs		0.5–3.5									[88]
InSb at 77 K		0.5–5.5			a few	10	very low				[86]

[91]. These are the simplest avalanche photodiodes. They operate at relatively low voltages and are useful between about 0.4 and 0.8 µm. Similar Ge n+–p diodes have been constructed [30] with a guard ring that terminates in a mesa structure in order to reduce the surface leakage current. These Ge diodes are useful as fast diodes with gain at wavelengths between 0.5 and 1.5 µm. Both the small area Si and Ge avalanche photodiodes can resolve optical pulses with widths of 130 ps between their 50-percent rise and fall time points. The current gain–bandwidth products are 100 and 60 GHz, respectively, thus indicating that current gains of 100 and 60 are possible in a detection system with 1-GHz bandwidth.

Two different structures have been developed for Si avalanche photodiodes for operation at wavelengths between 0.8 and about 1 µm. In order to achieve high quantum efficiencies these diodes have wide space charge layers. One structure, the n+–ν–p+ structure [89] [see Fig. 10(f)] has a wide high field region in which carrier multiplication can take place. Because the electric fields are relatively low in this case, a very low hole ionization coefficient results [84]. These diodes exhibit low excess noise [77] but will have only a low current gain–bandwidth product [81], [82]. In the n+–p–π–p+ structure [90], [91] [Fig. 10(c)], the multiplication is concentrated to the narrow n+–p region whereas the wide π region acts as a collection region for photon excited carriers. This narrow multiplication region will result in a high gain–bandwidth product but results also in higher excess noise. The noise is higher because, at the high fields in this narrow multiplication region, the ionization

coefficients for holes are not much lower than the ionization coefficients for electrons [84].

Silicon avalanche photodiodes with high current gain and relatively low excess noise are thus available for the wavelength range between 0.4 and 0.9 µm. Germanium avalanche photodiodes have higher excess noise and higher leakage current if not cooled, but they are excellent detectors for 1.06 µm and can be used up to over 1.5 µm as well [93]. As an example, the application of a Ge avalanche diode to the demodulation of small signals at 6 GHz on a 1.15-µm laser beam is illustrated in Fig. 12. As can be seen, the signal from the diode increases as M^2 with multiplication and the noise from the avalanche diode as M^3. The best operating point (M_{opt}) with the largest S/N ratio and the highest sensitivity to weak light signals is reached if the carrier multiplication is adjusted so that the noise of the avalanche diode is about equal to the amplifier noise.

Results for GaAs and InSb diodes indicate that avalanche photodiodes with almost no excess noise (only a factor of two) and high gain that is not limited by current gain–bandwidth products will be forthcoming in the foreseeable future. The InSb avalanche diodes will extend the range of solid-state diodes with internal current gain up to 5.6 µm.

Avalanche carrier multiplication has further been observed in a number of additional materials, among them InAs [88], but no practical devices have been reported.

Extension of avalanche photodetection to wavelengths beyond the bandedge of a semiconductor material is possible in metal–semiconductor junctions as photoemission of

107

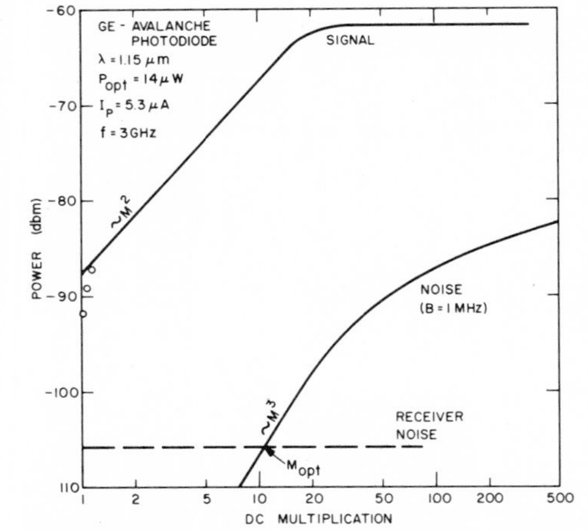

POWER (dbm)

GE - AVALANCHE
PHOTODIODE
λ = 1.15 μm
P_{opt} = 14 μW
I_p = 5.3 μA
f = 3 GHz

SIGNAL

NOISE
(B = 1 MHz)

RECEIVER
NOISE

M_{opt}

DC MULTIPLICATION

Fig. 12. Signal and noise power output of germanium photodiode in 1-MHz band at 3 GHz. Optimum operating point (M_{opt}) for best S/N ratio and sensitivity is indicated.

carriers from the metal contact can be combined with ionization of carriers within the high field region of the junction. Only small quantum efficiencies can, however, be expected. For palladium–silicon Schottky barrier diodes quantum efficiencies of 0.75 percent and avalanche multiplications > 200 have been observed at 1.3 μm [94].

C. Photoconductive Detectors

Electron–hole pairs created in a bulk semiconductor by the absorption of optical radiation increase the conductivity of the material during the lifetime of the pairs. This effect, termed photoconductivity [95], provides a basic mechanism for the detection of optical signals, since upon application of a bias voltage this conductivity modulation can be translated into a modulation of the current that flows through the output circuit.

For a simple dc biased photoconductor with ohmic contacts, whose conductance is dominated by one type of excited carrier, the optically induced small signal peak ac current is [16], [31]

$$i_S = \eta \frac{q m P_{opt}}{h\nu} \frac{\tau}{T_r} \frac{\exp(-j\phi)}{(1 + \omega^2\tau^2)^{\frac{1}{2}}} \qquad (16)$$

where τ = mean carrier lifetime, $T_r = L^2/\mu V$ = carrier transit time, μ = carrier mobility, L = interelectrode spacing, V = bias voltage, and $\phi = \tan^{-1}(\omega\tau)$.

The factor $M = (\tau/T_r) \exp(-j\phi)(1 + \omega^2\tau^2)^{-1/2}$ is the photoconductive gain and indicates the number of carriers which cross the photoconductor for each absorbed photon. Practical application of photoconductors generally requires this gain to be maximized consistent with frequency response requirements by operating the photoconductor at the highest bias voltage compatible with breakdown and extraneous noise considerations. The fundamental tradeoff in photoconductors between the current gain $M = \tau/T_r$ and

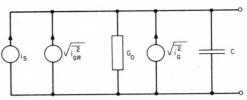

Fig. 13. Small signal equivalent circuit of photoconductive detector.

the speed of response τ is indicated by the existence of a current gain–bandwidth product [96], [97]

$$MB = \frac{1}{2\pi T_r} \leq \frac{1}{2\pi\tau_{rel}}. \qquad (17)$$

The maximum value of this current gain–bandwidth product is independent of the carrier lifetime τ and limited as indicated by the dielectric relaxation time $\tau_{rel} = \rho\varepsilon$ [95], where ε = permitivity and ρ = resistivity of the photoconductor.

Evaluation of an optical detection system that utilizes photoconductors requires in addition consideration of the differential conductance G and capacitance C of the photoconductor and of the different noise sources indicated in the small signal equivalent circuit of Fig. 13. The differential conductance is given by [98]

$$G_0 = \frac{Aq\mu n}{L} \bigg|_{\text{if determined by doping}}$$
$$= \frac{\mu\tau(I_{ph} + I_B)}{L^2} \bigg|_{\text{if determined by illumination}} \qquad (18)$$

where $(I_{ph} + I_B)/q \sim$ number of photons absorbed per unit time due to both signal (I_{ph}) and background (I_B) radiation.

The principal mean-square noise current generators of a photoconductor are [95]

1) generation–recombination noise [99], [100]

$$\overline{i_{GR}^2} = 4q\left(\frac{\tau}{T_r}\right)\frac{I_0 B}{1 + \omega^2\tau^2} \qquad (19)$$

with $I_0 = (\tau/T_r)(I_{ph} + I_B)$ being the dc light induced output current of the photoconductor.

2) thermal noise due to the differential conductance G_0 of the photoconductor at temperature T_G

$$\overline{i_G^2} = 4kT_G G_0 B. \qquad (20)$$

The S/N and sensitivity considerations of Section II can be applied to detection systems with photoconductors. The generation–recombination noise of (19) can be brought into the form $2q(I_{ph} + I_B)|M(\omega)|^2 F \cdot B$ with $|M(\omega)|^2 = M_0^2/(1 + \omega^2\tau^2)$ and $F = 2$. The thermal noise due to the output circuit with a load conductance G_L, if any, at temperature T_L and amplifier input conductance G_A and effective amplifier noise temperature T_A

$$\overline{i_L^2} = 4kT_L G_L B$$

and

$$\overline{i_A^2} = 4kT_A(G_0 + G_L)B \qquad (21)$$

can be combined with the thermal noise of the photoconductor into the single thermal noise source used in (2)–(7).

The equivalent circuit [31] of Fig. 13 takes account of the main features of the photoconductor. It should, however, be kept in mind, that in practical photoconductors material parameters often change with bias voltage, operating temperature, and optical power. For example, if the differential conductance G_0 of a photoconductor is too low, the carrier transit time T_r becomes shorter than the dielectric relaxation time τ_{rel} at relatively low bias voltages. This limits the current gain–bandwidth product [95] and leads to space charge limited currents [95]. Saturation of the current gain is often observed at high bias voltages [111]. This is because the carrier velocities cease to increase as $v = \mu E$ with electric field E but saturate at a scattering limited velocity v_{sat} thus leading to a voltage independent transit time $T_r = L/v_{sat}$ and gain $M = v_{sat}\tau/L$. In several extrinsic photoconductors an increase of the carrier lifetime τ (as shown in Fig. 14 for Ge:Cu and Ge:Hg(Sb) [102]) and of the $\mu\tau$ product [104] and impact ionization have been observed with rising bias fields. Nonohmic contacts lead to carrier depletion regions and low current gain [95]. In certain materials, such as in CdSe [105], the current gain and the carrier lifetime depend strongly on light intensity, typically decreasing like the square root of the optical power at high light intensities [16].

In many photoconductors both types of excited carriers will contribute to the photoconductivity and be able to move out through the contacts thus leading to a much lower photoconductive gain than indicated by (16). For this case and for nonohmic contacts, microwave bias has been explored as a potential means to increase the current gain and the gain–bandwidth product [106]. The effect of RF bias with a sufficiently high frequency is to effectively "trap" carriers in the bulk of the photoconductor, so that they are not swept out to the electrodes for recombination, but instead move back and forth contributing to conductivity, until volume recombination takes place. Although impressive results have been reported [101], [107], practical application of ac biasing might be somewhat limited because of complicated circuitry and because of the noise that is induced through instabilities in the microwave bias source.

In the design of photoconductive direct detection systems one strives to reach the generation–recombination noise limited sensitivity. This is obtained by the use of high gain to overcome the thermal noise of the photoconductor and output circuit. Once sensitivity is limited by background radiation induced g–r noise, a further increase in detection sensitivity is still possible through the use of aperturing, which narrows the field of view of the photoconductor, and by spectral filtering and cooling [5]–[7].

To obtain a flat frequency response and to minimize phase distortion within the bandwidth of operation B, the photoconductive lifetime should be as short as $\tau \leq 1/2\pi B$. Operation in the rolloff region $(\tau > 1/2\pi B)$ is not acceptable in most communications applications. On the other hand,

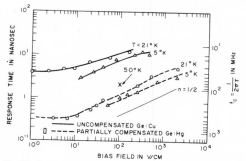

Fig. 14. Response time of photoconductive Ge:Cu and Ge:Hg(Sb) (After Buczek and Picus [102]).

the lifetime should not be shorter than that dictated by the bandwidth, in order to optimize the current gain. Once the value of τ is established the MB product can be optimized by minimizing the transit time, which is accomplished by the use of high dc bias voltages and small detector dimensions in the direction of current flow. In addition, sufficient speed of response must be provided by the output circuit to satisfy the condition $R_{eq}C \leq 1/2\pi B$. The value of C includes the capacitance of the photoconductor, its mount, and the output circuit which must thus be kept as low as possible. Given the value of B and C, a value of R_L is chosen equal to or lower than the photoconductor resistance to satisfy the previously mentioned condition. For wide signal bandwidths this loading leads to reduced sensitivity in direct detection systems since the photoconductive gain is now insufficient to overcome receiver thermal noise. However, by the use of heterodyne detection, photoconductors can yield high sensitivity and wide bandwidth in combination by using sufficient local oscillator power to overcome the thermal noise contributions.

1) Characteristics of Infrared Photoconductors: The spectral sensitivity for a number of photoconductors (and photovoltaic detectors) is shown in Fig. 15. Direct detection sensitivity is given in terms of detectivity D^*, the commonly used figure of merit for infrared detectors [5], D^* is in units of cm·Hz$^{1/2}$·W^{-1}, it is normalized to a sensitive area = 1 cm^2 and a postdetection bandwidth $B = 1$ Hz. D^* is related to the noise equivalent power NEP ($S/N = 1$ and $m = 1$ in (4)) by $D^* = A^{1/2}B^{1/2}/$NEP. The values for D^* given in Fig. 15 are only representative and are given mostly for a 60° field of view. The highest values of D^* are usually reached by the use of high purity materials with long time constants and high resistances.

D^* is commonly measured at audio modulation frequencies using a matched load resistance. This is satisfactory for radiation sensing applications of infrared detectors, where speed is usually secondary to sensitivity. The advent of optical communications and other laser applications gave impetus to more detailed investigations of detector speed and to the development of infrared detectors with wide bandwidth capabilities.

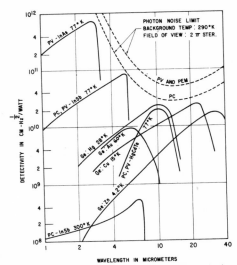

Fig. 15. Spectral dependence of detectivity for photoconductive and photovoltaic infrared detectors. Representative values are given [5], [6], [119].

TABLE IV

INFRARED PHOTOCONDUCTIVE DETECTORS

Material	Maximum Temperature for Background Limited Operation	Test Temperature (K)	Long Wavelength Cutoff (50%) (μm)	Peak Wavelength λ_m (μm)	Absorption Coefficient (cm^{-1})	Quantum Efficiency η	Resistance (Ω)	$D^*_{\lambda_m}$ ($cm \cdot Hz^{1/2}/W$)	Approximate Response Time (seconds)	Reference
InAs		195	3.6	3.3	$\sim 3 \times 10^3$			3×10^{11}	5×10^{-7}	[101]
InSb	110	77	5.6	5.3	$\sim 3 \times 10^3$	0.5–0.8	10^3–10^4	6×10^{10} -1×10^{11}	5×10^{-6}	[4], [108], [118], [119]
Ge : Au	60	77	9	6	~ 2	0.2–0.3	4×10^5	3×10^9–10^{10}	3×10^{-8}	[108], [110], [111], [119], [120]
Ge : Au(Sb)	60	77	9	6			10^6	6×10^9	1.6×10^{-9}	[102], [119]
Ge : Hg	35	4.2 27	14 14	11 10.5	~ 3 ~ 4	0.2–0.6 0.62	1–4×10^4 1.2×10^5	7×10^9 -4×10^{10} 4×10^{10}	3×10^{-8} -10^{-9}	[108], [113]–[115], [119], [120] [115]
Ge : Hg(Sb)	35	4.2	14	11			5×10^5	1.8×10^{10}	$3 \times 10^{-10} - 2 \times 10^{-9}$ $3 \times 10^{-10} - 3 \times 10^{-9}$	[102], [119] [102]
Ge : Cu	17	4.2 20	27	23	~ 4	0.2–0.6	2×10^4	2–4×10^{10}	$3 \times 10^{-9} - 10^{-8}$ $4 \times 10^{-9} - 1.3 \times 10^{-7}$	[108], [119], [120]
Ge : Cu(Sb)	17	4.2	27	23			2×10^5	2×10^{10}	$<2.2 \times 10^{-9}$	[119], [122]
$Hg_{1-x}Cd_xTe$ $x = 0.2$		77	14	12	$\sim 10^3$	0.05–0.3	60–400 20–200	10^{10} 6×10^{10}	$<10^{-6}$} $<4 \times 10^{-6}$}	[4], [119], [123]–[125], [133]
$Pb_{1-x}Sn_xTe$ $x = 0.17$–0.2		77 4.2	11 15	10 14	$\sim 10^4$		42 52	3×10^9 1.7×10^{10}	1.5×10^{-7}} 1.2×10^{-6}}	[4], [74], [126]

Table IV lists the properties of infrared photoconductors in further detail. For wavelengths up to 3.5 to 5.7 μm, the use of cooled intrinsic InAs and InSb is indicated. For the longer wavelengths, specifically 10.6 μm, extrinsic and intrinsic photoconductors are available and will now be discussed.

a) *Extrinsic infrared photoconductors:* Extrinsic infrared photoconductors [4]–[6], [108], [109] rely on optical excitation of holes from impurity centers such as Au[110], [111], Cu [4], [112], [113], Hg [113]–[115], or Cd [116], into the valence band of a p-type semiconductor. Ge is used principally as the host crystal but Ge-Si [108] and Si [117] have also been investigated. In order to keep competing thermal excitation low, extrinsic photoconductors are cooled to temperatures between 77 and 4.2°K (see Table IV). Since optical absorption constants in the wavelength

PROCEEDINGS OF THE IEEE, OCTOBER 1970

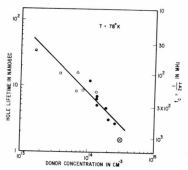

Fig. 16. Dependence of hole lifetime on antimony compensation in Ge:Au (after Bratt [128]). The bottom point is from [120].

TABLE V

Summary of Results on Infrared Heterodyne Detection

Material	Detection Mode*	Laser Wavelength (μm)	Measured Noise Equiv Power† (W/Hz)	Measured Quantum Noise Factor† (dB)	Quantum Efficiency	Calculated Noise Equiv Power (W/Hz)	Temperature (°K)	Response Time (Seconds)	Bandwidth	Dynamic Resistance (Ω)	Ref.
InAs	PV	3.39	1.25×10^{-19}	3.3			300				[140]
			$\sim 3 \times 10^{-19}$	7.1	0.25		300				[141]
InSb	PC	0.6328 $\}$ 3.39 $\}$			0.5		77	10^{-8}		300	[142]
Ge:Au	PC	3.39			0.08		77	10^{-8}			[143]
Ge:Hg	PC	10.6	1.73×10^{-19}	9.7	0.5	7.5×10^{-20}	4.2	3.3×10^{-9}	50 MHz		[4], [144]
Ge:Hg(Sb)	PC	3.39					4.2	$< 3.7 \times 10^{-10}$	>450 MHz		[4], [102]
Ge:Cu	PC	10.6	$1.3 \times 10^{-19} S$ $7 \times 10^{-20} Q$	8.4 5.7	0.5	7.5×10^{-20}	4.2	2×10^{-9}			[4], [18], [145]
Ge:Cu(Sb)	PC	10.6	7.5×10^{-20}	6	0.56	6.7×10^{-20}	4.2	2×10^{-10}	1.5 GHz	1.2×10^3	[4], [17], [146]
$Hg_{1-x}Cd_xTe$	PC	10.6	7×10^{-20}	5.7			77	$0.5 - 2.5 \times 10^{-7}$	1–10 MHz	200	[133]
$Hg_{1-x}Cd_xTe$	PV	10.6	2.2×10^{-19}	10.7	0.09	2.1×10^{-19}	77		>50 MHz	190	[138]
$Pb_{1-x}Sn_xSe$	PV	10.6	$2.5 \times 10^{-19} Q$ $1.0 \times 10^{-18} S$	11.2 17.3	0.085	2.2×10^{-19}	77	2×10^{-8}		1.5	[4], [18]

* PV = photovoltaic; PC = photoconductive
† Quantum noise limit $h\nu = 1.87 \times 10^{-20}$ and 5.85×10^{-20} W/Hz at 10.6 and 3.39 μm, respectively.
Q = sensitivity calculated using LO-generated noise only.
S = measured overall receiver sensitivity.

range of extrinsic carrier excitation are generally low (typically of the order of 1 to 10 cm^{-1}), [127], the requirement for high quantum efficiency leads to detector dimensions in the direction of light incidence of several millimeters.

Although Ge:Hg and Ge:Cu have drastically different cutoff wavelengths of 14 and 27 μm, respectively, their D^* at 10.6 μm are close [113]. Time constants for uncompensated samples are typically 10^{-7} to 3×10^{-9} seconds, with

peak quantum efficiency in the 0.2 to 0.6 range [102], [113]–[115], [119], [120]. The usefulness of Ge:Au (and Ge:Au(Sb)) at 10.6 μm is limited by low quantum efficiency ($< 10^{-2}$) at this wavelength but it is used in the laboratory because it operates at liquid N_2 temperature.

For wide-band applications, short carrier lifetimes can be obtained in impurity-activated germanium by compensation with antimony donors [102], [112], [119], [128], [129]. The measured shortening of the hole lifetime with increasing compensation by donors in p-type Ge:Au is shown in Fig. 16 [128]. Compensation initially has only a small effect on D^* as the data in Table IV for Ge:Au(Sb), Ge:Hg(Sb), and Ge:Cu(Sb) show. Compensation increases the resistance of the detector [129], thereby reducing the direct detection circuit bandwidth unless a low R_L with a concomitant reduction in sensitivity is used. The usable degree of compensation has a further limit in that quantum efficiency decreases in proportion to $(N_A - N_D)$, where N_A, N_D = acceptor and donor concentrations, respectively [129]. N_A typically has a value of the order of $10^{16} \mathrm{cm}^{-3}$ [108]. The usefulness of compensation is indicated in Table IV (and in Table V) where measured lifetime values as short as 3×10^{-10} and 2×10^{-10} seconds are given for Ge:Hg(Sb) and Ge:Cu(Sb), respectively. [4], [17], [102].

b) *Intrinsic infrared photoconductors:* Intrinsic photoconductors utilize band-to-band excitation. Consequently, the light absorption coefficient is high, typically 10^3 to 10^4 cm^{-1}, so that the dimension in the direction of light incidence need only be a few micrometers. Intrinsic infrared photoconductors are also listed in Table IV. In addition to the III–V photoconductors InAs and InSb with response to a few micrometers, narrow bandgap intrinsic detectors have been fabricated from mixed crystals $\mathrm{Hg}_{1-x}\mathrm{Cd}_x\mathrm{Te}$ [4], [119], [123] and $\mathrm{Pb}_{1-x}\mathrm{Sn}_x\mathrm{Te}$ [4], [74], [126]. These detectors can be tailored to the desired wavelength by varying the molar content x [4], [130]. $\mathrm{Hg}_{1-x}\mathrm{Cd}_x\mathrm{Te}$ detectors have been fabricated for operation at wavelengths from 1 to 30 μm [131], [132]. Emphasis has been on the 8- to 13-μm atmospheric window. Here their technical importance comes about because they operate at higher temperatures, that is 77° to beyond 110°K [4], [125] than the extrinsic photoconductors. As Tables IV and V show, response times as short as 50 and 15 ns have been obtained in photoconductive $\mathrm{Hg}_{1-x}\mathrm{Cd}_x\mathrm{Te}$ and $\mathrm{Pb}_{1-x}\mathrm{Sn}_x\mathrm{Te}$, respectively [4], [131]–[133].

Background limited (BLIP) detectivity has been achieved in $\mathrm{Hg}_{1-x}\mathrm{Cd}_x\mathrm{Te}$ [125]. Although the possible design of fast photoconductive $\mathrm{Hg}_{1-x}\mathrm{Cd}_x\mathrm{Te}$ detectors has been considered in detail [134], [135] photovoltaic $\mathrm{Hg}_{1-x}\mathrm{Cd}_x\mathrm{Te}$ and $\mathrm{Pb}_{1-x}\mathrm{Sn}_x\mathrm{Te}$ detectors are demonstrating greater capability in meeting the requirements of MHz bandwidth optical communication systems [8]. Bandwidths of 30 MHz to greater than 50 MHz have been obtained in both direct detection [136], [137] and heterodyne detection [138] modes.

IV. Applications of Photodetectors to Various Optical Communication Systems

The most important parameter in the choice of a photodetector is the wavelength of the optical communication system. The current state of the art in high efficiency high power lasers leads to four principal possible system wavelength regions:

1) 4880 Å—Argon, and 5300 Å—doubled YAG
2) 8500–9000 Å—GaAs
3) 1.06 μm—YAG
4) 10.6 μm—CO$_2$.

Although the type of the detector chosen for each system will depend upon many factors, the characteristics of well-suited detectors for each system will now be discussed.

A. Argon and Doubled YAG Laser Systems

Optical communication systems utilizing argon or doubled YAG lasers are likely to use photomultipliers in a direct detection mode of operation because of high current gain and low dark current. These wavelengths are very near to the peak response of S-20 or S-25 photosurfaces. Slight changes in the process used to form these photosurfaces can yield small changes in the wavelength of maximum response, thus allowing photomultipliers to be optimized for these systems.

Parameters affecting the type of photomultiplier to be employed are modulation bandwidth and optical signal power available at the receiver. Systems designed for modulation frequencies up to 100 MHz may employ conventional electrostatically focused photomultipliers with the number of secondary emission stages determined by the received signal power and desired (S/N) ratio. Systems with a bandwidth of several hundred MHz will require a photomultiplier in which both transit time dispersion in the electron multiplier and output capacitance have been minimized. The use of a high-voltage accelerating grid between each pair of dynodes, together with a coaxial anode structure such as employed in the RCA C70045 photomultiplier is one method for achieving the necessary modulation bandwidth. The use of high gain GaP dynode material reduces both the excess shot noise and required number of stages and it would be desirable for this application.

Very high data rate systems with modulation bandwidths of several GHz impose the most difficult requirements upon the photomultiplier. High interstage voltages must be used in the electron multiplier to minimize transit time dispersion due to initial velocity distributions and the spatial variation of electric fields must be minimized to achieve uniform interstage electron transit times. The magnetically focused crossed-field photomultiplier, such as the Sylvania 502, can provide a useful bandwidth of several GHz because of both extremely low transit time dispersion and a well-matched signal output circuit. The use of high-gain dynodes would again be very desirable to reduce excess shot noise.

Silicon avalanche photodiodes are also useful at these wavelengths. Their quantum efficiency is higher than in photomultipliers and the response and current gain mechanism are sufficiently fast. However, unless cooled, noise and dark current are larger thus limiting their sensitivity for direct detection of weak signals. The higher quantum efficiency of silicon diodes would make them useful in heterodyne detection systems where the preservation of optical frequency and phase information and rejection of interference may be of importance.

B. GaAs Systems

Optical communication systems utilizing the GaAs (8400–9000 Å) laser operate at wavelengths where both photomultipliers and solid-state photodiodes may be utilized. Small light-weight systems designed for short range communication may employ silicon photodiodes, while systems requiring maximum sensitivity will utilize photomultipliers or silicon avalanche photodiodes.

Photomultipliers with an S-25 or one of the new photosurfaces such as GaAs:Cs_2O offer reasonable sensitivity at these wavelengths. Wide bandwidths may be achieved by incorporating these cathodes into fast photomultiplier structures. For silicon photodiodes and avalanche photodiodes, a tradeoff exists between quantum efficiency and bandwidth as mentioned in Section III-B. However, quantum efficiencies of 90 percent have been achieved for speeds of response of 10 ns [65]. Extensive development activity in many laboratories on both GaAs:Cs_2O and other type III-V photoemitters and on silicon avalanche photodiodes is expected to result in practical photomultipliers and avalanche photodiodes with high sensitivity in this wavelength region.

C. YAG Systems

Photomultipliers are presently not very useful for the demodulation of 1.06-μm Nd:YAG laser radiation. This is because even the best developmental infrared photocathodes have quantum efficiencies of only 0.8 percent at 1.06 μm [43]. Silicon photodiodes which combine quantum efficiencies and response times of 75 percent and 25 ns or 45 percent and 10 ns are available [65]. Photomultipliers and Si avalanche photodiodes have recently been compared theoretically as detectors for laser pulses of 10^{-7} to 10^{-9} duration at 1.06 μm [139] and it has been found that Si avalanche diodes can detect one to two orders of magnitude lower pulse energies than the presently available photomultipliers. However, avalanche photodiodes with optimized properties for 1.06 μm are not yet available. Detection systems for 1.06 μm with signal bandwidths in excess of 100 MHz would thus use germanium avalanche photodiodes which, in order to reduce dark current, might be cooled to about 250°K.

D. CO_2 Laser Systems

Both photoconductive and photovoltaic detectors are useful for the detection of 10.6-μm radiation. Their characteristics were given in Tables II and IV. Direct detection is useful in 10.6-μm systems where simplicity but not high sensitivity are desired. Heterodyne detection is preferred because it combines high sensitivity with wide bandwidths.

Heterodyne detection is most practical at the longer wavelengths due to such factors as: sensitivity improvement due to reduced quantum noise (see (7)), easier alignment of signal and local oscillator, and greater diffraction-limited acceptance angle for a given aperture size. The CO_2 laser has proven capable of single-frequency stabilized operation as is desired for the LO [133].

Sensitivities in coherent detection approaching the theoretical quantum noise limit have been demonstrated in a variety of photodetectors. The results obtained are summarized in Table V. Results are listed for 10.6 μm (CO_2) and 3.39 μm (HeNe). Sensitivities within 3 to 10 dB of the quantum noise limit ($h\nu = 1.87 \times 10^{-20}$ W/Hz at 10.6 μm) have been demonstrated. Experimental sensitivity values in Table V are marked as to whether they take into account the total receiver noise (S), or only the LO-generated noise (Q). With improved heterodyne receiver design, the quantum noise limit will be more closely approached in the future. In the case of photovoltaic mixed crystal materials, particularly HgCdTe, heterodyne sensitivity is being increased as improved quantum efficiency elements become available.

As Table V shows, receiver sensitivity near 10^{-19} W/Hz simultaneously with a bandwidth exceeding 1 GHz has been obtained using highly compensated photoconductive Ge:Cu(Sb) with a time constant of 200 ps [5]. The generation–recombination noise spectrum of this mixer element measured up to 4 GHz is shown in Fig. 17, when illuminated with sufficient LO power to depress mixer resistance from 5×10^5 to 1.2×10^3 ohms. Due to the available conversion gain, heterodyne operation is obtained well beyond the 750-MHz rollover frequency. The measured variation of the heterodyne noise equivalent power ($S/N = 1$) with dc bias, and hence conversion gain, is shown in Fig. 18. The behavior shown is consistent with (7), which for a photoconductive mixer with high dark resistance and low background irradiation can be written [17]

$$P_{Smin} = \frac{S}{N} B \left[\frac{2h\nu}{\eta} + \frac{k(T_G + T_A)}{G} \right] \tag{22}$$

where the available conversion gain due to the local oscillator is given by

$$G = \frac{\eta q V}{2h\nu} \left(\frac{\tau}{T_r} \right) \frac{1}{1 + \omega^2 \tau^2}. \tag{23}$$

Note that for this photoconductive case G is independent of LO power variations since mixer conductance is directly proportional to P_{LO} (18). Expressions similar to the preceding ones but for the photodiode heterodyne case are given in [138].

For systems with bandwidths up to 100 MHz, uncompensated extrinsic photoconductors have demonstrated sensitive heterodyne performance [4], [18], [144], [145]. Photoconductive HgCdTe has given performance to 10 MHz [133]. Photovoltaic HgCdTe has emerged as a strong competitor offering in combination good frequency response, operation at moderate cryogenic temperatures, and low power dissipation [138]. As problems in mixed crystal fabrication are overcome, the use of 10.6-μm photodiode mixers will be extended to higher modulation frequencies.

Heterodyne operation requires the use of a laser local oscillator which is offset by the IF frequency, and frequency-stabilized by means of a circuit such as Fig. 19(a). The LO could be a CO_2 or possibly a PbSnTe laser [142]. In systems where significant Doppler shifts occur between transmitter and receiver, the use of the receiver configuration of Fig.

113

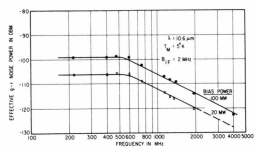

Fig. 17. Measured g–r noise spectrum of Ge:Cu(Sb) mixer [17].

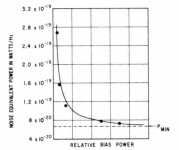

Fig. 18. Measured dependence of sensitivity of 10.6 μm heterodyne receiver on dc bias power (after [4] and [17]). (Relative bias power has linear scale.) As bias power is increased generation–recombination noise swamps out all other receiver noise contributions.

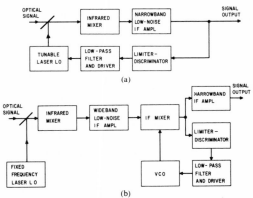

Fig. 19. Methods of frequency tracking in optical heterodyne receivers. (a) Optical tracking. (b) IF tracking.

19(b) is useful, and has been demonstrated at 10.6 μm [148] in a quantum noise limited 1.3-GHz bandwidth heterodyne receiver. Here frequency tracking is accomplished at IF and the information bandwidth established in the narrow-band second IF amplifier.

The optical heterodyne detector behaves simultaneously as a receiver and an antenna [33], [149]–[151]. This prop-erty has been used to combine infrared heterodyne elements to form spatially coherent arrays at 10.6 μm with directional properties exactly analogous to those obtained with multi-feed microwave antennas [146]. These coherent arrays are capable of providing a multiplicity of resolution elements in the object field of the optical receiver for such purposes as spatial acquisition and tracking [8], [152].

Pyroelectric detectors with response times as short as a few nanoseconds have been reported [153]. They can be operated at room temperature, but such thermal detectors have limited sensitivity in both direct detection [4], [154], [155] and heterodyne detection [156]–[158] modes. High speed detection using metal point contacts with a time con-stant of 3×10^{-14} seconds has also recently been reported [159].

V. CONCLUSIONS

It is evident that there is a large number of photodetectors and detection techniques available for use in optical com-munication systems. The correct choice of a particular de-tection technique and photodetector depends mainly on the wavelength of operation, the information bandwidth of the signal, and on the sensitivity desired for the de-modulation of weak signals. Direct detection is preferred for wavelengths at which photomultipliers and avalanche photodiodes with internal current gain, wide electrical bandwidth, and low excess noise are available.

Photomultipliers and silicon avalanche photodiodes are useful in the ultraviolet, visible, and near infrared to about 1 μm, germanium avalanche diodes extend this range to be-yond 1.5 μm. Photomultipliers provide high-gain and low-excess noise. Specially constructed photomultipliers have achieved gains of 10^4 for baseband widths up to several GHz. Avalanche photodiodes also combine gain and high speed of response as manifested by current gain–bandwidth products of up to 100 GHz. The multiplication process in avalanche diodes, however, results in higher excess noise. But despite this excess noise, current gain in both photo-multipliers and avalanche photodiodes allows realization of direct detection systems with wide bandwidths and con-siderably higher sensitivities to weak light signals as com-pared to detection systems without gain that are limited by thermal noise of the detector and output circuit. Although quite desirable, the development of avalanche photodiodes with response at longer wavelengths from materials like InAs, InSb, and even mixed crystals is still at an early stage. Photoconductors with wide bandwidths have been de-veloped for the infrared region, but the lack of sufficient in-ternal current gain limits their performance in direct de-tection systems.

Heterodyne detection in both photoconductors and photodiodes is most successfully used at infrared wave-lengths where no fast detectors with sufficient internal cur-rent gain are available. For 10.6 μm, sensitivities close to the theoretical quantum noise limit ($h\nu B$) have been realized in heterodyne detection systems with bandwidths approach-ing 1 GHz.

ACKNOWLEDGMENT

The authors wish to thank D. Bode and P. Bratt of Santa Barbara Research Center; E. Sard, B. Peyton, and F. Pace of AIL; for supplying useful data and suggestions for this paper.

REFERENCES

[1] L. K. Anderson and B. J. McMurtry, "High-speed photodetectors," *Proc. IEEE*, vol. 54, pp. 1335–1349, October 1966.

[2] W. R. Pratt, *Laser Communication Systems*. New York: Wiley, 1969.

[3] M. Ross, *Laser Receivers*. New York: Wiley, 1967.

[4] A. Beer and R. Willardson, Eds., *Semiconductors and Semimetals*, vol. 5. *Infrared Detectors*. The following chapters are of special interest: P. Kruse, "Indium antimonide photoconductive and PEM detectors,"; D. Long and J. Schmit, "Hg$_{1-x}$Cd$_x$Te and closely related alloys as intrinsic IR detector materials,"; I. Melngailis and T. Harman, "Single crystal lead–tin chalcogenides,"; R. Keyes and T. Quist, "Low-level coherent and incoherent detection in the infrared,"; M. Teich, "Coherent detection in the infrared,"; F. Arams, E. Sard, B. Peyton, and F. Pace, "Infrared heterodyne detection with gigahertz IF response,"; E. Putley, "Pyroelectric detectors,"; H. Sommers, "Microwave biased photoconductive detector."

[5] P. W. Kruse, L. D. McGlauchlin, and R. B. McQuistan, *Elements of Infrared Technology*. New York: Wiley, 1962.

[6] R. Hudson, *Infrared System Engineering*. New York: Wiley, 1969.

[7] T. M. Quist, "Copper-doped germanium detectors," *Proc. IEEE* (Letters), vol. 56, pp. 1212–1213, July 1968.

[8] G. McElroy, McAvoy, H. Richard, T. McGunigal, and G. Schiffner, "Carbon dioxide laser communication systems for near earth applications," to be published, and H. Plotkin and J. Randall, "Systems aspects of optical space communications," to be published.

[9] A. M. Johnson, "Square law behavior of photocathodes at high light intensities and high frequencies," *IEEE J. Quantum Electron.*, vol. QE-1, pp. 99–101, May 1965.

[10] B. J. McMurtry and A. E. Siegmann, "Photomixing experiments with a ruby optical maser and a traveling-wave microwave phototube," *Appl. Opt.*, vol. 1, pp. 51–53, January 1962.

[11] B. M. Oliver, "Signal-to-noise ratios in photoelectric mixing," *Proc. IRE* (Correspondence) vol. 49, pp. 1960–1961, December 1961.

[12] H. A. Hans, C. H. Townes, and B. M. Oliver, "Comments on noise in photoelectric mixing," *Proc. IRE* (Letters), vol. 50, pp. 1544–1546, June 1962.

[13] B. M. Oliver, "Thermal and quantum noise," *Proc. IEEE*, vol. 53, pp. 436–454, May 1965.

[14] S. Jacobs and P. Rabinowitz, "Optical heterodyning with a CW gaseous laser," in *Quantum Mechanics III*, P. Grivet and N. Bloembergen Eds. New York: Columbia University Press, 1964, pp. 481–487.

[15] G. Lucovsky, M. E. Lasser, and R. B. Emmons, "Coherent light detection in solid-state photodiodes," *Proc. IEEE*, vol. 51, pp. 166–172, January 1963.

[16] O. Svelto, P. D. Coleman, M. DiDomenico, and R. H. Pantell, "Photoconductive mixing in CdSe single crystals," *J. Appl. Phys.*, vol. 34. pp. 3182–3186, November 1963.

[17] F. R. Arams, E. W. Sard, B. J. Peyton, and F. P. Pace, "Infrared 10.6 micon heterodyne detection with gigahertz IF capability," *IEEE J. Quantum Electron.*, vol. QE-3, pp. 484–492, November 1967; see also *IEEE Spectrum*, vol. 5, p. 5, June–July 1968.

[18] M. C. Teich, "Infrared heterodyne detection," *Proc. IEEE*, vol. 56, pp. 37–46, January 1968.

[19] J. E. Geusic and H. E. D. Scovil, "A unidirectional traveling-wave optical maser," *Bell Syst. Tech. J.*, vol. 41, pp. 1371–1397, July 1962.

[20] H. Kogelnik and A. Yariv, "Considerations of noise and schemes for its reduction in laser amplifiers," *Proc. IEEE*, vol. 52, pp. 165–172, February 1964.

[21] F. Arams and M. Wang, "Infrared laser preamplifier system," *Proc. IEEE* (Correspondence), vol. 53, p. 329, March 1965.

[22] G. C. Holst and E. Snitzer, "Detection with a fiber laser preamplifier at 1.06 μ," *IEEE J. Quantum Electron.*, vol. QE-5, pp. 319–320, June 1969.

[23] W. H. Louisell and A. Yariv, "Quantum fluctuations and noise in parametric processes. I," *Phys. Rev.*, vol. 124, pp. 1646–1654, December 1961.

[24] J. E. Midwinter and J. Warner, "Up-conversion of near infrared to visible radiation in lithium-metaniobate," *J. Appl. Phys.*, vol. 38, pp. 519–523, February 1967.

[25] J. Warner, "Photomultiplier detection of 10.6 μm radiation using optical up-conversion in proustite," *Appl. Phys. Lett.*, vol. 12, pp. 222–224, March 1968.

[26] G. D. Boyd, T. J. Bridges, and E. G. Burkhardt, "Up-conversion of 10.6 μ radiation to the visible and second harmonic generation in HgS," *IEEE J. Quantum Electron.*, vol. QE-4, pp. 515–519, September 1968.

[27] D. A. Kleinman and G. D. Boyd, "Infrared detection by optical mixing," *J. Appl. Phys.*, vol. 40, pp. 546–566, February 1969.

[28] Y. Klinger and F. Arams, "Infrared 10.6 micron CW up-conversion in proustite using an Nd:YAG laser pump," *Proc. IEEE*, vol. 57, pp. 1797–1798, October 1969.

[29] G. Lucovsky and R. B. Emmons, "High frequency photodiodes," *Appl. Opt.*, vol. 4, pp. 697–702, June 1965.

[30] H. Melchior and W. T. Lynch, "Signal and noise response of high speed germanium avalanche photodiodes," *IEEE Trans. Electron Devices*, vol. ED-13, pp. 829–838, December 1966.

[31] M. DiDomenico, Jr., and O. Svelto, "Solid-state photodetection: A comparison between photodiodes and photoconductors," *Proc. IEEE*, vol. 52, pp. 136–144, February 1964.

[32] G. Lucovsky, R. B. Emmons, B. Harned, and J. K. Powers, "Detection of coherent light by heterodyne techniques using solid state photodiodes," in *Quantum Electronics III*. P. Grivet and N. Bloembergen, Eds. New York: Columbia University Press, 1964, pp. 1731–1738.

[33] A. E. Siegman, "The antenna properties of optical heterodyne receivers," *Proc. IEEE*, vol. 54, pp. 1350–1356, October 1966.

[34] A. J. Bahr, "The effect of polarization selectivity on optical mixing photoelectric surfaces," *Proc. IEEE* (Correspondence), vol. 53, p. 513, May 1965.

[35] O. DeLange, "Optical heterodyne detection," *IEEE Spectrum*, vol. 5, pp. 77–85, October 1968.

[36] A. H. Sommer and W. E. Spicer, *Photoelectronic Materials and Devices*. Princeton, N. J.: Van Nostrand, 1965.

[37] A. H. Sommer, *Photoemissive Materials*. New York: Wiley, 1968.

[38] J. J. Scheer and J. van Laar, "GaAs–Cs—a new type of photoemitter," *Solid-State Commun.*, vol. 3, pp. 189–193, August 1965.

[39] A. A. Turnbill and G. B. Evans, "Photoemission from GaAs-Cs-O," *Brit. J. Appl. Phys.*, ser. 2, vol. 1, pp. 155–160, February 1968.

[40] J. J. Uebbing and R. L. Bell, "Improved photoemitters using GaAs and InGaAs," *Proc. IEEE* (Letters), vol. 56, pp. 1624–1625, September 1968.

[41] R. L. Bell and J. J. Uebbing, "Photoemission from InP-Cs-O," *Appl. Phys. Lett.*, vol. 12, pp. 76–78, February 1968.

[42] B. F. Williams, "InGaAs-CsO, a low work function (less than 1.0 eV) photoemitter," *Appl. Phys. Lett.*, vol. 14, pp. 273–275, May 1969. Also, B. F. Williams *et al.*, *Appl. Phys. Letters*, to be published.

[43] H. Sonnenberg, "InAsP-Cs$_2$O, a high efficiency infrared photocathode," *Appl. Phys. Lett.*, vol. 16, pp. 245–246, March 1970.

[44] R. E. Simon, A. H. Sommer, J. J. Tietjen, and B. F. Williams, "New high-gain dynode for photomultipliers," *Appl. Phys. Lett.*, vol. 13, pp. 355–357, November 1968.

[45] R. C. Miller and N. C. Wittwer, "Secondary emission amplification at microwave frequencies," *IEEE J. Quantum Electron.* vol. QE-1, pp. 49–59, April 1965.

[46] M. B. Fisher and R. T. McKenzie, "A traveling-wave photomultiplier," *IEEE J. Quantum Electron.*, vol. QE-2, pp. 322–327, August 1966.

[47] W. Shockley and J. R. Pierce, "A theory of noise for electron multipliers," *Proc. IRE*, vol. 26, pp. 321–332, March 1938.

[48] R. M. Matheson, "Recent photomultiplier developments at RCA," *IEEE Trans. Nucl. Sci.*, vol. NS-11, pp. 64–71, June 1964.

[49] J. R. Kerr, private communication.

[50] D. E. Persyk, "New photomultiplier detectors for laser applications," *Laser J.*, pp. 21–23, November–December 1969.

[51] W. W. Gaertner, "Depletion-layer photoeffects in semiconductors," *Phys. Rev.*, vol. 116, pp. 84–87, October 1959.

[52] D. E. Sawyer and R. H. Rediker, "Narrow base germanium photodiodes," *Proc. IRE*, vol. 46, pp. 1122–1130, June 1958.

[53] R. P. Riesz, "High-speed semiconductor photodiodes," *Rev. Sci. Instr.*, vol. 33, pp. 994–998, September 1962.

[54] L. K. Anderson, "Photodiode detection," *Proc. 1963 Symp. on Optical Masers,* Polytechnic Press, pp. 549–563.

[55] W. M. Sharpless, "Evaluation of a specially designed GaAs Schottky-barrier photodiode using 6328 Å radiation modulated at 4 GHz," *Appl. Opt.,* vol. 9, pp. 489–494, February 1970.

[56] M. V. Schneider, "Schottky barrier photodiode with antireflection coating," *Bell Syst. Tech. J.,* vol. 45, pp. 1611–1638, November 1966.

[57] United Detector Technology (manufacturers data).

[58] H. Melchior, M. P. Lepselter, and S. M. Sze, "Metal-semiconductor avalanche photodiode," presented at IEEE Solid-State Device Res. Conf., Boulder, Colo., June 1968.

[59] R. D. Baertsch and J. R. Richardson, "An Ag-GaAs Schottky barrier ultraviolet detector," *J. Appl. Phys.,* vol. 40, pp. 229–235, January 1969.

[60] J. R. Richardson and R. D. Baertsch, "Zinc sulfide Schottky barrier ultraviolet detectors," *Solid-State Electron.,* vol. 12, pp. 393–397, May 1969.

[61] L. K. Anderson, P. G. McMullin, L. A. D'Asaro, and A. Goetzberger, "Microwave photodiodes exhibiting microplasma-free carrier multiplication," *Appl. Phys. Lett.,* vol. 6, pp. 62–64, February 1965.

[62] E. Labate, private communication.

[63] R. J. McIntyre and H. C. Sprigings, "Multielement silicon photodiodes for detection at 1.06 microns," presented at the Conf. on Prep. Control of Electron. Mat., Boston, Mass., August 1966.

[64] H. C. Sprigings and R. J. McIntyre, "Improved multielement silicon photodiodes for detection of 1.06 μm," presented at the Int. Electron Devices Meet., Washington, D. C., October 1968.

[65] RCA, Montreal, Canada (manufacturers data).

[66] D. P. Mathur, R. J. McIntyre, and P. P. Webb, "A new germanium photodiode with extended long-wavelength response," presented at the Int. Electron Device Meet., Washington, D. C., October 1968.

[67] B. R. Pagel and R. L. Petritz, "Noise in InSb photodiodes," *J. Appl. Phys.,* vol. 32, pp. 1901–1904, October 1961.

[68] H. Protschka and D. C. Shang, "InSb photodiodes with high reverse breakdown voltage," presented at the Int. Electron Device Meet., Washington, D. C., October 1967.

[69] A. Kohn and J. Schlickman, "1–2 micron (HgCd)Te photodetectors," *IEEE Trans. Electron Devices,* vol. ED-16, pp. 885–890, October 1960.

[70] M. Rodot, C. Vérié, Y. Marfaing, J. Besson, and H. Lebloch, "Semiconductor lasers and fast detectors in the infrared (3 to 15 microns)," *IEEE J. Quantum Electron.,* vol. QE-2, pp. 586–593, September 1966.

[71] C. Vérié and J. Ayas, "Cd$_x$Hg$_{1-x}$Te infrared photovoltaic detectors," *Appl. Phys. Lett.,* vol. 10, pp. 241–243, May 1967.

[72] I. Melngailis and A. R. Calawa, "Photovoltaic effect in Pb$_x$Sn$_{1-x}$Te diodes," *Appl. Phys. Lett.,* vol. 9, pp. 304–306, October 1966.

[73] J. F. Butler, A. R. Calawa, I. Melngailis, J. O. Dimmock, "Laser action and photovoltaic effect in Pb$_{1-x}$Sn$_x$Se diodes," *Bull. Am. Phys. Soc.,* vol. 12, p. 384, March 1967.

[74] I. Melngailis, "Laser action and photodetection in lead–tin chalcogenides," *J. de Physique,* vol. 29, colloque C4, supplement au no. 11–12, pp. C4–84–C4–94, November–December 1968.

[75] K. M. Johnson, "High-speed photodiode signal enhancement at avalanche breakdown voltage," *IEEE Trans. Electron Devices,* vol. ED-12, pp. 55–63, February 1965.

[76] A. S. Tager, "Current fluctuations in a semiconductor (dielectric) under conditions of impact ionization and avalanche breakdown," *Sov. Phys.—Solid State,* vol. 6, pp. 1919–1925, February 1965.

[77] R. J. McIntyre, "Avalanche multiplication noise in semiconductor junctions," *IEEE Trans. Electron Devices,* vol. ED-13, pp. 164–175, January 1966.

[78] H. Melchior and L. K. Anderson, "Noise in high speed avalanche photodiodes," presented at the Int. Electron Device Meet., Washington, D. C., October 1965.

[79] R. D. Baertsch, "Noise and ionization rate measurements in silicon photodiodes," *IEEE Trans. Electron Devices* (Correspondence), vol. ED-13, p.987, December 1966.

[80] W. T. Read, "A proposed high-frequency negative resistance diode," *Bell Syst. Tech. J.,* vol. 37, pp. 401–446, March 1958.

[81] R. B. Emmons and G. Lucovsky, "The frequency response of avalanching photodiodes," *IEEE Trans. Electron Devices,* vol. ED-13, pp. 297–305, March 1966.

[82] R. B. Emmons, "Avalanche-photodiode frequency response," *J. Appl. Phys.,* vol. 38, pp.3705–3714, August 1967.

[83] J. J. Chang, "Frequency response of PIN avalanche photodiodes," *IEEE Trans. Electron Devices,* vol. ED-14, pp. 139–145, March 1967.

[84] C. A. Lee, R. A. Logan, J. J. Kleimack, and W. Wiegmann, "Ionization rates of holes and electrons in silicon," *Phys. Rev.,* vol. 134, pp. A761–A773, May 1964.

[85] S. Donati and V. Svelto, "The statistical behavior of the avalanche photodiode," *Alta Freq.,* vol. 37, pp. 476–486, May 1968.

[86] R. D. Baertsch, "Noise and multiplication measurements in InSb avalanche photodiodes," *J. Appl. Phys.* vol. 38, pp. 4267–4274, October 1967.

[87] W. T. Lindley, R. J. Phelan, C. M. Wolfe, and A. G. Foyt, "GaAs Schottky barrier avalanche photodiodes," *Appl. Phys. Lett.,* vol. 14, pp. 197–199, March 1969.

[88] G. Lucovsky and R. B. Emmons, "Avalanche multiplication in InAs photodiodes," *Proc. IEEE* (Correspondence), vol. 53, p. 180, February 1965.

[89] R. J. Locker and G. C. Huth, "A new ionizing radiation detection concept which employs semiconductor avalanche amplification and the tunnel diode element," *Appl. Phys. Lett.,* vol. 9, pp. 227–230, September 1966.

[90] H. Ruegg, "An optimized avalanche photodiode," *IEEE Trans. Electron Devices,* vol. ED-14, May 1967.

[91] J. R. Biard and W. N. Shaunfield, "A model of the avalanche photodiode," *IEEE Trans. Electron Devices,* vol. ED-14, pp. 233–238, May 1967.

[92] W. N. Shaunfield and J. R. Biard, "A high-speed germanium photodetector for 1.54 microns," presented at the Solid-State Sensors Symp., Minneapolis, Minn., September 1968.

[93] W. T. Lynch, "Elimination of the guard ring in uniform avalanche photodiodes," *IEEE Trans. Electron Devices,* vol. ED-15, pp. 735–741, October 1968.

[94] F. D. Shepherd, Jr., A. C. Yang, and R. W. Taylor, "A 1 to 2 μm silicon avalanche photodiode," *Proc. IEEE* (Letters), vol. 58, pp. 1160–1162, July 1970.

[95] A. Rose, *Concepts in Photoconductivity and Applied Problems.* New York : Interscience, 1963.

[96] A. Rose, "Maximum performance of photoconductors," *Helv. Phys. Acta,* vol. 30, pp. 242–244, August 1957.

[97] R. W. Reddington, "Gain-bandwidth product of photoconductors," *Phys. Rev.,* vol. 115, pp. 894–896, August, 1959.

[98] P. D. Coleman, R. C. Eden, and J. N. Weaver, "Mixing and detection of coherent light in a bulk photoconductor," *IEEE Trans. Electron Devices,* vol. ED-11, pp. 488–497, November 1964.

[99] R. E. Burgess, "The statistics of charge carrier fluctuations in semiconductors," *Proc. Phys. Soc. London,* (Gen.), vol. 69, pp. 1020–1027, October 1956.

[100] K. M. Van Vliet, "Noise in semiconductors and photoconductors," *Proc. IRE,* vol. 46, pp. 1004–1018, June 1968.

[101] H. Sommer and E. K. Gatchell, "Demodulation of low-level broadband optical signals with semiconductors," *Proc. IEEE,* vol. 54, pp. 1553–1568, November 1966.

[102] C. Buczek and G. Picus, "Far infrared laser receiver investigation," Hughes Res. Labs., Malibu, Calif., A. F. Rept. AFAL-TR-68-102, May 1968.

[103] A. Yariv, C. Buczek, and G. Picus, "Recombination studies of hot holes in mercury-doped germanium," *Proc. Int. Conf. Phys. of Semicond.* (Moscow), vol. 9–10, p. 500, July 1968.

[104] R. L. Williams, "Response characteristics of extrinsic photoconductors," *J. Appl. Phys.,* vol. 40, pp. 184–192, January 1969.

[105] M. DiDomenico and L. K. Anderson, "Microwave signal-to-noise performance of CdSe bulk photoconductive detectors," *Proc. IEEE,* vol. 52, pp. 815–822, July 1964.

[106] H. Sommer and W. Teutsch, "Demodulation of low-level broadband optical signals with semiconductors; pt. II analysis of photoconductive detector," *Proc. IEEE,* vol. 52, pp. 144–153, February 1964.

[107] C. Sun and T. E. Walsh, "A solid-state microwave-biased photoconductive detector for 10.6 μm," *IEEE J. Quantum Electron.,* vol. QE-5, pp. 320–321, June 1969.

[108] H. Levinstein, "Extrinsic detectors," *Appl. Optics,* vol. 4, pp. 639–647, June 1965.

[109] E. Putley, "Far infrared photoconductivity," *Phys. Status Solidi,* vol. 6, pp. 571–614, September 1964.

[110] Johnson and H. Levinstein, "Infrared properties of gold in germanium," *Phys. Rev.,* vol. 117, pp. 1191–1203, March 1960.

PROCEEDINGS OF THE IEEE, OCTOBER 1970

[111] W. Beyer and H. Levinstein, "Cooled photoconductive infrared detectors," *J. Opt. Soc. Am.*, vol. 49, pp. 686–692, July 1959.

[112] G. S. Picus, "Carrier generation and recombination processes in copper-doped germanium photoconductors," *J. Phys. Chem. Solids*, vol. 23, pp. 1753–1761, 1962.

[113] D. Bode and H. Graham, "Comparison of performance of copper-doped germanium and mercury-doped germanium detectors," *Infrared Phys.*, vol. 3, pp. 129–137, September 1963.

[114] S. Borrello and H. Levinstein, "Preparation and Properties of mercury doped germanium," *J. Appl. Phys.*, vol. 33, pp. 2947–2950, October 1962.

[115] Y. Darviot, A. Sorrentino, B. Joly, and B. Pajot, "Metallurgy and physical properties of mercury-doped germanium related to the performance of the infrared detector," *Infrared Phys.*, vol. 7, pp. 1–10, March 1967.

[116] P. Bratt, W. Engeler, H. Levinstein, A. Mac Rae, and J. Pehek, "A status report on infrared detectors," *Infrared Phys.*, vol. 1, pp. 27–38, March 1961.

[117] R. Soref, "Extrinsic IR photoconductivity of Si doped with B, Al, Ga, P, As or Sb," *J. Appl. Phys.*, vol. 38, pp. 5201–5209, December 1967; see also, "Coherent homodyne detection at 10.6 μm with an extrinsic photoconductor," *Electron. Lett.*, vol. 2, pp. 410–411, November 1966.

[118] F. D. Morten and R. E. J. King, "Photoconductive indium antimonide detectors," *Appl. Opt.*, vol. 4, pp.659–663, June 1965.

[119] Manufacturers' Data (Santa Barbara Research Center, Texas Instruments, Honeywell, Philco-Ford, Mullard Raytheon, Société Anonyme de Télécommunications).

[120] D. Bode and P. Bratt, private communication.

[121] T. Bridges, T. Chang, and P. Cheo, "Pulse response of electrooptic modulators and photoconductive detectors at 10.6 μm," *Appl. Phys. Lett.*, vol. 12, pp. 297–300, May 1968.

[122] J. T. Yardley and C. B. Moore, "Response times of Ge : Cu infrared detectors," *Appl. Phys. Lett.*, vol. 7, pp. 311–312, December 1965.

[123] J. Schlickman, "Mercury cadmium telluride intrinsic photodetectors," *Proc. Electrooptic. Syst. Conf., Industr. Sci. Conf. Mgmt, Inc.*, (Chicago, Ill.), pp. 289–309, September 1969.

[124] P. W. Kruse, "Photon effects in Hg$_{1-x}$Cd$_x$Te," *Appl. Opt.*, vol. 4, pp. 687–692, June 1965.

[125] B. Bartlett, D. Charlton, W. Dunn, P. Ellen, M. Jenner, and M. Jervis, "Background limited photoconductive HgCdTe detectors for use in the 8–14 micron atmospheric window," *Infrared Phys.*, vol. 9, pp. 35–36, 1969.

[126] I. Melngailis and T. Harman, "Photoconductivity in single-crystal Pb$_{1-x}$Sn$_x$Te," *Appl. Phys. Lett.*, vol. 13, pp. 180–183, September 1968.

[127] W. Kaiser, R. Collins, and H. Fan, "Infrared absorption in p-type germanium," *Phys. Rev.*, vol. 91, pp. 1380–1381, September 1953.

[128] P. Bratt, "Photoconductivity in impurity-activated germanium," Ph.D. dissertation, University of Syracuse, Syracuse, N. Y., January 1965.

[129] T. Vogl, J. Hansen, and M. Garbuny, "Photoconductive time constants and related characteristics of p-type gold-doped germanium," *J. Opt. Soc. Am.*, vol. 51, pp. 70–75, January 1961.

[130] J. Schmit and E. Stelzer, "Temperature and alloy compositional dependences of energy gap of Hg$_{1-x}$Cd$_x$Te," *J. Appl. Phys.*, vol. 40, pp. 4865–4869, November 1969.

[131] W. Saur, "Long wavelength mercury-cadmium telluride photoconductive infrared detectors," *Infrared Phys.*, vol. 8, pp. 255–258, 1968.

[132] A. Kohn and J. Schlickman, "1–2 micron (Hg, Cd)Te photodetectors," *IEEE Trans. Electron Devices*, vol. ED-16, pp. 885–890, October 1969.

[133] H. Mocker, "A 10.6 μ optical heterodyne communication system," *Appl. Opt.*, vol. 8, pp. 677–684, March 1969.

[134] D. Breitzer, E. Sard, B. Peyton, and J. McElroy, "G–R noise for auger band-to-band processes," to be published; see also: J. H. McElroy, S. C. Cohen, and H. E. Walker, "First and second summary design report ATS-F laser communication experiment infrared mixer and radiation cooler subsystem," Goddard Space Flight Center, Greenbelt, Md., Rep. X-524-69-227, 1969.

[135] D. Long, "On generation-recombination noise in infrared detection materials," *Infrared Phys.*, vol. 7, pp. 169–170, September 1967.

[136] A. Sorrentino, "IR detectors, developments at SAT," presented at Infrared Symp. Royal Radar Establishment, Malvern, England, April 1969.

[137] E. D. Hinkley and C. Fried, private communication.

[138] B. Peyton, E. Sard, R. Lange, and F. Arams, "Infrared 10.6-μm photodiode heterodyne detection," this issue, pp. 1769–1770.

[139] R. J. McIntyre, "Comparison of photomultipliers and avalanche photodiodes for laser applications," *IEEE Trans. Electron Devices*, vol. ED-17, pp. 347–352, April 1970.

[140] J. Hanlon and S. Jacobs, "Narrow-band optical heterodyne detection," *Appl. Opt.*, vol. 6, pp. 577–578, March 1967.

[141] F. E. Goodwin and M. E. Pedinoff, "Application of CCl$_4$ and CCl$_2$:CCl$_2$ ultrasonic modulators to infrared optical heterodyne experiments," *Appl. Phys. Lett.*, vol. 8, pp. 60–61, February 1966.

[142] E. N. Fuls, "Optical frequency mixing in photoconductive InSb," *Appl. Phys. Lett.*, vol. 4, pp. 7–8, January 1964.

[143] R. A. Wood, "Gold-doped germanium as an infrared high-frequency detector," *J. Appl. Phys.*, vol. 36, pp. 1490–1491, April 1965.

[144] C. Buczek and G. Picus, "Heterodyne performance of mercury-doped germanium," *Appl. Phys. Lett.*, vol. 11, pp. 125–126, August 1967.

[145] M. Teich, R. Keyes, and R. Kingston, "Optimum heterodyne detection at 10.6 microns in photoconductive Ge : Cu," *Appl. Phys. Lett.*, vol. 9, pp. 357–360, November 1966.

[146] F. Pace, R. Lange, B. Peyton, and F. Arams, "Infrared heterodyne receiver array with wide IF bandwidth for CO$_2$ laser systems," to be published.

[147] E. Hinkley, T. Harman, and C. Freed, "Optical heterodyne detection at 10.6 microns of the beat frequency between a tunable PbSnTe diode laser and a CO$_2$ gas laser," *Appl. Phys. Lett.*, vol. 13, pp. 49–51, July 1968.

[148] W. Chiou, T. Flattau, B. Peyton, and R. Lange, "High-sensitivity infrared heterodyne receivers with gigahertz IF bandwidth Pt. III: Packaged 10 μm heterodyne 1.3 GHz bandwidth tracking receiver," *IEEE Spectrum*, vol. 7, p. 5, January 1970 (Advertisement for AIL, Div. Cutler-Hammer, Deer Park, N. Y.).

[149] A. E. Siegman, "A maximum-signal theorem for the spatially coherent detection of scattered radiation," *IEEE Trans. Antennas Propagat.*, vol. AP-15, pp. 192–194, January 1967.

[150] V. J. Corcoran, "Directional characteristics in optical heterodyne detection process," *J. Appl. Phys.*, vol. 36, pp. 1819–1825, June 1965.

[151] W. Read and D. Fried, "Optical heterodyning and noncritical angular alignment," *Proc. IEEE* (Letters), vol. 51, p. 787, December 1963.

[152] H. A. Bostick, "A carbon dioxide laser radar system," *IEEE J. Quantum Electron.*, vol. QE-3, p. 232, June 1967.

[153] A. Glass, "Investigation of electrical properties of Sr$_{1-x}$Ba$_x$Nb$_2$O$_6$ with special reference to pyroelectric detection," *J. Appl. Phys.*, vol. 40, pp. 4699–4713, November 1969.

[154] M. Kimmit, J. Ludlow, and E. Putley, "Use of pyroelectric detector to measure Q-switched CO$_2$ laser pulses," *Proc. IEEE* (Letters), vol. 56, p. 1250, 1968.

[155] A. Hadni, R. Thomas, and J. Perrin, "Response of triglycine sulfate pyroelectric detector to high frequencies (300 KHz)," *J. Appl. Phys.*, vol. 40, pp. 2740–2745, 1969.

[156] E. Leiba, "Heterodynage optique avec un detecteur pyroelectrique," *C. R. Acad. Sci.* (Paris), vol. 268, ser. B, pp. 31–33, January 1969.

[157] R. L. Abrams and A. M. Glass, "Photomixing at 10.6 μ with strontium barium niobate pyroelectric detectors," *Appl. Phys. Lett.*, vol. 15, pp. 251–253, October 1969.

[158] S. Eng and R. Gudmundsen, "Theory of optical heterodyne detection using pyroelectric effect," *Appl. Opt.*, vol. 9, pp. 161–166, January 1970.

[159] V. Danen, D. Sokoloff, A. Sanchez, and A. Javan, "Extension of laser harmonic-frequency mixing techniques into the 9μ region with an infrared metal-metal point-contract diode," *Appl. Phys. Lett.*, vol. 15, pp. 398–401, December 1969.

9

Reprinted from *J. Opt. Soc. Amer.*, **50**(11), 1058–1059 (1960)

Proposal of the Detectivity D^{**} for Detectors Limited by Radiation Noise†

R. Clark Jones

Research Laboratory, Polaroid Corporation, Cambridge, Massachusetts

(Received July 18, 1960)

A new kind of detectivity, called D^{**}, is proposed for cells that have a detectivity that is limited by radiation noise. The detectivity D^{**} takes specific account of the solid angle Ω, from which external radiation can reach the responsive element. Usually, the external radiation is produced by objects at room temperature. D^{**} is also appropriate for use with detectors that are non-Lambertian for reasons other than cooled radiation shields: for example, detectors that are immersed in a high-index medium, or detectors that are closely associated with a lens.

INTRODUCTION

DURING the last few years, an ever-increasing number of refrigerated detectors have been described that have a detectivity limited by the radiation noise of the steady incident radiation.

Such detectors usually contain refrigerated shields to reduce the amount of room-temperature radiation that strikes the responsive element. As would be expected for the radiation noise limited detector, the detectivity D^* of these cells increases as the aperture in the refrigerated shield is made smaller.

In this paper we describe a modified kind of detectivity that is denoted by D^{**} and is the measured detectivity, not only referred to a bandwidth of 1 cps and an area of 1 cm², but also referred to an (effective weighted) solid angle of π steradians.

Let D be the measured detectivity (in reciprocal watts) of a detector with the effective area A (in sq cm) when measured with the noise equivalent bandwidth Δf (in cps). Then the detectivity D^* is defined[1] by

$$D^* = (A\Delta f)^{\frac{1}{2}} D \qquad (1)$$

and is measured in the unit: cm-(cps)$^{\frac{1}{2}}$/w. D^* is pronounced "D-star."

The detectivity D^{**} (pronounced "D-double-star") is defined by

$$D^{**} = (\Omega/\pi)^{\frac{1}{2}} D^* \qquad (2)$$

or equivalently by

$$D^{**} = (A\Omega\Delta f/\pi)^{\frac{1}{2}} D, \qquad (3)$$

where Ω is the effective weighted solid angle that the responsive element sees through the aperture in the radiation shield. D^{**} is measured in the same unit as D^*.

If the detector has circular symmetry and the solid angle can be represented as a cone with the half-angle θ_0, the relation between Ω and θ_0 is

$$\Omega = \pi \sin^2\theta_0. \qquad (4)$$

This is as far as the reader need go if he is interested in the broad outline of the definition of D^{**}. In the next section we consider the precise definition of the area A and the solid angle Ω.

† Supported by the Geophysics Research Directorate.
[1] R. Clark Jones, Proc. IRIS 2, No. 1, 9–12 (1957).

DEFINITIONS OF A AND Ω

It has always been an awkward part of the detector art to provide a suitable and general definition of the effective detector area A. It is just as awkward to provide a suitable and general definition of the solid angle Ω. The following discussion is restricted to detectors with a flat surface.

Let $R(x,y)$ denote the responsivity in volts per watt of the detector when the radiation falls normally on a very small area centered at the point x, y on the surface of the detector. Then, provided x, y is a rectangular Cartesian coordinate system, the effective area A can be defined by

$$A = \int\int R(x,y)dxdy/R_{max}, \qquad (5)$$

where R_{max} is the maximum value of $R(x,y)$.

This definition of A does not always give a reasonable answer, but it is the best general definition of which I know.

We let $R(x,y,\theta,\varphi)$ denote the responsivity in volts per watt when the radiation falls on a very small area centered at the point x, y and arrives from the direction specified by the angles θ and φ, where θ is the polar angle and φ the azimuthal angle. Then the effective weighted solid angle Ω may be defined by

$$\Omega = \int\int dxdy \int_0^{\pi/2} \cos\theta \, \sin\theta d\theta$$
$$\times \int_0^{2\pi} R(x,y,\theta,\varphi)d\varphi/[AR_{max}(0,0)], \qquad (6)$$

where $R_{max}(0,0)$ is the maximum value of $R(x,y,0,0)$.

It is scarcely contemplated that anyone would ever go to the trouble of measuring $R(x,y,\theta,\varphi)$ and then performing the integration indicated. Nevertheless, the above quadruple integral does indicate the nature of the solid angle that is relevant to the definition of D^{**}.

The solid angle Ω is an *effective* solid angle; this is indicated by the fact that it is an average over the total solid angle and over the entire plane of the detector. The solid angle Ω is a *weighted* solid angle. This means that the solid angle in a given direction is weighted by the *projected* area of the detector. This weighting is taken care of by the presence of the factor $\cos\theta$ in the above integral.

To illustrate the use of the above equation, let us consider a hypothetical detector that has uniform responsivity for radiation with θ less than θ_0, and zero responsivity otherwise. Then

$$\Omega = 2\pi \int_0^{\theta_0} \cos\theta \, \sin\theta d\theta \qquad (7)$$

or

$$\Omega = \pi \sin^2\theta_0, \qquad (8)$$

and thus Eq. (4) has been confirmed. For this detector, one has

$$D^{**} = D^* \sin\theta_0. \qquad (9)$$

Most actual detectors do not have circular symmetry; even if the aperture in the shield is circular, the responsive element itself is usually not. But to the extent that the symmetry is an approximation to circular, a reasonable estimate of Ω may be obtained by use of the above equation.

The ordinary, unweighted solid angle ω of a cone of half-angle θ_0 is given by

$$\omega = 2\pi \int_0^{\theta_0} \sin\theta d\theta \qquad (10)$$

or

$$\omega = 2\pi(1 - \cos\theta_0). \qquad (11)$$

We can define a Lambertian detector as one whose responsivity is independent of the angles θ and φ. Thus, a Lambertian detector corresponds to $\theta_0 = 90°$. For this special case, one has

$$\Omega = \pi \qquad (12)$$

$$\omega = 2\pi. \qquad (13)$$

The difference corresponds to the fact that Ω is weighted by the factor $\cos\theta$, whereas ω is not so weighted. One has also

$$D^{**} = D^*. \qquad (14)$$

for a Lambertian detector.

DISCUSSION

There is nothing profound in the definition of D^{**}. The writer offers the definition and notation in the hope that the proposal will promote uniformity in the concepts and notation in the field of infrared detectors.

It is interesting that as early as 1947, a figure of merit quite similar to D^{**}, was proposed by Golay.[2] The figure of merit m, proposed by Golay, involves the factors $A^{\frac{1}{2}}$ and $\sin\theta_0$.

[2] Marcel J. E. Golay, Rev. Sci. Instr. **18**, 347–356 (1947).

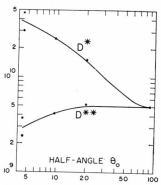

FIG. 1. The upper curve shows the detectivity D^*, and the lower curve shows the detectivity D^{**}, both in cm-$(cps)^{\frac{1}{2}}$/w. The abscissa is the half-angle θ_0 (in degrees) of the cone of room temperature radiation that strikes the cooled detector. Both curves are for the same detector. The two curves represent different ways of presenting the same information. The relation between D^{**} and D^* in the figure is $D^{**} = D^* \sin\theta_0$.

The thinking processes of some persons require that every concept shall have a name. For these persons, I suggest the following name for D^{**}: the detectivity in the reference condition F. The detectivity D^* has already been named the detectivity in the reference condition E.[3]

To illustrate the use of the detectivity D^{**}, there is shown in Fig. 1 a plot of both D^* and D^{**} for a detector in which the half-angle θ_0 of room-temperature radiation was varied from 4.5° to 90° with the same cooled detector.[4] The figure shows clearly that whereas D^* increases markedly as the half-angle θ_0 is reduced, the detectivity D^{**} is substantially independent of the angle θ_0 for large angles and drops slightly at the smaller angles.

Figure 1 illustrates the fact that with a radiation limited detector the detectivity D^{**} is an invariant method of rating the performance.

D^{**} is recommended for cooled detectors that have a detectivity limited by radiation noise. The fact that D^{**} drops at the smaller angles indicates that the internal noise of the detector begins to be important as the amount of external radiation drops.

D^{**} is also suitable for use with detectors, cooled or not, that are non-Lambertian for any reason. An example of a non-Lambertian room-temperature detector is a thermistor bolometer in optical contact with a germanium or arsenic trisulfide lens.

[3] R. Clark Jones, Proc. Inst. Rad. Engr. **47**, 1495–1502 (1959).
[4] R. M. Talley, T. H. Johnson, and D. F. Bode, Proc. IRIS 4, No. 4, 256–259 (1959).

III
Photon Detectors

Editors' Comments on Paper 10

10 Griffin, Hubbard, and Wald: *The Sensitivity of the Human Eye to Infra-Red Radiation*

The reader may be surprised to find a paper on the human eye included in a book on infrared detectors. Its selection was deliberate, because very few people seem to know that the response of the human eye extends to beyond $1\,\mu$m. Paper 10 is interesting not only for the measurements of the spectral response of the human eye that it contains but also for its very detailed description of how such measurements are made. The measurements reported extend to a wavelength of $1.05\,\mu$m, where the eye's response is 3×10^{-13} times its peak value. Griffin, Hubbard, and Wald calculate that at wavelengths beyond $1.15\,\mu$m the response of the eye will be less than that of the skin, so radiation of this and longer wavelengths will be more readily felt as heat by the skin than perceived as light by the eye. More information on the performance of the human eye will be found in References 1 through 3.

George Wald is Professor of Biology at Harvard University, Cambridge, Massachusetts, and was the recipient of the Nobel Prize in Physiology and Medicine in 1967.

It is interesting to note that snakes of two large families, the pit vipers and the boas, are equipped with sensitive infrared receptors with which they sense fractional-degree differences in their surrounding environment [4, 5]. There also seems to be evidence that moths communicate by means of infrared [6].

References

1. Walraven, P. L., and H. J. Leebeek, "Foveal Sensitivity of the Human Eye in the Near Infrared," *J. Opt. Soc. Amer.,* **53,** 765 (1963).
2. Leavitt, G. A., "Spectral Response of the Eye," *Opt. Spectra,* **5,** 28 (Sept. 1971).
3. Rose, A., *Vision–Human and Electronic,* Plenum Press, New York (1973).
4. Harris, J. F., and R. I. Gamow, "Snake Infrared Receptors: Thermal or Photochemical Mechanism?" *Science,* **172,** 1252 (1971).
5. Gamow, R. I., and J. F. Harris, "The Infrared Receptors of Snakes," *Sci. Amer.,* **228,** 94 (May, 1973).
6. Hsiao, H. S., and C. Susskind, "Infrared and Microwave Communication by Moths," *IEEE Spectrum,* **7,** 69 (Mar. 1970).

10

Reprinted from *J. Opt. Soc. Amer.*, **37**(7), 546–554 (1947)

The Sensitivity of the Human Eye to Infra-Red Radiation

Donald R. Griffin, Ruth Hubbard, and George Wald*

The Biological Laboratories of Harvard University, Cambridge, Massachusetts

(Received March 12, 1947)

The spectral sensitivity of human vision has been measured in the near infra-red, in two areas of the dark adapted eye: the central fovea (cones) to 1000 mμ, and a peripheral area, in which the responses are primarily caused by rods, to 1050 mμ. In both cases the estimates of spectral sensitivity are based upon determinations of the visual thresholds for radiation passing through a series of infra-red filters. By successive approximation, sensitivity functions were chosen which were consistent with the observed thresholds.

The spectral sensitivity of the fovea determined in this way is consistent with previous measurements of Goodeve on the unfixated eye. At wave-lengths beyond 800 mμ the periphery becomes appreciably more sensitive than the fovea. This tendency increases at longer wave-lengths, so

that at the longest wave-lengths studied, the radiation appeared colorless at the threshold and stimulated only rods.

Lengthening the exposure time increases the sensitivity of the peripheral retina relative to the fovea. Our measurements involved exposures of 1 second and fields subtending a visual angle of 1 degree. With shorter exposures or smaller fields the fovea is favored, so that under such circumstances the fovea may become more sensitive than the periphery well into the infra-red.

At 1050 mμ the sensitivity of the peripheral retina is only 3×10^{-13} times its maximum value at 505 mμ. A computation shows that by 1150 or 1200 mμ radiation should be more readily felt as heat by the skin than seen as light by the eye.

IT is customary to set the limits of the visible spectrum at about 400 and 750 mμ, and to consider that shorter or longer wave-lengths are invisible ultraviolet or infra-red. However, recent measurements have shown that with sufficient energy the eye can be stimulated by wave-lengths well outside of these limits. Goodeve (1936) has measured the visual threshold at wave-lengths as long as 900 mμ,[1] and has discussed the visibility of ultraviolet radiations down as low as 312.5 mμ (1934).

In this paper we shall describe new measurements in the near infra-red which extend to 1000–1050 mμ, and in which fixation of the eye was adequately controlled so that data could be obtained on the sensitivity of both foveal cones and peripheral rods. Previous work had shown that from 550 to 700 mμ the rods decline in sensitivity more rapidly than the cones and that in the far red the fovea is the more sensitive

portion of the retina (Abney and Watson, 1916; Wald, 1945b). Our measurements show that at longer wave-lengths this tendency is reversed and that beyond 800 mμ, under our conditions of observation, the rods are appreciably more sensitive than the cones. At the threshold of the whole dark-adapted eye, therefore, infra-red lights are seen only with the peripheral retina, and they appear colorless.

APPARATUS AND METHODS

Our method of obtaining sensitivity functions for the fovea and periphery in the infra-red was similar to that used by Dresler (1937) in the ordinary visible spectrum. It was based upon determinations of the visual transmissions of several filters—the ratios of the transmitted to the incident luminous flux. For this purpose we measured the visual threshold of the dark-adapted eye, using each filter in turn, and varying the intensity of the stimulus by means of a neutral optical wedge.

Such thresholds are inversely proportional to the visual transmissions of the filters. The visual transmission of each filter could also be calculated from its physical properties and an assumed sensitivity function for the eye. By successive approximation the assumed sensitivity function was adjusted until the calculated visual transmissions were brought into agreement with the

* This research was performed under contract with the Army Engineer Board. An account of it was presented at the Cleveland meeting of the Optical Society in March, 1946 (Griffin, Hubbard, and Wald, 1946).

[1] A recent paper by N. I. Pinegin (1945) has been received since our work was completed. Visual thresholds were measured at 655, 700, 800, 900, and 950 mμ, using as stimulus the exit slit of a monochromator viewed through a 9× eye-piece. The angular subtense of the test field is not stated. The results agree closely with Goodeve's foveal sensitivity function, and the color of the 950 mμ radiation was reported as "an unsaturated red" at threshold, so that it seems clear that foveal vision was involved.

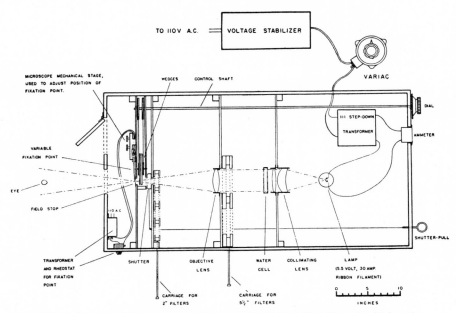

Fig. 1. Top view of adaptometer used to measure visual thresholds in the red and infra-red.

measurements. The final corrected function represents the sensitivity of foveal cones and peripheral rods from 700 mμ to 1000 and 1050 mμ, respectively (see Fig. 4 and Table V).

Our threshold measurements were made with a specially constructed adaptometer shown in schematic top-view in Fig. 1. A ribbon-filament tungsten lamp was used as source, and an image of the filament was focused on the retina. An f:1.6 lens was used to concentrate the necessary energy to enable the observer to see the filament through the very dense infra-red filters which we employed. A water cell 2 centimeters thick excluded long wave-length infra-red radiation, and provision was made for inserting the infra-red filters in a collimated portion of the light path. The light was focused after passing through water cell and infra-red filter to form an image of the filament at a field stop which could be varied in size to control the area of retina stimulated. Just before reaching this focus the light passed successively through 2-inch filters, if these were wanted, a photographic shutter to control exposure time, and a calibrated neutral optical wedge to control the intensity.

A movable fixation point similar to that described by Wald (1945a) was located very close to the plane of the field stop. For most of our measurements the image of the filament was viewed directly from a distance of 14 inches; but in some cases a magnifying ocular was used to increase the area of retina stimulated and incidentally to concentrate more radiation into the eye. A chin-and-forehead rest fixed the observer's head.

The optical wedges were made from photographic plate by graded exposure, and they permitted continuous variation of intensity over a range of about 1,000,000 times. Their construction was identical with those described by Wald (1945a) and they were calibrated similarly, by means of a Beckman spectrophotometer. The calibration was carried far enough into the infra-red to make certain that the calibration curve was accurate for all wave-lengths used in our threshold determinations.

The greatest care had to be exercised to avoid any leakage of light, either between the partitions or to the outside. The dark-adapted eye is about 10^{11} times more sensitive at 500 mμ than at 944 mμ, the wave-length at which some of our filters transmitted their greatest luminous flux. Leaks of white light which ordinarily would be negligible may in this case be many times more effective in stimulating the eye than the infra-red radiation which the apparatus is intended to deliver.

Before making threshold measurements the observer's eye was aligned with the optic axis of the instrument by means of the adjustable chin-and-forehead rest. He was allowed to become adequately dark-adapted and was then asked to fixate upon the dim fixation point while flashes of light were presented by means of the shutter. We always approached the threshold from below, i.e., by setting the intensity initially at a sub-threshold level and raising it until the observer reported seeing a flash. Final determinations were ordinarily made with intensity intervals of approximately 0.07 log unit. Several such readings were made, and a series was considered to be satisfactory when the individual measurements showed no systematic drift and varied within a range of not more than 0.2–0.3 log unit. The value assigned to the threshold was the average of such a series of readings. Our observers ranged in age from 17 to 40 and most

of them were men. All had 20/20 vision, either with their unaided eyes or with correcting lenses.

To isolate the infra-red radiation for these measurements, we employed a graded series of infra-red filters with exceptionally sharp cut-offs, manufactured by the Polaroid Corporation (Blout et al., 1946). In the present paper these filters are arbitrarily designated with the notation EBH and a number. They provided bands of radiation which were limited on the short wave-length side by the cut-off of the filter, and on the long wave-length side by the steeply falling sensitivity of the eye. They had maximum spectral visual transmissions, as finally determined, at 835, 873, 908, 928, and 944 mμ. A 685 mμ filter (Jena RG8 plus Corning 3961) was used as a standard reference red; since it transmitted a band from 660 to 750 mμ its visual transmission could be calculated from the familiar sensitivity function for the ordinary visible spectrum. The visual transmissions of the other filters were always computed relative to this red reference filter. To fill out the spectral range we also used a Corning 2600 filter, with maximum spectral visual transmission at 725 mμ. Figure 2 shows the spectral (radiant) transmissions of five of these filters, measured with a Beckman spectrophotometer.

In calculating the visual transmission of a filter, one multiplies together at convenient wave-length intervals the following three quantities:

(1) The spectral (radiant) transmission of the filter, t_λ,
(2) The relative spectral energy distribution of the source R_λ, and,
(3) The relative spectral sensitivity of the eye, K_λ.[2]

These are all pure numbers, the spectral transmission being the ratio of transmitted to incident radiation or illumination at each wave-length and the spectral energy emission and spectral sensitivity being expressed as fractions of the maximal values of those functions.

The product $R_\lambda t_\lambda K_\lambda$ expresses the relative luminous flux or relative visual effectiveness of the radiation transmitted at wave-length λ from

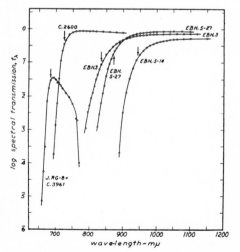

FIG. 2. Spectral radiant transmissions of five of the filters used in the measurement of visual thresholds in the red and infra-red. Arrows show the wave-lengths of maximum visual transmission for each filter.

[2] The terms, *spectral luminosity* and *spectral sensitivity*, are used interchangeably to denote the reciprocal of the relative energy at each wave-length needed to produce a constant visual effect. The term *luminosity* directs attention primarily to the radiation, *sensitivity* to the properties of the eye.

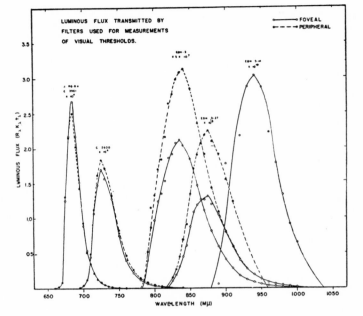

FIG. 3. Spectral visual transmissions of the five filters, the radiant transmissions of which are shown in Fig. 2. The visual transmissions are different for foveal (cone) and peripheral (rod) vision; and these differences increase toward the infrared, as the peripheral retina grows more sensitive relative to the fovea. Source: blackbody radiation at 3200°K.

a given type of source. It is this product which we call the spectral visual transmission, to distinguish it from the total visual transmission defined below.[3] It also, of course, is a pure number. At shorter wave-lengths this product is reduced by the low values of t_λ; at long wave-lengths it is diminished by the rapidly decreasing values of K_λ.

In Fig. 3 are plotted the values of $R_\lambda t_\lambda K_\lambda$ for the same five filters shown in Fig. 2. (The values for K_λ used in Fig. 3 are those obtained at the end of the investigation.) Figure 3 shows the bands of luminous flux passed by each filter, and it is clear that most of the effective luminous flux is contained within a sharply peaked band 50 to 100 mμ in width.

The total luminous flux transmitted by each filter is given by $\int R_\lambda t_\lambda K_\lambda d\lambda$, while the flux

incident on the filter, i.e., that characteristic of the unfiltered source, is given by $\int R_\lambda K_\lambda d\lambda$. The visual transmission of the filter, T_v, is the ratio of the transmitted to the incident flux, or:

$$T_v = \frac{\int R_\lambda t_\lambda K_\lambda d\lambda}{\int R_\lambda K_\lambda d\lambda}.$$

Thus one could compute T_v provided that he knew R_λ, t_λ, and the sensitivity function, K_λ. Our problem, however, was the converse one: given values of T_v from the threshold measurements, to find a sensitivity function yielding calculated values of T_v which agree with the measurements.

We first computed trial values of T_v for each filter, using the values of t_λ shown in Fig. 2, the appropriate values of R_λ from the radiant energy spectrum of a blackbody at 3200°K, and a sensitivity function derived from Goodeve's previous measurements in the infra-red (1936). These computed values of T_v were then compared with those obtained from the threshold measure-

---

[3] The total visual transmission or luminous transmission of a filter is given by the expression $T_v = \int R_\lambda t_\lambda K_\lambda d\lambda / \int R_\lambda K_\lambda d\lambda$. It would seem obvious, therefore, to define the visual transmission at each wave-length as $(R_\lambda t_\lambda K_\lambda / R_\lambda K_\lambda)$; but this would equal simply the radiant transmission, t_λ, which is, of course, the same whether measured by eye or with a thermopile (cf. Committee on Colorimetry, 1944, p. 256). Such a definition of spectral visual transmission would be useless. We have chosen instead the very useful expression for this term indicated above.

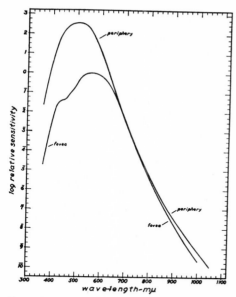

FIG. 4. Relative spectral sensitivity of the dark adapted fovea and peripheral retina. The spacing of these curves is appropriate for 1° test fields fixated within the fovea or 8° above the fovea, and exposed for 1 second. Below 750 mµ, the foveal curve is a composite function based on the original data of a number of workers; the peripheral function is from Wald (1945b). To these the data for the far red and infra-red have been joined.

ments. Corrections could then be applied to the sensitivity function, if necessary, until a function was obtained which produced agreement between observed and calculated values of T_v.

FOVEAL SENSITIVITY IN THE INFRA-RED

The measurements reported by Goodeve (1936) apparently were essentially foveal.[4] Two principal observers and an unspecified number of others were employed, and the visual threshold was measured by changing the wave-length, with the source constant. The image observed was the exit slit of a monochromator, a rectangle about forty times as high as it was wide and subtending a visual angle of 6.2°×0.15°. The eye was dark-adapted, fixation was purposely avoided, and the infra-red stimulus was reported to have various colors from red to yellow. Since Goodeve's sensitivity function appeared to have

[4] As noted above, this is true also of the work of Pinegin (1945).

been obtained largely with foveal vision, we used it tentatively to calculate visual transmissions of our filters. In so doing we joined Goodeve's function to a preferred foveal sensitivity function for the visible spectrum.[5]

We also obtained measured values of T_v for several filters by means of visual threshold determinations, using a one degree circular field fixated foveally and exposed for one second. The average thresholds of 39 to 54 observers yielded values for the visual transmissions of filters which are shown in Table I. This table also shows calculated values of T_v for the same filters.

The differences between observed and calculated values are small and show no systematic trend. We may conclude that Goodeve's function is an adequate description of foveal sensitivity in the infra-red. Since the computation of the visual transmission of the densest filter (No. EBH 5–27) required values of K_λ at wavelengths as long as 1000 mµ, it was necessary to extrapolate Goodeve's sensitivity function beyond 900 mµ. This offers no difficulties since it

TABLE I. Observed and calculated visual transmissions, foveal.

Observations with a 1-degree field, fixated within the fovea, and exposed for 1 second. Calculated visual transmissions, (T_v), are based on foveal values of relative sensitivity (K_λ) taken from Fig. 4, and on the energy distribution of a black body at 3200°K. λ_{max} is the wave-length of maximum spectral visual transmission of each filter (Fig. 3). In the designations of the filters, J = Jena, C = Corning, and EBH designates a special group of infra-red filters manufactured by the Polaroid Corporation.

Filter	JRG-8 plus C 3961	C 2600	EBH 3	EBH 5-27
λ_{max}	685	725	835	873
No. of observers	54	48	39	54
I. Calc. log T_v relative to 685 mµ filter	0.00	+0.03	−3.30	−3.90
II. Obs. log T_v relative to 685 mµ filter	0.00	−0.09	−3.23	−3.91
III. log $\dfrac{\text{calculated}}{\text{observed}}$ values (I–II)	0.00	+0.12	−0.07	+0.01

[5] This foveal sensitivity curve, shown in Fig. 4, is a composite of the original measurements of Hyde and Forsythe (1915), Coblentz and Emerson (1918–19), Gibson and Tyndall (1923), Jainski (1938), Fedorov et al. (1940), and Wald (1945b). Its derivation will be described in detail elsewhere. The large depression and minor inflections in the region 440–520 mµ mark the absorption bands of xanthophyll, the pigment of the human yellow patch (Wald, 1945b).

is linear when plotted on a frequency basis. This extrapolated portion of the foveal function yielded a calculated visual transmission in good agreement with the measurements.

The complete foveal sensitivity function for the red and infra-red is shown in Fig. 4, joined to the foveal function for the ordinary visible spectrum discussed above. The values themselves are given in Table V.

PERIPHERAL SENSITIVITY IN THE INFRA-RED

When calculating the visual transmissions of filters for peripheral vision, we assumed as a starting point that the peripheral sensitivity function would parallel Goodeve's curve which had proved satisfactory for the fovea. Table II shows the results of such computations, together with the observed values of visual transmission of the same four filters listed in Table I. We can see at once that there are considerable discrepancies between observed and calculated values and that the sensitivity function must therefore be corrected.

The discrepancies in Table II show that the periphery was more sensitive than the assumed function had predicted, and this was also evident from a direct comparison of foveal and peripheral thresholds. Such a comparison is presented in Table III, which reveals a progressive increase in sensitivity of the periphery as compared with the fovea as one goes to longer wave-lengths.

TABLE II. Observed and calculated visual transmissions, peripheral.

Observations made with a 1-degree field, fixated 8 degrees above the fovea and exposed for one second. Source at 3200°K; 2 cm of water in light path. Otherwise as in Table I.

Filter	JRG-8 plus C 3961	C 2600	EBH 3	EBH 5-27
λ_{max}	685	725	838	874
No. of observers	54	48	39	54
I. Obs. log T_v relative to 685 mμ filter	0.00	+0.02	−3.07	−3.68
II. Calc. log T_v based on Goodeve's sensitivity function	0.00	+0.01	−3.27	−4.01
III. Discrepancy (I minus II)	0.00	+0.01	+0.20	+0.33
IV. Calc. log T_v based on corrected sensitivity function	0.00	+0.09	−3.10	−3.62
V. Discrepancy (I minus IV)	0.00	−0.07	+0.03	−0.06

TABLE III. Variation, with respect to wave-length, of the ratio between foveal and peripheral thresholds in the red and infra-red.

Measurements with the adaptometer described in text, using a 1-degree field and 1-second exposures. Peripheral fields fixated 8 degrees above the fovea.

Filter	JRG-8 plus C 3961	C 2600	C 2540 plus C 2600	EBH 3	EBH 5-27	EBH 5-36
λ_{max}	685	725	762	835	873	908
No. of observers	54	48	10	39	54	4
log $\dfrac{\text{peripheral threshold}}{\text{foveal threshold}}$	+0.04	−0.08	−0.08	−0.16	−0.20	−0.34

The ratios between foveal and peripheral thresholds shown in Table III were used to correct the assumed peripheral function; the corrected sensitivity curve thus obtained yielded the calculated values of T_v listed in the fourth row of Table II. These corrected values are in satisfactory agreement with the measurements, the discrepancies being 0.07 log unit or smaller and showing no consistent trend.

We were able to extend the measurements of visual thresholds with the peripheral retina further into the infra-red than was possible with the fovea, both because of the greater sensitivity of the periphery and because larger retinal areas could be stimulated. In order to accomplish this we used a magnifying ocular between the observer's eye and the stimulus field so that we obtained a 7 degree field centered 8 degrees above the fovea. With this arrangement we could measure visual thresholds with filters having their maximum spectral visual transmissions at 908, 928, and 944 mμ; the results of these measurements are shown in Table IV.

In order to calculate the visual transmissions of these denser filters it was necessary to extend the corrected peripheral sensitivity function described above to 1050 mμ. The resulting function yielded calculated values of T_v which showed close agreement with the measurements. We feel justified, therefore, in presenting this function in Fig. 4 and in Table V as the best available description of the sensitivity of the peripheral retina from 700 to 1050 mμ. In Fig. 4 it has been joined to a peripheral rod sensitivity function for the ordinary visible spectrum (Wald, 1945b).

It should be noted in Fig. 4 that the foveal function is almost completely fused with the peripheral between 675 and 740 mμ. This may

TABLE IV. Observed and calculated visual transmissions, peripheral.

Data from 11 observers, using a 7-degree field, fixated peripherally and exposed for 1 second.

Filter	EBH 5-36	EBH 5-8	EBH 5-14
λ_{max}	908	928	944
I. Obs. log T_r relative to 873 mμ filter	−0.75	−1.18	−1.44
II. Calc. log T_r relative to 873 mμ filter, based on corrected sensitivity function	−0.76	−1.14	−1.46
Discrepancy (I minus II)	+0.01	−0.04	+0.02

appear to contradict a previous statement that in the far red the fovea is more sensitive than the periphery by about 0.2 log unit (cf. Wald 1945b). The difference between Fig. 4 and the earlier data lies in the exposure times employed. The earlier work was done with 1/25 second exposures, whereas the present measurements involved one second exposures, since we wished to obtain the lowest possible thresholds in order to push the determinations as far as possible into the infra-red.

In a special series of measurements designed to test this point we compared 1-second exposures with 1/25-second flashes using both the adaptometer described in this paper and also the instrument employing a mercury arc which was described earlier (Wald 1945a). With both instruments the foveal and peripheral thresholds at 685 or 691 mμ were within a few hundredths of a log unit of each other with 1 second exposures, while with 1/25 second exposures the fovea was more sensitive, on the average, by 0.25 log unit.

In addition to the effect of exposure time, one must also consider that with fields smaller than 1 degree the fovea is favored relative to the periphery. That is, in fields of this size the foveal threshold rises much more slowly than the peripheral threshold as the diameter of field is reduced below 1° (cf., for example, Graham and Bartlett, 1939). Consequently the greater sensitivity of the periphery in the infra-red may well disappear when either smaller test fields or shorter exposure times are used. This is probably why Goodeve and others who have studied the visibility of infra-red radiation have not reported the increasing sensitivity of the

rods relative to the cones, as one goes to longer wave-lengths.

THE SPECTRAL LIMITS OF VISION

In closing, it is interesting to inquire whether there is any end to the process of extending the spectral range of human vision by employing more and more intense sources of radiation. One could probably make threshold determinations beyond 1050 mμ under the proper conditions,

TABLE V.

Spectral sensitivity ($K\lambda$) in the red and infra-red. Values selected as described in the text, for 1° test fields, exposed for 1 second, and fixated either within the fovea or 8° above the fovea in the wholly dark-adapted eye. The foveal values beyond 750 mμ are identical with those of Goodeve (1936). All spectral sensitivities are stated relative to the maximum sensitivity of the fovea, here taken as 1 (log$K\lambda$ =0.0); on this basis the maximum sensitivity of the peripheral area, at 505 mμ, is 363 (log$K\lambda$ =2.56).

Wave-length (mμ)	1° foveal field		1° field fixated 8° above fovea	
	log$K\lambda$	$K\lambda$	log$K\lambda$	$K\lambda$
700	$\overline{3}.55$	3.55×10^{-3}	$\overline{3}.53$	3.4×10^{-3}
710	$\overline{3}.23$	1.7	$\overline{3}.23$	1.7
720	$\overline{4}.92$	8.3×10^{-4}	$\overline{4}.92$	8.3×10^{-4}
730	$\overline{4}.61$	4.1	$\overline{4}.63$	4.3
740	$\overline{4}.31$	2.0	$\overline{4}.34$	2.2
750	$\overline{4}.02$	1.04	$\overline{4}.06$	1.15
760	$\overline{5}.71$	5.1×10^{-5}	$\overline{5}.78$	6.0×10^{-5}
770	$\overline{5}.43$	2.7	$\overline{5}.51$	3.2
780	$\overline{5}.15$	1.4	$\overline{5}.25$	1.8
790	$\overline{6}.87$	7.4×10^{-6}	$\overline{6}.99$	9.8×10^{-6}
800	$\overline{6}.59$	3.9	$\overline{6}.73$	5.4
810	$\overline{6}.34$	2.2	$\overline{6}.48$	3.0
820	$\overline{6}.08$	1.2	$\overline{6}.23$	1.7
830	$\overline{7}.83$	6.8×10^{-7}	$\overline{7}.99$	9.8×10^{-7}
840	$\overline{7}.57$	3.75	$\overline{7}.76$	5.75
850	$\overline{7}.34$	2.2	$\overline{7}.53$	3.4
860	$\overline{7}.11$	1.3	$\overline{7}.32$	2.1
870	$\overline{8}.88$	7.5×10^{-8}	$\overline{7}.11$	1.3
880	$\overline{8}.65$	4.5	$\overline{8}.90$	7.9×10^{-8}
890	$\overline{8}.45$	2.8	$\overline{8}.71$	5.1
900	$\overline{8}.23$	1.7	$\overline{8}.52$	3.3
910	$\overline{8}.03$	1.07	$\overline{8}.35$	2.2
920	$\overline{9}.82$	6.6×10^{-9}	$\overline{8}.17$	1.5
930	$\overline{9}.62$	4.2	$\overline{8}.00$	1.0
940	$\overline{9}.43$	2.7	$\overline{9}.82$	6.6×10^{-9}
950	$\overline{9}.24$	1.7	$\overline{9}.65$	4.5
960	$\overline{9}.05$	1.1	$\overline{9}.48$	3.0
970	$\overline{10}.87$	7.4×10^{-10}	$\overline{9}.31$	2.0
980	$\overline{10}.69$	4.9	$\overline{9}.14$	1.4
990	$\overline{10}.51$	3.2	$\overline{10}.98$	9.55×10^{-10}
1000	$\overline{10}.34$	2.2	$\overline{10}.82$	6.6
1010			$\overline{10}.66$	4.6
1020			$\overline{10}.51$	3.2
1030			$\overline{10}.35$	2.2
1040			$\overline{10}.19$	1.55
1050			$\overline{10}.04$	1.1

129

but eventually one would reach a point beyond which the eye ceases to be a specialized sense organ which is more sensitive to radiant energy than the general body surface.

The human skin can respond to electromagnetic radiation, both visible and infra-red, by giving rise to the sensation of warmth, provided that the intensity of the radiation is sufficiently great. This is why infra-red radiation is often called radiant heat.[6] The most extensive data on the threshold of the skin to heat radiation are those of Hardy and Oppel (1937). They found that the threshold is inversely proportional to the area stimulated, over a wide range of areas. Their values for the threshold ranged from 0.019 g cal./cm²/sec. for 0.2 cm² of the skin of the forehead to 0.00015 g cal./cm²/sec. for the entire posterior body surface of a naked man.

It is probably unprofitable to attempt a direct comparison between the retina and the skin, since the density of receptor cells in the skin is low and variable, compared with the retina, and because the skin is not equipped with a refracting system to concentrate the radiation upon a discrete area. It is possible, however, to compare the eye as a whole with the general body surface, as represented by the undifferentiated skin. In this case we are interested in the energy falling on the cornea on the one hand and on the skin on the other.

Hecht, Shlaer, and Pirenne (1942) give the absolute threshold of human rod vision at 510 mμ under optimal conditions as about 3×10^{-10} ergs delivered in a 1/1000-second flash. This is equivalent to 1.85×10^{-14} g cal./cm²/sec. falling on the cornea. However, it is well known that the threshold intensity could have been reduced by a factor of about 100 if exposures of 1 second or longer were used instead of the 1/1000-second flash (Graham and Margaria, 1935). This would give a threshold for the rods at 510 mμ of

1.85×10^{-16} g cal./cm²/sec. From Fig. 4 we see that at 1050 mμ the retina is less sensitive than this by a factor of 3.3×10^{12} which indicates a threshold at this wave-length of 0.00061 g cal./cm²/sec. This is slightly higher than Hardy and Oppel's threshold for the entire posterior body surface of a man, but considerably lower than their values for small areas of skin. Figure 4 shows that in the infra-red the sensitivity of the eye falls by a factor of ten when the wave-length increases by about 60 mμ, so that by 1150 mμ one would expect the sensitivity of the eye to drop below that of even small areas of skin.

Another factor which operates in favor of the skin is the absorption spectrum of water itself. This is negligible at wave-lengths shorter than 950 mμ, but at 1150 mμ according to Hartridge and Hill (1915), all but 15.9 percent of the radiation entering the eye is absorbed by the water of the ocular media before reaching the retina, while at 1200 mμ only about 7.9 percent reaches the rods or cones. Thus it seems probable that the absorption of water, together with the rapidly falling sensitivity of the eye, sets a limit at about 1150 or 1200 mμ beyond which incident radiation is more readily felt as heat than seen as light.

BIBLIOGRAPHY

W. de W. Abney and W. Watson, "The threshold of vision for different coloured lights," Phil. Trans. Roy. Soc. London, **A216**, 91 (1915).
E. R. Blout, W. F. Amon, Jr., R. G. Shepherd, Jr., A. Thomas, C. W. West and E. H. Land, "Near infra-red transmitting filters," J. Opt. Soc. Am. **36**, 460 (1946).
W. W. Coblentz and W. B. Emerson, "Relative sensibility of the average eye to light of different colors and some practical applications to radiation problems," Bull. Bur. Stand. **14**, 167 (1918–1919).
Committee on Colorimetry, "The psychophysics of color," J. Opt. Soc. Am. **34**, 245 (1944).
A. Dresler, "Beitrag zur Photometrie farbiger Lichtquellen, insbesondere zur Frage des Verlaufs der spektralen Hellempfindlichkeit," Das Licht **7**, 81–85 (1937).
N. T. Fedorov, V. I. Fedorova, A. G. Plakhov, and L. O. Seletzkaya, "A new determination of the relative luminosity curve," J. Physics Acad. Sci. U.S.S.R. **3**, 5 (1940).
K. S. Gibson and E. P. T. Tyndall, "Visibility of radiant energy," Sci. Pap. Bur. Stand. **19**, 131 (1923).
C. F. Goodeve, "Vision in the ultra-violet," Nature **134**, 416–417 (1934).
C. F. Goodeve, "Relative luminosity in the extreme red," Proc. Roy. Soc. London **A155**, 664–683 (1936).
C. H. Graham and N. R. Bartlett, "The relation of size of stimulus and intensity in the human eye: II. Intensity thresholds for red and violet light," J. Exper. Psychol. **24**, 574 (1939).
C. H. Graham and R. Margaria, "Area and the intensity-time relation in the peripheral retina," Am. J. Physiol. **113**, 299–305 (1935).

---

[6] Thomas Young (1802) stated the basic relation between light and heat radiation very soon after the discovery of the infra-red by Herschel in 1800: ". . . the affections of heat may perhaps hereafter be rendered more intelligible to us; at present, it seems highly probable that light differs from heat only in the frequency of its undulations or vibrations; those undulations which are within certain limits, with respect to frequency, being capable of affecting the optic nerve, and constituting light; and those which are slower, and probably stronger, constituting heat only; . . . ." The present experiments extend into the curious region in which radiation may be regarded with equal justice as light or as radiant heat.

D. R. Griffin, R. Hubbard, and G. Wald, "The sensitivity of human rod and cone vision to infra-red radiation," J. Opt. Soc. Am. **36**, 360 (1946).

J. D. Hardy and T. W. Oppel, "Studies in temperature sensation. Part III. The sensitivity of the body to heat and the spatial summation of the end organ responses," J. Clin. Investigation **16**, 533–540 (1937).

H. Hartridge and A. V. Hill, "Transmission of infra-red by the media of the eye and the transmission of radiant energy by Crookes and other glasses," Proc. Roy. Soc. London **B89**, 58 (1915).

S. Hecht, S. Shlaer, and M. H. Pirenne, "Energy, quanta, and vision," J. Gen. Physiol. **25**, 819–840 (1942).

E. P. Hyde and W. E. Forsythe, "The visibility of radiation in the red end of the visible spectrum," Astrophys. J. **42**, 285 (1915).

P. Jainski, "Die Spektrale Hellempfindlichkeit des menschlichen Auges und ihre Bedeutung für die Lichtmesstechnik," Diss. Tech. Hochschule, Berlin, 1938.

N. I. Pinegin, "Absolute sensitivity of the eye in the infrared spectrum," C. R. Acad. Sci. U.R.S.S. **47**, 627 (1945).

G. Wald, "The spectral sensitivity of the human eye. I. A spectral adaptometer," J. Opt. Soc. Am. **35**, 187–196 (1945a).

G. Wald, "Human vision and the spectrum," Science **101**, 653–658 (1945b).

T. Young, "On the theory of light and colours" (Bakerian Lecture), Phil. Trans. Roy. Soc. London **92**, Part I, 12 (1802).

Editors' Comments on Papers 11 Through 13

11   **Cashman:**   *Film-Type Infrared Photoconductors*

12   **Johnson, Cozine, and McLean:**   *Lead Selenide Detectors for Ambient Temperature Operation*

13   **Bode, Johnson, and McLean:**   *Lead Selenide Detectors for Intermediate Temperature Operation*

The next three papers describe the characteristics of film-type photoconductive detectors. The designation "film-type" refers to the fact that these detectors consist of a thin polycrystalline film that is produced either by vacuum evaporation or chemical deposition. Although films of many materials show limited response in the infrared, only those of the lead salts have, thus far, proved to be useful as detectors. In Paper 11, Cashman discusses techniques for producing film-type detectors and discusses the interrelationships between processing details and such detector parameters as spectral response, time constant, signal and noise characteristics, resistance, linearity of response, and the effect of cooling. The explanation of the importance of oxygen in the production process is of particular interest because it was Cashman himself who first deduced its importance. His application of these findings was the key to successful wartime production of film-type thallous sulfide detectors; oxygen sensitizing is, even today, the most crucial step in the production of film-type detectors.

Robert J. Cashman is a professor of physics at Northwestern University, Evanston, Illinois.

Lead selenide detectors optimized for operation at 77°K have been available for many years. Papers 12 and 13 describe developments in the technology of lead selenide that have led to the fabrication of detectors optimized for use in the temperature range from 195°K to as high as 373°K. For the below-ambient portion of this temperature range, thermoelectric or carbon-dioxide-activated systems offer simple and inexpensive ways to provide the required cooling of the detector. At 195°K, lead selenide is partially background limited and appropriate radiation shielding can double the value of D^*. There are many situations, such as those often found in process control instrumentation, where detectors must operate at elevated temperatures, yet it is quite rare to find published data on the performance of detectors under such conditions. The authors are to be commended for including it here. Additional information on lead salt detectors is contained in References 1 through 17.

Thomas H. Johnson is Manager, Space Focal Plane Assemblies, and Donald E. Bode is Manager, Detector Research and Development Department, both at the Santa Barbara Research Center, Goleta, California.

References

1. Moss, T. S., "Photoconductivity," *Repts. Progr. Phys.*, **28**, 15 (1965).
2. Dalven, R., "A Review of the Semiconductor Properties of PbTe, PbSe, PbS, and PbO," *Infrared Phys.*, **9**, 141 (1969).

132

3. Lovell, D. J., "The Development of Lead Salt Detectors," *Amer. J. Phys.*, **37,** 467 (1969).
4. Hammar, G. W., "Highly Sensitive Lead Sulfide Surfaces and the Method of Manufacture," U.S. Patent No. 2,917,413, Dec. 16, 1959.
5. Cooperstein, R., "Method of Forming a Photosensitive Layer of Lead Sulfide Crystals on a Glass Plate," U.S. Patent No. 3,018,236, Jan. 8, 1962.
6. Autrey, E. A., "Production of Infrared Detector Patterns," U.S. Patent No. 3,356,500, Dec. 5, 1967.
7. Markov, M. N., and E. P. Kruglyakov, "Zonal Sensitivity of PbS Photoconductors," *Optics and Spectroscopy*, **9,** 284 (1960).
8. Bisbee, J., and S. I. Moody, "Lead-Salt Film Detector Parameters," *Infrared Phys.*, **3,** 185 (1963).
9. Humphrey, J. N., "Optimum Utilization of Lead Sulfide Infrared Detectors Under Diverse Operating Conditions," *Appl. Optics*, **4,** 665 (1965).
10. Rogovoy, I. D., "The Absolute Spectral Sensitivity of Type FS-A Commercial Polycrystalline Lead Sulfide Photoresistors," *Sov. J. Opt. Tech.*, **33,** 257 (1966).
11. Feldner, E., et al., "Some Aspects of the Influence of Temperature on the Photoconductive Properties of PbS Layers," *Infrared Phys.*, **8,** 161 (1968).
12. Astaf'yev, A. I., and G. K. Kholopov, "The Voltage Sensitivity of Lead Sulfide Photoresistors as a Function of Background Illumination," *Sov. J. Opt. Tech.*, **36,** 655 (1969).
13. Belyakova, V. V., L. N. Biller, and S. I. Freyvert, "Heterojunction Photocell, Sensitive in the Near Infrared," *Sov. J. Opt. Tech.*, **39,** 134 (1972).
14. Bode, D. E., "Lead Selenide Infrared Detectors," *Proc. Natl. Electronics Conf.*, **19,** 630 (1963).
15. McLean, B. N., "Method of Production of Lead Selenide Photodetector Cells," U.S. Patent No. 2,997,409, Aug. 22, 1961.
16. Spencer, H. E., "Chemically Deposited Lead Selenide Photoconductive Cells," U.S. Patent No. 3,121,022, Feb. 22, 1964. (See also No. 3,121,023, same title and date.)
17. Alekseyev, A. M., "The Effect of Unmodulated Illumination on the Sensitivity of PbTe Resistors," *Sov. J. Opt. Tech.*, **35,** 742 (1968).

II

Reprinted from *Proc. IRE,* **47**(9), 1471–1475 (1959)

Film-Type Infrared Photoconductors*

R. J. CASHMAN†

INTRODUCTION

PHOTOCONDUCTIVITY was first observed and correctly interpreted in 1873 by Smith.[1] Selenium rods, which served as high-resistance elements in circuits designed for testing long submarine cables, were found to decrease in resistance in the presence of light. Smith's discovery provided a fertile field of investigation for several decades, though most of the effort was of doubtful quality. By 1927, over 1500 articles and 100 patents had been listed on photosensitive selenium.[2] Selenium has a spectral response which extends slightly into the infrared to about 9000 A.

The first infrared photoconductor of high responsivity was developed by Case in 1917. He discovered that a substance composed of thallium and sulfur exhibited photoconductivity.[3] Later he found that the addition of oxygen greatly enhanced the response.[4] During the next fifteen years, many laboratories throughout the world engaged in research on photoconductivity and related phenomena.[5] In spite of all this activity, certain undesirable characteristics persisted in photoconductors. Instability of resistance in the presence of light or polarizing voltage, loss of responsivity due to overexposure to light, high noise, sluggish response and lack of reproducibility seemed to be inherent weaknesses. About 1930, the appearance of the Cs-O-Ag phototube, with its more stable characteristics, discouraged to a great extent further development of photoconductive cells until about 1940.

During World War II, the thallous sulfide cell was subjected to intensive development in the United States,[6] and its former shortcomings were largely overcome. Moreover, the detectivity was increased by almost two orders of magnitude. The cell was extensively used in communication and identification systems.[7–9]

Lead-sulfide photoconductors were brought to the manufacturing stage of development in Germany about 1943. They were first produced in the United States at Northwestern University in 1944[10] and, in 1945, at the Admiralty Research Laboratory in England.[11] Since the war, there has been extensive work on these cells and also on lead-selenide and lead-telluride types in the United States and other countries throughout the world.

Other substances which in film form at room temperature show infrared response in the 1- to 2-micron region include Ag_2S, MoS_2, In_2S_3, Bi_2S_3, SnS, In_2Se_3, and In_2Te_3. A great many films of other materials exhibit photoconductivity when cooled.[12] To date, none of these has shown as much promise as the lead-salt photoconductors.

PROCESSING OF PHOTOCONDUCTIVE FILMS

Thallous sulfide and lead telluride films are prepared by vacuum evaporation onto an insulating substrate, generally glass. Lead-sulfide and lead-selenide films may be prepared by vacuum evaporation[10,12,13] and also by chemical deposition.[13] Lead sulfide films are deposited chemically from a solution of lead acetate and thiourea made basic with sodium hydroxide. Lead selenide films may be deposited chemically from a solution of lead acetate and selenourea. In both methods, oxygen plays a role in producing a photoconductive film. This is accomplished in the evaporation method by controlled oxygen pressures and elevated substrate temperatures during film formation, or by oxygen-temperature treatments after the film is formed. In the chemical method, oxygen or oxygen-containing compounds are trapped in the polycrystalline film during film deposition. Subsequent baking in oxygen or vacuum may optimize the sensitization process.

* Original manuscript received by the IRE, June 26, 1959.
† Northwestern University, Evanston, Ill.
[1] W. Smith, "The action of light on selenium," *J. Soc. Tel. Engrs.,* vol. 2, p. 31; 1873.
[2] M. F. Doty, "Selenium, List of References, 1817–1925," New York Public Library, New York, N. Y.; 1927.
[3] T. W. Case, U. S. Patent No. 1,301,227; April 22, 1919.
[4] ———, U. S. Patent No. 1,316, 350; September 16, 1919.
———, "The thalofide cell—a new photoelectric substance," *Phys. Rev.,* vol. 15, p. 289; 1920.
[5] F. C. Nix, "Photoconductivity," *Rev. Mod. Phys.,* vol. 4, p.723; 1932.
R. J. Cashman, "Development of Sensitive Thallous Sulfide Photoconductivity Cells for Detection of Near Infrared Radiation," OSRD 1325, PB 27332; March 17, 1943. Available from Office of Technical Services, U. S. Dept. of Commerce, Washington, D. C. This report contains 176 references to photoconductors and related phenomena.
[6] R. J. Cashman, "Development of Stable Thallous Sulfide Photoconductive Cells for Detection of Near Infrared Radiation," OSRD 5997, PB 27354; October 31, 1945. Available from Office of Technical Services, U. S. Dept. of Commerce, Washington, D. C.
———, "Photodetectors for ultraviolet, visible and infrared radiation," *Proc. NEC,* vol. 2, p. 171; 1946.
———, "New photoconductive cells," *J. Opt. Soc. Am.,* vol. 36, p. 356; 1946.
U. S. Patent Nos. 2,448,517 and 2,448,518; September 7, 1948.

[7] J. M. Fluke and N. E. Porter, "Some developments in infrared communications components," PROC. IRE, vol. 34, pp. 876–883; November, 1946.
[8] W. S. Huxford and J. R. Platt, "Survey of near infrared communication systems," *J. Opt. Soc. Am.,* vol. 38, p. 253; 1948.
[9] V. K. Zworykin and E. G. Ramberg, "Photoelectricity and Its Applications," John Wiley and Sons, Inc., New York, N. Y., ch. 3; 1949.
[10] R. J. Cashman, "Development of Sensitive Lead Sulfide Photoconductive Cells for Detection of Intermediate Infrared Radiation," OSRD 5998, ATI 25740; October 31, 1945. Available from ASTIA Document Service Center, Dayton 2, Ohio. U. S. Patent No. 2,448,516; September 7, 1948.
[11] L. Sosnowski, J. Starkiewicz and O. Simpson, "Lead sulfide photoconductive cells," *Nature,* vol. 159, p. 818; 1947.
[12] R. A. Smith, F. E. Jones, and R. P. Chasmar, "The Detection and Measurement of Infrared Radiation," Oxford University Press, Oxford, Eng., p. 137; 1957.
[13] T. S. Moss, "Lead salt photoconductors," PROC. IRE, vol. 43, pp. 1869–1881; December, 1955.

FILM STRUCTURE

The films formed by either of the above processes are of the order of 1 micron thick and consist of a composite of microscopic crystallites. X-ray and electron diffraction studies indicate that the crystallites have dimensions of the order of 0.1 micron and are separated by intercrystalline barriers of the order of 10 A thick. For the lead-salt films, the barriers are oxidation products such as PbO and PbO·PbSO₄.

ROLE OF OXYGEN

Optical absorption[14] and electrical conductivity[15] measurements on single crystals of the lead salts have shown that the energy gap of the single crystal corresponds to the long wavelength limit of photoconductivity of the film. Thus, the photoelectric process in which free carriers are generated by absorption of photons involves a main-band transition in which an electron is raised from the full band to the conduction band, thereby leaving a hole in the full band. The role that oxygen plays in the photoconductive process therefore must be found in the recombination of charge carriers. Recombination processes are discussed generally in terms of the carrier lifetime of importance.[13,16-18] Three models will be discussed briefly.

In the *intrinsic-carrier model*,[16] oxygen acts as a *p*-type compensator impurity to balance the *n*-type impurities in the film. Both the lifetime and the number of free holes and electrons are equal. Recombination occurs between electrons and holes either directly or via recombination centers. Maximum response is obtained by minimizing the electron and hole densities by compensation and by maximizing the lifetime of hole-electron pairs.

The *minority-carrier model*[16,19] visualizes the film as a composite of microscopic *n-p-n* junctions. The crystallites are *n*-type with a thin layer of *p*-type material in between, presumably produced by the oxidation treatment. Barrier modulation plays an important role in this model. The diffusion of minority carriers across the *p-n* junctions results in lowering the space-charge barrier at the junction, thereby allowing more current to flow across the junction. This sort of secondary amplification is similar to *p-n* hook multiplication in a *p-n-p-n* junction transistor. A long minority-carrier

lifetime and a low minority-carrier charge density are again necessary for high responsivity.

The *majority-carrier model*[20,21] depends on the presence of minority-carrier traps in the film which are due to oxygen or oxygen-containing molecules. The traps may be of either the surface or the bulk type. The free-minority carrier created initially by the photon, when trapped, leaves the majority carrier free to conduct. The increase in the majority-carrier lifetime is proportional to the time the minority carrier spends in the trap. High responsivity is obtained by optimizing the ratio of the majority-carrier lifetime to the majority-carrier density. Secondary amplification is also possible through lowering of the intercrystalline-barrier potential by the trapped minority carriers.

Many of the properties of photoconducting films have been explained by the last two models. It is not unlikely that both these processes and also possibly other mechanisms are present in the various photoconductive films.

SPECTRAL RESPONSE

One of the most important characteristics of a photodetector is its spectral response. Fig. 1 shows the absolute spectral response of thallous sulfide, lead sulfide,[22] lead selenide,[23] and lead telluride[23] films at room temperature. In agreement with recent practice,[24]

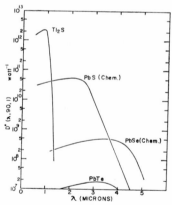

Fig. 1—Absolute spectral response of some infrared photoconductors at room temperature.

[14] W. Paul, D. A. Jones and R. V. Jones, "Infrared transmission of galena," *Proc. Phys. Soc. B*, vol. 64, p. 528; 1951.
 M. A. Clark and R. J. Cashman, "Transmission and spectral response of PbS, PbSe and PbTe," *Phys. Rev.*, vol. 85, p. 1043; 1952.
 A. F. Gibson, "The absorption spectra of single crystals of PbS, PbSe and PbTe," *Proc. Phys. Soc. B*, vol. 65, p. 378; 1952.
[15] W. W. Scanlon, "Interpretation of Hall effect and resistivity data in PbS and similar binary compound semiconductors," *Phys. Rev.*, vol. 92, p. 1573; 1953.
[16] A. Rose, E. S. Rittner, and R. L. Petritz in "Photoconductivity Conference," John Wiley and Sons, Inc., New York, N. Y.; 1956.
[17] A. Rose, "Performance of photoconductors," PROC. IRE, vol. 43, pp. 1850–1869; December, 1955.
[18] J. N. Humphrey and R. L. Petritz, "Photoconductivity of lead selenide," *Phys. Rev.*, vol. 105, p. 1736; 1957.
[19] J. C. Slater, "Barrier theory of photoconductivity of PbS," *Phys. Rev.*, vol. 103, p. 1631; 1956.

[20] R. H. Harada and H. T. Minden, "Photosensitization of PbS films," *Phys. Rev.*, vol. 102, p. 1258; 1956.
[21] R. L. Petritz, "Theory of photoconductivity in semiconductor films," *Phys. Rev.*, vol. 104, p. 1508; 1956.
[22] R. F. Potter and D. H. Johnson, "The Status of the Uniformity of Photodetectors in the U.S.A.," presented at 2nd Annual Natl. IRIS Meeting, Boston, Mass.; September 17, 1958.
[23] H. Levinstein, W. Beyen, P. Bratt, W. Engeler, L. Johnson, and A. MacRae, "Infrared Detectors Today and Tomorrow," presented at 2nd Annual Natl. IRIS Meeting, Boston, Mass.; September 17, 1958.
[24] R. C. Jones, "Phenomenological description of the response and detecting ability of radiation detectors," paper 4.1.1., this issue, p. 1495.

the detectivity D^* has been plotted as ordinate $(D^* = (A\Delta f/1\ \mathrm{cm}^2 \cdot 1\ \mathrm{cps})^{1/2} \cdot 1/\mathrm{NEP}$, where A is the sensitive area, Δf is the amplifier bandwidth, and NEP is the noise equivalent power). The quantities in parenthesis $(\lambda, 90, 1)$ refer to the test conditions; λ refers to monochromatic detectivity; 90, to the chopping frequency in cycles per second; 1, to the amplifier bandwidth in cycles per second.

Lead sulfide cells are now available from a number of manufacturers, and the spectral response may vary somewhat from that shown in Fig. 1. The peak detectivity may be higher by about a factor of 5 than that shown, but may be lower at 4 microns by a factor of 10 or more. Evaporated lead-sulfide films show greater variations in their spectral responses than do the chemically-deposited types.

The absolute spectral response of some cooled photoconducting films is shown in Fig. 2. The detectors are liquid nitrogen cooled, except for the lead selenide evaporated layer which is cooled with a dry-ice-alcohol mixture. This type of film is thinner than is used ordinarily and has a smaller long-wavelength cutoff than the chemically-deposited type of lead-selenide film[23] shown. The two curves for lead telluride films[23] give an indication of the "spread" in response to be expected among these detectors. The 900-cps chopping frequency used for these data is approximately the frequency which gives an optimum signal-to-noise ratio and hence optimum detectivity.

Lead sulfide cells from different sources may show considerable variation in their spectral response characteristics when cooled. This is due in part to the fact that a cell processed to give optimum detectivity at room temperature may not have the optimum detectivity possible when operated at dry-ice or liquid-nitrogen temperatures. Conversely, a cell processed to give optimum detectivity at dry-ice or liquid-nitrogen temperature may have a poor detectivity when operated at room temperature. When cooled-type lead-sulfide cells are so processed, their peak detectivity at dry-ice or liquid-nitrogen temperatures does not differ greatly from that of the uncooled cell of high detectivity shown in Fig. 1. The position of the peak and the long wavelength tail, however, shift to longer wavelengths as the temperature is lowered. If the wavelength is expressed in electron volts, the shift is about 4×10^{-4} ev/°C. The detectivity D^* of a liquid-nitrogen cooled cell may be as high as 10^{10} cm/watt at 4 microns or 100 times greater than that of the uncooled cell.

TIME CONSTANT

The speed of response of a photodetector is determined in two ways. The first consists of exposing the detector to square-wave radiation pulses and observing the rise and decay of the signal. The second consists of varying the chopping frequency of the incident radiation and observing the resulting changes in the signal. If the rise and decay are exponential and can be repre-

sented by a single time constant, τ, the two methods agree and the dependence of the signal S on the frequency, f, is given by the relation: $S = S_0(1 + 4\pi^2 f^2 \tau^2)^{-1/2}$. By observing the frequency, f_1, at which the signal voltage drops 3 db, one can obtain τ from the relation: $2\pi f_1\tau = 1$. For the fast detectors ($\tau = 1$ to 10 μsec), this method may not be feasible due to the difficulty of modulating sources at high frequency. Some detectors, particularly cooled types, may exhibit more than one time constant, in which case the above expression for S is not valid.

The time constant of thallous-sulfide cells varies from about 300 μsec to 1 msec. The rise time is generally about one-half the decay time. Chemically-prepared PbS cells with a spectral response similar to that shown in Fig. 1 have a response time of 150–500 μsec. The rise time is generally faster than the decay time. Evaporated types are generally faster, ranging from 10 to 150 μsec. These are commonly used in motion picture sound reproduction.[25] Uncooled PbSe cells, both chemically-deposited and evaporated, and uncooled PbTe cells are very fast with time constants in the 1–10-μsec range.

Dry-ice-cooled evaporated PbSe cells with a spectral response like that shown in Fig. 2 have a 10–20-μsec time constant. Liquid nitrogen-cooled PbSe and PbTe cells generally exhibit two time constants and sometimes more. There is a fast component of 10–30 μsec and a slow component of the order of a millisecond. The most desirable of these cells are those in which the slow component contributes only a small percentage to the signal. The time constant of the PbTe detectors also frequently exhibits a wavelength dependence. The response to wavelengths of 1.5 μ and shorter is slower than for the longer wavelengths. Cooled PbS cells have time constants ranging from about 0.5 to 3.0 msec.

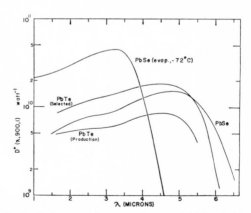

Fig. 2—Absolute spectral response of some cooled infrared photoconductors. The lead-telluride and chemical lead-selenide cells are cooled to −195°C.

[25] R. J. Cashman, "Lead sulfide photoconductive cells for sound reproduction," *J. Soc. Mot. Pic. Engrs.*, vol. 49, p. 342; 1947.

Noise

Present-day thin-film photoconductors exhibit several types of noise. These include Johnson or Nyquist noise, current noise which is of two types, flicker and generation-recombination noise, and radiation noise.

Johnson noise may often be made unimportant in practice by increasing the polarizing current until current noise predominates. Flicker noise is generally of the 1/f type, in which the noise power varies inversely with frequency. For some detectors, the noise power varies less rapidly than 1/f. Generation-recombination noise, which is due to fluctuations in the number of charge carriers in the crystallites, has the same frequency dependence as the signal response. Radiation noise, due to fluctuations in the rate of emission and absorption of background radiation, and which sets a fundamental limit on detectivity, provides one of the mechanisms by which generation-recombination noise is produced. The responsivity of thin-film photoconductors has not developed to the point where radiation noise overrides the other sources of noise.[26]

Thallous-sulfide and lead-telluride detectors exhibit a 1/f noise-power spectrum as do most evaporated lead-sulfide types. Chemically-deposited lead-sulfide cells generally are dominated by 1/f noise below 100 cps and by generation-recombination noise at higher frequencies. Chemically-deposited lead-selenide detectors have a noise power which decreases with increasing frequency but generally not as rapidly as 1/f. Occasionally the noise behavior is similar to that of chemically-deposited lead-sulfide detectors. Evaporated lead-selenide detectors are dominated at low frequencies by 1/f noise and by generation-recombination noise at higher frequencies.[27] Noise data for a detector of this type are shown in Fig. 3.

A quantitative expression may be derived for genera-tion-recombination noise based on shot-noise theory and a one-carrier model.[28] If Ns is the ratio of generation-recombination noise power to thermal or Johnson noise power, then

$$Ns = \frac{e}{kT}\left(\frac{V}{d}\right)^2 \frac{\mu\tau}{1 + 4\pi^2 f^2 \tau^2}$$

where V is the potential difference applied across the detector, d is the electrode spacing, μ the charge carrier mobility and τ the carrier lifetime. The Hall mobility of photoconducting PbSe films has been measured[29] and at room temperature has the value of 10 cm²/volt sec. The carrier lifetime is taken as the measured time constant of these films which is 1 microsecond at room temperature. Substitution of values in the above equation gives a value of 15 for the noise-power ratio, which is in good agreement with the experimental value shown in Fig. 3. The agreement is probably somewhat fortuitous as the model is probably over-simplified. Moreover, experimental values may be as low as 3 for "quiet" cells and as high as 100 for "noisy" cells. The validity of the frequency dependence could not be determined at high frequencies because of equipment limitations.

Resistance

Uncooled lead-sulfide and evaporated lead-selenide cells have dark resistances of around 1 megohm, while chemically-deposited lead-selenide cells have values of about 50,000 ohms. Thallous-sulfide cells range from 1 to 10 megohms in dark resistance.

Liquid-nitrogen-cooled lead-telluride cells have quite high, dark resistances—50 to 1000 megohms. At this temperature, chemically-deposited lead-sulfide cells have resistances in the range of 10 to 20 megohms; those of evaporated types are around 100 megohms, and those of chemically-deposited lead-selenide cells are 2 to 10 megohms.

Other Cell Parameters

Cell users are frequently confronted with problems which bear on other cell characteristics than those discussed. One of these is the determination of the *optimum bias voltage*. Ordinarily, the signal response and noise of a cell increase linearly with cell bias voltage until a value is reached where either the signal increases less than linearly or the noise increases more than linearly. Further increase in bias potential beyond this optimum value results in a decreased signal-to-noise ratio.

Linearity of response to intensity of irradiation is important where the cell is used in quantitative flux measurements. Distortion-free sound reproduction also requires cells with a linear response. The maximum inten-

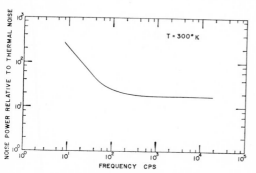

Fig. 3—Noise-power spectrum of an evaporated lead-selenide photoconductor.

[26] F. L. Lummis and R. L. Petritz, "Noise, time constant and Hall studies on lead sulfide photoconductive films," *Phys. Rev.*, vol. 105, p. 502; 1957.
[27] R. J. Cashman, C. R. Betz, and L. J. Rose, unpublished data.

[28] A. van der Ziel, "Shot noise in semiconductors," *J. Appl. Phys.*, vol. 24; 1953.
[29] K. G. Halvorsen, "The Hall Effect in Lead Sulfide and Lead Selenide as a Function of Temperature," Ph.D. dissertation, Northwestern University, Evanston, Ill.; 1951. (Unpublished.)

sity beyond which nonlinearity sets in is not the same for all detectors, but is less for those of high responsivity than for those of low responsivity. If a cell is operating in the nonlinear range, the output signal depends on the image spot size on the cell. Lacking actual signal vs intensity of irradiation data for a given cell, it is still possible to estimate its linear range of response if detectivity data are known. From these data, the noise equivalent intensity, NEI, may be determined for a 1-cps bandwidth. It has been found empirically that cells are linear with intensity from NEI to 10^5 NEI. For many cells the range is greater.

In operation, there is *power dissipation* in the sensitive layer of the cell due to the incoming flux and I^2R heating. The maximum-safe-power dissipation for cells operated at room temperature is generally considered to be 100 milliwatts per cm² of cell area. A related subject is the maximum ambient temperature allowable during storage or operation that will not cause damage to the cell. Detector manufacturers will supply information on this subject and on most of the characteristics which have been discussed. Through control of processing parameters and by selection, manufacturers are able to supply cells with a wide range of characteristics.

12

Reprinted from *Appl. Optics,* **4**(6), 693–696 (1965)

Lead Selenide Detectors for Ambient Temperature Operation

T. H. Johnson, H. T. Cozine, and B. N. McLean

Chemically deposited lead selenide detectors have excellent detectivities in the 1-μ to 4.8-μ region even when operated at room temperature. Recent improvements in processing have yielded D^* (4.0 μ, 10 kc/sec, 1) greater than 1 × 10^{10} cm(cps)$^{1/2}$/W at 300°K. Performance data for these detectors are presented in this paper covering an operating temperature range of 256°K to 373°K. The high level of detectivity, short time constant (<3 μsec), and high-temperature storage capability attest to the superiority of these detectors for use in the 3.3-μ to 4.8-μ region when little or no cooling is available.

Introduction

The purpose of this paper is to describe recent advances in the technology of lead selenide (PbSe) detectors for near ambient temperature operation. This development has resulted in PbSe detectors with 4.0-μ detectivity as much as a factor of 5 greater than previously available.[1] This paper will present the electrical and optical characteristics of these detectors and compare these properties with other types of detectors for the same spectral region.

It is believed that an awareness of the detectivity achievable with PbSe over the temperature range of 240°K to 323°K, coupled with the use of radiation-to-space or thermoelectric cooling, will stimulate an interest in the utilization of this type of detector for infrared systems and devices.

Lead Selenide Detector Fabrication

Ambient PbSe detectors are fabricated utilizing the same general techniques employed in the preparation of all chemically deposited film detectors. Consequently, they can be prepared as immersed or plate-type detector configurations with element size ranging from approximately 2.5 × 10^{-5} cm^2 to 6.0 cm^2.

Lead Selenide Detector Characteristics

The spectral performance of PbSe detectors at various operating temperatures is illustrated in Fig. 1. For a photoconductive device with a $\lambda_{1/2}$ of 4.5 μ, the theoretical limit of detectivity is approximately 2 × 10^{11} cm(cps)$^{1/2}$/W. Thus, it can be seen that the PbSe at 300°K is slightly more than an order of magnitude below this value.

The authors are with the Santa Barbara Research Center, a subsidiary of Hughes Aircraft Company, Goleta, California.
Received 12 February 1965.

Figure 1 also illustrates the shift of the long-wavelength edge with operating temperature. The $\lambda_{1/2}$ value can be seen to vary at the rate of approximately 6 × 10^{-3} μ/°K. It is interesting to note that significant

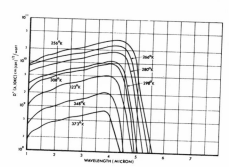

Fig. 1. PbSe spectral detectivity vs temperature.

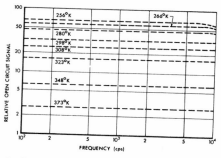

Fig. 2. Dependence of PbSe detector signal response on operating temperature.

detectivity, i.e., in excess of 1×10^9 cm(cps)$^{1/2}$/W at 4.4 μ, is possible at a temperature as high as 323°K.

The frequency response of the PbSe for the same range of operating temperatures is shown in Fig. 2. Note that the response at the lowest temperature shown, 256°K, has not dropped more than 1 dB, indicating a time constant at this temperature of less than 10 μsec. Data received from NOL/Corona[2] have shown a time constant of the PbSe at an operating temperature of 296°K to be < 3 μsec.

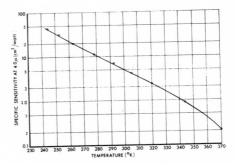

Fig. 6. PbSe (4.0-μ) specific sensitivity (S_1) vs temperature.

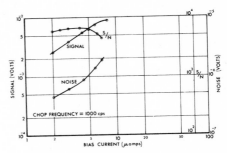

Fig. 7. Determination of optimum bias, PbSe temperature 256°K.

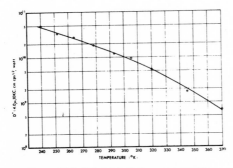

Fig. 3. Dependence of PbSe detector noise spectrum on operating temperature.

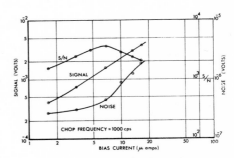

Fig. 8. Determination of optimum bias, PbSe temperature 298°K.

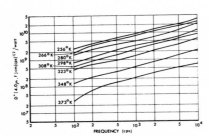

Fig. 4. PbSe (4.0-μ) detectivity vs frequency.

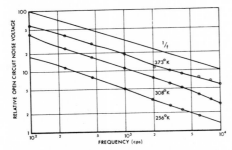

Fig. 5. PbSe (4.0-μ) detectivity vs temperature.

The noise voltage spectra for a PbSe detector are shown in Fig. 3 for three operating temperatures: 256°K, 308°K, and 373°K. In all cases, the 1/f noise dominates to 10 kc/sec. Thus, the detectivity of the PbSe from 373°K down to at least 250°K approaches its maximum value at a frequency of 10 kc/sec or higher. These results are shown graphically in Fig. 4. The reduction and/or elimination of 1/f noise in these detectors would undoubtedly result in detectivities closer to the theoretical limit.

The variation of the 4.0-μ detectivity with temperature over the range of 230°K to 370°K is shown in Fig. 5. The significant reduction of the 4.0-μ detectivity at temperatures above $\sim$320°K reflects the spectral shift of the peak to shorter wavelengths.

A plot of responsivity at 4.0 μ as a function of operating temperature is shown in Fig. 6. Responsivity in this case is expressed as S_1 (cm²/W) or specific sensitivity and is defined as

$$S_1 = \frac{V}{JE\left[\dfrac{4R_e R_L}{(R_e + R_L)^2}\right]},$$

where

V = rms value of fundamental component of the signal voltage (V),

J = rms value of the fundamental component of the heat signal (W/cm²),

E = bias voltage applied to series combination of detector and lead resistor (V),

R_e = detector dark resistance (MΩ),

R_L = load resistance (MΩ).

The relationship of signal and noise values as a function of bias current for four different operating temperatures is shown in Figs. 7–10. It can be seen that optimum detector bias currents lie in the range of 5 μA to 20 μA at these temperatures. Typical dark

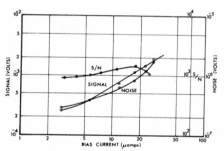

Fig. 9. Determination of optimum bias, PbSe temperature 323°K.

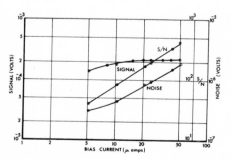

Fig. 10. Determination of optimum bias, PbSe temperature 373°K.

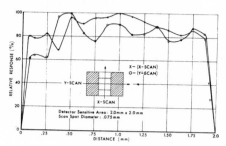

Fig. 11. PbSe responsivity profile.

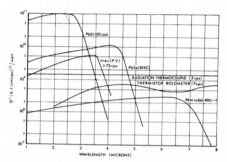

Fig. 12. Spectral detectivity comparison for various detectors at 298°K.

resistance of a square element PbSe detector will vary from 30 MΩ to 0.5 MΩ as the temperature increases from 250°K to 373°K.

The degree of the uniformity of response that can be expected with this type of PbSe detector is shown in Fig. 11. A response profile was obtained on a PbSe detector with a 2 × 2 mm² sensitive area, using a 75-μ diameter scanning spot. Two passes were made, one parallel to the electrodes and one perpendicular to the electrodes. The lowest response observed was a value of 60% of the peak.

Aging tests of this type of PbSe detector at a storage temperature of 343°K for a period of 240 h have resulted in detectivity changes of approximately ±10%. Data obtained on an earlier type PbSe detector had shown the ambient PbSe detectors capable of being stored for approximately 200 days at this temperature without degradation of electrical characteristics.

There are only four photodetectors that can be considered for operation in the intermediate infrared range at normal ambient temperatures. These are PbS, InAs, InSb, and PbSe. A spectral detectivity comparison of these detectors is shown in Fig. 12. The data for the PbS and PbSe detectors were obtained from Santa Barbara Research Center detectors, whereas the curves for the InSb and InAs were obtained from refs. 3 and 4, respectively. For comparison, the performances of a radiation thermocouple[3] and a thermistor bolometer[5] are given. Each of the curves is plotted for the particular device operated at its optimum chopping frequency. The PbSe detector is clearly superior to any of the other detectors over the range of 3.5 μ to 4.7 μ.

Conclusions

Lead selenide detectors for near room temperature operation offer a number of advantages, including high detectivity in the 4.0-μ range, high responsivity, and a short time constant. They are readily available in a wide variety of sizes and shapes, and have been successfully produced in large quantities.

In comparison with currently available detectors that can operate at 300°K, the PbSe is far superior in the detection of radiation in the 4.0-μ range.

References

1. U.S. NOL Rept. No. 474 (3 December 1959).
2. U.S. NOL/Corona Data Nos. 2094 and 2095 (5 November 1964), unpublished.
3. P. W. Kruse, L. D. McGlauchlin, and R. B. McQuistan, *Elements of Infrared Technology* (John Wiley, New York, 1962), p. 424, Figs. 10, 11(a).
4. U.S. NOL/Corona Rept. No. 588 (1 July 1963).
5. U.S. NOL Rept. No. 570 (1 October 1962).

13

Reprinted from *Appl. Optics*, **4**(3), 327–331 (1965)

Lead Selenide Detectors for Intermediate Temperature Operation

D. E. Bode, T. H. Johnson, and B. N. McLean

At about 195°K, PbSe detectors have a much higher detectivity in the 3.5-μ to 5.2-μ wavelength region than any other type of infrared detector. Detectors with a D^* (4.4 μ, 2500 cps) greater than 3×10^{10} cm (cps)$^{1/2}$/W are standard. These detectors are partially background limited at 195°K and can be improved to at least 6×10^{10} cm (cps)$^{1/2}$/W by appropriate radiation shielding. When used at room temperature (295°K), these films exhibit peak detectivity at about 3.8 μ and about one-half detectivity at 4.4 μ. Although the absolute magnitude of the peak detectivity at 295°K is considerably less than that at 195°K, it is better than a thermal detector, such as a bolometer or thermocouple. In addition, the time constant at 295°K is less than 2 μsec.

I. Introduction

The purpose of this paper is to describe recent advances in detector technology for the 3-μ to 5-μ region. PbSe detectors optimized for operation near liquid nitrogen temperature (77°K) have been available for many years. Recent improvements in the chemistry of PbSe films have lead to the development of infrared detectors optimized for high performance at temperatures near that of dry ice (195°K). This paper will present the electrical and optical characteristics of these films and compare these properties with other types of photon detectors for the same spectral region. Practical application of these detectors requires suitable techniques for refrigeration. Various cryogenic techniques will also be discussed.

II. Lead Selenide Film Preparation

Lead selenide films can be prepared by either vacuum evaporation of bulk lead selenide, or by chemical deposition. The films reported here were all prepared by chemical deposition. The fundamental chemical reaction in this process is the reaction between selenourea and lead acetate. The films are generally deposited on glass, ceramic strontium titanate, or single crystal strontium titanate. The single crystal strontium titanate is used when the immersion of the high index of refraction film on a lens is required. In all cases, evaporated gold electrodes are used to make ohmic contact to the PbSe film and help define the sensitive area. Various techniques such as sawing, sandblasting, scribing, and photoetching are then used to delineate a specific sensitive element. Elemental sizes ranging from 25 μ on a side up to 1 sq cm have been achieved. Rectangular areas suitable for optimum coupling to the exit beam of a spectrometer are frequently used.

After deposition, electroding, and area delineation, the films are sensitized by heat treatment in an oxidizing atmosphere. Variation in the sensitization technique results in a film optimized for either low-temperature operation (LTO) or intermediate-temperature operation (ITO). The LTO-type films are used for infrared detection in the temperature range below about 145°K. The ITO films are superior to the LTO films above about 145°K, as will be discussed later. The preparation of intricate detector arrays is far simpler using lead salt films than single crystal semiconductors. The various processing steps described above are pictured in Fig. 1, for the case of a seven-element array. Step 1 shows the glass substrate, whose diameter is 4.8 mm. Step 2 shows an evaporated gold layer on this substrate. Step 3 shows the same layer after photoetching to delineate the individual electrodes. Step 4 shows the substrate after chemical deposition of PbSe including registration marks. Step 5 shows the result of photoetching the PbSe film to delineate the individual areas. Step 6 shows the attachment of leads by microwelding techniques. The six square sensitive areas in this configuration have the dimensions 1 × 1 mm.

Fig. 1. Array processing steps.

The authors are with the Santa Barbara Research Center (subsidiary of Hughes Aircraft Company), Goleta, California.

Received 29 June 1964.

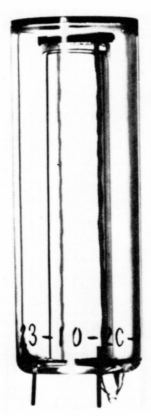

Fig. 2. Vacuum dewar for the LTO and ITO detectors.

nitrogen transfer from a storage reservoir, or by open cycle Joule-Thomson expansion of compressed nitrogen, air, or argon gas, depending on the particular application. In cases where indefinite standby and refrigeration at the flip of a switch is desired, a closed cycle compressor such as that shown in Fig. 4 can be used. This compressor is designed for airborne refrigeration of infrared detectors by Joule-Thomson expansion of nitrogen, argon, or air in the cold finger of the detector dewar.

Recent improvements in thermoelectric materials and techniques have provided an auxiliary method for the electrical refrigeration of $3-\mu$ to $5-\mu$ infrared detectors to temperatures as low as $200°K$. Figure 5 shows a complex array of lead-selenide detector elements mounted on a two-stage Peltier cooler designed to operate at $-40°C$.

There are other techniques for cooling ITO-type PbSe films that are particularly suited to space applications. One of these techniques involves the use of a solid cryogen, such as solid CO_2. Another technique is to use radiation-to-space cooling.

III. PbSe Detector Characteristics

The spectral performance of PbSe detectors is compared with many other types of detectors in Fig. 6. Figure 6 represents the best state-of-the-art detectors prepared at the Santa Barbara Research Center. With the exception of InAs (which is a photovoltaic detector) all of the detectors shown are photoconductive. The data is reported for a field of $2-\pi$ sr looking at $300°K$ radiation. The theoretical curve is that of an ideal background limited infrared

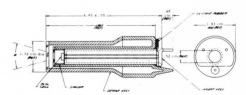

Fig. 3. SBRC style 120 dewar flask.

Refrigerated infrared detectors are not very useful to the average user unless they are coupled to some type of cooled heat sink. The common method of cooling this detector is to mount the sensitive element on the end of the cold finger of a dewar flask as shown in Fig. 2. This particular flask is made of standard Pyrex glass with an antireflection-coated silicon window. This type of evacuated dewar is particularly recommended for ITO PbSe. ITO PbSe films are quite stable in vacuum. Figure 3 shows a dewar design that is particularly suited for use with LTO PbSe. In this case, the envelope has outer- and inner-coated silicon windows, and the element is isolated from the vacuum. This practice helps ensure the stability of the electrical characteristics of the LTO films. Since the outer dewar does not contain a detector element, it can be vacuum-baked at very high temperatures ensuring indefinite vacuum integrity.

LTO-type detectors, in packages as shown in Figs. 2 or 3, can be cooled by direct liquid nitrogen, liquid-

Fig. 4. Closed-cycle cooler. (The over-all length of the cylinder is about 20 cm.)

144

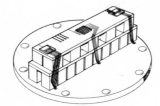

Fig. 5. Layout of thermoelectric cooler plus arrays.

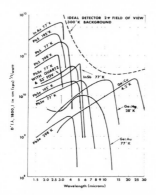

Fig. 6. Best state-of-the-art summary of all types of detectors being made at SBRC.

photoconductor (BLIP detector) with a 100% quantum efficiency.

In the limit, a photoconductive detector exhibits generation noise produced by the statistical fluctuations in the background photon flux as well as statistical fluctuations in the recombination mechanism. A photovoltaic detector only exhibits generation noise, consequently, its theoretical limit is $\sqrt{2}$ times that shown for the ideal BLIP detector.

Figure 7 shows the spectral detectivity of an ITO PbSe detector operated at four different temperatures. This figure shows how the peak spectral detectivity approaches the theoretical limit on cooling to 193°K. The spectral peak D^* then decreases as the temperature is lowered to 145°K. This is caused by the shift in the spectral edge to longer wavelengths which results in the deleterious capture of more effective photons from the 295°K background.

Figure 8 shows the frequency response of a typical ITO PbSe detector at the same four temperatures (room temperature, Freon 12, Freon 13, and Freon 14). Note that the photoconductive time constant increases significantly as the temperature is reduced and the signal approaches the background limit. The noise spectra of the ITO PbSe detector are shown in Fig. 9 for these same temperatures. In all cases, the $1/f$ noise dominates to 10 kc/sec. The detectivity of most detectors at 193°K approaches its maximum value

at 10 kc/sec as the ultimate signal rolloff of 6 dB/octave tends to overpower the noise rolloff of 3 dB/octave. The presence of $1/f$ noise in these detectors is probably one of the chief reasons that these detectors do not come closer to the theoretical limit.

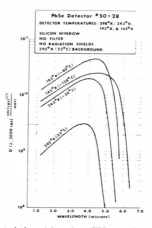

Fig. 7. Spectral detectivity of an ITO type for several temperatures.

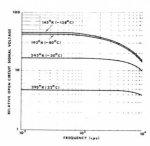

Fig. 8. Dependence of PbSe detector signal response on operating temperature.

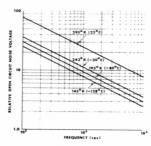

Fig. 9. Dependence of ITO PbSe detector noise spectrum on operating temperature.

145

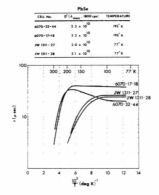

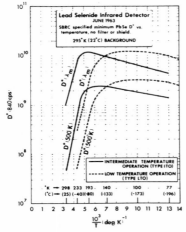

Fig. 10. Temperature dependence of photoconductive lifetime for ITO and LTO PbSe.

the number of measurements of time constant below 77°K has been quite limited. The other significant point in this graph is the temperature shift in the edge resulting from the different degree of film oxidation. This shift from LTO to ITO has increased the operating temperature by about 70°K.

Figure 11 shows the detectivity of an LTO compared to an ITO type as a function of the reciprocal of the absolute temperature. In this case, both D^* (500°K) and D^* (λ_{max}) are shown for comparison. The general behavior is similar to photoconductive lifetime behavior shown in Fig. 10.

As the magnitude of effective background radiation is reduced the effective time constant is increased. This in turn gives rise to an increase in detectivity. However, the increase in detectivity does not follow a simple BLIP theory, because the mechanism of photoconductivity in lead-salt films is far more complicated than the mechanism of photoconductivity in extrinsic germanium detectors, such as Ge:Cu and Ge:Hg.

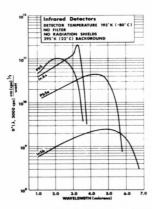

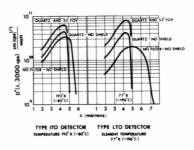

Fig. 12. Effect of filters and radiation shields on PbSe detectors. The quartz filter is 0.23 mm thick. FOV-detector field of view; 295°K background; silicon window peaked at 3.9 μ.

Fig. 11. Lead selenide infrared detector. SBRC specified minimum PbSe D^* vs temperature, no filter or shield. 295°K (22°C) background. The dependence of reflectivity on operating temperature.

Figure 10 shows the photoconductive time constant of two ITO-type PbSe detectors #6070–17–18 and #6070–32–44 plotted against the reciprocal temperature compared to two LTO-type PbSe detectors. The low temperature behavior of the two ITO units represents the sort of spread one observes between different depositions. The exact reason for the large decrease in time constant with decreasing temperature in the case of 6070–32–44 is not completely understood. Presumably, the increase in background photon flux with the shift in the spectral edge has a more severe effect on 6070–32–44 than it does on 6070–17–18. Wide variations in spread of the LTO-type PbSe detectors have not been observed to date; however,

Fig. 13. Spectral detectivity comparison of the ITO PbSe detector with InAs, PbS, and InSb.

Further discussion of these complications is beyond the scope of this paper.

There are two basic ways to reduce the magnitude of background radiation. One method is to employ cold-radiation shield within the detector dewar which restricts the detector field of view to the necessary optics. The other method is to use a cold spectral filter and restrict the upper wavelength cutoff to a lower desired value and, in this way, reduce the number of background photons coming from the ambient. Figure 12 shows the improvement that can be made by operating a typical LTO PbSe detector at 77°K. Note that, although the peak detectivity of the unshielded LTO type is less than the unshielded ITO type, the improvement in D^* by the LTO-type detector through filtering and shielding is about a factor of two greater than the ITO type. Further background shielding has little effect on the detectivity of ITO PbSe at 195°K. However, further background shielding continues to improve the detectivity of LTO PbSe at 77°K. As one reduces the background radiation below a specific level, operating temperatures below 77°K are necessary to observe further improvement in D^*. A D^* (λ_{max}) of 1.8×10^{11} cm (cps)$^{1/2}$W has been achieved for an LTO-type PbSe detector with a background level of about 10^{14} effective photons/cm^2

sec. However, in this case, a temperature below 77°K was necessary.

There are four detectors that can be considered for operation in the 3-μ to 5-μ region at dry ice temperature. These detectors are PbS, PbSe, InAs, and InSb. Figure 13 shows a spectral detectivity comparison of these four detectors. PbS crosses PbSe at about 3.2 μ, InAs falls below PbSe at about 3.5 μ, and the InSb detectivity exceeds that of PbSe for wavelengths beyond 5.7 μ.

IV. Conclusions

Lead-selenide detectors for intermediate temperature operation offer a number of cryogenic advantages. Detectors with high detectivity in the 3-μ to 5-μ region can now be cooled by a variety of techniques such as: (1) radiation to space cooling, (2) thermoelectric cooling, (3) solid CO_2 cooling, (4) Freon 13 or 14 refrigeration. Refrigeration efficiencies are far higher for this type of detector. The size, weight, and cost of closed cycle refrigeration is certainly reduced for these detectors.

In comparison with all known detectors that can operate at 195°K, ITO PbSe is far superior in the detection of radiation in the 3.5-μ to 5.2-μ wavelength range.

Editors' Comments on Papers 14 Through 19

This group of papers describes the development and the characteristics of intrinsic infrared detectors. In an intrinsic, or pure, semiconductor the incident photon produces an electron-hole pair, consisting of the electron raised to the conducting state and the hole at the site vacated by the electron. These charge carriers can be separated by a bias voltage applied across the detector, in which case the detector functions in the photoconductive mode, or they can be separated by the field that exists across a *p-n* junction, in which case the detector functions in the photovoltaic mode. The separation of the charge carriers can also be accomplished by an applied magnetic field, in which case the detector functions in the photoelectromagnetic (PEM) mode. In Paper 14 Rieke, DeVaux, and Tuzzolino give a basic introduction to intrinsic detectors covering their history, physics, and characteristics.

Lloyd H. DeVaux is Head, Detectors Section, Hughes Research Laboratories, Malibu, California.

Pruett and Petritz, in Paper 15, provide a detailed treatment of the characteristics of photovoltaic indium antimonide detectors and of the factors that must be considered in using them as circuit elements. This is one of the first papers that was written with the needs of the circuit designer in mind. The analysis of the requirements for a preamplifier and how best to achieve them is particularly noteworthy.

George R. Pruett is Manager, Infrared Detector Research, Texas Instruments, Inc., Dallas, Texas.

Photoconductive indium antimonide detectors have enjoyed a wide acceptance despite the fact that their ultimate detectivity is 40 percent less than with the same material used in the photovoltaic mode. Morten and King, in Paper 16, describe details of the manufacturing process for photoconductive indium antimonide, develop a simple theory to explain the performance of these detectors, present performance data over the temperature range from room temperature down to 77°K, and finally show several of the possible embodiments this versatile detector can assume. Of particular interest is the 30-element linear array shown in Figure 8.

F. D. Morten is Development Manager, Mullard Ltd., Southampton, England.

Intrinsic detectors have a long-wavelength cutoff that is determined by the width of the energy gap between the valence and conduction bands of the detector material. The width of the energy gap is a measure of the minimum energy required to produce electron-hole pairs in the material. Intrinsic detectors available today all require energies greater than 0.18 eV, which corresponds to a long-wavelength cutoff of about 7 μm. In Paper 17, Putley describes various hot-electron effects that extend the response of intrinsic photoconductors to the millimeter portion of the spectrum. Hot-electron effects depend on the fact that very long wavelength radiation can be absorbed directly by the free carriers of a semiconductor. If the semiconductor is cooled sufficiently, this absorption causes a change in the energy distribution of the electrons within the conduction band which, in turn, changes the mobility of the electrons. The result is a change in conductivity produced by a change in mobility, rather than the change in carrier concentration that is observed at shorter wavelengths. As Putley shows, the effect can be enhanced by the application of an external magnetic field. Additional information on indium antimonide detectors is contained in References 1 through 10.

The search for materials having intrinsic energy gaps small enough to permit operation at 10 μm and beyond has led to tri-metal detectors that are made from mixed compounds of mercury telluride/cadmium telluride and of lead telluride/tin telluride. By varying the ratio of the two components, the energy gap of the compound can be varied so as to give long-wavelength cutoffs ranging from 5 to 15 μm. From a system designer's standpoint these are exciting developments, since these detectors need be cooled only to 77°K. All other currently available photon detectors for the 8–13 μm region require cooling to much lower temperatures.

In Paper 18, Bartlett et al. announce the achievement of background-limited performance with photoconductive mercury cadmium telluride detectors cooled to 77°K. The reader who wishes to compare the values of $D*$ found in Table 1 of this paper with values elsewhere is cautioned that the values reported here are for a shielded detector having a field of view of 60°. Note also from Table 1 how changes in the relative proportions of the constituents comprising the detector shifts the wavelength of peak response from 9.5 to 12 μm.

B. E. Bartlett is Materials Development Group Leader, Mullard Ltd., Southampton, England.

In Paper 19, Rolls and Eddolls describe lead tin telluride photovoltaic diodes having detectivities that are very close to the background limit. Previous work on detectors of this type indicated that there was a problem with excess $1/f$ noise. Results reported here show that if the detectors are back-biased to the zero volts point on the I–V characteristic, the $1/f$ noise is negligible down to frequencies as low as 70 Hz. Owing to the high index of refraction of the lead tin telluride material, antireflection coatings can increase the value of $D*$ by 50 percent or more. Further information on tri-metal detectors is contained in References 11 through 23. Miscellaneous intrinsic materials are covered in References 24 through 26.

D. V. Eddolls is a Research Scientist at the Allen Clark Research Center, Caswell, Towcester, Northamptonshire, England.

References

1. Butter, C. D., "Some Properties of Cooled InSb Photoconductive Infrared Detectors," *Infrared Phys.*, **3**, 207 (1963).
2. Williams, D. B., "Effect of Reduced Background Radiation on Photoconductive InSb Detector Arrays," *Infrared Phys.*, **5**, 57 (1965).
3. Morten, F. D., and R. E. J. King, "Multi-Element Infra-Red Detectors for High Information Rate Systems," *Infrared Phys.*, **8**, 9 (1968).
4. Vinogradova, G. N., and V. Ye. Il'in, "Distribution of Sensitivity over the Surfaces of Indium Antimonide Photocells," *Sov. J. Opt. Tech.*, **35**, 487 (1968).
5. Klein, C. A., "On Photosaturation of Intrinsic Infrared Detectors," *Appl. Optics*, **8**, 1897 (1969).
6. Kruse, P. W., "Indium Antimonide Photoconductive and Photoelectromagnetic Detectors," Chap. 2 in Willardson, R. K., and A. C. Beer (eds.), *Semiconductors and Semimetals*, Vol. 5, *Infrared Detectors*, Academic Press, New York (1970).
7. Breckenridge, R. G., and W. Oshinsky, "Photoconductive Cell," U.S. Patent No. 2,793,275, May 21, 1957.
8. Mesecke, C. M., "Infrared Detector with *p-n* Junctions in Indium Antimonide," U.S. Patent No. 3,139,599, June 30, 1964.
9. Beaupre, H. J., and C. M. Mesecke, "Infrared Detector and Method of Making Same," U.S. Patent No. 3,128,253, Apr. 7, 1964.
10. Lucovsky, G., R. B. Emmons, and H. Altemose, "Infrared Photomixer Diodes," *Infrared Phys.*, **4**, 193 (1964).
11. Kruse, P. W., "Photon Effects in $Hg_{1-x}Cd_xTe$," *Appl. Optics*, **4**, 687 (1965).
12. Saur, W., "Long Wavelength Mercury–Cadmium Telluride Photoconductive Infrared Detectors," *Infrared Phys.*, **8**, 255 (1968).
13. Kohn, A. N., and J. J. Schlickman, "1–2 Micron (Hg, Cd) Te Photodetectors," *IEEE Trans. Electron Devices*, **ED-16**, 885 (1969).
14. Stelzer, E. L., J. L. Schmit, and O. N. Tufte, "Mercury Cadmium Telluride as an Infrared Detector Material" *IEEE Trans. Electron Devices*, **ED-16,** 880 (1969).
15. Long, D. L., and J. L. Schmit, "Mercury–Cadmium Telluride and Closely Related Alloys," Chap. 5 in Willardson, R. K., and A. C. Beer (eds.), *Semiconductors and Semimetals*, Vol. 5, *Infrared Detectors*, Academic Press, New York (1970).
16. Aldrich, N. C., and J. D. Beck, "Performance of S-192 (Hg, Cd) Te Arrays," *Appl. Optics*, **11,** 2153 (1972).
17. Halpert, H., and B. L. Musicant, "N-Color (Hg, Cd) Te Photodetectors," *Appl. Optics*, **11,** 2157 (1972).
18. Long, D., "Uniformity of Infrared Detector Parameters in Alloy Semiconductors," *Infrared Phys.*, **12**, 115 (1972).
19. Soderman, D. A., and W. H. Pinkston, "(Hg, Cd) Te Photodiode Laser Receivers for the 1–3 μ Spectral Region," *Appl. Optics*, **11**, 2162 (1972).
20. Ameurlaine, J., J. Coester, and H. Hofheimer, "Breakthrough in Detectors: Photovoltaic HgCdTe," *Opt. Spectra*, **7**, 28 (Oct. 1973).
21. Melngailis, I., and T. C. Harman, "Single Crystal Lead–Tin Chalcogenides," Chap. 4 in Willardson, R. K., and A. C. Beer (eds.), *Semiconductors and Semimetals*, Vol. 5, *Infrared Detectors*, Academic Press, New York (1970).
22. Rolls, W. H., et al., "Lead–Tin Telluride Photodiode Arrays for 8 to 14 Micron Detection," *Proc. Conf. Infra-Red Techniques, IERE Conf. Proc. 22*, London (1971), p. 75.
23. Joseph, A. S., "Heterojunction PbSnTe Detectors Solve IR System Problems," *Electro-Opt, Syst. Design*, **5**, 24 (Oct. 1973).

24. Wrobel, J. S., and H. Levinstein, "Photoconductivity in InSb–GaSb and InAs–GaAs Alloys," *Infrared Phys.,* **7,** 201 (1967).
25. Prince, M. B., "Narrowband Self-Filtering Detectors," Chap. 3 in Willardson, R. K., and A. C. Beer (eds.), *Semiconductors and Semimetals,* Vol. 5, *Infrared Detectors,* Academic Press, New York (1970).
26. Sherring, C. W., and D. V. Eddolls, "Free Carrier Screening in a Microwave Biased InAs Photoconductor," *Infrared Phys.,* **11,** 203 (1971).

Reprinted from *Proc. IRE*, **47**(9), 1475–1478 (1959)

Single-Crystal Infrared Detectors Based Upon Intrinsic Absorption*

F. F. RIEKE†, L. H. DeVAUX†, AND A. J. TUZZOLINO†

UNTIL the early 1950's, the development of infrared photodetectors revolved principally around polycrystalline films of PbS, PbSe, or PbTe, deposited either by evaporation or from solution. The success of semiconductor physics based on single-crystal techniques, culminating in the invention and development of the transistor, was an incentive to apply similar methods to the development of infrared detectors.

Photodetectors based on bulk crystals of germanium were described by Shive[1] in 1950. Although his devices had attractive features, the spectral response was restricted to the region covered by the intrinsic absorption of germanium, namely, wavelengths less than 1.7 microns.

It was evident that to develop a detector with response extending farther into the infrared, it would be necessary either 1) to "sensitize" Si or Ge to the infrared by introducing impurities that were normally unionized, but could be ionized by low-energy quanta, or 2) to employ a semiconductor with an energy gap considerably smaller than that of germanium.

Approach 1) has resulted in the development of the doped-germanium family of detectors, which is described elsewhere in this issue. Approach 2) is the subject of the present discussion.

The long wavelength limit to intrinsic absorption and photoconductivity, λ_c, in a semiconductor is inversely proportional to the energy gap E_g:

$$\lambda_c E_g = 1.24 \text{ ev} \cdot \text{microns.}$$

If one, somewhat arbitrarily, sets the goal of developing a detector with response extending at least as far into the infrared as that of lead sulfide ($\lambda_c \geq 3$ microns), one requires a semiconductor with energy gap of 0.4 ev or less; the number of such semiconductors now known is small. Of these, only indium antimonide has been extensively developed.

The primary consideration in improving the performance of a detector employing photoconductive phenomena is to minimize the rate at which free carriers are generated spontaneously within the active material. This requirement applies regardless of the chemical species or physical form of the semiconductor, or of the particular mode of operation. The generation rate affects the performance in two ways. First, through its inverse corollary, the lifetime of excess carriers, it determines the magnitude of the electrical response to the photogeneration of carriers. Secondly, through its effect on fluctuations in the total number of carriers, it influences the internal noise of the detector. The two effects combine in such a way that an increased generation rate tends to decrease the attainable signal-to-noise ratio. The significance of excess-carrier lifetimes has

* Original manuscript received by the IRE, June 26, 1959.
† Chicago Midway Labs., Chicago, Ill.
[1] J. A. Shive, *Bell Lab. Rec.*, vol. 28, p. 8; 1950.

been emphasized by Rose[2] and by Petritz.[3] The subject of noise in photoconductors has been reviewed recently in the PROCEEDINGS.[4]

Generation of carriers occurs by way of two mechanisms, one thermal, the other optical. Thermal generation results from the transfer of energy from lattice vibrations to electrons, either within the bulk of the semiconductor or at the surface. Surface generation and recombination of carriers is particularly important in detectors that depend upon intrinsic optical absorption, which is characterized by absorption coefficients of the order of magnitude of 10^4 cm^{-1}. In consequence of this strong absorption, the photoeffects originate primarily in a shallow surface layer, at most a few microns deep, from which the excess carriers can readily diffuse to the surface. Thermal generation is catalyzed by particular kinds of impurities and imperfections and is thus subject to some measure of control in the growth of the crystal and in the preparation of its surface. It generally can be greatly reduced by lowering the temperature of the crystal, and it is usually for this reason that many detectors exhibit enhanced performance when they are cooled.

The optical generation of carriers occurs through the absorption of radiation to which the detector is sensitive, and is a necessary accompaniment to the useful sensitivity of the detector. The generation of carriers by ambient radiation, however, lowers the signal-to-noise ratio. Ambient radiation can be minimized by placing the sensitive elements within a cooled cavity having an aperture that admits radiation from only the useful field of view, but such shielding will be rather ineffective if the performance of the detector is limited by thermal generation of carriers.

The single-crystal detectors that have been developed are of three types: photoconductor, *p-n* junction, and photoelectromagnetic (PEM).

PHOTOCONDUCTOR

A photoconductive detector employing simple photoconductivity is constructed by cementing a thin section of semiconductor to a backing, and attaching current leads. The backing is important in dissipating heat generated in the sample by the bias current, which is advantageous to make as large as possible without raising the temperature of the sample unduly. If the semiconductor is to be cooled, it is placed within the vacuum space of a Dewar flask.

Photoconductive detectors using indium antimonide have been described by Avery, Goodwin and Rennie.[5]

Data on the performance of a typical cell[6] is given in Table I.

Photoconductive detectors have also been made using single-crystal tellurium cooled to 77°K. A noise equivalent power of 10^{-10} watts has been obtained for 500°K blackbody radiation at 85 cps with a 5-cps bandwidth. A minimum NEP of 10^{-11} watts occurs at a wavelength of 3.5 microns with a cutoff at 4.2 microns. The detector has an area of $1 \times \frac{1}{2}$ mm^2, a resistance of approximately 1000 ohms and a time constant of 30 to 50 microseconds.[7] In terms of detectivity D^{*}[8] we thus have:

$$D^{*}(500°K, 85 \text{ cps}) = 1.6 \times 10^9 \text{ cm cps}^{1/2} \text{ watt}^{-1},$$

$$D^{*}(3.5u, 85 \text{ cps}) = 1.6 \times 10^{10} \text{ cm cps}^{1/2} \text{ watt}^{-1}.$$

TABLE I

CHARACTERISTICS OF COOLABLE InSb PHOTOCONDUCTIVE CELLS (SENSITIVE AREA 1.4 MM^2, TIME CONSTANT LESS THAN 4×10^{-7} SECOND)

Temperature (°K)	90	195	249	292
Cell resistance (Ω)	20 K	1.8 K	490	120
D^{*}(cm cps$^{1/2}$ watt^{-1}) (λ_{peak}, 800 cps)	4.5×10^9	1.7×10^9	3×10^8	1.7×10^8
Wavelength of peak sensitivity (μ)	5.6	5.6	6.2	6.7
$\lambda_{1/2}(\mu)$	5.85	6.0	6.65	7.2
Responsivity (V/w)	1300	400	30	1

P-N JUNCTION

The theory and principles of operation of *p-n* junction detectors have been discussed elsewhere.[9,10] Photosensitive junctions can be produced by methods used in transistor technology, such as growth from the melt, vapor-phase diffusion, and alloying.

Grown *p-n* junctions are formed in a single-crystal bar of semiconductor with the *p-n* junction near the center. The junction plane is perpendicular to the major axis of the bar and is illuminated near its edge. The sensitive area is a small region surrounding the intersection of the junction plane with the surface of the bar and its magnitude is given by the product of the bar width and the sum of the minority carrier diffusion lengths. Diffused junctions can be formed by producing a thin *p*-type layer on an *n*-type semiconductor. Light is incident perpendicular to the resulting junction plane. The sensitive area can be varied by protecting a portion of the *p*-layer of the size and shape desired and etching away the remaining *p*-layer. *P-n* junction detectors sensitive in the intermediate infrared region generally require cooling to relatively low temperatures (~77°K). The detector can be operated with bias (photodiode) or without (photovoltaic), depending on which mode of operation leads to higher sensitivity.

[2] A. Rose, in "Photoconductivity Conference," John Wiley and Sons, New York, N. Y., pp. 3–48; 1955.
[3] R. L. Petritz, "Title," *Ibid.*, pp. 49–77.
[4] K. M. van Vliet, "Noise in semiconductors and photoconductors," PROC. IRE, vol. 46, pp. 104–118; June, 1958.
[5] D. G. Avery, D. W. Goodwin and A. E. Rennie, *J. Sci. Instr.*, vol. 34, p. 394; 1957.

[6] D. W. Goodwin, *Ibid.*, p. 368.
[7] G. Suits, private communication.
[8] D^{*} is defined as $\sqrt{A \Delta f}/$NEP where the NEP is measured with bandwidth Δf, the nature of the radiation and the chopping frequency being indicated by the quantities in parentheses.
[9] L. P. Hunter, "Handbook of Semiconductor Electronics," McGraw-Hill Book Co., Inc., New York, N. Y.; 1956.
[10] E. S. Rittner, "Photoconductivity Conference," John Wiley and Sons, New York, N. Y., p. 215; 1955.

Most of the development in the intermediate infrared region has been devoted to InSb *p-n* junctions. Detectors using grown junctions have been developed at Chicago Midway Laboratories.[11] Diffused-junction types have been developed and are manufactured by the Philco Radio Corporation and Texas Instruments, Inc. Grown-junction detectors have an inherently small sensitive area, whereas diffused-junction types can be made larger and of various shapes. The characteristics of the detectors are given in Table II.

TABLE II

CHARACTERISTICS OF INSB GROWN- AND DIFFUSED-JUNCTION
DETECTORS

	CML, Grown-Junction Detector	Philco, Diffused-Junction Detector[12]	Texas Instruments Diffused-Junction Detector[13]
Temperature (°K)	77°K	77°K	77°K
Cell resistance (Ω)	22 K	50–200	1200–1500
Area (cm²)	3.2×10^{-4}	4.1×10^{-2}	4.16×10^{-2}
Time constant (μsec)	<1.8	<1	~0.1
Wavelength of peak sensitivity (μ)	5.5	5.3	5.35
D^* (cm cps$^{1/2}$ watt^{-1})	9.4×10^9 (500°K, 90 cps)	2.7×10^9 (500°K, 800 cps)	5×10^9 (500°K, 1000 cps)
Cutoff $\lambda\frac{1}{2}(\mu)$	~5.7	~5.5	~5.5

PHOTOELECTROMAGNETIC (PEM) TYPE

The use of the PEM effect[15,16] leads to another type of single-crystal photodetector. When electrons and holes produced at the front surface of a semiconductor by light diffuse downward, a magnetic field applied perpendicular to this initial diffusion current will bend the electron and hole currents in opposite directions The resulting separation of charge gives rise to a photovoltage.

The strength of the PEM effect is determined by the product of the carrier mobility and the magnetic-field strength. For practicable magnetic fields, a semiconductor of high mobility such as InSb must be employed to achieve useful sensitivity. No bias current is required and the detector can be operated at room temperature. The detectors have been developed and manufactured by many agencies.

A typical room-temperature PEM detector made at Chicago Midway Laboratories has a peak noise-equivalent power of 8×10^{-10} watts at 6.2 microns with a cutoff wavelength at 7 microns. The detector has an area of 1.8×10^{-2} cm² and a resistance of 33 Ω. The detectivity $D^*(6.2 \, \mu, \, 90 \, \text{cps})$ is thus 1.9×10^8 cm cps$^{1/2}$ watt^{-1}. Although the time constant of the cell has not been measured directly, minority carrier-lifetime measurements indicate a value of less than 1 microsecond for the time constant.

In conclusion, we shall attempt to assess what has been done and what remains to be done. One can estimate how closely present-day intrinsic photoconductors can approach the radiation limit by introducing the noise figure, defined as the ratio of the detectivity limited by radiation fluctuations to peak detectivity. In principle, the noise figure $(\text{NF}) = (\tau^*/\tau)^{1/2}$, where τ^* is the radiation lifetime and τ is the actual lifetime.[3] Values of D^* and the noise figure of InSb and Te detectors are given in Table III.

The best cooled InSb cells have a noise figure of ten or less, and therefore the ultimate possibilities seem to

TABLE III

VALUES OF D^* AND THE NOISE FIGURE (NF) FOR INSB AND TE DETECTORS

	Photoconductive			Photovoltaic			PEM		
	T°K	$D^*(\lambda_{peak}, 800 \text{ cps})$ (cm cps$^{1/2}$ watt^{-1})	NF	T°K	$D^*(5.4 \, \mu, 90 \text{ cps})$ (cm cps$^{1/2}$ watt^{-1})	NF	T°K	$D^*(6.2 \, \mu, 90 \text{ cps})$ (cm cps$^{1/2}$ watt^{-1})	NF
InSb	90	4.5×10^9	10	78	2×10^{10}	5	292	1.9×10^8	300
	195	1.7×10^9	50						
	292	1.7×10^8 ($3.5 \, \mu$, 85 cps)	300						
Te	78	1.6×10^{10}	27						

[11] G. R. Mitchell, A. E. Goldberg, and S. W. Kurnick, *Phys. Rev.*, vol. 97, p. 239; 1955.

[12] M. E. Lasser, P. Cholet, and E. C. Wurst, Jr., *J. Opt. Soc. Am.*, vol. 48, p. 468; 1958.

[13] R. L. Petritz, private communication.

[14] The value of D^* for CML and Philco presented in the table is characteristic of their better detectors. The value for Texas Instruments is an average value.

[15] S. W. Kurnick and R. N. Zitter, *J. Appl. Phys.*, vol. 27, p. 278; 1956.

[16] C. Hilsum and I. M. Ross, *Nature*, vol. 179, p. 146; 1957.

have been approached. In this case, the future possibilities lie in improving design and reproducibility in manufacture. For the Te detector, the noise figure of 27 indicates that some improvement should be possible for this cooled detector. The uncooled InSb cells all have noise figures of the order of 300. For InSb at room temperature, $(\tau^*/\tau) = 10.$[13] The reason for the large noise figure is not evident. It would appear that in this case a very

154

great improvement in uncooled detectors is within the realm of possibility. Whether or not the same can be said for other single-crystal semiconductors must wait until results of lifetime measurements become available.

No other small-gap semiconductor has been so highly developed as InSb. There would probably not be much point in developing other materials having a cutoff wavelength ~6 microns. Detectors having cutoffs ~4 microns which might compete with PbS have not been extensively developed. The single-crystal-cooled Te de tector discussed above has a cutoff at 4.2 microns, but further development on this detector will be necessary to determine whether or not it will compare with PbS. In addition, some work has been done on InAs where photoconductive, PEM, and photovoltaic responses have been reported.[17,18] Useful sensitivity is obtained out to 4 microns.

Little work has been done on intrinsic photoconductors having cutoff wavelengths ~10 microns. Some work has been done on HgTe-CdTe alloys in which appreciable sensitivity has been observed as far out as 13 microns.[19] Further development will be necessary before a comparison can be made of these intrinsic alloy detectors with the doped-Ge family of detectors.

[17] R. M. Talley and D. P. Enright, *Phys. Rev.*, vol. 95, p. 1092; 1954.
[18] C. Hilsum, *Proc. Roy. Phys. Soc. B*, vol. 70, p. 1011; 1957.
[19] W. D. Lawrence, *et al.*, *J. Chem. Phys. of Solids*, to be published.

Reprinted from *Proc. IRE*, **47**(9), 1524–1529 (1959)

Detectivity and Preamplifier Considerations for Indium Antimonide Photovoltaic Detectors*

G. R. PRUETT AND R. L. PETRITZ†

INTRODUCTION

THE indium antimonide single-crystal photovoltaic detector[1] shows promise of reaching the ultimate limit of detectivity, termed the *Blip* condition of operation.[2] The detector has a lower impedance than the lead-salt detectors, and special attention must be given to the associated preamplifier. The detector signal, noise and detectivity characteristics as well as the preamplifier requirements will be discussed. Equations will be derived to show the conditions that must be fulfilled by the preamplifier in order to utilize the full detectivity of the detector. Equations which will simplify problems in design are derived, and a specific circuit is included with experimental data.

VOLTAGE-CURRENT CHARACTERISTICS

These detectors are prepared from single crystals of n-type indium antimonide. Diffusion or alloy techniques are used to create a thin p-type layer on the surface, and contact is made to the surface and base regions; thus, the model is a simple p-n junction with radiation incident normal to the junction. A more complete discussion of this type of detector is given elsewhere in this issue.[1]

In thermal equilibrium a p-n junction has voltage-current characteristics[3] given by

$$I_d = I_s(e^{qV/\beta kT} - 1), \qquad (1)$$

where I_d is the junction current, I_s is the saturation current, q is the electronic charge, V is the drop in potential across the junction of the diode, k is Boltzmann's constant, T is the equilibrium temperature of the diode, and β is a factor[3] which is unity for an ideal diode, but is found to be greater than unity in cooled indium-antimonide junctions. Cooled indium antimonide appears to be similar to silicon at room temperature and to cooled germanium where the main generation of carriers occurs in the space-charge region of the junction,

and where β is expected to be of the order of two.[4]

When radiation from a background with a temperature higher than the equilibrium temperature of the diode falls onto the detector, the voltage-current equation for the diode becomes[5]

$$I = -I_{sc} + I_s(e^{qV/\beta kT} - 1). \qquad (2)$$

I_{sc} is the current induced by the incident background radiation. A derivation of (2) with an expression for I_{sc} in terms of basic semiconductor parameters is given by Rittner[5] in a treatment of the solar battery.

I_{sc} can be expressed in terms of the background-radiation flux as

$$I_{sc} = q\eta_r J_r A, \qquad (3)$$

where

$$J_r = \int_0^{\lambda_i} J_\lambda d\lambda. \qquad (4)$$

J_r is the radiation flux expressed in photons per second per square centimeter of area, A is area in square centimeters, η_r is the quantum efficiency with respect to background radiation, and $\lambda_i = hc/E_i$ is the long wavelength cutoff of the detector, corresponding to the energy gap, E_i. In indium antimonide, the energy gap at 77°K is 0.23 ev corresponding to a long wavelength cutoff of 5.35 microns. The radiation flux, J_r, and the current density, j_r, for an indium antimonide photodiode with an equilibrium temperature of 77°K while looking at 300°K radiation are

$$J_r = 2 \times 10^{16} \text{ photons/cm}^2 \text{ sec},$$

and

$$j_r = qJ_r = 3.2 \times 10^{-3} \text{ amp/cm}^2.$$

An additional path for current in diffused, alloyed, or grown photodiodes is a conductance, G_s, which is in shunt with the diode. When G_s is incorporated in the voltage-current equation, the latter becomes

$$I = -I_{sc} + I_s(e^{qV/\beta kT} - 1) + G_s V. \qquad (5)$$

* Original manuscript received by the IRE, April 27, 1959; revised manuscript received, July 8, 1959.

† Central Res. Labs., Texas Instruments Inc., Dallas, Texas.

[1] A general discussion of indium antimonide infrared detectors is given by F. F. Rieke, L. H. DeVaux, and A. J. Tuzzolini, "Single crystal infrared detectors based upon intrinsic absorption," paper 3.3.5, this issue, p. 1475.

[2] For a discussion of the *Blip* (background-limited infrared photodetector) detector, see R. L. Petritz, "Fundamentals of infrared detectors," paper 3.3.1, this issue, p. 1458.

[3] J. L. Moll, "The evolution of the theory of the voltage-current characteristics of p-n junctions," PROC. IRE, vol. 46, pp. 1076–1082; June, 1958.

[4] C. T. Sah, R. N. Noyce, and W. Shockley, "Carrier generation and recombination in p-n junctions and p-n junction characteristics," PROC. IRE, vol. 45, pp. 1228–1243; September, 1957.

[5] E. S. Rittner, "Electron processes in photoconductors," in "Photoconductivity Conference," R. G. Breckenridge, B. R. Russell, and E. E. Hahan, eds., John Wiley and Sons, Inc., New York, N. Y., pp. 250–257; 1956.

Typical voltage-current curves for indium antimonide photodiodes are shown in Fig. 1. In curve A, the detector is at thermal equilibrium at 77°K with no excess background radiation; curve B shows the effect of adding 300°K radiation while the diode is held at 77°K, and curve C shows the effect of still more background radiation. Note that I_{sc} can be taken directly from Fig. 1 at the intercept with the current axis. The slope of the curve in the reverse biased condition would approach zero for an ideal photodiode, but in the actual photodiode this slope gives the value for the shunt conductance G_s. By plotting $\log (I+I_{sc}-G_sV)$ vs V for the forward-bias condition, the values of β and I_s can be determined. The value of β has varied from 2 to 3.8 for the photodiodes measured.

The impedance of the Norton equivalent circuit is found by differentiating (5) with respect to V while holding J_s constant:

$$\Delta I = (G_d + G_s)\Delta V, \qquad (8)$$

where

$$G_d = \frac{qI_s}{\beta kT}\, e^{qV/\beta kT}, \qquad (9)$$

$$R_d = \frac{1}{G_d}, \quad \text{and} \quad R_s = \frac{1}{G_s}. \qquad (10)$$

At low frequencies (7) and (8) indicate the Norton equivalent circuit that is shown in Fig. 2.

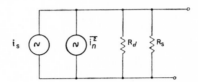

Fig. 2—Norton's equivalent circuit for indium-antimonide photodiode.

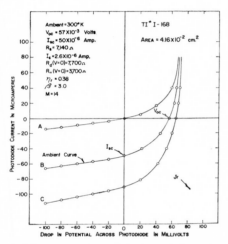

Fig. 1—Voltage-current characteristics for an indium-antimonide photodiode under several conditions of ambient radiation.

SMALL-SIGNAL PROPERTIES

Response of the photodiode to a small radiation signal, ΔJ_s, in the presence of the background radiation, is obtained from (5) as

$$\Delta I = \left[-\frac{dI_{sc}}{dJ_s} + \frac{q}{\beta kT} I_s e^{qV/\beta kT}\frac{dV}{dJ_s} + G_s\frac{dV}{dJ_s} \right]\Delta J_s. \quad (6)$$

The small signal short-circuit current generator for a Norton representation is found by holding V constant:

$$i_s = \Delta I]_{V=\text{const}} = -\frac{dI_{sc}}{dJ_s}\Delta J_s = -q\eta_s A\Delta J_s, \quad (7)$$

where η_s is the quantum efficiency with respect to the signal radiation. The designation η_s is used to distinguish it from η_r because of the possible difference in the spectral character of the signal radiation and background radiation.

NOISE PROPERTIES

The fundamental noise associated with an ideal narrow-base diode ($\beta = 1$) can be expressed[6] in terms of shot noise for each current:

$$\frac{\overline{i^2}}{\Delta f} = 2q[I_{sc} + I_s(e^{qV/kT} + 1)]. \qquad (11)$$

When current is generated in the space-charge region ($\beta > 1$), (11) must be modified according to Schneider and Strutt,[7] who have recently made a study of this problem in silicon. When β is considered, the shot noise equation for the photodiode is

$$\frac{\overline{i^2}}{\Delta f} = 2q\left[I_{sc} + \frac{I_s}{\beta}(e^{qV/\beta kT} + 1) \right]. \qquad (12)$$

This equation is correct at thermal equilibrium ($I_{sc}=0$, $V=0$) since

$$\frac{\overline{i^2}}{\Delta f}\bigg]_{\substack{V=0 \\ I_{sc}=0}} = \frac{4qI_s}{\beta} = 4kTG_d, \qquad (13)$$

in agreement with the Nyquist result. Eq. (12) is valid for small bias near the $V=0$ condition but needs modification for large forward bias. Since the bias condition near $V=0$ is of the greatest interest for this discussion, (12) will be used throughout this paper.

[6] A recent review and extensive bibliography of the theory of shot noise in *p-n* junctions is by A. van der Ziel, "Noise in junction transistors," PROC. IRE, vol. 46, pp. 1019–1038; June, 1958.
[7] B. Schneider and M. J. O. Strutt, "Theory and experiments on shot noise in silicon *p-n* junction diodes and transistors," PROC. IRE, vol. 47, pp. 546–554; April, 1959.

The shunt conductance has a Nyquist noise,

$$\frac{\overline{i_s^2}}{\Delta f} = 4kTG_s. \tag{14}$$

Experiment has indicated that the main source of $1/f$ noise is the current which passes through the shunt conductance. An empirical equation for the $1/f$ noise can be written as

$$\frac{\overline{i_f^2}}{\Delta f} = \frac{1}{f}\left[k_1 G_s^2 V^2 + k_2(I - G_s V)^2 + k_3 I^2\right], \tag{15}$$

where

$$k_1 \gg k_2 \text{ and } k_3.$$

The values for k_1, k_2, and k_3 must be determined experimentally, since no quantitative theory of $1/f$ noise exists.

The equation for the total short-circuit noise generator is obtained from (12), (14), and (15).

$$\frac{\overline{i_n^2}}{\Delta f} = 2q\left[I_{sc} + \frac{I_s}{\beta}(e^{qV/\beta kT} + 1)\right] + 4kTG_s$$
$$+ \left[\frac{k_1 G_s^2 V^2 + k_2(I - G_s V)^2 + k_3 I^2}{f}\right]$$
$$= N_n. \tag{16}$$

SIGNAL-TO-NOISE RATIO AND DETECTIVITY

When (7) and (16) are combined, the signal-to-noise ratio for the photodiode can be expressed as

$$\frac{i_s}{\sqrt{\overline{i_n^2}}} = \frac{qn_s A \Delta J_s}{\sqrt{\Delta f}(N_n)^{1/2}}. \tag{17}$$

Examination of (17) and (16) show that the S/N will be near a maximum when $V = 0$, and this has been verified by experiment. The chief reason for this is because the $1/f$ noise term due to the shunt conductance disappears at this bias. Another point is that the condition

$$I_{sc} \gg \frac{I_s}{\beta}(e^{qV/\beta kT} + 1) \tag{18}$$

is desirable, since I_{sc} reflects noise due to the background radiation, while the I_s term contains the lattice-induced noise.[2] Conditions for (18) can be obtained if the forward-bias state is avoided, since I_{sc} is usually much greater than $2I_s$.

The specific detectivity of the detector, D^*_d, is defined[8] as

$$D_d^* = \frac{(S/N_d)\sqrt{A\Delta f}}{E_s A \Delta J_s}, \tag{19}$$

where E_s is the energy per photon of the signal radiation, and N_d is the detector noise. This is a measure

[8] For a discussion of the specific detectivity see R. C. Jones, "Phenomenological description of the response and detecting ability of radiation detectors," paper 4.1.1, this issue, p. 1495.

of S/N per unit of incident power, normalized to 1 cm² area and unit bandwidth.

When (17) is substituted into (19), and using (16),

$$D^*_d = \frac{qn_s}{E_s\left\{2q\left[j_{sc} + \dfrac{j_s}{\beta}(e^{qV/\beta kT} + 1)\right] + \dfrac{4kTG_s}{A}\right.}$$
$$\overline{\left. + \dfrac{1}{fA}[k_1 G_s^2 V^2 + k_2(I - G_s V)^2 + k_3 I^2]\right\}^{1/2}}, \tag{20}$$

where

$$j_{sc} = \frac{I_{sc}}{A} \text{ and } j_s = \frac{I_s}{A}.$$

If

$$\eta_s = 1, \quad \eta_r = 1, \quad j_{sc} \gg \frac{j_s}{\beta}(e^{qV/\beta kT} + 1),$$

and there is no $1/f$ noise, then this equation becomes that of the ideal radiation detector discussed elsewhere in this issue,[2] that is, the *Blip* (background-limited infrared photodetector):

$$D^*_{Blip} = \frac{1}{E_s\sqrt{2J_r}}. \tag{21}$$

Note that $\sqrt{2}$ appears in (21) whereas $\sqrt{4}$ appears in (46) of reference 2. This is because the photodiode registers only generation noise of the background radiation; the recombination of hole-electron pairs occurs in the base material and does not contribute to the noise.

D^*_{Blip} for indium antimonide facing hemispherical 300°K background radiation is

$$D^*_{Blip}(500° \text{ K}) = 24 \times 10^9 \text{ watts}^{-1},$$

and

$$D^*_{Blip}(\lambda_i = 5.35\mu) = 13.6 \times 10^{10} \text{ watts}^{-1}.$$

DETECTOR NOISE FIGURE

A quantitative measure of the performance of a photodetector is the noise figure, F_d, defined[2] by

$$F_d = \frac{(S/N)^2_{Blip}}{(S/N)_d^2} = \frac{(D^*_{Blip})^2}{(D^*_d)^2}. \tag{22}$$

Substituting (20) and (21) into (22), one finds

$$F_d = \frac{1}{\eta_s^2}\left(\frac{N_n}{2qI_r}\right), \tag{23}$$

where Nn is defined in (16) and

$$I_r = qJ_r A.$$

Note that for a detector which is not limited at all by radiation noise, F_d depends on the quantum efficiency as $1/\eta_s^2$, but for detectors which see radiation noise, (23) can be written

$$F_d = F_d \text{ (quantum eff)} \cdot F_d \text{ (noise)}, \tag{24}$$

where

$$F_d \text{ (quantum eff)} = \frac{\eta_r}{\eta_s^2},$$

and

$$F_d \text{ (noise)} = 1 + \cfrac{2q\dfrac{I_s}{\beta}(e^{qV/\beta kT} + 1) + 4kTG_s + \cfrac{k_1G_s^2V^2 + k_2(I - G_sV)^2 + k_3I^2}{f}}{2qI_{sc}} \quad (25)$$

For radiation signals of spectra $0 \leq \lambda_i \leq 5.35\mu$, η_s can be taken equal to η_r so that

$$F_d \text{ (quantum eff)} = \frac{1}{\eta_s}. \quad (26)$$

For the example of Fig. 1,

$$F_d \text{ (quantum eff)} = \frac{1}{0.38} = 2.6, \text{ or } 4.2 \text{ db.}$$

To calculate F_d(noise) precisely, the parameters specified in (25) must be determined. Of these, only k_1, k_2, and k_3 require direct noise measurements; all the other parameters can be taken from the voltage-current curves as previously discussed. It is of interest to calculate F_d (noise) at $V = 0$, where the major source $(k_1G_s^2V^2)$ of $1/f$ noise is zero. The $1/f$ noise associated with the terms containing k_2 and k_3 is neglected. Then

$$F_d \text{ (noise)}\big]_{\substack{V=0 \\ k_3=0}} = 1 + \frac{1}{M}, \quad (27)$$

where

$$M = \frac{qI_{sc}R_{\parallel}}{2kT} = 75I_{sc}R_{\parallel}. \quad (28)$$

$$1/R_{\parallel} = Gs + Gd$$

Eq. (27) is useful since it gives the lower limit of F_d (noise) for $V = 0$. The presence of $1/f$ noise will cause F_d to be larger than this value.

For the example of Fig. 1,

$$F_d \text{ (noise)}\big]_{V=0} = 15/14, \text{ or } 0.3 \text{ db;}$$

thus the detector should be only 0.3 db from a 100 per cent radiation noise-limited condition if there is no $1/f$ noise present.

The total theoretical noise figure for the detector is

$$F_d = 2.6(15/14) = 2.8, \text{ or } 4.5 \text{ db.}$$

This indium antimonide detector should be only 4.5 db from the fundamental limit of detectivity; that is

$$D^*_d(500° \text{ K}) = \frac{D^*_{Blip}}{\sqrt{2.8}} = 14 \times 10^9 \text{ watts}^{-1}$$

With detectors having such high theoretical performance available, requirements on the preamplifier should

is sufficiently stringent that the performance of the photodiode is not degraded by the preamplifier.

EFFECTIVE NOISE FIGURE OF PREAMPLIFIER

When the detector is connected to a preamplifier, the problem becomes one of determining how the noise figure of the preamplifier affects the detectivity of the detector. The usual amplifier noise figure, F_a, is defined relative to the Nyquist noise of a resistor, R, at room temperature. The amplifier contribution to the noise can be represented in an equivalent circuit diagram as a series noise generator,

$$\overline{e_a^2} = 4kTR(F_a - 1)\Delta f, \quad (29)$$

as shown in Fig. 3(a).

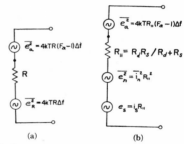

(a)   (b)

Fig. 3—Equivalent circuit for preamplifier input (a), and for detector-preamplifier combination (b).

When the indium antimonide detector is connected in place of the Nyquist noise resistor, a different situation exists: the detector is cooled, and the source noise is not that of a pure resistance but that of the detector. If the equivalent circuit of the detector, Fig. 2, is converted to a voltage equivalent circuit as in Fig. 3(b) and added in series with the amplifier noise generator, an effective noise figure, F^*_a, which has meaning relative to the cooled indium antimonide detector, can be defined. The effective noise figure of the amplifier can now be expressed in the same formal manner as the conventional noise figure,

$$F^*_a = \frac{(S/N)_d^2}{(S/N)_{d+a}^2} = \frac{(D^*_d)^2}{(D^*_{d+a})^2} = \frac{(N_{d+a})^2}{(N_d)^2}. \quad (30)$$

D^*_{d+a} is the detectivity of the detector including the amplifier noise:

$$D^*_{d+a} = \frac{S\sqrt{A\Delta f}}{N_{d+a}E_sA\Delta J_s}, \quad (31)$$

where N_{d+a} includes both the noise of the detector and preamplifier.

Substituting (16) and (29) into (30), and using the circuit of Fig. 3(b),

$$F^*_a = 1 + \frac{4kT_a(F_a + 1)}{R_\| N_n} \qquad (32)$$

is obtained for the effective noise figure of the preamplifier. For indium antimonide detectors at the $V=0$ condition and assuming $k_2 = k_3 = 0$, (32) becomes

$$F_a^* = 1 + \frac{T_a(F_a + 1)}{T_d(1 + M)}, \qquad (33)$$

where $T_d = 77°$K and $T_a = 300°$K. If the data from Fig. 1 are used along with $F_a = 2$ (3 db) as an example, F^*_a is 1.27 or 1.0 db. This means that the detectivity as measured through this amplifier (D^*_{d+a}) is a factor of $\sqrt{1.27}$ less than the actual detectivity of the detector (D^*_d), as can be seen from the definition of F^*_a in (30):

$$D^*_{d+a}(500°\text{K}) = \frac{D^*_d}{\sqrt{1.29}} = 12.5 \times 10^9 \text{ watts}^{-1}.$$

To minimize the noise introduced by the amplifier, I_{sc} and $R_\|$ need to be as large as possible, and F_a should be as nearly unity as possible. A recent increase in detector impedance as a result of improved detector technology has helped in its direct effect in (33) as well as in its effect on transistor preamplifiers, which have better noise figures with source impedances of a few thousand ohms rather than a few hundred ohms.

Noise Figure of Detector-Amplifier Combination

The over-all noise figure of the detector with the amplifier can be defined as

$$F_{d+a} = \frac{(S/N)_{Blip}^2}{(S/N)_{d+a}^2} = \frac{(D^*_{Blip})^2}{(D^*_{d+a})^2}$$
$$= \frac{(D^*_{Blip})^2}{(D^*_d)^2} \times \frac{(D^*_d)^2}{(D^*_{d+a})^2}, \qquad (34)$$

or

$$F_{d+a} = F_d \cdot F^*_a, \qquad (35)$$

where F_d is given by (23) and F^*_a by (32). Using the example from Fig. 1,

$$F_{d+a} = 4.5 \text{ db} + 1.0 \text{ db} = 5.5 \text{ db (or the ratio, 3.54)}, \qquad (36)$$

and

$$D^*_{d+a}(500° \text{ K}) = \frac{D^*_{Blip}}{\sqrt{3.54}} = 12.5 \times 10^9 \text{ watts}^{-1}, \qquad (37)$$

in agreement with the results of the previous section. This theoretical value for D^*_{d+a} is very satisfactory and indicates that contemporary indium antimonide photodetectors constitute very sensitive infrared detectors which can be connected directly into transistor preamplifiers without the use of a coupling transformer.

Measuring Technique and Preamplifier Application

Transistorized preamplifiers have been constructed, and measurements have been taken to investigate the above conclusions. A simple preamplifier circuit using a selected germanium transistor (2N185) is shown in Fig. 4. The characteristics of the amplifier as a function of frequency are shown in Fig. 5, and the dependence of F_a on source resistance is shown in Fig. 6. At present the noise figure for alloyed germanium transistors is superior to that of other transistors when operated in the "starved" condition[9] (low collector currents). Silicon transistors can be obtained with noise figures between 3 and 6 db.

Two methods are used for biasing to the zero-voltage condition. A choke can be placed across the photodiode to provide a dc short while maintaining a high impedance for ac signals, or the detector can be back-biased to the condition of zero voltage across the diode as shown in Fig. 4. The latter method has been used for making the experimental measurements. R_b should be large compared to R_s so that no Nyquist noise is introduced by the biasing resistor. If R_b is not large compared to R_s, the above formulas are still applicable provided R_s is considered to be the parallel combination of R_b and R_s.

The experimental values for D^*_{d+a} (500°K) are measured with the transistor preamplifier shown in Fig. 4 coupled directly into a standard amplifier. Signal and noise measurements are taken from a narrow-band wave analyzer, and the radiation signal is obtained from a standard calibrated 500°K blackbody chopped at 1000 cps. Both noise voltage and signal voltage are measured with the detector biased to the zero dc voltage condition with a 300°K ambient radiation. Thus D^*_{d+a} as defined in (31) is measured experimentally without transformers.

The value of D^*_{d+a} for the detector of Fig. 1, as measured under the condition noted above was

$$D^*_{d+a}(500°\text{K})]_{V=0} = 8.7 \times 10^9 \text{ watts}^{-1}, \qquad (38)$$
$$\text{f=1000 cps}$$

and

$$F^*_{d+a}(500°\text{K}) = \frac{24}{8.7} = 2.76, \text{ or } 8.8 \text{ db}. \qquad (39)$$

When the value from (39) or (38) is compared to the theoretical value from (36) or (37), it is found that the experimental value of D^*_{d+a} is 3.3 db below the theoretical value. The difference is due to the neglect of $1/f$ noise in calculating D^*_{d+a} (theoretical). In the measurement of a large number of detectors, results such as those of (38) and (39) are typical. A difference of 3 to 6

---

[9] R. D. Middlebrook and C. A. Meed, "Transistor ac and dc amplifiers with high input impedance," *Semiconductor Products*, vol. 2, pp. 26–35; March, 1959.

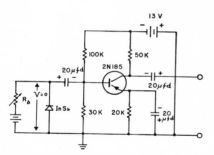

Fig. 4—Circuit diagram for transistor preamplifier with an indium antimonide photodiode back-biased to the $V=0$ operating point.

db between the experimental and the theoretical value of D^*_{d+a} is common and is attributable to $1/f$ noise in the detectors.

Summarizing the values obtained for the example of Fig. 1, it is found that

$$D^*_{d+a}(\text{exp}) = 8.7 \times 10^9 \text{ watts}^{-1},$$

and

$$F^*_{d+a}(\text{exp}) = 8.8 \text{ db}$$

of which

$$F_d \text{ (quantum eff)} = 4.2 \text{ db},$$
$$F_d \text{ (noise other than } 1/f) = 0.3 \text{ db},$$
$$F_d \text{ (}1/f \text{ noise)} = 3.3 \text{ db},$$
$$F^*_a = \underline{1.0 \text{ db}}$$
$$8.8 \text{ db}.$$

Conclusion

Contemporary indium antimonide detectors can be used to nearly their full potentialities directly with germanium or silicon transistor preamplifiers; no coupling transformer or choke is required. With further increase in detector impedance and I_{sc}, the amplifier noise contribution can be reduced to negligible proportions for even relatively noisy preamplifiers. The formulas which have been derived for calculation of the noise figure of the detector and the preamplifier separately should be useful in the design of both preamplifiers and detectors. These formulas should also provide guidance in detector and preamplifier research and development.

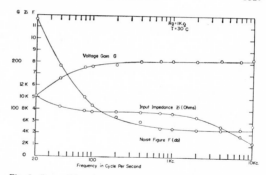

Fig. 5—Preamplifier characteristics as a function of frequency for an input of 1000 ohms at 30°C.

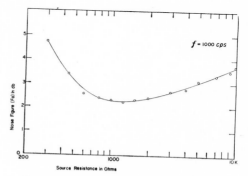

Fig. 6—Preamplifier noise figure as a function of source resistance at 1000 cps.

Acknowledgment

The authors wish to acknowledge the contributions to this subject of W. A. Craven, Jr., and R. H. Genoud, who kindly gave us a preprint of their paper, "Characterization of Indium-Antimonide Photodetectors," presented before the Optical Society of America on April 3, 1959. We would also like to acknowledge the aid given by the personnel at Texas Instruments Inc. who were associated with the development and production of infrared detectors, especially B. R. Pagel and R. Fewer, who designed the transistor preamplifiers, and L. G. Sloan, who supervised the construction of the detectors used in collecting the experimental data.

Reprinted from *Appl. Optics*, **4**(6), 659–663 (1965)

Photoconductive Indium Antimonide Detectors

F. D. Morten and R. E. J. King

The manufacture of InSb photoconductive detectors is briefly described. A simplified design theory is given, followed by a description of typical performance between room temperature and 77°K. The practical embodiment of photocells is described, together with the design of arrays of detector elements.

I. Introduction

The semiconducting properties of InSb were first investigated by Blum *et al.*,[1] and by Welker.[2] Measurement of the forbidden energy gap as 0.18 eV, at room temperature, indicated a potential infrared detector material. Photoconductive detectors operating at room temperature have been described by Avery *et al.*[3] and the low-temperature photoconductive properties studied by Goodwin[4] and by Kurnick and Zitter.[5] Today, photoconductive and photoelectromagnetic detectors operating at room temperature, and photoconductive and photovoltaic detectors operating at low temperatures are available and are proving to be the most suitable detectors in the wavelength region 3–7 μ for many applications.

In this paper we aim to give an account of the present state of development of InSb photoconductive detectors, together with some of the theoretical background which leads to an understanding of the design requirements and limitations. Photovoltaic InSb detectors and InSb photoconductive detectors in the 100–1000 μ wavelength range are not considered since these are dealt with in other papers in this issue.

II. Manufacture

The basic requirement for a photoconductive detector of the InSb type is, as will be seen from the next section, material of very high purity into which small and carefully controlled amounts of a particular impurity can be introduced. Such material also must have a high carrier mobility and also a high value for the excess carrier lifetime. The detector elements made from such materials must be thin. Such thin layers can be made by evaporation[6] as in some types of lead-salt detectors, or by etching of single crystal material. The

The authors are with Associated Semiconductor Manufacturers Ltd., Mullard-Southampton Works, Southampton, England.
Received 18 December 1964.

latter method leads to a better control of material properties with existing techniques, and has been used almost exclusively for making InSb photoconductive detectors.

The preparation of InSb single crystals of the required purity has been described by Hulme and Mullin,[7] King and Bartlett,[8] and others.[9] Indium and antimony are separately purified before being combined in a closed vessel at 750°C. The resulting polycrystalline material is then zone-refined to give material with a net donor concentration less than 3×10^{14} cm^{-3}. Crystals may be grown either by pulling or by horizontal growth; the horizontal growth technique gives larger volumes of constant purity. For photoconductive devices a p-type dopant such as germanium is often used.[10]

The cutting and lapping of InSb may be carried out by methods similar to those used for germanium or silicon, but regard must be paid to the greater brittleness of the material.

The preparation of thin specimens may be carried out by the electrochemical process described by Kurnick and Zitter.[5] Specimens of the desired shape, with a thickness of about 200 μ, have indium-based contacts applied. They are then electrolytically etched in an ethylene glycol–nitric acid mixture. This produces an oxide which stops the process when material of thickness 10–20 μ has been removed. This process is repeated until the required thickness is reached. As surface conductance and recombination can be important, the etching process must be carefully controlled.

The specimens are fragile and consequently are usually mounted by means of a suitable adhesive on a substrate, which may be part of a dewar in a cooled cell.

III. Design Theory for InSb Photoconductive Detectors

In this section, a simple theory of intrinsic detector performance applicable to InSb is developed.

A. Responsivity

It is convenient to consider the open circuit signal[11] as a measure of responsivity. This can be shown,[12] by con-

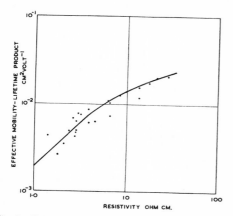

Fig. 1. Variation of effective mobility-lifetime product with resistivity at 77°K for *p*-type InSb.

sidering the change in conductance on illumination, to be given by

$$V_s = \frac{IQe(\mu_n\tau_n + \mu_p\tau_p)R^2W}{(1+\beta)L},$$ (1)

where

I = dc bias current,
Q = signal photons per cm² per sec,
e = electronic charge,
μ_n, μ_p = electron and hole mobilities,
τ_n, τ_p = electron and hole lifetimes,
R = specimen resistance,
W = specimen width,
L = specimen length,
β is a term involving surface and bulk properties and specimen dimensions. β can be made negligible with the appropriate surface treatment and will be neglected hereafter.

This expression may be simplified in two extreme cases: (*a*) for temperatures above about 150°K when $\tau_n \approx \tau_p$ and $\mu_n \gg \mu_p$,

$$V_s = \frac{IQe\mu_n\tau_nR^2W}{L},$$ (2)

(*b*) for *p*-type material at low temperatures in which electron trapping occurs (Laff and Fan),[13]

$$V_s = \frac{IQe\mu_p\tau_pR^2W}{L}.$$ (3)

B. Noise

The noise may be due to five main causes: (a) Johnson noise, (b) generation-recombination (G–R) noise, (c) background noise, (d) $1/f$ noise, and (e) *environmental* noise (cooling, mechanical, or optical).

The theory and causes of $1/f$ noise are not well understood. It is therefore not possible to introduce it into a simple theory of detector performance, but it must be realized that $1/f$ noise will in general appear. At

77°K, it is likely to dominate the noise for frequencies below 200–1000 cps depending on the treatment of the element. Environmental noise is dependent on the conditions under which the photocell is operated and, while it cannot be ignored in practice, will not be considered further.

Johnson noise is given in the form of an open circuit voltage, as

$$V_{n_1} = \sqrt{4kTR\Delta f}.$$ (4)

Generation-recombination noise can be calculated for a single-level model, and written, for frequencies small compared with the reciprocal of the lifetime, in the form

$$V_n = \frac{IR^{1/2}\sqrt{4e\mu\tau\Delta f}}{L},$$ (5)

where the resistance R is controlled by the thermally generated carriers from the single impurity level. Background noise can be written in a similar form to (5), if the conductance is due to background generated carriers alone.

C. Detectivity

It is convenient to consider three conditions.

a. At or Near Room Temperature

Under these conditions with practical exciting currents, the Johnson noise (current independent) dominates G–R noise (proportional to current). The area-normalized detectivity is, from Eqs. (2) and (4):

$$D^* = \frac{IR^{1/2}e\mu_n\tau_n}{\sqrt{4kTWL}}\frac{\lambda}{hc}$$ (6)

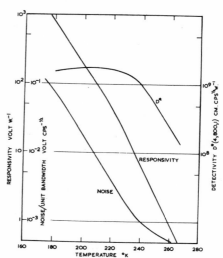

Fig. 2. Variation of responsivity, noise, and detectivity with temperature between 180°K and 270°K. (4-μ signal radiation, 3.5 mm × 0.6 mm specimen, 10 mA current.)

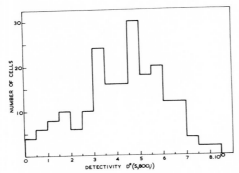

Fig. 3. Distribution of detectivity at 5 μ for two hundred detectors at 77°K.

and rises with current. Writing the wattage dissipation in the element per unit area $I^2R/WL = C^2$, and $R = \rho L/wt$, where ρ is the resistivity and t the thickness of the element,

$$D^* = \frac{C \ e\mu_n\tau_n}{\sqrt{4kT}} \frac{\rho}{t} \frac{\lambda}{hc}. \qquad (7)$$

Thus, to obtain a high detectivity, the material must be thin and the thermal impedance to the mounting low.

b. Generation-Recombination Noise

Generation-recombination noise becomes more prominent at low temperatures. Under these conditions, the detectivity is theoretically independent of current, and from Eqs. (3) and (5) we have

$$D^* = \sqrt{\frac{e\mu_p\tau_p RW}{4L}} \frac{\lambda}{hc}. \qquad (8)$$

If surface conductance is neglected, $R = \rho L/Wt$ and

$$D^* = \sqrt{\frac{e\mu_p\tau_p}{4} \frac{\rho}{t}} \frac{\lambda}{hc}. \qquad (9)$$

c. Background Noise

To approach the background noise limit, thermal generation must be small compared to the background generation, i.e., the conductance due to thermal carriers must be less than $Q_B e\mu_n\tau_n/t$, or $pe\mu_p < Q_B e\mu_p\tau_p/t$, i.e.,

$$pt/\tau_p < Q_B. \qquad (10)$$

The lifetime τ is expected to vary with carrier concentration. Figure 1 shows the variation of the effective mobility-lifetime product with resistivity [determined from photoconductivity measurements using Eq. (1)] for *p*-type InSb. Over a wide range we can take $\mu\tau = 2 \times 10^{-3}\rho$ cm²/V.

For radiation over 2π steradian from a 300°K blackbody, and between 0 and 5.2 μ (the cut-off of InSb at 77°K), $Q_B = 1.6 \times 10^{16}$ photons/(cm² sec). Inserting this value into Eq. (10), and using $\mu\tau = 2 \times 10^{-3} \rho$ cm²/V and $\rho = (pe\mu)^{-1}$, we have

$$\rho^2/t > 2 \times 10^5 \ \Omega^2 \ \text{cm}.$$

This is satisfied by, for example,

$$\rho > 20 \ \Omega \ \text{cm}, \ t = 20 \ \mu,$$

$$\rho > 10 \ \Omega \ \text{cm}, \ t = 5 \ \mu.$$

The above is only an approximate theory in that, for example, it assumes the same lifetime to apply for both optical and thermal generation which may not be the case in the presence of trapping, but it gives a guide in designing detectors and indicates the type of material and thicknesses required. It also ignores surface conductance and recombination both of which must be minimized to give good performance (see refs. 14 and 15).

IV. Performance

A. Room Temperature Detectors

Detectors operating at or near room temperature can be made so as to be limited by Johnson noise. The low resistivity of InSb at this temperature normally used, a surface resistance of typically 5Ω per square. There therefore may be difficulty in reaching the amplifier noise levels required to make full use of the detector unless transformer coupling is used.

Detectivities at peak wavelength, D^* (6 μ, 800, 1) of 2.5×10^8 cm cps$^{1/2}$/W may be achieved. The responsivity of a 1-mm² detector at current for optimum detectivity may be 0.5 V/W. The low lifetime (about 0.05 μsec) makes these detectors suitable for applications for which high speed of operation is desirable.

B. Detectors Operating Down to −80°C

A number of methods of cooling are suitable for temperatures down to −80°C (thermoelectric, Freon, etc.), and there is thus particular interest in this range. Figure 2 shows typical variation of detectivity, responsivity, and noise with temperature.

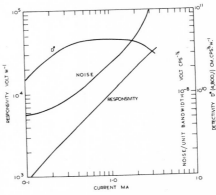

Fig. 4. Variation of responsivity, noise, and detectivity with current at 77°K. (1 mm × 1 mm specimen, 4-μ signal radiation.)

C.   Detectors Operating at 77°K

Operated with liquid nitrogen or liquid air as coolant, InSb photoconductors are of great interest. Devices can be made which approach the background-limited detectivity. Also output levels obtainable are high and electronic problems in the system are reduced. Figure 3 shows a distribution of detectivity for a sample of two hundred detectors of various shapes and sizes, showing a median detectivity D^* (5 μ, 800, 1) of 4.3 $\times$ 10^{10} cm cps$^{1/2}$/W. Responsivities of 2 $\times$ 10^4 V/W for a 1-mm^2 element at optimum signal-to-noise ratio are obtained, and the resistance is typically 2000 Ω per square.

Time constants determined from photoconductive decay or from noise spectra are normally in the range 5–10 μsec. Lower time constants can be obtained by the use of lower resistivity material but only at the expense of detectivity [Eq. (9)], unless particular care is taken to obtain thin specimens; this may not always be possible with large elements.

Figure 4 shows a typical variation of detectivity, responsivity, and noise with current for a 1-mm^2 element. The spectral variation of detectivity is shown in Fig. 5 in which a comparison is made between detectors operated 300°K, 233°K, and 77°K.

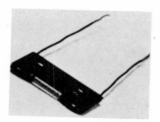

Fig. 6.   Room temperature InSb photoconductive detector with sensitive area 6 mm $\times$ 0.5 mm.

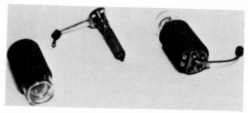

Fig. 7.   77°K InSb photoconductive detector, 4 mm $\times$ 4 mm element, with Joule-Thomson cooler.

V.   Practical Embodiments

A.   Room Temperature Detectors

Detectors operating at room temperature require a good heat sink. In normal ambient conditions the element requires no protection and a simple arrangement as shown in Fig. 6 can be used.

B.   Detectors Operating between Room Temperature and −80°C

In this range either thermoelectric or liquid cooling may be used. It is normally necessary to provide a vacuum envelope for the detector to prevent condensation and icing. Windows may be made of sapphire (which, however, limits the long wave cut-off slightly) or of silicon. Liquid coolants may be used in the type of dewar described below for liquid nitrogen.

C.   Detectors Operating at Liquid Nitrogen Temperature

Three types of construction may be used depending on the source of coolant. The simplest consists of a dewar with capacity sufficient to give the operating time required, which, if desired, may be used with a forced liquid transfer system.[16] An alternative and convenient method of cooling is with a Joule-Thomson liquefier; in this case the dewar must be made with a precision bore to match the liquefier. Figure 7 shows a dewar and liquefier* suitable for this type of operation.

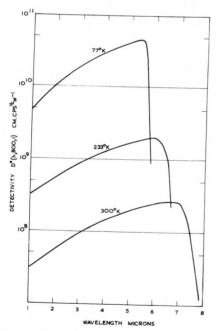

Fig. 5.   Typical spectral variation of detectivity for detectors operating at 300°K, 233°K, and 77°K.

* Manufactured by the Hymatic Engineering Co., Redditch, Worcs., England.

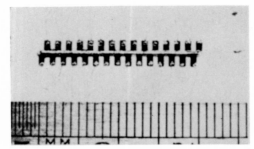

Fig. 8. 77°K InSb photoconductive array of thirty staggered elements.

D. Element Shapes

To achieve uniformity of current flow, which is necessary for uniformity of signal and high detectivity, rectangular specimens with contacts along the sides are normally desirable. Within this limitation, single element detectors from 0.5 mm × 0.5 mm to 8 mm × 8 mm can be made. At the larger size, some sacrifice of detectivity has to be made, as the detectors cannot be made so thin as those of smaller size. At 77°K, the resistance of a square detector is usually about 2 kΩ, a value suitable for use with transistor amplifiers. At higher temperatures, the resistance is lower and if a large square detector is required it may be more desirable to fabricate this from a number of filaments geometrically in parallel, but electrically in series.

Similar techniques may be used to make arrays. Figure 8 shows a thirty-element array for cooling to 77°K with the elements staggered to give minimum dead area along the line of the array. The uniformity of such arrays at constant current can be ±20% in responsivity and ±12% in detectivity.

VI. Conclusion

The availability of single-element InSb photoconductive detectors has already made its impact in infrared instrumentation and in a number of applications. The use of arrays will lead to the ability to make more sensitive and faster picture-forming instruments, and raises the possibilities of a more flexible approach to instrumentation and detection systems. InSb detectors fulfill this function in the wavelength range 3.5–7 μ.

The authors wish to thank the directors of Associated Semiconductor Manufacturers Limited, and the Ministry of Aviation for permission to publish this paper.

References

1. A. N. Blum, N. P. Mokrovski, and A. R. Regel, Zh. Tekhn. Fiz. **21**, 237 (1951).
2. H. Welker, Z. Naturforsch. **7a**, 744 (1952).
3. D. G. Avery, D. W. Goodwin, and A. E. Rennie, J. Sci. Instr. **34**, 394 (1957).
4. D. W. Goodwin, Report of the Meeting on Semi-conductors, London Physical Society, April 1956, p. 137; J. Sci. Instr. **34**, 367 (1957).
5. S. W. Kurnick and R. W. Zitter, J. Appl. Phys. **27**, 278 (1956).
6. J. C. M. Brentano and J. D. Richards, Phys. Rev. **94**, 1427 (1954).
7. K. F. Hulme and J. B. Mullin, Solid State Electron. **5**, 211 (1962).
8. R. E. J. King and B. E. Bartlett, Philips Tech. Rev. **22**, 217 (1961).
9. S. Liang, *Compound Semiconductors* (Reinhold, New York, 1962), Vol. 1, p. 227.
10. R. W. Cunningham, E. E. Harp, and W. M. Bullis, Rept. of Internatl. Conf. on Semi-conductors, Exeter, Inst. of Phys. and Phys. Soc., 1962, p. 732.
11. *Electronics for Spectroscopists* (Hilger and Watts, London, 1959), p. 9.
12. T. S. Moss, L. Pincherle, and A. M. Woodward, Proc. Phys. Soc. **B66**, 745 (1953).
13. R. A. Laff and H. Y. Fan, Phys. Rev. **121**, 53 (1961).
14. G. K. Eaton, R. E. J. King, F. O. Morten, A. T. Partridge, and J. G. Smith, J. Phys. Chem. Solids **23**, 1473 (1962).
15. R. K. Mueller and R. L. Jacobson, J. Appl. Phys. **35**, 1524 (1964).
16. J. A. Jamieson, R. H. McFee, G. N. Plass, R. H. Grube, R. G. Richards, *Infrared Physics and Engineering* (McGraw-Hill, New York, 1963), p. 183.

Reprinted from *Appl. Optics*, **4**(6), 649–657 (1965)

Indium Antimonide Submillimeter Photoconductive Detectors

E. H. Putley

The nature of photoconductive effects at wavelengths longer than 100 μ is discussed with particular reference to n-type InSb in (a) zero magnetic field, (b) magnetic fields too low to produce magneto-optical resonance absorption, and (c) large magnetic fields. The performance of practical detectors using these effects is described and is compared with that of other submillimeter detectors.

I. Introduction

In most photoconductive detectors the incident radiation produces a change in conductivity by raising electrons from the valence band of a semiconductor to the conduction band or by freeing electrons or holes from impurity centers. Detectors employing these processes thus will have a long-wave threshold determined by the energy required to free the electrons or holes. For practical intrinsic photoconductors the threshold occurs at a wavelength less than 10 μ. The ionization energy of the shallowest levels in Ge corresponds to a wavelength longer than 100 μ, so that detectors based on photoionization of impurities in Ge (extrinsic photoconductors) can be used for wavelengths as long as 120 μ. The only known semiconductor in which shallower levels can be introduced in a reasonably controlled way is InSb. However, as we shall see, this requires the use of a magnetic field, and we shall find that the type of extrinsic photoionization process which occurs in Ge is of much smaller importance than other photoconductive processes which will now be discussed.

The principal new process is one which leads to photoconductivity extending into the microwave region of the spectrum. It is well known that at long wavelengths the free carriers in a semiconductor absorb radiation efficiently. Since this absorption involves only intraband transitions one would not at first sight expect this process to produce a significant change in conductivity and therefore a photoconductive effect. When the temperature of the semiconductor is sufficiently low, however, the absorbed radiation causes an appreciable change in the distribution in energy of the electrons within their band. If this occurs, the average mobility of the electrons will change. Hence there will be a change in conductivity produced by the change in mobility rather than by a change in the carrier concentration as in the photoconductive effects observed at

The author is at the Royal Radar Establishment, Malvern, England.

Received 18 December 1964.

shorter wavelengths; this is in fact another example of a hot electron effect. Semiconductors with high mobilities are required so that suitable materials are n-type InSb or Ge cooled to liquid helium temperature. The effect has been observed in both of these materials but InSb has been studied most extensively in this connection and has been used mainly to provide a photoconductive detector.

The behavior of this hot electron photoconductivity is markedly affected by the application of a magnetic field. At the wavelength at which cyclotron resonance occurs there is a sharp peak in the free-carrier absorption. In Ge a corresponding peak in the photoconductive effect has been observed. The position of this peak can be varied by adjusting the magnetic field so that a narrow-band tunable detector, in principle, could be made. With InSb the behavior is more complicated because of a magnetic freeze-out effect of the free carriers (see below) but with sufficiently large magnetic fields the effect has been observed and has been used to make a practical detector (see below). At lower fields a general enhancement of the submillimeter photoconductive effect is observed.

In the next five sections the effects referred to in this introduction will be discussed in more detail while in the last two sections the performance of practical detectors based on these effects will be described.

II. Hydrogenic Impurity Centers in Semiconductors

The shallowest impurity centers in Ge or Si are those associated with substitutional group III or group V impurities. In these atoms the number of electrons in the outermost shell differs by one from the four in Ge or Si. The extra electron (or hole) is balanced by an extra unit of charge on the atomic core. The system of free carrier plus fixed core charge can be treated as a hydrogen atom embedded in the dielectric medium of the Ge or Si lattice. Application of the quantum theory of the hydrogen atom to this system gives a good first approximation to the properties of the impurity center. The ionization energy and the Bohr radius are given by

$$\epsilon = 13.6(m^*/mK^2) \text{ eV}, \tag{1}$$

$$a = 5.29 \times 10^{-9}(m^*/m)K \text{ cm}, \tag{2}$$

where m^* is the effective mass of the free carrier and K the dielectric constant. When applied to Ge these expressions show that the ionization energy should be about 10^{-2} eV and the Bohr radius about 4.2×10^{-7} cm. The value for the ionization energy agrees reasonably with the measured values, although a more exact theory is required to account for the difference between different impurities.

The impurities will behave in this simple way only if their concentration is sufficiently small. For Ge the value of the Bohr radius indicates that interaction should become appreciable at concentration of the order of 10^{17} cm^{-3}. Experimentally, it is found desirable to use a concentration of not more than about 10^{15} cm^{-3} if serious interaction effects are to be avoided. At higher concentrations the ionization energy is reduced and when the concentration reaches 10^{19} cm^{-3} the impurity levels merge with the bands and the behavior at low temperatures is metallic.

Equations (1) and (2) show that to obtain smaller ionization energies semiconductors are required with smaller effective masses or larger dielectric constants but that in these semiconductors the critical concentration above which interaction between neighboring impurities becomes serious will be smaller. For instance, for InSb the effective mass for electrons is about 0.013 **m** which is appreciably smaller than for Ge. The dielectric constant is approximately the same but Eq. (1) indicates an ionization energy of about 0.001 eV, corresponding to a wavelength of the order of 1000 μ. Now p-type impurities in InSb are observed to behave as hydrogenic centers, and there is no reason to suppose that n-type impurities do not. The latter would seem to be ideal for observing submillimeter photoconductivity. Unfortunately, Eq. (2) shows that the Bohr radius is increased to 5.7×10^{-6} cm. Therefore, by comparison with Ge we expect the behavior to be metallic for concentrations greater than 10^{14} cm^{-3} and that a concentration of less than 10^{13} cm^{-3} is required before the interaction between neighboring impurities becomes negligible. Unfortunately, in the purest InSb so far examined the impurity concentration is greater than 10^{14} cm^{-3}. It is not surprising, therefore, that at low temperatures its behavior is metallic. There are other semiconductors which either on account of small effective masses or large dielectric constants might have smaller impurity ionization energies than Ge. Possible materials are InAs, PbTe, or HgTe. The purity of these materials is, however, considerably inferior to that of InSb. No material is known which has the required properties and which can be prepared purer than InSb. It is for this reason that it has not been possible so far to produce extrinsic photoionization photoconductivity beyond the limit achieved with doped Ge, without the use of a magnetic field. (See Sec. VI.)

III. Free-Carrier Absorption

To discuss the order of magnitude expected for the free-carrier absorption the classical Drude theory of high-frequency conductivity is still adequate.

The conductivity varies with frequency as

$$\sigma = \sigma_0(1 + \omega^2\tau_e^2)^{-1}, \tag{3}$$

where σ_0 is the zero-frequency conductivity and τ_e the electron scattering time; so that for an electron conductor,

$$\sigma_0 = ne^2\tau_e/m^*, \tag{4}$$

where n is the free electron concentration. Similar expressions apply for hole conduction and for mixed conduction.

If the frequency is so high that the displacement current is large compared with the conduction current, the optical absorption coefficient α is given by the equation

$$\alpha = 4\pi\sigma/c(K)^{1/2}. \tag{5}$$

This shows that at low frequencies, where $\omega\tau_e < 1$, α will be independent of frequency, but at high frequencies, ($\omega\tau_e \gg 1$), α will vary as ω^{-2} or as λ^2. Thus any effect which depends upon free-carrier absorption should be optimal at long wavelengths but should fall off as λ^2 at short wavelengths. It is for this reason that the submillimeter InSb photodetectors do not show a long-wave threshold like the photoionization detectors but show a falling off in performance as the wavelength is reduced below 1 mm.

The order of magnitude of the free-carrier absorption in the n-type InSb used in the submillimeter detectors can be estimated from Eqs. (3)–(5). At 4°K (and zero magnetic field) in typical material $n = 5 \times 10^{13}$ cm^{-3} and the mobility $\mu = e\tau_e/m^* = 10^5$ cm^2 V^{-1} sec^{-1}. Also, $\tau_e = 8.5 \times 10^{-13}$ sec so that $\omega\tau_e = 1$ when $\lambda = 1.6$ mm. Then when $\lambda = 1$ mm the absorption coefficient α will be 22 cm^{-1} while at $\lambda = 100$ μ it will have fallen to 0.30 cm^{-1}. These values are considerably smaller than those found in intrinsic photoconductors but they compare favorably with the values found in extrinsic Ge photoconductors. The optimum thickness t for a detector satisfies the condition $\alpha t \sim 1$ which indicates that a thickness of the order of 1 mm would be a convenient one.

IV. Hot Electron Effects

The free carriers in a semiconductor exchange energy with the lattice and with blackbody radiation and, when in thermal equilibrium, will have a mean energy of the order of kT. If an electric field (either static or alternating) is applied, the carrier will receive energy from this source. A steady state will be set up in which this energy is passed to the lattice and then to the surroundings of the sample. Under most circumstances the coupling between the electrons and the lattice is so strong that, in the steady state, the energy of the electrons will not be significantly greater than the thermal equilibrium value. However, in pure high-mobility semiconductors (such as Ge or InSb) at low temperatures, the coupling between the electrons and the lattice becomes so weak that even for quite small electric fields ($\backsim$ 1 V/cm) the steady-state energy of the free carriers is appreciably greater than the thermal equilibrium value. Since the change in the lattice temperature is negligible, this effect manifests itself as a departure from Ohm's law. The carrier mobility depends upon the mean energy of the carriers so that increasing this energy produces a change in mobility similar to that produced by an increase in temperature. Using static fields this hot electron effect was first observed in Ge and subsequently has been observed also in InSb. The electrons can draw energy from a high-frequency field as well as from a static one, so that, by applying a combination of static and alternating fields, changes in the level of the alternating field can be detected through changes in the static current-voltage characteristic of the sample. This is the basis for the hot electron photoconductive effect.

For frequencies for which $\omega\tau_e < 1$ the effect will be independent of frequency, but when $\omega\tau_e$ becomes appreciably greater than 1 the effect falls off due to decrease in the absorption of the radiation. For InSb, as we have seen, this decrease begins to take place when the wavelength of the incident radiation becomes less than 1 mm.

V. Hot Electron Photoconductivity

The characteristics of this form of photoconductivity have been considered in some detail by Kogan[1] and by Rollin.[2] For small electric fields the deviation from Ohm's law varies quadratically with electric field E so that we can write

$$\sigma = \sigma_0(1 + \beta E^2). \qquad (6)$$

At somewhat larger fields (but at fields sufficiently small that avalanche ionization does not occur) higher powers of E become important. It is convenient to redefine β by writing

$$\beta = (1/\sigma)[d\sigma/d(E^2)]. \qquad (7)$$

Kogan[1] has shown that under open circuit conditions the photoconductive responsivity R is given by

$$R = \beta V/v\sigma, \qquad (8)$$

where V is the applied voltage and v the volume of the sample. The corresponding response time τ is given by

$$\tau = (^3/_2\mathbf{k}/\mathbf{e})\beta(dT/d\mu). \qquad (9)$$

Thus, by measuring the current-voltage characteristics and the temperature dependence of the mobility, the photoconductive responsivity and time constant can be estimated. Figure 1 shows the results of such an estimate. From the measured values of σ, β, and $d\sigma/dT$ ($= ne d\mu/dT$) the responsivity (in arbitrary units) and the time constant are calculated. The results show that the responsivity passes through a well-defined maximum at the optimum field and that at this point the response time is about 0.4 μsec. Although the

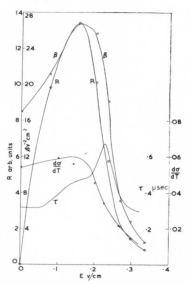

Fig. 1. Performance of a hot electron detector calculated from measured values of β and $d\sigma/dT$.

responsivity has been expressed in arbitrary units the actual value can be calculated by assuming values for the dimensions of the sample. If the volume $v = 5 \times 10^{-3}$ cm³ and the applied voltage = $^1/_2E$ then Eq. (8) gives $R = 650\ V/W$. The value found for the response time agrees well with measured values but the calculated responsivity is somewhat larger than the measured value. This is to be expected since this calculation of the responsivity assumes that we have complete absorption of the incident radiation, which never can be achieved in practice.

VI. Behavior in a Magnetic Field

When a sufficiently large magnetic field is applied to a semiconductor the conduction band will be split into a series subbands or Landau levels. In the plane perpendicular to the direction of the magnetic field the motion of electrons will be in orbits described with the cyclotron resonance frequency

$$\nu_c = eB/2\pi m^*, \qquad (10)$$

where B is the magnetic flux density, and the energy associated with motion in this plane will be quantized into a series of levels separated by $\mathbf{h}\nu_c$. The position of the energy minimum of the lowest Landau band will be raised by $^1/_2\mathbf{h}\nu_c$ above the normal band minimum on account of the zero-point energy of the cyclotron motion. Considering a unit volume of material and neglecting spin, the energy density of states in the nth Landau level is

$$N_n(\epsilon)d\epsilon = \pi\mathbf{h}\nu_c(2m^*/\mathbf{h}^2)^{3/2}[\epsilon - (n + {}^1/_2)\mathbf{h}\nu_c]^{-1/2}d\epsilon. \quad (11)$$

169

If the condition $\mathbf{h}\nu_c \gg \mathbf{k}T$ is satisfied and if integrating (11) over the energy range $^1/_2\mathbf{h}\nu_c$ to $^3/_2\mathbf{h}\nu_c$ shows that enough states are available to accommodate all the available electrons, then only the lowest Landau level will be occupied. This situation (known as "the extreme quantum limit") is difficult to achieve with most materials, but because electrons in the conduction band of InSb have such a small effective mass ($\sim$0.013 $\mathbf{m}$ near the bottom of the band) this condition can be achieved with the available material ($n \sim 5 \times 10^{13}$ cm^{-3}) with magnetic fields of only about 1000 Oe and at temperatures not greater than 4°K. It is therefore to be expected that even moderate magnetic fields will have a marked effect upon the properties of InSb at low temperatures.

As pointed out in Sec. II, in even the purest available n-type InSb the impurity levels are merged with the conduction band. The situation is changed by the application of a magnetic field. When a field greater than a certain critical value is applied, impurity levels appear below the conduction band so that when the temperature is lowered below 4°K the majority of the conduction electrons fall into them.[3] The critical field depends upon the purity of the sample; in the material used for submillimeter detectors it is about 3000–4000 Oe. The position of these energy levels can be found by Hall effect measurements; above the critical field their depth below the conduction band minimum varies approximately linearly with magnetic flux density B and for $B < 10$ kG the ionization energy corresponds to a wavelength longer than 1 mm.

When a magnetic field is applied hot electron effects are still significant.[4] At temperatures so high that carrier freeze-out does not occur the magnetic field reduces the mobility μ and hence affects the quantity β (see Sec. V). At lower temperatures the behavior is more complicated. Initially deviation from Ohm's law due to carrier heating effects occur; on raising the applied electric field to a certain point avalanche ionization occurs and the electric field becomes independent of the current until all the electrons frozen out by the magnetic field are returned to the conduction band.

In Sec. V calculation of the parameters of the hot electron photoconductive effect from the current voltage characteristics are described. This procedure can be applied equally well when a magnetic field is present. Typical values for the quantities appearing in Eqs. (8) and (9) are: $\sigma = 3.5 \times 10^{-4}\,\Omega^{-1}$ cm^{-1}; $\beta = 0.3\ V^{-2}$ cm^2, $V = 0.2$ V, and $dT/d\mu = 2.7 \times 10^{-4}$. These values were observed at a temperature of 1.3°K and a transverse magnetic field of 6 kOe on the same specimen as considered in Sec. V. The values obtained for the responsivity and time constant were

$$R = 3 \times 10^4 \text{ V/W and } \tau = 1.0 \times 10^{-7} \text{ sec.}$$

Comparing these results with those obtained in zero magnetic field shows that the effect of the field is to increase the responsivity and to reduce the time constant.

The calculation of the hot electron photoconductivity in a magnetic field assumes that the free electrons are the principal absorbing centers. Since with tempera-

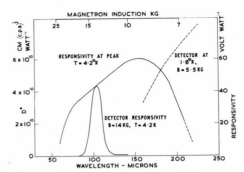

Fig. 2.   Performance of a narrow-band detector mounted in the cryostat shown in Fig. 4.

tures below 2°K and magnetic fields greater than 5 kOe the majority of the electrons will have fallen from the conduction band into the impurity levels, this assumption needs careful scrutiny. Measurement of the optical transmission as a function of magnetic field[5] has shown that at a wavelength of 1 mm the absorption of an electron in an impurity level is very small compared with that of a free electron. The magnetic field reduces the absorption of a sample by removing free electrons but a practical photoconductive detector can be designed in which this does not seriously affect the performance until the magnetic field approaches 10 kOe. As we shall see in the next section, application of a magnetic field does in fact produce the improvements in performance predicted by this calculation.

Measurement of the transmission at about 0.2-mm wavelength[5] showed that the free-carrier absorption had fallen and that when a magnetic field of about 7 kOe was applied the absorption increased. At this wavelength the absorption of an electron attached to an impurity center is greater than that of a free carrier. Since this wavelength marks the effective short-wave limit of the InSb submillimeter detector, it appears that over the working range of this device free carrier absorption is the dominant absorption process.

When magnetic fields larger than about 7 kOe are applied the photoconductive effect at long wavelengths is reduced, but near the wavelength corresponding to cyclotron resonance a narrow peak occurs in the photoconductive responsivity. This peak has a bandwidth of about 10% at 4°K or 5% below 2°K, and by increasing the magnetic field to 76 kOe the wavelength can be reduced to 26 μ. Figure 2 then shows results obtained with a field of 14 kOe. A similar effect has been observed in Ge where cyclotron resonance absorption can be used to enhance the hot electron photoconduction effect at short wavelengths[6]; in this case, the bulk of the electrons will occupy the impurity levels. Consider the energy level diagram shown in Fig. 3. If the radius of a cyclotron orbit is small compared with the Bohr radius it is possible for an electron attached to an impurity center to rotate about the magnetic field. Its state can then be represented by one of the impurity levels drawn under the higher Landau bands. The radius a_c of the cyclotron orbit is given by (a_c in centimeters and B in Gauss)

$$a_c = \left[\frac{(n + \frac{1}{2})\mathbf{h}}{\pi \mathbf{e} B} \right]^{1/2} = 3.63 \times 10^{-4}(n + \frac{1}{2})^{1/2}B^{-1/2}. \quad (12)$$

Comparing this with Eq. (2) for the Bohr radius shows that the condition for impurity cyclotron resonance absorption is easily satisfied. Hasegawa and Howard[7] have shown that under these conditions the resonance absorption will correspond to the transition from the impurity ground state to the first impurity Landau level, as indicated by the vertical arrow in Fig. 3. This process alone will not lead to a photoconductive effect, but emission of a phonon will permit a nonvertical transition into the lowest Landau conduction band and hence produce a change in conductivity. In principle, if the energy band structure is known sufficiently precisely it will be possible to distinguish between the absorption process depicted in Fig. 3 and true cyclotron resonance absorption by measuring accurately the wavelength at resonance. At present, the band structure is not known sufficiently accurately for this to be possible.

VII. Performance of InSb Submillimeter Detectors

A. With Zero or Weak Magnetic Fields

The behavior in zero field and in fields up to about 7–10 kOe will be discussed in this section. At these fields resonant effects are unimportant. Photoconductive effects under these conditions were first observed by Putley.[8,9] Further investigations have been made by several other workers, principally Rollin and co-workers,[2,10] Lifshitz *et al.*,[11] Wallis and Shenker,[12] Danilychev and Osipov,[13] Bratt *et al.*,[14] Besson *et al.*,[15] and Nad and Oleinikhov.[16]

Figure 4 is a diagram of the type of cryostat used by the author both for studying photoconductive effects and also as a practical detector. Radiation is condensed upon the sample by means of a light pipe. Cooled filters of black polythene and crystal quartz are fitted to remove the bulk of the room temperature radiation; the magnetic field is provided by a superconducting niobium solenoid. Glass dewars are used which when filled with about 1.8 l of liquid helium enable the detector to be operated for up to 11 h at 1.8°K.

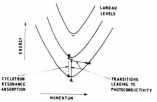

Fig. 3. Energy-level diagram showing process responsible for photoconductivity in large magnetic field.

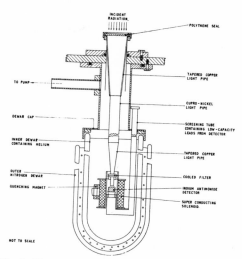

Fig. 4. Diagram of cryostat used for studying submillimeter photoconductivity.

Figure 5 shows how the responsivity varies with magnetic field and with sample current at a wavelength of 0.2 mm. This shows that application of a magnetic field produces a large increase in responsivity but that an optimum field exists above which the responsivity falls. This behavior agrees with the calculation of the effect of a small magnetic field on the hot-carrier photoconductivity as given in the last section.

Figure 6 shows how the responsivity varies with wavelength for various magnetic fields. At zero magnetic field the effect falls approximately as λ^2 at the shorter wavelengths, as would be expected if free-carrier absorption determined the size of the effect. At the optimum magnetic field the responsivity falls approximately linearly with wavelength. At first this was interpreted as evidence for a simple photoionization process but it is probably the fortuitous result of the combination of free-carrier absorption and a very broad impurity cyclotron resonance absorption.

Figure 6 shows mainly measurements of responsivity for wavelengths shorter than 1 mm; measurements out to 8 mm show, however, that the variation of responsivity with magnetic field is similar to that shown in Fig. 6 at 1 mm. There is no evidence for a long-wave threshold nor does the value of the optimum magnetic field fall as the wavelength is increased. Other workers have obtained somewhat different results at wavelengths of several millimeters. Thus Danilychev and Osipov[13] find that with their most sensitive specimen the responsivity is largest in zero magnetic field. Similar results have been found by Bratt *et al.*[14] but the results of Wallis and Shenker[12] are similar to our own. These discrepancies might be caused by differences in

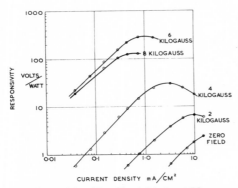

Fig. 5. Responsivity of broad-band detector at 0.2-mm wavelength and 1.8°K showing dependence upon magnetic field and detector current.

the apparatus used by different workers, but, more probably, are caused by differences in the samples of InSb used. Strong photoconductivity is only shown by material of the highest purity that present techniques can produce. The performance attained has not been closely correlated so far with other properties of the material. Comparable, but not identical, results have been obtained at the Royal Radar Establishment from material produced by Mullards (Southampton), Texas Instruments Ltd. (Bedford), Cominco (Trail), and the Naval Research Laboratory (Washington). Nevertheless, differences in material seem the most likely reason for the difference between the behavior found and that reported from Raytheon (U.S.A.) and from the Lebedev Institute (Moscow).

The calculations of Sec. V [see Fig. (1)] show that the response time should increase at first as the applied electric field is increased, but at a field somewhat greater than that which optimizes the responsivity the response time should fall. It also appears that, for small electric fields, applying a magnetic field should reduce the response time. Measurements by Shaw and Putley (Fig. 7) (private communication and ref. 4) for applied electric fields up to the value for optimum responsivity behave in the way expected. At the optimum electric field the response time is about 0.2–0.5 μsec.

Measurement of the noise equivalent power of the InSb detector when operated at the optimum magnetic field for broad-band operation shows that the principal source of noise is that from the room temperature amplifier. With the device operating at 1.8°K the resistance can lie between about 1 kΩ and 100 kΩ. To make full use of the response time of the device it is desirable to choose elements of resistance not greater than about 5 kΩ. With higher resistances difficulties are encountered in obtaining an adequate frequency response for the input stage of the amplifier. A resistance of 5 kΩ at 2°K will produce the same Johnson noise as a resistance of 33 Ω at room temperature, and the best

available amplifiers have equivalent noise resistance of 100–200 Ω. Therefore, the contribution of Johnson noise from the detector will be small compared with the amplifier noise. It has not been possible so far to produce a detector in which the radiation noise is greater than the Johnson noise. Putley[17] has calculated that using a cryostat similar to that shown in Fig. 4, and assuming that the cooled filter cuts out all room temperature radiation at wavelengths shorter than 100 μ, a noise equivalent power of 6×10^{-14} W would be attainable if only radiation noise were significant. For this condition to be achieved the Johnson noise in the element must be smaller than the radiation noise output from the detector; that is,

$$2(\mathbf{k}Tr)^{1/2} < R\Delta W,$$

where r is the element resistance, R its responsivity, and ΔW is the rms radiation power falling on the detector. Thus,

$$9.1 \times 10^{-12}(r)^{1/2} < 1000 \times 6 \times 10^{-14},$$

so that $r < 40$ Ω.

The assumed responsivity is only achieved in elements with resistance of more than 1000 Ω. Therefore, background-limited operation cannot yet be achieved with this detector.

So long as the amplifier is the main source of noise, as high a responsivity as possible is required. The maximum detector resistance is determined by the input circuit bandwidth requirements rather than by noise considerations. The magnetic field then can be looked on as a means for increasing the responsivity without sacrificing the response time. When major improvements in amplifier performance are forthcoming the use of a magnetic field may no longer be advantageous. This is not likely to occur until an amplifier has been

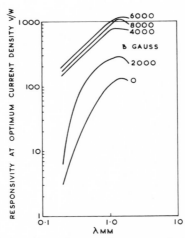

Fig. 6. Dependence of responsivity upon wavelength and magnetic field at 1.8°K and at optimum current for each field.

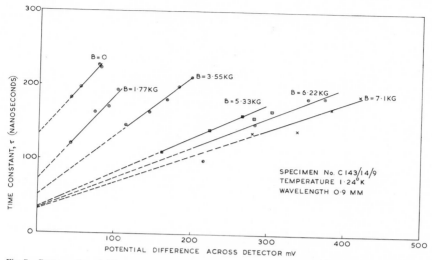

Fig. 7.   Response time at 0.9-mm wavelength and 1.24°K as a function of applied electric and magnetic fields.

developed which can operate in liquid helium. One step in this direction which gives a worthwhile improvement in detectivity, but at the sacrifice of frequency response, is the use of a helium-cooled step-up transformer.[10]   No magnetic field is applied but the resistance is stepped up from about 10 Ω to 100 kΩ. The transformer must be tuned at a frequency of about 1 kc/sec, and thus the effective time constant is increased to the order of 1 msec. Table I summarizes the performance obtained with the various methods of using the InSb submillimeter detector.

B.   With Large Magnetic Fields

As discussed in Sec. VI, the narrow-band tunable photoconductive effect becomes prominent when fields of greater than 10 kOe are applied. Figure 2 shows the performance obtained with a sample mounted in a cryostat similar to that shown in Fig. 4 but with the niobium solenoid replaced by a niobium-zirconium one.[18] The device was operated at 4°K to reduce the rise in resistance of the InSb element. The fall-off in performance near 50 μm is due to the reststrahlen band in InSb. With a solenoid producing about 58 kG the detector has been used at 32 μ. When operated at 4°K the bandwidth of the response is about 5–10% of the resonant wavelength, but on reducing the temperature to 1.8°K the bandwidth is reduced to about 3%. The bandwidth is not therefore so narrow that the detector can replace a high-resolution spectrometer but it has found application where a very high rejection of higher order grating harmonics was necessary. One notable advantage of this detector is that it does not itself respond to harmonics of its resonance frequency. Direct measurement of the response time with a large magnetic field have not been made but results obtained when

using this detector to study the radiation from a thetatron discharge show that its response time is probably not longer than 0.1 μsec.[19]   The characteristics of the tuned detector also are included in Table I.

VIII.   Comparison with other Submillimeter Detectors

There are now several detectors available for the submillimeter region of the spectrum (see article by Smith on p. 631, also ref. 20).   The three principal factors which influence the choice of detector for a given application are (1) detectivity, (2) response time, and (3) convenience.   When the highest available detectivity is required, a detector cooled with liquid helium is almost essential.   The only exception to this is that at the longer wavelengths a superheterodyne receiver may be used.   At wavelengths less than 2 mm the construction of this type of receiver becomes increasingly difficult and the tuning range is restricted.   In addition to the InSb detector there are three types of helium-cooled bolometer available, using as their sensing elements a tin film held near the superconducting transition, a piece of carbon composition resistor material, and a piece of Ge doped with Ga or In (see article by Smith on p. 631).   Reliable absolute standards do not exist at present in the submillimeter region of the spectrum, but by making comparative measurements the detectivity of the broad-band (low magnetic field) InSb detector has been compared with that of various bolometers.   One of the difficulties in making this comparison is that the detectors should be used with similar optical arrangements to eliminate transmission losses. For example, the transmission efficiency of the light pipe and filters shown in Fig. 4 is about 40%.   The data given in Table I which gives performance figures for

Table I.   Performance of InSb Submillimeter Detectors

Detector	Sensitive area (mm²)	Operating temperature (°K)	Noise equivalent power per unit bandwidth (W/cps)	Response time (sec)	Operating wavelength (μ)
Wide-band detector with magnetic field $\sim$7 kG	5	1.5	2×10^{-11} 1×10^{-11} 5×10^{-12}	2×10^{-7}	200 500 1000
Wide-band detector zero magnetic field but with cooled transformer	15	4	5×10^{-13}	10^{-3}	500–8000
Narrow band InSb detector	25	4	1×10^{-11} 1.5×10^{-11}	$\sim 10^{-7}$	150 100

some InSb detectors are not corrected for this. When comparisons are made under similar conditions, the broad-band InSb detector with a magnetic field of about 6–7 kOe has been found to be about 40–50 times more sensitive than a Golay cell at 1-mm wavelength and at about 500 μ the performance is comparable with that of the superconducting bolometer (see p. 683). At wavelengths longer than about 300 μ the performance appears better than that of the carbon-flake bolometer and of the Ge bolometer we so far have been able to examine. At shorter wavelengths than this the performance of the InSb detector falls below that of the best bolometers. Both with the InSb detector and with the bolometers the noise of the amplifier determines the detectivity. The InSb detector can detect modulation of higher frequency than is possible with the bolometers and this gives it an advantage over the latter since the amplifier noise generally will be reduced by such use of a higher frequency. Apart from this, the superconducting bolometer is a fairly delicate instrument and requires careful setting up. The other helium-cooled detectors are of comparable convenience in use.

The remarks in the last paragraph relate to the InSb detector operated in a magnetic field. This is a fast detector, and where high speed is not required, as in many spectroscopic applications, it is possible to exchange speed for sensitivity. One method of doing this is to remove the magnetic field and to step up the detector's resistance (about 10 Ω) by means of a helium-cooled transformer. This arrangement gives about an order of magnitude improvement in detectivity at 1 mm over that attained using the optimum magnetic field. The transformer is tuned at about 1 kc/sec so that this arrangement is only suitable for use with chopped radiation. At wavelengths near 1 mm the detectivity of this arrangement should compare favorably with that of the best cooled bolometers. The performance will fall off somewhat more rapidly at shorter wavelengths than it does when a magnetic field is used, but should still be comparable with the best bolometers at 200 μ.

When a short response time is required the only alternative to the InSb detector used with a magnetic field is the point contact rectifier used either as a detector or as a mixer in a superheterodyne receiver. The response time of the InSb detector is not significantly shorter than 0.2 μsec; the response time of a microwave crystal rectifier is much shorter than this. Superheterodyne receivers have been built for 2-mm wavelength with a noise equivalent power of 7×10^{-15} W/cycle, but they do not appear to have been built for shorter wavelengths. Where they are available their performance is superior both in terms of detectivity and response time. The crystal rectifier has been used as a detector at wavelengths somewhat less than 500 μ, but the detectivity falls off rapidly as the wavelength is reduced. Thus for wavelengths shorter than 2 mm the detectivity of the InSb detector is superior to that of the point contact rectifier.

The author wishes to thank M. F. Kimmitt and his other colleagues at the Royal Radar Establishment, Malvern, for many valuable discussions. He also would like to thank D. H. Martin of Queen Mary College, London, for several interesting conversations on the relative merits of different detectors.

References

1. S. M. Kogan, Fiz. Tverd. Tela **4**, 1891 (1962); Soviet Phys.—Solid State **4**, 1396 (1963).
2. B. V. Rollin, Proc. Phys. Soc. (London) **77** 1102 (1961).
3. R. J. Sladek, J. Phys. Chem. Solids **5**, 157 (1958).
4. E. H. Putley, Proc. Internl. Conf. on the Physics of Semiconductors (Dunod, Paris, 1964).
5. E. H. Putley, Physica Status Solidii **6**, 571 (1964).
6. D. W. Goodwin and R. H. Jones, J. Appl. Phys. **32**, 2056 (1961).
7. H. Hasegawa and R. E. Howard, J. Phys. Chem. Solids **21**, 179 (1961).
8. E. H. Putley, Proc. Phys. Soc. (London) **76**, 802 (1960).
9. E. H. Putley, J. Phys. Chem. Solids **22**, 241 (1961).
10. M. A. Kinch and B. V. Rollin, Brit. J. Appl. Phys. **14**, 672 (1963).
11. T. M. Lifshitz, S. M. Kogan, A. N. Vystavkin, and P. G. Melnik, Zh. Eksperim. i Teor. Fiz. **42**, 959 (1962); Soviet Phys.–JETP **15**, 661 (1962).
12. R. F. Wallis and H. Shenker, N.R.L. Rept. 5996 (29 August 1963).
13. V. A. Danilychev and B. D. Osipov, Fiz. Tverd. Tela **5**, 2369 (1963); Soviet Phys.—Solid State **5**, 1724 (1964).
14. P. R. Bratt, P. P. Debye, G. F. Giggey, and S. Koozekanani, Rept. RADC-TDR-63-199 (April 1963).
15. J. Besson, R. Cano, M. Matteoli, R. Papoular, and B. Philippeau, L'Onde Electrique **45**, 107 (1965).
16. F. Ya Nad and A. Ya. Oleinikhov, Fiz. Tverd. Tela **6**, 2064 (1964); Soviet Phys.—Solid State **6**, 1629 (1965).

174

17. E. H. Putley, Infrared Phys. **4,** 1 (1964).
18. M. A. C. S. Brown and M. F. Kimmitt, Brit. Commun. and Electron. **10,** 608 (1963); Infrared Physics, in press (1965).
19. M. F. Kimmitt and G. B. F. Niblett, Proc. Phys. Soc. (London) **82,** 938 (1963).
20. E. H. Putley, Proc. IEEE **51,** 1412 (1963).

Reprinted from *Infrared Phys.*, **9**(1), 35–36 (1969), with the permission of Microforms International Marketing Corporation as exclusive copyright licensee of Pergamon Press journal back files.

Background Limited Photoconductive HgCdTe Detectors for Use in the 8–14 Micron Atmospheric Window

B. E. BARTLETT, D. E. CHARLTON, W. E. DUNN, P. C. ELLEN, M. D. JENNER, and M. H. JERVIS

(*Received 13 November 1968*)

THIS letter describes the achievement of background limited performance in HgCdTe photoconductive detectors operated at 77°K.

Extrinsic photon detectors sensitive in the 8–14 μm atmospheric window have been available for some years. They have the disadvantage that they have to be cooled to low temperatures, for example, to less than 50°K for Hg : Ge and to less than 20°K for Cu : Ge. Thus considerable interest has arisen over the possibility of using intrinsic detector materials at the more easily obtainable temperature of liquid air (77°K), for this atmospheric window.

The two main detector materials that have been studied are $Hg_{1-x}Cd_xTe$, first suggested by Lawson *et al.*[1] in 1959, and $Pb_{1-x}Sn_xTe$.[2]

Table 1 shows the performance of some recent photoconductive detectors made during this study. All the detector elements have been fabricated from single crystal n type Hg Cd Te grown by the vertical Bridgman

TABLE 1. PERFORMANCE DATA ON NINE PHOTOCONDUCTIVE HgCdTe DETECTORS

Detector number	Area L × W (mm²)	Resistance (Ω)	Responsivity (λp) (V/W)	Noise per unit bandwidth at 800 Hz (VHz$^{-\frac{1}{2}}$)	Noise Equivalent resistance (Ω)	Response time (μ sec)	Peak wavelength λp(μm)	D* (λp, 800, 1) (cm Hz$^{\frac{1}{2}}$/W)
N 182	0·22 × 0·25	38	8400	2·2 × 10⁻⁹	290	1·2	10·0	9·0 × 10¹⁰
N 193	0·33 × 0·37	23	5900	2·3 × 10⁻⁹	300	5·5	9·5	9·0 × 10¹⁰
N 180	0·27 × 0·27	22	4700	1·5 × 10⁻⁹	130	2·5	9·5	8·5 × 10¹⁰
N 172	0·14 × 0·25	33	16000	4·5 × 10⁻⁹	1200	1·5	10·0	6·8 × 10¹⁰
N 174	0·25 × 0·25	35	11000	3·3 × 10⁻⁹	670	0·8	9·5	8·0 × 10¹⁰
N 173	0·25 × 0·25	23	6100	3·0 × 10⁻⁹	560	1·0	9·5	5·1 × 10¹⁰
N 190	0·50 × 0·50	40	21000	1·8 × 10⁻⁸	18000	6·0	9·5	5·7 × 10¹⁰
N 157	0·25 × 0·25	110	25000	1·2 × 10⁻⁸	8500	1·8	12·0	5·1 × 10¹⁰
N 154	0·25 × 0·25	43	16000	1·0 × 10⁻⁸	6000	1·1	12·0	3·9 × 10¹⁰

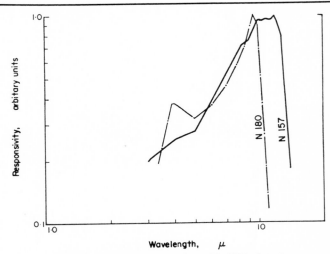

FIG. 1. Spectral responses of two HgCdTe detectors.

technique.[3] The detector elements are between 20 μm and 50 μm thick and are mounted by means of a suitable adhesive on to the substrate of a vacuum encapsulation. The field of view of the detector element is restricted to 60 deg. with a cooled radiation shield. The window of the encapsulation is silicon which has been bloomed with ZnS to give a peak transmittance at approximately 10 microns. The detectors can be cooled to 77°K by either using a Joule–Thompson cooler or bulk liquid nitrogen.

$D^*(\lambda p)$BLIP lies between $9\cdot6 \times 10^{10}$ cm Hz$^{\frac{1}{2}}$/W at 8 μm and $7\cdot0 \times 10^{10}$ cm Hz$^{\frac{1}{2}}$/W at 14 μm. The measured performances in Table 1 are close to these figures.

The spectral responses are very close to that expected for ideal photon detectors modified by the transmission characteristics of the bloomed silicon window, Fig. 1.

The detectors have a low resistance, typically in the range 20–200 Ω for square elements. Despite this, the noise is sufficiently high for the detector to be used with conventional amplifiers with little degradation of performance. In the early stages of the programme detector performance was seriously degraded by $1/f$ noise, extending to beyond 10 kHz. This has now been largely overcome as is shown in Fig. 2, where the noise spectrum of one of our later detectors is given.

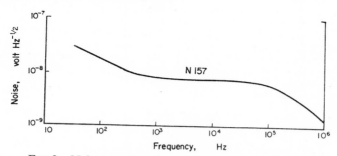

FIG. 2.   Noise spectrum of HgCdTe detector number 157.

This noise spectrum exhibits $1/f$ noise up to 500 Hz and generation recombination (g–r) noise at higher frequencies. The time constants deduced from noise spectra agree very closely with the measured response times of the detectors.

The performance of the detectors reported in this letter show that it is possible to obtain background limited performance with single element photoconductive HgCdTe detectors operated at 77°K.

Acknowledgements—We wish to thank our colleagues who have contributed to the success of this work and the Directors of Mullard Limited for permission to publish this letter. This work is supported by the Department of Co-ordination of Valve Development, Ministry of Defence, and the Royal Radar Establishment, Malvern.

Mullard Limited
Mullard Southampton Works
Millbrook, Southampton

B. E. Bartlett
D. E. Charlton
W. E. Dunn
P. C. Ellen
M. D. Jenner
M. H. Jervis

REFERENCES

1. Lawson, W. D., S. Nielson, E. H. Putley and A. S. Young, *J. Phys. Chem. Solids* **9**, 325 (1959).
2. Calawa, A. R., T. C. Harman, M. Finn and P. Youtz, *Trans. Metal. Soc. AIME* **242**, 374 (1968).
3. Bartlett, B. E., J. Deans and P. C. Ellen. *Int. Conf. Crystal Growth*, Birmingham July (1968).

Reprinted from *Infrared Phys.*, **13**(2), 143–147 (1973)

HIGH DETECTIVITY Pb$_x$Sn$_{1-x}$Te PHOTOVOLTAIC DIODES

W. H. ROLLS and D. V. EDDOLLS

Allen Clark Research Centre, Caswell, Towcester, Northamptonshire, England

(*Received* 30 *October* 1972)

Abstract—Pb$_x$Sn$_{1-x}$Te diodes have been made with detectivities up to 10^{11} cm Hz$^{\frac{1}{2}}$W^{-1} at 10·6 μm and 77°K. Diodes with a peak response at 12 and 13 μm have detectivities greater than 50 per cent of the 180° F.O.V. background limited value. Measurements from 70 Hz to 50 kHz have shown that 1/f noise is negligible if the diode is biased to the zero volts point on the I–V characteristic. The same result will be obtained by operating the diode into the d.c. short circuit input of a virtual earth current amplifier. The advantages of this mode of operation for thermal imaging applications and CO$_2$ laser detection are discussed.

1. INTRODUCTION

Pb$_x$Sn$_{1-x}$Te PHOTOVOLTAIC diodes are high sensitivity, fast detectors for the 8–14 μm window. Rolls and Eddington[1] have previously reported D^* (11·7, 800, 1) values up to $2·6 \times 10^{10}$ cm.Hz$^{\frac{1}{2}}$ W^{-1} for diodes with an antireflection coating. However, the detectivity at 800 Hz was limited by 1/f noise. This research note describes a study of the effect of bias on noise. This has shown that if the diodes are back-biased to the zero volts point on the I–V characteristic, 1/f noise is negligible at least at frequencies down to 70 Hz. The electrical and optical performance has been measured for diodes with a peak wavelength response in the range 10·6–13 μm. Detectors with an antireflection coating have D^*-values close to the background limit. Even without an antireflection coating D^*-values are greater than 50 per cent of the background limit.

Pb$_x$Sn$_{1-x}$Te is grown on PbTe substrates by vapour phase epitaxy.[2] This technique gives high purity, single crystal layers of uniform thickness and composition. The as-grown layers are p-type with a thickness of 200–300 μm. After removing the substrate by etching, a p–n junction is induced by a carefully controlled annealing process.[3] Mesa diodes are fabricated by conventional photoengraving techniques.

2. DETECTOR PERFORMANCE

Table 1 lists the performance of detectors with peak response at 10·6, 12 and 13 μm. The authors believe that the detectivity of 10^{11} cm Hz$^{\frac{1}{2}}$ W^{-1} measured for diode number 94/13 is the highest value achieved with any detector at 10·6 μm and 77°K. The spectral response of this detector is shown in Fig. 1. A zinc selenide antireflection coating acted as a cooled filter and reduced the short wavelength response. The total noise in the detector was measured over the frequency range 70 Hz to 50 kHz using a Brookdeal nanovolt pre-amplifier and a Marconi wave analyser. Biasing the diode to the zero volts point on the I–V characteristic reduced the 1/f noise to a negligible value. It is shown in the next section that the remaining noise sources should then be Johnson noise in the diode incremental

TABLE 1. PERFORMANCE OF LEAD–TIN TELLURIDE DETECTORS

No.	R_0 (Ω)	A (cm^{-2})	R (v/w)	λ_p (μm)	F.O.V.	D^* (λ_p, 800 Hz, 1 Hz) (cm Hz$^{\frac{1}{2}}$ W^{-1})
94/13	400	$2 \cdot 8 \times 10^{-3}$	3150	10·6	60°	10^{11} (measured)
						9×10^{10} (calculated)
96/9	190	4×10^{-3}	700	10·6	180°	$2 \cdot 6 \times 10^{10}$
96/22	135	$2 \cdot 8 \times 10^{-3}$	535	10·6	60°	$2 \cdot 7 \times 10^{10}$
87/9	62	$2 \cdot 5 \times 10^{-3}$	290	12·1	180°	$2 \cdot 6 \times 10^{10}$
89/16	65	$2 \cdot 5 \times 10^{-3}$	320	13·0	180°	$2 \cdot 7 \times 10^{10}$
89/11	58	$2 \cdot 1 \times 10^{-3}$	260	13·0	180°	$2 \cdot 0 \times 10^{10}$
96/10	160	$2 \cdot 8 \times 10^{-3}$	610	10·6	180°	$2 \cdot 6 \times 10^{10}$
86/30	58	$2 \cdot 5 \times 10^{-3}$	214	11·2	180°	$2 \cdot 0 \times 10^{10}$
87/10	35	$2 \cdot 5 \times 10^{-3}$	206	11·8	180°	$2 \cdot 5 \times 10^{10}$

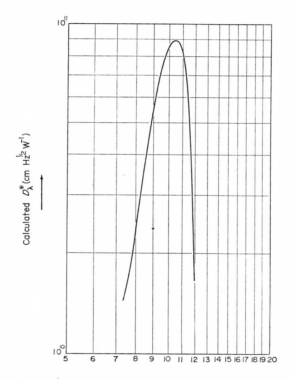

FIG. 1. Spectral response of Pb$_x$Sn$_{1-x}$Te diode 94/13.

slope resistance, R_0, and shot noise on the photocurrent, I_{sc}. The detectivity of a diode of area A and quantum efficiency η can then be calculated from the expression

$$D^* = \frac{\eta q \lambda A^{\frac{1}{2}}}{hc \left[\dfrac{4kT}{R_0} + 2qI_{sc} \right]^{\frac{1}{2}}} = \frac{R_c A^{\frac{1}{2}}}{\left[\dfrac{4kT}{R_0} + 2qI_{sc} \right]^{\frac{1}{2}}} \, \text{cmHz}^{\frac{1}{2}} \, \text{W}^{-1}$$

(1)

where R_c is the current responsivity in A W^{-1}.

The calculated D^* for diode 94/13 was 9×10^{10} cm Hz$^{\frac{1}{2}}$ W^{-1}, in reasonable agreement with the measured value of 10^{11} cm Hz$^{\frac{1}{2}}$ W^{-1}. Equation (1) was used to obtain the other results of Table 1. In all cases the slope resistance, responsivity and short circuit photo-current were measured with the detector biased to eliminate $1/f$ noise. The dependence of $1/f$ noise on bias is discussed below.

3. EFFECT OF BIAS ON $1/f$ NOISE

The I–V characteristic of an unilluminated diode ideally passes through the origin. It is well known that the incidence of radiation displaces the I–V characteristic parallel to the current axis. The diode operating point will depend on the applied bias and the d.c. input impedance of the following amplifier. Figure 2a shows the I–V characteristics for a typical diode. The total noise in two diodes was measured as a function of reverse bias at 250 Hz. These results are shown in Figs. 2b and 2c. With the diode coupled into a voltage amplifier and no applied bias the operating point was A (Fig. 2a) on the characteristic. As reverse bias was increased the operating point moved along the characteristic through B. The total diode noise decreased as the bias point moved from A to B, exhibited a minimum at B, and increased again as more reverse bias was applied. When the diode was biased at B, that is the zero volts point on the I–V characteristic, no $1/f$ noise was observed at frequencies down to 70 Hz. This was the lowest frequency at which measurements were made.

These results can be explained by considering the noise sources in a junction diode. In an ideal, unilluminated diode the reverse saturation current is due to the random thermal generation of electron-hole pairs and consequent diffusion of minority carriers across the junction. This just balances the forward current due to the thermal excitation of majority carriers across the junction, and the net current flow is zero. When this condition prevails the diode noise is Johnson noise in the incremental slope resistance R_0. In a real diode at zero volts the total noise will still be Johnson noise in the slope resistance as the diode is in thermal equilibrium with its surroundings. However, the slope resistance will now be the series and parallel combination of the junction resistance, the contact resistance and other

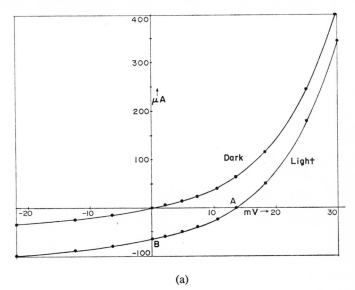

(a)

W. H. ROLLS and D. V. EDDOLLS

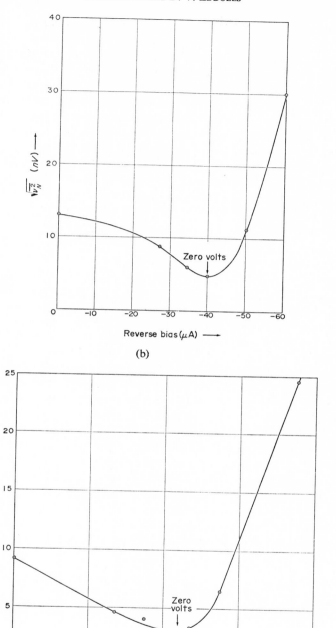

(b)

(c)

FIG. 2. Effect of bias on noise in $Pb_xSn_{1-x}Te$ diodes. (a) Measured I–V characteristics of a typical diode. (b) Noise vs bias for diode 94/13. (c) Noise vs bias for diode 94/16.

parasitic impedances associated with surface states and dislocations. Noise in the diode junction arises from the saturation and forward currents and generation–recombination in the space charge region. When the diode is illuminated the photocurrent will show full shot noise. However, provided the diode is still biased to the zero volts point on the I–V characteristic there will be no potential drop across the parasitic impedances and the reverse saturation current will still balance the forward current. The diode noise will now be Johnson noise in the slope resistance plus the uncorrelated shot noise on the photocurrent.

If either forward or reverse bias is applied to the junction, the resultant voltage will produce current flow in the parasitic impedances. It is known that current flow in surface states and contacts can generate $1/f$ noise. Further, Bess[4] and Morrison[5] have proposed that this noise might be associated with dislocations. Clearly operation anywhere on the I–V characteristic except at zero volts may give rise to $1/f$ noise. The higher the applied voltage, and therefore the current, the greater the noise level, in agreement with the results of Figs. 2b and 2c.

DISCUSSION

To minimize $1/f$ noise it is necessary to minimize current flow in the parasitic impedances associated with the diode. This can be done by biasing the diode to the zero volts point on the I–V characteristic or by working into the d.c. short circuit input of a current amplifier. For $Pb_xSn_{1-x}Te$ diodes the second approach offers two further advantages. These are faster response speed without degradation of D^*, and, for monolithic arrays of detectors, improved uniformity of responsivity. Owing to the high dielectric constant in $Pb_xSn_{1-x}Te$ the response time is determined by the RC time constant of the diode and its associated circuitry. Hence operating the diode as a current generator into a low input impedance amplifier will significantly reduce the detector rise time. This result is of particular importance in the development of detectors for laser radar applications. In practice $1/f$ noise is relatively unimportant in broad band applications and reverse bias could be used to increase sensitivity. In arrays of detectors for thermal imaging it is desirable to reduce the spread in responsivity as much as possible since this eases the requirements on signal processing. The current responsivity of a photovoltaic diode is proportional to the quantum efficiency, η. This varies by less than 2:1 across an epitaxial layer and as little as ±15 per cent across a ten element array. By contrast the voltage responsivity is proportional to the product ηR_0. Since R_0 is affected by surface properties and can vary by up to 4:1 across an array it is clear that a worthwhile improvement can be obtained by operating the diodes into current amplifiers.

Acknowledgements—The authors wish to thank B. L. H. Wilson for helpful criticism of the manuscript, C. J. Rogers, who grew the epitaxial layers and diffused the *p–n* junctions, and R. S. Simkins, who made the diodes. This work was carried out in part under a C.V.D. contract sponsored by Dr. Ellis (R.A.E.) and Dr. Elliot (R.R.E.), and is published by permission of the U.K. Ministry of Defence (Procurement Excutive) and the directors of the Plessey Company.

REFERENCES

1. ROLLS, W. H. and R. J. EDDINGTON, *Infrared Phys.* **10**, 71 (1970).
2. ROLLS, W. H., R. LEE and R. J. EDDINGTON, *Solid-St. Electron.* **13**, 75 (1970).
3. CALAWA, A. R., T. C. HARMAN, M. FINN and P. YOULZ, *Trans. AIME*, **242**, 374 (1968).
4. BESS, L., *Phys. Rev.* **91**, 1569 (1953).
5. MORRISON, S. R., *Phys. Rev.* **104**, 619 (1956).

Editors' Comments on Papers 20 Through 22

20 Levinstein: *Extrinsic Detectors*

21 Bode and Graham: *A Comparison of the Performance of Copper-Doped Germanium and Mercury-Doped Germanium Detectors*

22 Buczek and Picus: *Heterodyne Performance of Mercury Doped Germanium*

The papers in this group describe extrinsic, or doped, detectors whose principal area of application is in the wavelength region beyond 10 μm. In an extrinsic detector a pure host material is doped with an impurity whose energy states lie within the forbidden band of the host. These impurity states may lie just below the conduction band of the host or just above its valence band. Either way, they provide a small energy gap that requires very little energy from the incident photon in order to generate a charge carrier. Extrinsic detectors may have long-wavelength cutoffs extending to about 400 μm but, because of the small energy gaps involved, they generally require cooling to temperatures of 4 to 30°K.

Paper 20 by Levinstein serves as an introduction to this section. In it he describes the characteristics of dopants that can be used in extrinsic detectors, analyzes the factors that determine their time constants, describes the conditions that must be met for background-limited operation, and offers some interesting information on the problems of packaging detectors.

Henry Levinstein, a professor of physics at Syracuse University, Syracuse, New York, has the unique distinction that his former graduate students hold important positions with virtually all U.S., and many foreign, manufacturers of detectors.

For many years copper-doped and mercury-doped germanium detectors have been among the most popular photon detectors for use in the 8 to 13 μm atmospheric window. Bode and Graham, in Paper 21, compare these detectors on the basis of (1) performance as a function of operating temperature, (2) the current–voltage characteristics and their influence on the accompanying circuit design, and (3) the effects of background radiation. An indication of the care taken by these authors is their correction to the measured values of spectral response that was occasioned by their use of a reference thermocouple that was not perfectly black. For further information on the problems of blackening thermal detectors, refer to Papers 28 and 29. Other papers that contain comparative performance data on detectors are listed in References 1 through 3. Further information on extrinsic detectors will be found in References 4 through 16.

Most of the early experiments on heterodyne detection with photoconductive detectors were made in the visible or near-infrared portions of the spectrum. Buczek and Picus, in Paper 22, were among the first to report measurements of the minimum detectable power and frequency response in the heterodyne mode for a mercury-doped germanium detector cooled to 4.2°K. The minimum detectable power was within a factor of 10 of the ideal photon limit. The frequency response at 10.6 μm showed a roll-off starting at about 50 MHz. In more heavily compensated detector material a frequency response flat to 450 MHz was observed. Additional work in this area is found in References 17 through 22.

Gerald Picus is Manager, Chemical Physics Department, Research Laboratories, Hughes Aircraft Company, Malibu, California.

References

1. Anderson, N. C., "Target-to-Background Performance of Various Detectors for a Wide Range of Operating Conditions," *Infrared Phys.,* **4,** 149 (1964).
2. Leftwich, R. F., "Comparison of InSb and HgCdTe in a Real-Time Scanning Infrared Camera," *Appl. Optics,* **9,** 1941 (1970).
3. Jaeger, M., A. Nordbryhn, and P. A. Stokseth, "Detection of Low Contrast Targets at 5 μm and 10 μm: A Comparison," *Appl. Optics,* **11,** 1833 (1972).
4. Levinstein, H., "Impurity Photoconductivity in Germanium," *Proc. Inst. Radio Engrs.,* **47,** 1478 (1959).
5. Schultz, M. L., "Silicon: Semiconductor Properties," *Infrared Phys.,* **4,** 93 (1964).
6. Dalven, R., "Germanium: Semiconductor Properties," *Infrared Phys.,* **6,** 129 (1966).
7. Burstein, E., "Infrared Detector," U.S. Patent No. 2,671,154, Mar. 2, 1954.
8. Burstein, E., "Germanium Far Infra-Red Detector," U.S. Patent No. 2,816,232, Dec. 10, 1967.
9. Schultz, M. L., and W. E. Harty, "Germanium Silicon Alloy Semiconductor Detector for Infrared Radiation," U.S. Patent No. 3,105,906, Oct. 1, 1963.
10. Darviot, Y., et al., "Metallurgy and Physical Properties of Mercury-Doped Germanium Related to the Performances of the Infrared Detector," *Infrared Phys.,* **7,** 1 (1967).
11. Baker, G., and D. E. Charlton, "Low Frequency Noise in Copper Doped Germanium Infrared Detectors Caused by Thermal Impedance Fluctuations," *Infrared Phys.,* **8,** 15 (1968).
12. Quist, T. M., "Copper-Doped Germanium Detectors," *Proc. IEEE,* **56,** 1212 (1968).
13. Moore, W. J., and H. Shenker, "A High-Detectivity Gallium-Doped Germanium Detector for the 40–120 μ Region," *Infrared Phys.,* **5,** 99 (1965).
14. Jeffers, W. Q., and C. J. Johnson, "Spectral Response of the Ge:Ga Photoconductive Detector," *Appl. Optics,* **7,** 1859 (1968).
15. Stillman, G. E., "Far-Infrared Photoconductivity in High-Purity Epitaxial GaAs," *Appl. Phys. Letters,* **13,** 83 (1968).
16. Feldman, P. D., and D. P. McNutt, "A Rocket-Borne Liquid Helium-Cooled Infrared Telescope," *Appl. Optics,* **8,** 2205 (1969).
17. Kurtin, S. L., G. S. Picus, and C. J. Buczek, "Wideband Optical Heterodyne Performance of Extrinsic Photoconductive Infrared Radiation Detectors," *Appl. Optics,* **9,** 1848 (1970).
18. Arams, F. R., et al., "Infrared 10.6-Micron Heterodyne Detection with Gigahertz IF Capability," *IEEE J. Quantum Electronics,* **QE-3,** 484 (1967).
19. Asnis, L. N., A. I. Vereshchaka, and Yu. V. Popov, "Heterodyne Detection with a Ge:Hg Photoresistor at High Frequencies," *Sov. J. Opt. Tech.,* **38,** 729 (1971).
20. Walker, B. J., and J. N. Crouch, Jr., "Internal Current Gain in Microwave-Biased Gallium-Doped Silicon 8–14 μm IR Detectors," *Proc. IEEE,* **57,** 2167 (1969).
21. Crouch, J. N., Jr., and B. J. Walker, "High Gain Microwave-Biased Mercury-Doped Germanium Photoconductors," *Infrared Phys.,* **11,** 129 (1971).
22. Teich, M. C., R. J. Keyes, and R. H. Kingston, "Optimum Heterodyne Detection at 10.6 μm in Photoconductive Ge:Cu," *Appl. Phys. Letters,* **9,** 357 (Nov. 1966).

Reprinted from *Appl. Optics*, **4**(6), 639–647 (1965)

Extrinsic Detectors

Henry Levinstein

The photoconductive response of intrinsic detectors may be extended to longer wavelengths through the addition of selected impurities to the semiconductor which has been properly purified. Although impurities in many semiconductors, such as germanium, silicon, indium antimonide, and germanium-silicon alloys have been investigated, only impurity-activated germanium detectors are commercially available. Germanium–gold (Ge:Au) detectors have a response to about 9 μ and require cooling between 60°K and 80°K; germanium–mercury (Ge:Hg) detectors have a response to 14 μ and should be cooled to about 30°K; germanium–copper (Ge:Cu) detectors have a response to 30 μ and require cooling to about 15°K. Their detectivity, which depends on the amount of background radiation falling on the detector, is of the order of 10^{10} cm(cps)$^{1/2}$W^{-1} at spectral peak for a 180° field of view and 300°K background radiation. Time constants are less than 1 μsec when the detectors are cooled properly.

I. Introduction

The long-wavelength threshold of intrinsic photoconductive detectors is determined by the energy gap between the valence and conduction band, i.e., the photon energy required to produce free electron-hole pairs in the semiconductor. The relation between energy gap (E) and threshold (λ_m) is given by the expression $\lambda_m(\mu) = 1.23/E(\text{eV})$. Most elements and compounds investigated to date have energy gaps either larger than 0.2 eV or negligibly small. Although the energy gap may be a function of temperature and certain other variables, such as pressure, most of these changes are not significant in extending the detector response into the 8–100 μ region. In order to fill the need for detectors in that region, material studies have proceeded in two directions: the formation of solid solutions from several compounds of which one is either a semimetal or a small gap semiconductor, and the addition of impurities to large energy gap materials. Since the former technique has a serious drawback in the difficulty of preparing materials of that type to the required purity, the latter technique is responsible for all photon detectors now in use in the 8–100 μ region. The energy required to liberate charge carriers from the impurities within a crystal is usually considerably less than the energy required to free charge carriers from atoms of the host lattice. Since detectors prepared from these materials owe their long wavelength response to the impurities, they are designated impurity-activated or extrinsic detectors. The success of the method depends entirely on the choice of the host lattice and the selection of the impurity. The host lattice must permit purification to the extent that there are fewer residual than intentionally added impurities. It must be possible to find impurities which have activation energies that are such as to provide the proper long-wavelength threshold and which can be added to the host lattice in high enough concentrations so that reasonable absorption will occur. Although the activation energies of many impurities in Ge,[1] Si,[2] InSb,[3] and GeSi[4] alloys have been investigated, all extrinsic detectors in general use are prepared with germanium as host lattice. However, GeSi alloys have a unique advantage in that a variation of the percentages of germanium and silicon in the alloy will vary its energy gap and with it the ionization energy of the impurities. In this way detectors with a variety of long-wavelength thresholds may be obtained, using only one particular impurity. Since it is difficult to prepare these alloys, and, also, since a large number of impurities with a great variety of energy gaps in germanium are known, GeSi alloy detectors have not been used extensively. Few impurity-activated silicon detectors have been constructed, and the use of impurity-activated indium antimonide detectors is limited to wavelengths longer than 100 μ.[5] Although a great many different types of impurity-activated detectors may be built to fill one specific need or another, only very few with broad application possibilities are commercially available items. These include Ge:Au, Ge:Cu, and Ge:Hg.

II. Germanium Detectors

A. Spectral Response

Germanium is a semiconductor which crystallizes in the diamond structure, each atom sharing one of its four valence electrons to form four covalent bonds, one

The author is at Syracuse University, Syracuse, New York.
Received 18 December 1964.

with each of its four nearest neighbors. The substitution of various impurities at normal germanium lattice sites introduces donor or acceptor electron energy levels into the forbidden zone of germanium. If a column V atom, such as arsenic, occupies the place of a germanium atom in the lattice, it shares four of its valence electrons to form complete bonds. The fifth electron is bound to the excess positive charge of the arsenic nucleus and moves around it in a hydrogenlike orbit. It is easily ionized to the conduction band. This type of an impurity is designated as an *n* type or donor impurity. Column III atoms (boron, for example) are *p* type or acceptor impurities. They complete only three of the four covalent bonds demanded by the germanium lattice. The fourth is completed when they accept an electron from the nearby germanium atom, leaving a positive hole which is bound to the negative column III ion. This hole is easily ionized to the valence band. Ionization energies, *E*, may be calculated on the basis of a hydrogenlike model in which the Coulomb potential is reduced by the dielectric constant, ϵ, of the medium and the electronic mass is replaced by the effective electron or hole mass, m^*, in the solid. Thus

$$E = \frac{2\pi^2 e^4}{h^2} \frac{m^*}{\epsilon^2},$$

where e is the electronic change and h is Planck's constant. The value of 0.01 eV for the ionization energy obtained from this expression is in fairly good agreement with experimental values which vary between 9.6 meV and 12.7 meV. Since most semiconductors do not possess the simple band structure required for this approximation, Luttinger and Kohn[6] have extended the model to take into account the more complex band structures, such as germanium. Calculations based on their theory yield values of 9.2 meV for the ionization of shallow donor impurities. When the theory is refined to include corrections owing to the localized nature of the bound states and the disruption of lattice bonding by the impurity atoms, improved results are obtained.

Impurities with such small ionization energies are referred to as shallow impurities. Since charge carriers are easily excited by lattice vibrations, thereby making them no longer available for the excitation by photons, the material must be cooled to temperatures in the vicinity of liquid helium if it is to be used as a detector.

When column II elements occupy lattice sites, only two of the four covalent bonds are completed. A completion of the bonds by electrons from neighboring atoms produces two holes, circulating about the impurity site. One might hope that the hydrogenlike impurity center model for column III and V impurities could be extended to heliumlike impurity centers for column II impurities. This would lead to two activation energies, one for the excitation of the first hole and a greater one for excitation of the second hole. Although two activation energies are actually observed for column II impurities, values calculated from simple theory deviate considerably from experimental results.

Calculations to obtain agreement, through corrections to this theory, have so far not been made. Experimental evidence indicates that for the three elements investigated (zinc, cadmium, mercury) the activation energy for the first hole and also for the second is proportional to the atomic mass of the impurity.[7] Since the activation energy for these materials is greater than for column III elements, the temperature at which the number of thermally excited charge carriers becomes significant is considerably higher than for column III elements. Detectors made from materials in this group thus require less cooling than those prepared from column III or V elements. Column I atoms (gold, copper, silver), when added substitutionally to the germanium lattice, complete only one of the four covalent bonds. According to the simple model, there should be three energy states associated with the excitation of holes formed by the completion of the four possible bonds. This behavior is verified by triple acceptor energy levels for copper and silver. Gold, on the other hand, possesses in addition to triple acceptors, another level[8,9] close to the valence band. This has been identified as a donor level, and Kaus[10] has attributed its existence to interstitial gold atoms. As one would expect from the preceding discussion of double acceptor impurities, the energies of the triple acceptors also differ considerably from those predicted by simple theory. Large differences are observed between ionization energies of the three elements (copper, silver, gold).

In addition to these impurities, transition metals and column VI elements have been investigated. Figure 1 shows a summary of the energy levels associated with some of the impurities in germanium. While detectors could be prepared using either *n*- or *p*-type impurities, all detectors in use today are made with *p*-type material. (Only column I, II, and III impurities are of detector interest at present.) The spectral response of all impurity-activated detectors exhibits two peaks, one with a threshold at 1.8 μ determined by the energy to free charge carriers from the germanium bonds, the other at longer wavelengths determined by the energy to free charge carriers from the particular impurity which has been intentionally added. The relative magnitude of the two peaks may be adjusted and depends on constructional details of the detector, such as density of

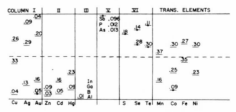

Fig. 1. Energy levels of impurities in germanium (energies above dotted line are measured from conduction band, below dotted line from valence band). Lines with circles indicate donor levels, those without circles acceptors.

impurity atoms, sample geometry, and surface preparation.

B. Time Constant

The response time or time constant of a detector is determined by the recapture time of the charge carriers which have been freed by incident photons. In the case of almost all impurity-activated Ge detectors which are in use, these charge carriers are holes and they are recaptured by singly charged negative impurity ions. Neutral impurities become negatively charged either when a hole is excited to the valence band by incident photons (from signal or background) or when donor atoms, which are present in the sample, give up electrons to the acceptor impurities. It is desirable that impurity-activated materials whose long-wavelength threshold is due to acceptor atoms have a certain density of such donor atoms. Let us assume that we deal with a detector containing about 10^{15} atoms/cm³ of a particular column I or II deep acceptor impurity. It is then highly probable that there would be, in addition, at least 10^{13} shallow acceptor impurities which cannot be removed and which would be ionized at the operating temperature. The large number of free holes thereby produced decreases the detector responsivity. To eliminate this action of these impurities, shallow donors whose electrons fill these holes are added. Since it is usually not possible to obtain an exact balance which would prevent the existence of any holes, the sample is slightly overcompensated (the density of shallow donors slightly exceeds that of shallow acceptors). Perhaps 10^{13}–10^{14} deep acceptor atoms thus become singly charged negative ions owing to overcompensation. This is in addition to a much smaller number of atoms which have been ionized by photons from signal and background radiation. Figure 2(a) shows a schematic diagram of the electronic processes of such a compensated sample.

The time constant, τ, is given by the expression $\tau = 1/Nv\sigma$, where N is the density of negatively charged ion sites, σ their capture cross section, and v the thermal velocity of the holes. If one assumes reasonable values for these parameters, such as $N \sim 10^{13}$/cm³, $\sigma \sim 10^{-13}$ cm², $v \sim 10^{7}$ cm/sec, τ turns out to be of the order of 10^{-7} sec. It is assumed in this discussion that the detector is cooled sufficiently so that there are no thermally freed holes. A control of τ can thus be obtained by varying the density of donor impurities. Impurity-activated detectors with values of τ between 10^{-6} sec and 10^{-8} sec have been constructed. Since the density of recombination centers is determined by the degree of compensation, the time constant is not dependent on the amount of background and signal radiation falling onto the detector. If this were not the case and if background radiation were to determine the density of negative ion sites, time constants would decrease with increasing background radiation. This is shown in Fig. 2(b) where the recombination centers are determined by the background radiation, a condition not usually obtained. These detectors are always operated in the photoconductive mode, the photosensi-

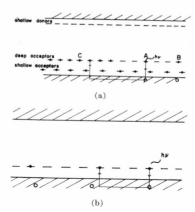

Fig. 2. Recombination of charge-carriers in a material containing deep *p*-type acceptor levels. (a) Overcompensated sample, recombination centers determined by compensating donor impurities. (b) Uncompensated sample, recombination centers determined by background radiation.

tive sample being connected in series with a load resistor and bias battery. The signal is taken across the load resistor and amplified as desired. Since the distributed capacitance of these detectors and associated circuit is of the order of 1–10 pF and resistances are as high at 10^{6} Ω, RC circuit time constants are of the order of 10^{-6} sec, making it difficult to take advantage of the shorter detector time constants. Only a reduction of the load resistance below the detector resistance decreases this time constant. Unfortunately, this results in an often undesirable lowering of detector signal.

In addition to the short time constant associated with the capture of holes by negative impurity ions, several other time constants have been observed in these detectors. The intrinsic time constant associated with radiation below 1.8 μ in germanium is governed by electron-hole recombination either directly or through recombination centers. Time constants from 1–10 μsec have been associated with that mechanism. If a response below 1.8 μ is not required, contributions from this intrinsic response may be eliminated through the use of a germanium filter which may be placed over the sensitive element or made an integral part of the element.

Time constants of several milliseconds have been observed in various other impurity-activated detectors in certain temperature ranges. Thus, when Ge:Au detectors are cooled considerably below the optimum of 60°K, such a time constant appears superimposed on the very short normal detector time constant. Ge:Hg detectors have also shown such behavior when they were not cooled to a sufficiently low temperature. Recent evidence tends to show that these time constants are a result of other impurities, especially copper, which have not been entirely removed.

C. Detectivity

The magnitude of the signal, S is given by the expression

$$S = \frac{ER_cR_L}{(R_c + R_L)^2}\frac{\Delta R_c}{R_c},$$

where R_c and R_L are detector and load resistance, respectively, E the emf of the bias battery, and ΔR_c the change in detector resistance, when exposed to radiation.

In terms of material parameters

$$\frac{\Delta R_c}{R_c} = \frac{\Delta n}{n} = \frac{\eta Q_s A \tau}{n},$$

where Δn is the change in the number of charge carriers n when exposed to radiation, η the number of charge carriers liberated per incident photon (usually less than one), Q_s the incident photon flux, and A the detector area.

Since devices of this sort are primarily used to measure low-intensity infrared radiation, one is interested not only in signal magnitude but also in the signal-to-noise ratio produced by a detector for a certain incident radiation flux. Two types of extrinsic detectors—$1/f$ noise and charge carrier fluctuation noise. $1/f$ noise is usually associated with the surface of a semiconductor.[11] Since here noise power varies inversely with frequency, it is significant only at the lower frequencies. Well-constructed detectors show appreciable $1/f$ noise only below 100 cps. Noise resulting from fluctuation in generation and recombination rates of charge carriers is inherent in all semiconductor detectors. It is given by the expression[12]:

$$V_n = 2\frac{ER_cR_L}{(R_c + R_L)^2}\left(\frac{\tau}{\langle n^2\rangle}\right)^{1/2}\left(\frac{\Delta f}{4\pi^2 f^2\tau^2 + 1}\right)^{1/2},$$

where f represents the frequency at which measurements are made and Δf the bandwidth of the amplifier. $\langle\Delta n^2\rangle$ is the variance in the charge-carrier concentration. When no signal is incident on the detector, the charge-carrier concentration is determined by the lattice vibrations and by background radiation photons with energy sufficient to free charge carriers. If the material is cooled sufficiently for the charge-carrier concentration to be determined solely by the background radiation, the detector is said to be background-limited, and noise arises from fluctuations in the arrival of background photons. $\langle\Delta n^2\rangle$ is then

$$\sum_\lambda n_\lambda\left(1 + \frac{1}{\exp(hc/\lambda kT) - 1}\right),$$

where h is Planck's constant, c the velocity of light, k Boltzmann's constant, and T the absolute temperature (°K). This reduces to n for background temperatures less than 300°K and wavelengths shorter than 20 μ. When the charge carriers are due to excitation by lattice vibrations, $\langle\Delta n^2\rangle = n$, and the noise is termed generation–recombination (G–R) noise.

In order to make possible a comparison among detectors, the S/N ratio per watt is normalized to a 1-cm² detector area and 1-cps bandwidth. This yields the parameter specific detectivity, D^*, used for all photon detectors and defined[13]:

$$D^* = \frac{S/N}{H_s}\left(\frac{\Delta f}{A}\right)^{1/2},$$

where H_s is the rms power density at the detector. Although this quantity is independent of circuit parameters E and R_L, their choice determines the proper operation of the detector. If R_L is too small, the signal voltage appearing across it may be smaller than the noise of the amplifier, and the device becomes amplifier noise limited. If R_L is too large, the time constant may be the RC circuit time constant rather than the detector time constant. Although both signal and noise voltage vary more or less linearly with applied bias voltage, each detector has a certain critical voltage above which noise voltage increases superlinearly and signal voltage sublinearly. Thus, both bias voltage and load resistance must be selected to fit the particular detector and the particular application. If one assumes a background-limited detector, the number of charge carriers is given by $n = A\eta Q_B \tau$, where Q_B is the background photon flux. If, furthermore, the limiting noise is radiation noise rather than $1/f$ noise,

$$D^* = \frac{1}{2}\left(\frac{\eta}{Q_B}\right)^{1/2}\frac{Q_s}{H_s}.$$

At a particular wavelength

$$H_s = (hc/\lambda)Q_s.$$

Thus, D_λ^*, the detectivity at a particular wavelength λ is

$$D^* = \frac{\lambda}{2hc}\left(\frac{\eta}{Q_B}\right)^{1/2}.$$

D^* at a particular wavelength may thus be enhanced by having the quantum efficiency as close to unity as

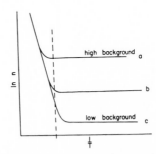

Fig. 3. The variation of charge-carrier concentration with temperature for various amounts of background radiation. Dotted line shows how a detector may be background-limited for one particular amount of background radiation, but not background-limited as the background is reduced.

possible and the background radiation as small as possible. Both approaches are being used. The quantum efficiency is being raised by minimizing radiation which is not absorbed. This could be radiation reflected from the front surface of the sample or radiation which on its course through the sample does not encounter an absorption center. Absorption may be enhanced by making the sample dimensions in the direction of the incident radiation greater or by making the impurity concentration as large as possible. Increased absorption may also be obtained by mounting the element in an integrating sphere which permits the nonabsorbed radiation to make several passes through the sample, thereby increasing the possibility for absorption. This is especially vital for material with impurities which cannot be incorporated to concentrations greater than $10^{15}/cm^3$. Background may be reduced by limiting the angular field of view of the detector to no larger a size than is required by the signal and by using a cooled filter which eliminates background radiation at wavelengths containing no signal information. Since a reduction in background radiation reduces the density of charge carriers freed by the background, the sample resistance increases until the density of thermally liberated charge carriers becomes dominant. This behavior is shown in Fig. 3. The sample must then be cooled to lower temperatures if the detector is to become background-limited again. Such a reduction in the number of charge carriers may lead to detectors with resistance of several hundred $M\Omega$ for the smallest fields of view.

D. Detector Construction

The preparation and constructional details of all impurity-activated Ge detectors are similar. The variations arise from different methods for adding impurities to the germanium, and the use of various types of detector envelopes to provide the proper cooling. Impurities may be added to the germanium either during crystal growth or, following crystal growth, by diffusion. The photosensitive sample is cut from the single crystal ingot to the desired dimensions, such as $1 \times 1 \times 3$ mm³, but frequently either larger or smaller. It is then lapped and etched. Contacts are attached either by the use of indium solder (for p-type materials) with a suitable flux (Divco 335, for example) or alloyed. Since the sensitive element must be cooled, the cold finger of the dewar container in which the sample is to be mounted frequently becomes one of the contacts. If the impurity concentration is less than $10^{16}/cm^3$, the sample is mounted in an integrating sphere which is cooled to the temperature of the sample. The integrating sphere also makes possible the attachment of limiting cooled apertures or filters. Depending on the cooling requirement, the sample is either mounted on the cold finger of a liquid nitrogen dewar [Fig. 4(a)], a liquid helium dewar [Fig. 4(b)], or at the cold finger of a Sterling or other closed cycle refrigerator. The detector envelope is provided with a window, transparent in the special region in which the detector is to be operated. During operation the dewar is main-

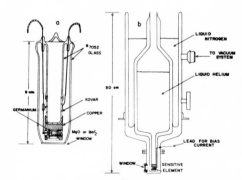

Fig. 4. Detector envelopes for: (a) a detector cooled to liquid nitrogen temperature, (b) a detector cooled to liquid helium temperature.

tained at pressure less than 10^{-6} mm of mercury to prevent frosting of the window and the formation of frost on the detector surface. Although it is desirable to maintain such a vacuum when the detector is not in use, this is not always essential.

Detector performance, especially the amount of $1/f$ noise and the maximum voltage which may be applied, depends on the preparation of the surface and the quality of the contacts. Well-established semiconductor techniques, as well as some trial and error, can improve results.

E. Germanium Detectors with Column I Impurities

Impurities investigated in that group include gold, silver, and copper. Because of an original interest in developing detectors with response in the 3–5 μ region, the gold impurity with comparatively deep energy levels was investigated. Ge:Au was the first impurity-activated germanium detector in use.[14,15] As the need for longer wavelength detectors arose, Ge:Cu detectors were fabricated[16] based on considerable experience of the behavior of copper impurities in germanium. Silver impurities in germanium have not been investigated thoroughly, because of the low solubility of silver in germanium.

Ge:Au

Gold is most conveniently added to germanium during crystal growth. Concentrations of gold atoms in germanium as high as 10^{16} have been reported. The energy required to free the least tightly bound hole is 0.16 eV. The reaction may be given by the expression, $Au^\circ + h\nu = Au^- + p^+$, where Au° represents neutral gold, $h\nu$ a photon, Au^- a singly charged gold ion, and p^+ the excited hole. This yields a detector with response to about 9 μ. Since the concentration of gold in germanium is comparatively low, a large fraction of the incident radiation is not absorbed in the detector element, and the element should be mounted in an integrating chamber. Desirable bias currents for the conventional detector element with a resistance of about 10^6 Ω are in the range 20–80 μA (fields $\sim$ 100 V/cm). Although much lower currents may yield an adequate D^*, they might produce electrical signals smaller than amplifier noise, thus making the system amplifier noise limited. One of the difficulties with the detector is that, in order to obtain background-limited operation, its temperature must be in the vicinity of 60°K. Under ideal conditions, detectivity at 78°K (liquid nitrogen) is lower by approximately a factor 3. In addition, when the detector is not background-limited, noise may be introduced by small fluctuations in coolant temperature, such as the bubbles in the liquid nitrogen. When this is remedied by mounting the element on a base with sufficient thermal capacity, a considerable lengthening of detector cooldown-time results. The recent introduction of Sterling cycle coolers which reach 60°K with ease should make this detector competitive with intrinsic detectors and superior to them if a response to 9 μ is required. Figure 5 shows the spectral response of a typical Ge:Au detector at 78°K and at 60°K.

As may be seen from Fig. 1, gold has three acceptor and one donor states. Although the hole ionization associated with the 0.16-eV level has been applied most frequently to detector construction, several other possibilities present themselves. If donor impurities are increased to the extent that they fill one of the holes per gold atom, then addition of further donors will begin to eliminate the second hole. The energy required to remove electrons to the conduction band from each of these doubly negatively charged gold atoms is 0.2 eV. This leads to an n-type detector with a response to 6 μ (Au$^-$ + $h\nu$ = Au$^-$ + n^-). Since this detector is inferior in every way to available intrinsic detectors (InSb, for example), it is no longer used. Alternately, if shallow acceptor impurity atoms are added to Ge:Au, the electron from the donor state will be lost to the acceptor atom, producing an additional hole. The energy to free this hole is 0.5 eV producing a detector with response to 24 μ (Au$^+$ + $h\nu$ = Au$^\circ$ + p^+). Cooling to about 20°K is required, however, to prevent thermal excitation of this hole. Since detectors with other impurities may be prepared more easily in that range, this level has been only of scientific interest.

Ge:Au:Cu

Copper is most conveniently added to germanium by diffusion. Copper atoms are deposited on the germanium surface (either by evaporation or plating from solution) and diffused into the sample by heating it for several hours just below the melting point of germanium. Copper, just as gold, possesses three acceptor states. The energy required to free the most weakly bound hole is 0.04 eV. This yields a detector with response to about 30 μ, the spectral peak occurring at about 24 μ. Since the energy required to remove the second hole is 0.37 eV, a detector using this energy level would respond to only 4 μ, a range where intrinsic detectors are more suitable. The small activation energy for the weakly

bound hole requires that the detector be cooled to about 15°K. Because of the difficulty of reaching this intermediate temperature, liquid helium is generally used. This requires a double dewar (Fig. 4) rather than the one used for Ge:Au. The field of view and spectral response of the detectors are often decreased to improve detectivity. Since the detector when operated at 4°K is background-limited even for very small amounts of background radiation, detectivity is improved by decreasing the angular field of view through the use of a limiting aperture or by limiting the spectral response through the uses of a cooled filter. Figure 5 shows a Ge:Cu detector with and without the use of a limiting filter.

Ge:Cu detectors show a nonlinear voltage–current curve, resistance decreasing for higher fields. Since this decrease is not accompanied by a reduction in D^*, this lower dynamic resistance with higher fields is frequently a very desirable feature. It decreases the RC time constant at no decrease in signal. Detectors with dynamic resistance as low as 3000 Ω have been constructed.[16,17] Thus the 0.1-μsec material time constants may actually be utilized.

F. Column II Detectors

Ge:Hg

Ge:Hg detectors were developed to fill the need for fast detectors in the 8–14 μ atmospheric window usable at temperatures considerably higher than the 4–20°K range.[7] Mercury is added to the germanium during crystal growth either by passing mercury vapor over the growing crystal or sealing the growing crystal in a quartz crucible containing mercury. Only the lower energy level of mercury in germanium (0.09 eV) is currently being used. Since the concentration of mercury in germanium seldom exceeds 10^{15} atoms/cm³,

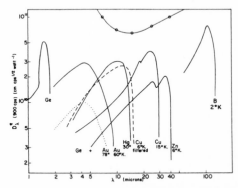

Fig. 5. Spectral response of typical impurity-activated germanium detectors: Ge:Au at 60°K and 78°K, Ge:Cu with and without filter, Ge:Hg and Ge:B. All detectors with the exception of Ge:B have a 60° field of view. Field of view of Ge:B is ~10°. The line with circles indicates peak detectivity of ideal detectors with a 60° field of view.

an integrating sphere is required. Cooling in the range 30–40°K is sufficient to produce background-limited operation. However, in that temperature range the time constant has two components, one less than 1 μsec, the other considerably longer. Cooling below 30°K is required to remove the slow component.[18] The magnitude of the slow component has been observed to vary with the density of copper recombination centers in germanium.[19] Other impurities as well as imperfections probably play a major part. Improved crystal growing techniques should eventually produce Ge:Hg detectors which can be operated in the 40–50°K temperature range. In the meanwhile, liquid neon or Sterling cycle mechanical refrigerators provide adequate cooling just below 30°K. Ge:Hg detectors, operated at 30°K, have a response similar to Ge:Cu, whose spectral response has been modified through the use of cooled filters to remove radiation beyond 14 μ, while the element is cooled to liquid helium temperature. Ge:Hg thus presents an advantage over the more easily prepared Ge:Cu only when used in the 30–40°K temperature range where copper impurities in germanium would be completely ionized. Resistances of the Ge:Hg detector are in the MΩ range. Voltage-current curves are much more linear than for Ge:Cu. Detectivity frequently remains fairly constant with increasing voltage until a critical voltage is reached above which detectivity decreases rapidly. Bode and Graham[17] note a slight increase in D^* with bias to the point where it decreases very rapidly. The region just above optimum bias is characterized by current oscillations whose frequency varies with bias voltage.

Detectors with a variety of sensitive areas have been constructed, but it has been observed that D^* values vary somewhat with area, smaller area detectors having larger D^*.[20] Figure 5 shows the spectral response of a typical Ge:Hg detector.

Ge:Zn

Ge:Zn was the first of the long wavelength ir detectors developed.[21] The ionization energy for the first hole is 0.03 eV. Thus the long-wavelength threshold occurs at 40 μ (Fig. 5). Zinc may be added to germanium during crystal growth up to concentrations of 10^{17} atoms/cm³. This makes possible the use of thinner samples without an integrating sphere. Since the temperature to prevent appreciable thermal ionization of the zinc centers is about 7–8°K, liquid helium must be used as the refrigerant. Element resistances of the order of 1 MΩ are usually obtained. The resistance remains fairly constant with increasing field to a value of about 20 V/cm. It may then either rise or decrease slowly to a value of about 60 V/cm above which resistance decreases rapidly, probably due to impact ionization. Since at this point the detectivity decreases rather drastically, the detector must be used below that critical value of bias field. The availability of Ge:Hg for the 8–14 μ region and Ge:Cu with response to 30 μ has reduced the need for this detector to the point where it is no longer commercially available.

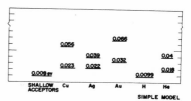

Fig. 6. Energy levels of impurities in indium antimonide.

Attempts have been made to develop a detector through excitation of the second hole of zinc which is attached to the zinc center with an energy of 0.09 eV. If the first hole is filled by electrons from shallow donor impurities, such as antimony, the material need only be cooled to the range between 40–50°K to prevent thermal ionization of the second hole. A detector, sensitive in the 8–14 μ region, is thus obtained. Detectors so far prepared show lower D^* than Ge:Hg, are more difficult to prepare, and are, therefore, used only rarely.

Ge:Cd

Ge:Cd detectors were developed before the advent of Ge:Hg to provide detectors in the 8–14 μ region which could be operated at higher temperatures than Ge:Cu or Ge:Zn. Their response extends to about 25 μ when they are cooled to 25°K (liquid hydrogen range).[22]

G. Column III or Group V Impurities

The addition of either shallow donor or acceptor impurities to Ge produces detectors with a response beyond 100 μ. The long-wavelength threshold varies somewhat with the impurity, occurring between 120 μ and 140 μ. The spectral peak is somewhere between 90 μ and 100 μ. Ge:Sb was used in one of the early detectors.[23] More recently Ge:B has been studied.[24] It requires cooling to about 4°K, has a spectral peak at 104 μ and a long-wavelength threshold at 140 μ (Fig. 5). Detectivity at spectral peak is about 10^{11} (cps)$^{1/2}$ W⁻¹. One of the difficulties with these materials is their low bias voltage before impact ionization sets in. Fields in excess of 3–4 V/cm, that is, more than 1–2 V across a typical detector element, make the detector useless. Because of a very low signal associated with low bias voltage, such detectors would be amplifier noise limited under normal use and would find competition from liquid helium-cooled germanium bolometers if their short time constants (10^{-7}–10^{-8} sec) are not a specific requirement.

III. Ge:Si Alloys

Good single crystals may be obtained when germanium and silicon are alloyed in all proportions. Increasing the amount of silicon in the alloy decreases the lattice constant without noticeable segregation or without the formation of a superlattice. At the same time the band gap changes from that of germanium at 0.7 eV to that of silicon at 1.2 eV. The band gap changes rapidly at first until about 15% Si in Ge:Si is reached, then more slowly as more and more Si is added. At the same time the ionization energy of impurities increases. It is, thus, possible to select an impurity which in germanium has too small an ionization energy (a threshold which is at longer wavelengths than required) and increase the ionization energy by the addition of Si. In the search for a detector sensitive in the 8–14 μ region, usable with a minimum amount of cooling, two types of impurities were considered: the 0.05-eV donor level of gold in germanium and the 0.09-eV acceptor level of zinc in germanium.[4] As is the case when the 0.09-eV level of zinc is used in germanium alone, donor impurities have to be introduced in sufficient amounts to fill holes excited from the 0.03-eV level. The addition of 7.5% Si to Ge produces a shift in spectral peak from 13 μ to 8 μ, 4.4% Si in a shift to 11 μ. When the latter material is cooled to about 48°K, a satisfactory 8–14 μ detector with peak detectivity of about 10^{10} cm (cps)$^{1/2}$ W^{-1} is obtained. Corresponding changes are produced when the 0.05-eV level of gold is used as the impurity. Since the growing of crystals from this alloy is more complicated than from germanium, detectors of this type are expected to find use only when the restrictions on spectral response are sufficiently severe so that no impurity-activated germanium detector may be adopted for this particular use.

IV. Impurity-Activated InSb Detectors

Energy levels in the forbidden gap of InSb may be obtained when an impurity atom replaces either indium or antimony. When an atom from column VI of the Periodic Table replaces antimony (column V) there is one more electron than the number needed to complete the bonds with neighboring atoms, thus producing *n*-type InSb. When this material is used in conjunction with a magnetic field, response to 8 mm may be obtained.[5] This detector is described elsewhere in this issue. When a column II impurity replaces indium in the lattice, an electron from one of the surrounding indium–antimony bonds is required to complete the bonds between the impurity and the closest lattice atoms. The resulting hole will circulate around the impurity if no thermal ionization takes place. An energy of 0.075 eV is required to excite the hole. Column I impurities, such as gold, silver, and copper, require two electrons to complete the bonds. Thus two holes circulate about the impurities very similar to column II impurities in germanium. Figure 6 shows the energies required to free both holes for the three impurities. It is interesting to note here that for the simple model of a heliumlike impurity site used to calculate impurity level energies, results are in better agreement than for the similar model in germanium.[3] Extrinsic response in each case has been so low that detectors with these materials as impurities have not as yet been constructed.

A. Impurities in Si

The energy levels of impurities in silicon have not been studied in as great detail as those in germanium.

Those investigated behave in much the same way as impurities in germanium.[2] In general, ionization energies are greater in silicon than for the same impurities in germanium. The simple theory for hydrogenlike impurity centers may be used to calculate ionization energies of shallow impurities, but deviations from the calculated values are greater than in germanium. No impurity-activated silicon detectors have been developed to date, probably because the work on germanium was started sooner and was successful. Furthermore, extrinsic silicon detectors cannot be expected to present a significant advantage over extrinsic germanium detectors.

The research on impurity-activated germanium detectors at Syracuse University has been supported by the Reconnaissance Division, Air Force Avionics Laboratory, Wright-Patterson Air Force Base, Ohio. The author is indebted to the late N. F. Beardsley and his successor, T. Pickenpaugh, at Wright-Patterson Air Force Base who have acted as project monitors. It has been their foresight, understanding, and cooperation which have led to uninterrupted support of this work, under ideal conditions, for the past eighteen years. He is also indebted for the valuable contributions of his present and former students at Syracuse University who were responsible for many of the detectors which have been developed and the physical interpretations of their characteristics.

References

1. R. Newman and W. W. Tyler, Solid State Phys. **8**, 49 (1959).
2. C. B. Collins and R. O. Carlson, Phys. Rev. **108**, 1409 (1957).
3. W. Engeler, H. Levinstein, and C. Stannard, Jr., J. Phys. Chem. Solids **22**, 249 (1961).
4. G. A. Morton, M. L. Schultz, and W. E. Harty, RCA Rev. **20**, 599 (1959).
5. E. H. Putley, J. Phys. Chem. Solids **22**, 241 (1961).
6. J. M. Luttinger and W. Kohn, Phys. Rev. **97**, 869 (1955).
7. S. Borrello and H. Levinstein, J. Appl. Phys. **33**, 2947 (1962).
8. W. C. Dunlap, Phys. Rev. **100**, 1629 (1955).
9. G. A. Morton, E. E. Hahn, and M. L. Schultz, in *Photoconductivity Conference*, R. J. Breckenridge et al., eds. (Wiley, New York, 1956).
10. P. E. Kaus, Phys. Rev. **109**, 1944 (1958).
11. A. U. MacRae and H. Levinstein, Phys. Rev. **119**, 62 (1960).
12. K. M. Van Vliet, Proc. Inst. Radio Engrs. **46**, 1004 (1958).
13. R. C. Jones, Proc. Inst. Radio Engrs. **47**, 1498 (1959).
14. M. L. Schultz and G. A. Morton, Proc. Inst. Radio Engrs. **43**, 1819 (1955).
15. L. Johnson and H. Levinstein, Phys. Rev. **117**, 1191 (1960).
16. H. D. Adams, W. J. Beyen, and R. L. Petritz, J. Phys. Chem. Solids **22**, 168 (1961).
17. D. E. Bode and H. A. Graham, Infrared Phys. **3**, 129 (1963).
18. C. C. Rousseau and G. Pruett, private communication.
19. J. Stannard, private communication.
20. Infrared Devices, T. I. Tech. Rept., Texas Instruments Inc. (1964).

21. E. Burstein, S. F. Jacobs, and G. S. Picus, Proc. Intern. Comm. for Optics (1959).

22. P. Bratt, W. Engeler, H. Levinstein, A. Mac Rae, and J. Pehek, Infrared Phys. **1**, 27 (1961).

23. S. J. Fry and C. F. Oliver, J. Sci. Instr. **36**, 195 (1959).

24. H. Shenker, W. J. Moore, and E. M. Swiggard, J. Appl. Phys. **35**, 3965 (1964).

Reprinted from *Infrared Phys.*, **3**(3), 129–137 (1963), with the permission of Microforms International Marketing Corporation as exclusive copyright licensee of Pergamon Press journal back files.

A COMPARISON OF THE PERFORMANCE OF COPPER-DOPED GERMANIUM AND MERCURY-DOPED GERMANIUM DETECTORS

D. E. BODE and H. A. GRAHAM

Santa Barbara Research Center, Goleta, California

(*Received* 10 *May* 1963)

Abstract—This paper compares the performance characteristics of copper-doped germanium with mercury-doped germanium. The detectivity of Ge:Cu falls off above about 20°K, while, that for Ge:Hg falls off rapidly above about 40°K. The dynamic resistance of Ge:Hg increases with increasing bias in the useful range of the device. The impact of nonflat thermocouples on the accuracy of spectral detectivity is discussed and the noise characteristics of these detectors as they relate to preamplifier design are considered.

INTRODUCTION

INFRARED technology in the 8 to 14 μ atmospheric window has advanced considerably during the last few years. Many of these achievements have been made possible by the advances in detector technology for this wavelength region. To date there are several extrinsic germanium detectors that have been developed for this wavelength region. In chronological order of development, these units are: Ge:Zn, Ge:Cu, Ge:Cd, and Ge:Hg. The most commonly used detectors in modern systems applications are copper-doped germanium and mercury-doped germanium.

The purpose of this paper is to compare the characteristics of copper-doped germanium with those of mercury-doped germanium. Both of these detectors commonly exhibit a detective efficiency within a factor of two of the theoretical limit. The detective efficiencies of both types have been reported to be as high as 80 per cent. Therefore, the decision as to which detector should be used for a particular application must be based upon criteria other than detectivity.

The first criterion is the performance of the detector as a function of operating temperature. The second consideration is the current-voltage characteristic; and the third factor to examine is the effect of background radiation on the various detector parameters. Other factors that might influence a decision are (a) the true spectral response, (b) the effective time constant, and (c) the magnitude of 1/f noise exhibited by the detector.

This paper will present the results of experimental measurements made on representative samples prepared at Santa Barbara Research Center.

TEMPERATURE DEPENDENCE

The measurement of detectivity as a function of temperature was performed in the following manner. The representative sample was soldered to its heat sink with indium and mounted into a copper integrating cavity similar to the method described by Levinstein and Borello.[1] A 20-cm length of 0·003-in. diameter Teflon coated copper-constantan thermocouple was cemented to the integrating cavity and soldered to the heat sink. A thin mylar spacer was inserted between the copper integrating cavity and the copper heat sink of the double dewar flask to provide a degree of thermal insulation. A 50-Ω carbon resistor was embedded in the integrating cavity and held in place with GE insulating varnish #7031 to serve as a resistance heater. Cold copper radiation shields were used to establish the desired field of view. Both liquid hydrogen and liquid helium were used as the cryogenic liquids to perform these tests. By applying electrical power to the heater resistance, a variation in temperature between about 9°K and 50°K was easily achieved. Thermocouple potentials were measured with a Leeds-Northrup Type K potentiometer using a liquid nitrogen reference junction.

Figure 1 shows the dependence of the detectivity of Ge:Cu #ZN7753–1–21 on temperature, using two different levels of background radiation. Note that, with a large field of view

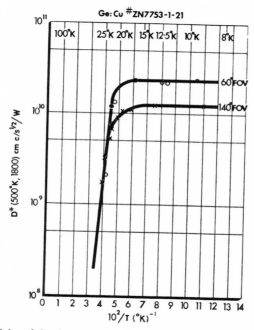

Fig. 1. The detectivity of Ge:Cu detector #ZN7753–1–21 as a function of temperature.

and a background radiation temperature of about 300°K, it is possible to operate this detector at temperatures up to 16°K. The D^* is down 3 dB at about 21°K. As the background radiation level is reduced, the maximum operating temperature must also be reduced.

If one considers that D^* is inversely proportional to the square root of the hole concentration (p) and that in the temperature range of thermal impurity activation

$$p = p_0 e^{-\epsilon/kT},$$

then since

$$D^* \propto p^{-1/2},$$

$$D^* \propto p_0^{-1/2} e^{\epsilon/2kT}.$$

The slope of the roll-off curve in Fig. 1 yields an activation energy of about 0·043 eV in general agreement with the accepted energy level of the copper acceptor impurity.

Figure 2 shows the dependence of the detectivity of Ge:Hg sample #10–8 on temperature for two different fields of view. In the case of the 150–degree field of view and a 300° background temperature, the $D*$ is down 3 dB at about 45°K. The $D*$ is effectively flat below about 36°K.

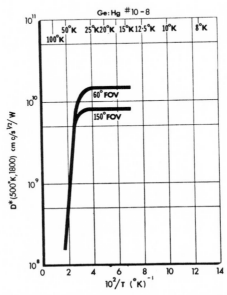

FIG. 2. The detectivity of Ge:Hg detector #10–8 as a function of temprature.

CURRENT-VOLTAGE CHARACTERISTICS

Neither Ge:Cu nor Ge:Hg exhibits a truly ohmic behavior. Consequently, it is important for the circuit engineer to recognize the type of nonohmic behavior that is characteristic of the particular detector.

Figure 3 shows the current-voltage characteristics of Ge:Cu #7753–1–21 for two different field of view. The measurements were made at about 11°K. Note that the current increases linearly with voltage for low voltages. As the voltage is increased, the current increases at a faster rate than the voltage. In other words, the resistance of the sample decreases with the increasing voltage. Although various explanations of the phenomenon have been offered, the effect is not completely understood at present.*

The signal/noise ratio of Ge:Cu is not critically dependent on the bias level. By proper adjustment of the bias, one can operate at a very low dynamic resistance without loss in responsivity from a fixed load. There are many advantages in maintaining a low dynamic impedance. The photoconductive lifetime of Ge:Cu is less than 10^{-7} sec. However, in order to utilize this short time constant in high-speed application, it is necessary to reduce the RC time constant of the detector–cooler–preamplifier combination. The capacitance of detector–dewar–preamplifier combination can be minimized only within practical limitations.

* For further discussion of this phenomenon, the interested reader can be referred to an article by G. Picus, *J. of Phys. & Chem. of Solids*, **23**, 1753 (1962).

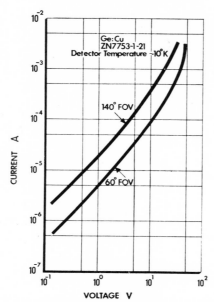

Fig. 3. The current-voltage characteristics of Ge:Cu detector #ZN7753–1–21.

A decrease in dynamic impedance is a distinct advantage. This decrease in dynamic impedance can be achieved by subjecting the Ge:Cu element to a high bias voltage.

Figure 4 shows the V-I characteristics of Ge:Hg #10–8 for two different fields of view. In this case, the resistance generally decreases slightly with increasing voltage and then passes through a point of inflection and increases. The optimum D^* generally occurs at the point where dV/dI is maximum. In contrast to Ge:Cu, the S/N ratio of Ge:Hg is more

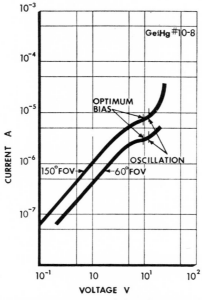

Fig. 4. The current-voltage characteristics of Ge:Hg detector #10–8.

sharply dependent on the bias level. As indicated in Fig. 4, the optimum bias voltage was nearly the same for the two different levels of background radiation. Optimum performance for Ge:Hg under strongly varying background conditions will require circuitry that maintains a fixed voltage across the detector element.

As indicated for the case shown in Fig. 4, some samples of Ge:Hg will exhibit severe current oscillations at voltage levels just above the point of optimum operation. The frequency of these oscillations is bias dependent. The oscillations are not necessarily stable for all bias levels and they often exhibit several modes. Cases have been observed where the oscillations appear quite sinusoidal on an oscillograph. Other cases exhibit more of a sawtooth appearance. Depending upon the particular sample and voltage level, the frequency of oscillation has been observed in the 10–kc/s to 50–kc/s range. The peak-to-peak amplitude of these oscillations can vary from several millivolts to several volts depending on the bias level. The mechanism for these oscillations is not understood at present.

It is important that the circuit engineer appreciate the existence of these oscillations. In the case of a constant current source, a reduced background radiation level would raise the detector resistance and increase the voltage drop across the detector. Figure 4 shows that an increase in bias voltage of several volts might drive the detector from optimum operation into oscillation.

SPECTRAL CHARACTERISTICS

The spectral detectivity of Ge:Cu is compared to Ge:Hg in Fig. 5. Both of these detectors were shielded to a 60° field of view. A KRS–5 window with a flat transmission of about 70

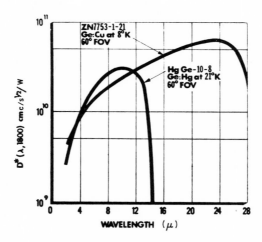

FIG. 5. The spectral detectivity of Ge:Cu #ZN7753–1–21 and Ge:Hg #10–8 for a 60° field of view. The background temperature was 298°K.

per cent over this spectral range was used. The data reported in Fig. 5 have not been corrected for this loss in window transmission. The two samples represented in Fig. 5 are not being represented as the best detectors that have been made. They are the same samples for which the temperature characteristics were determined, as previously described.

Although Ge:Cu was shown in Fig. 5 without any spectral alteration, it is possible to limit the performance of the detector to about 14 μ with the help of a cold filter. The standard technique is to use a 2·5-mm thick, cold Irtran 2 filter inside the dewar or cooling head. It is also possible to combine this with an 8-μ turn-on interference filter (multilayer coating on Ge) to provide a bandpass in the 8 to 14 μ region. Figure 6 shows the characteristics of a typical Ge:Cu detector utilizing an Irtran 2 cold cutoff filter and a room temperature interference type turn-on filter as the entrance window of the liquid helium dewar.

The spectral curves shown in Figs. 5 and 6 were obtained using a Perkin–Elmer Model

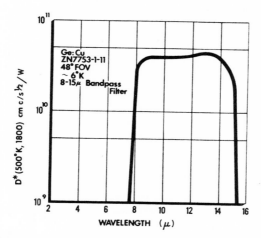

Fig. 6. The spectral detectivity of Ge:Cu #ZN7753–1–11 using an 8 to 15 micron bandpass filter and a 48°K field of view

112U infrared spectrometer and thermocouple TC–1 with the assumption that the thermocouple is flat with wavelength. However, it has been pointed out by Eisenman[2] of NOL-Corona that thermocouples do not have a flat response with wavelength beyond about 2 μ microns. There are also severe differences between thermocouples. Figure 7 shows the spectral detectivity of a typical Ge:Hg detector utilizing two different thermocouples (TC–1 and TC–2). The responsivity of TC–1 was about 70 per cent at 15 μ and the responsivity of TC–2 was 30 per cent at 15 μ compared to the value at 2 μ. The third curve in Fig. 7 represents the correction of curves TC–1 and TC–2 for thermocouple roll-off with wavelength. It should be noted that the absolute value of the true peak detectivity drops, and that the spectral peak is displaced to shorter wavelengths by this correction. The errors in detectivity introduced in Figs. 5 and 6 for the wavelength range to 15 μ by thermocouple TC–1 are not serious and are almost lost in the noise error of the measurement. However, the errors are more severe for copper-doped germanium, since its response extends to 28 μ. But, corrected Ge:Cu spectral results have not been presented because accurate thermocouple calibration data beyond 15 μ are not available at this time.

This section has limited its discussion to Ge:Cu and Ge:Hg materials as separate entities. For the sake of completeness, it should be mentioned that Ge:Hg:Cu can exist with varying magnitudes of copper and mercury concentrations. It is therefore possible to observe the superposition of the two spectral characteristics. Extremely good detectors of the mixed variety can be made. However, thermal ionization of the copper acceptors limits the use of this type of detector to temperatures below that which is possible for Ge:Hg without copper.

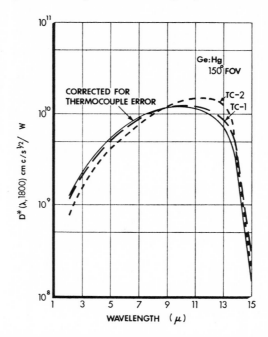

FIG. 7. An example of the errors in spectral response that can result with the assumption that the reference thermocouple is black.

NOISE LEVELS

In designing the proper preamplifier for use with an infrared detector, the electronic design engineer is generally interested in the noise characteristics of the device under optimum bias conditions. Figure 8 shows a comparison of the open circuit r.m.s noise voltage levels of Ge:Hg and Ge:Cu as a function of frequency for two different background conditions at maximum bias for optimum performance.

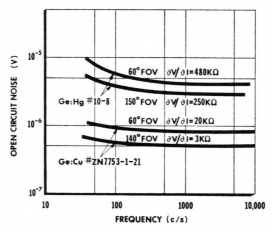

FIG. 8. The noise characteristics of Ge:Cu #ZN7753–1–21 and Ge:Hg #10–8.

The noise level of either Ge:Cu or Ge:Hg is flat above about 200 c/s. One can express the level of the observed generation-recombination noise looking at 300°K in relation to the Johnson–Nyquist noise of an equivalent wire-wound resistor at 300°K with a resistance equal to the observed dynamic resistance of the detector. Table 9 shows the comparison of the two different detectors at optimum bias for two levels of background radiation.

TABLE 1. Noise characteristics at optimum bias

Detector	Field of view (degrees)	Dynamic resistance (Ω)	Open circuit noise (V)	Equivalent Nyquist noise (V)	V_n/V_{th}	Noise figure (dB)
Ge:Cu ZN7753–1–21	140	3K	5.4×10^{-7}	1.65×10^{-8}	33	30
Ge:Cu ZN7753–1–21	60	20K	7.8×10^{-7}	4.25×10^{-8}	18	25
Ge:Hg 10–8	150	250K	3.1×10^{-6}	1.46×10^{-7}	21	26
Ge:Hg 10–8	60	480K	4.0×10^{-6}	2.03×10^{-7}	20	26

Amplifier bandpass = 5.6 c/s

For both Ge:Hg and Ge:Cu the noise figures are comparable. The noise figures are sufficiently high that one should not experience any serious difficulty in designing a preamplifier that will provide a detector noise-limited system for the ordinary audio-frequency system. However, if high-speed performance (the order of 1 Mc/sec) is desired, it may be necessary to load down the amplifier to 10 kΩ or less to compensate for a relatively high detector–dewar–preamplifier capacitance. In this event, the resultant noise figure from Ge:Hg may be considerably less than it is for the lower-impedance Ge:Cu.

CONCLUSION

The choice of the proper detector for a given application depends upon several preliminary considerations. The considerations that dictate this decision are:

1.   Desired spectral range
2.   Desired operating temperature
3.   Background temperature range
4.   Effective field of view for the detector
5.   Operating frequency and amplifier bandpass

In many cases the decision is simple. For example, if operation to 28 μ is mandatory, Ge:Cu is the only choice. If operation at 30°K over the full 8 to 14 μ range is desired, Ge:Hg is mandatory. However, if operation at 20° K or lower is considered, the decision is not so simple. For example, if high frequency performance is desired, one must also minimize the RC time constant of the detector–dewar–preamplifier combination. By proper design, the capacitance of the leads into the dewar and the input capacitance of the preamplifier can be minimized. In addition, it is necessary to have a low resistance detector to achieve minimum RC time constant and maximum responsivity. Since Ge:Cu exhibits a decreasing dynamic resistance with increasing bias, while Ge:Hg exhibits an increasing resistance with increasing bias, a decision in favor of Ge:Cu might be the wisest choice. When either can be used, the added ease of preparing Ge:Cu is an additional argument in favor of Ge:Cu.

Acknowledgements—The authors wish to express their appreciation to R. L. Burdick for his most useful assistance in materials preparation and to R. M. Talley for his interesting and valuable discussions.

REFERENCES

1. BORELLO, S. R., and H. LEVINSTEIN, *J. Appl. Phys.* 33, 2947 (1962).
2. EISENMAN, W. L., Private Communication, to be published in *J. Opt. Soc. Amer.* about May 1963.

Erratum

Page 134, Fig. 6, the label on the ordinate should read: "D*(λ, 1800) cm c/s$^{\frac{1}{2}}$/W."

22

Reprinted from *Appl. Phys. Letters*, **11**(4), 125–126 (1967)

HETERODYNE PERFORMANCE OF MERCURY DOPED GERMANIUM*

Carl J. Buczek and Gerald S. Picus
Hughes Research Laboratories
Malibu, California
(Received 7 July 1967)

Using optical heterodyne techniques, the response time of a mercury doped germanium detector operated at 4.2°K has been found to be 3.3 nsec and the signal-to-noise ratio measured at 15 MHz is within 4 dB of the quantum noise limit.

The minimum detectable signal and frequency response of a mercury doped germanium detector operating at liquid helium temperature have been measured. The detector was fabricated by the Santa Barbara Research Center and mounted in a specially designed dewar incorporating a 50 Ω coaxial transmission line passing through the helium reservoir to within 0.25 in. of the sensitive element. The frequency response was measured by modulating the output of a 10.6 μ CO₂ laser with a GaAs electro-optic modulator.[1] The results of this measurement are presented in Fig. 1, where it is evident that there is a roll-off in the response at about 50 MHz. This corresponds to a free carrier lifetime in the material of 3.3 nsec. Detectors fabricated from

more heavily compensated mercury doped germanium have shown flat responses (measured at 3.39 μ) up to frequencies of 450 MHz ($\tau < 3.7 \times 10^{-10}$ sec).

The heterodyne technique used to measure the minimum detectable signal at 10.6 μ is shown schematically in Fig. 2. Two tunable single-frequency CO₂ lasers are accurately superimposed to produce a beat frequency signal at the detector. The two lasers use intracavity NaCl flats as output couplers and combining beam splitters. The upper laser serves as the local oscillator laser, and the lower serves as the signal laser. The beams from the two lasers are aligned initially by maximizing the beat signal amplitude with all irises open. Irises 2 and 3 (both ~2 mm in diam) are used to control the intensity of the local oscillator beam. Iris 1 (~0.3 mm in diam) serves as an attenuator for the signal laser;

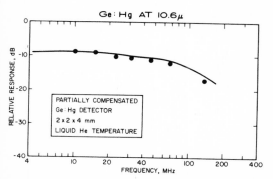

Fig. 1. Frequency response of mercury doped germanium infrared detector at 10.6 μ.

*This research was supported partially by the Electronic Technology Division, Air Force Avionics Laboratory, Air Force Systems Command, Wright-Patterson Air Force Base, Ohio.

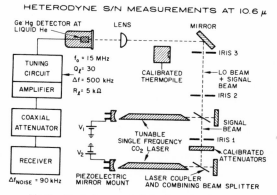

HETERODYNE S/N MEASUREMENTS AT 10.6 μ

Fig. 2. Block diagram of 10.6-μ heterodyne system.

the distance from iris 1 to 2 is such that the diffracted signal beam diameter is about five times the diameter of iris 2. Calibrated attenuators made from many sheets of polyethylene film serve to decrease the signal intensity further. The calibrated thermopile is used to measure the power of the local oscillator and the signal power at high levels. The superimposed local oscillator and signal beam are focused on the detector with a germanium lens antireflection coated for 10 μ.

The receiver used to make the signal and noise measurements was a spectrum analyzer adjusted for a 90-kHz noise bandwidth. The detector was operated with a wide field of view, and the background photon flux corresponded to 8 mW of 10.6-μ radiation. So that the detector noise generated by the local oscillator signal would be the predominant source of noise at 15 MHz, the local oscillator power was adjusted to be 18 mW at the detector. At this value of local oscillator power the local-oscillator-generated noise exceeded the thermal noise of the detector load resistor and any excess noise in the amplifiers. The results of the measurements are shown in Fig. 3. The highest signal power was measured directly with the thermopile. Subsequent values of signal power were obtained by using the calibrated attenuators. The experimental signal-to-noise ratio at the detector lies approximately 4 dB below the theoretical curve which is given by

$$(S/N)_{\text{power}} = \frac{P_s \eta}{2h\nu\Delta f_n}$$

with the quantum efficiency η approximated by $\frac{1}{2}$.

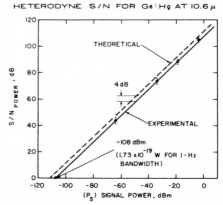

HETERODYNE S/N FOR Ge:Hg AT 10.6 μ

Fig. 3. Signal to noise ratio for mercury doped germanium detector at 10.6 μ in heterodyne mode of operation.

The signal and rms noise were equated visually on the spectrum analyzer; hence, about a 3-dB uncertainty exists in the noise value. The experimentally determined minimum detectable signal is found to be −108 dBm in a noise bandwidth of 90 kHz which corresponds to 1.73×10^{-19} W for a 1-Hz bandwidth. This result for Ge:Hg gives a minimum detectable signal within a factor of 10 of the ideal photon limit and is in accord with the lower frequency measurements made previously on Ge:Cu.[2]

[1] A. Yariv, C. A. Mead, and J. V. Parker, *IEEE J. Quantum Electronics* **QE-2**, 243–245 (1966).

[2] M. C. Teich, R. J. Keyes, and R. H. Kingston, *Appl. Physics Letters* **9**, 357–360 (1966).

IV
Thermal Detectors

Editors' Comments on Papers 23 Through 29

By long-standing tradition, thermal detectors have been characterized by their ability to operate satisfactorily without cooling and by their essentially uniform response across very large regions of the spectrum. The papers in this section describe a cross section of the available thermal detectors and, in addition, two deal with the very important problem of determining the true blackness of a detector. More recent developments that have led to cooled thermal detectors are described in the following section.

Paper 23, by DeWaard and Wormser, contains a review and a comparison of the characteristics of commercially available thermocouple-, pneumatic-, and bolometer-type detectors [1, 2]. The discussion of the pneumatic, or Golay, detector is of interest since it is, in essence, a gas thermometer and a lineal descendent of Herschel's thermometer [3–6]. The 100-element mosaic of thermistor bolometers is certainly one of the earliest disclosures of a multielement array. The authors report that individual elements give performance comparable to that attained with single-element detectors.

Russell DeWaard is a Senior Physicist with Barnes Engineering Co., Stamford, Connecticut.

Thermal detectors have response times that are several orders of magnitude longer than those of photon detectors. It is not surprising, therefore, that a large fraction of the literature on thermal detectors is concerned with methods of shortening their response times. A fine example is Paper 24 by Hornig and O'Keefe, who give detailed design criteria for fast thermopiles of high detectivity. A particularly valuable feature of the paper is the discussion of thermoelectric materials and the derivation of a thermoelectric figure of merit. The characteristics of thermoelectric materials that are suitable for detector use are shown in Table 1 along with their figures of merit. It is evident from these data that semiconductors are the most promising materials for use in thermocouple and thermopile detectors. The thermopiles constructed by Hornig and O'Keefe use a blackened gold foil of 0.1 μm thickness fastened to a thermoelectric couple having a bismuth–antimony alloy for one conductor and a bismuth–tin alloy for the other. The performance of these detectors is compared with that of contemporary detectors in Figures 1 and 2 of Paper 37.

Donald F. Hornig was Special Assistant for Science and Technology to the President of the United States from 1964 to 1969 and is currently the President of Brown University, Providence, Rhode Island.

In contrast to the techniques discussed in the previous paper, thermocouples and thermopiles (two or more thermocouples connected in series electrically) can be made by vacuum deposition techniques. In Paper 25, Astheimer and Weiner describe the fabrication of thermocouples consisting of bismuth and antimony junctions vacuum-deposited on a heat sink, one junction being in good thermal contact with the sink by conduction while the other is thermally isolated from it by a thick insulating layer. The principles of this type of construction were first reported in 1934 by Harris and Johnson [7]. The development of the improved techniques described here was spurred by the need for improved thermopiles for use in space-borne infrared systems. The paper includes examples of multielement arrays containing as many as 120 elements. For more information on thermocouple and thermopile detectors see References 8 through 12.

Robert W. Astheimer is Vice President-Engineering, Barnes Engineering Co., Stamford, Connecticut.

The thermistor bolometer was developed at Bell Telephone Laboratories during World War II [13]. Since that time it has become the most widely used of all thermal detectors for military and space systems. In Paper 26, Wormser describes how a thermistor (*therm*ally sensitive res*istor*) element is made from a thin flake of sintered semiconductor material chosen to have a large temperature coefficient of resistance. Characteristics and applications data are given for several different types of thermistor bolometers. For more information on bolometers, see References 14 through 21.

Eric M. Wormser is a Technical Consultant in Stamford, Connecticut.

The pyroelectric effect is observed in ferroelectric materials which are characterized by a permanent electric polarization with a strong temperature dependence. A change in the temperature of the material, such as might result from incident radiation, changes the polarization, which, in turn, changes the potential difference across the material. Pyroelectric detectors act like an electrical capacitance and their impedance drops with increasing frequency. As a consequence, their frequency response is flat up to frequencies that are much higher than those commonly observed with other types of thermal detectors. In Paper 27, Baker, Charlton, and Lock describe developments in high-performance pyroelectric detectors fabricated from strontium barium niobate, lithium sulfate, and alanine-doped triglycine sulfate. Their data show that for chopping frequencies above 20 Hz the pyroelectric detector can give a value of D^* higher than that of any other uncooled thermal detector. For more on pyroelectric detectors, see References 22 through 32.

G. Baker is Development Group Leader, Mullard Ltd., Southampton, England.

The materials used in thermal detectors are not in themselves good absorbers, so it is necessary to blacken them by applying some sort of an absorbing coating. An ideal black coating would have (1) uniformly high absorptance at all wavelengths, (2) negligible thermal capacity, (3) high thermal conductivity, and (4) no adverse effect on the electrical properties of the detector. In addition, application of the coating must not expose the detector to possible mechanical damage [33]. Obviously, no such coating exists and those that are available must represent a compromise.

Having found a suitable coating, against what does one compare it to verify its blackness? Papers 28 and 29, by Eisenman et al., describe the best solution yet developed, the *black radiation detector*. This detector was developed at the JSIRSETP[1] as a standard of blackness for use in determining the spectral response of other detectors. Their first detector had an effective absorptance of 0.990 out to a wavelength of 15 μm, while the most recent version has an effective absorptance of 0.995 out to a wavelength of 45 μm. Anyone naive enough to think that all thermal detectors have a response that is independent of wavelength, i.e., that they are perfectly black, should study the detector response data given in these papers. They show, for example, that the response of commercially available thermocouples may be only one-third as much at 35 μm as it is at short wavelengths and that the response of thermistor bolometers may be highly variable with wavelength. Additional measurements of this type are shown in Figure 2 of Paper 1 and in Figure 7 of Paper 21. Other pertinent information will be found in References 34 through 38.

References

1. Firth, J. R., and L. B. Davies, "A Comparison Between Pyroelectric, Pneumatic, and Thermocouple Detectors," *Proc. Conf. Infra-Red Techniques, IERE Conf. Proc. 22,* London (1971), p. 227.
2. Gillham, E. J., "Radiometry from the Viewpoint of the Detector," Chap. 2 in Drummond, A. J. (ed.), *Advances in Geophysics,* Vol. 14, *Precision Radiometry,* Academic Press, New York (1970).
3. Golay, M. J. E., "Radiation Detecting Device," U.S. Patent No. 2,557,096, June 19, 1951.
4. Golay, M. J. E., "Theoretical Considerations in Heat and Infra-Red Detection, with Particular Reference to the Pneumatic Detector," *Rev. Sci. Instr.,* **18,** 347 (1947).
5. Golay, M. J. E., "The Theoretical and Practical Sensitivity of the Pneumatic Infra-Red Detector," *Rev. Sci. Instr.,* **20,** 816 (1949).
6. Hennerich, K., W. Lahmann, and W. Witte, "The Linearity of Golay Detectors," *Infrared Phys.,* **6,** 123 (1966).
7. Harris, L., and E. A. Johnson, "The Technique of Sputtering Sensitive Thermocouples," *Phys. Rev.,* **45,** 635 (1934).
8. Cartwright, C. H., and J. Strong, "Vacuum Thermopiles and the Measurement of Radiant Energy," Chap. 8 in Strong, J. (ed.), *Procedures in Experimental Physics,* Prentice-Hall, Englewood Cliffs, N.J., (1938).
9. Stevens, N. B., "Radiation Thermopiles," Chap. 7 in Willardson, R. K., and A. C. Beer (eds.), *Semiconductors and Semimetals,* Vol. 5, *Infrared Detectors,* Academic Press, New York (1970).
10. Fastie, W., "Ambient Temperature Independent Thermopiles for Radiation Pyrometry," *J. Opt. Soc. Amer.,* **41,** 823 (1951).
11. Stafsudd, O., and N. Stevens, "Thermopile Performance in the Far Infrared," *Appl. Optics,* **7,** 2320 (1968).

---

[1]Joint Services Infrared Sensitive Element Testing Program. For further information see the Editors' Comments on Paper 6.

12. Stevens, N. B., and D. D. Errett, "Detector Having Radiation Collector Supported on Electrically Insulating Thermally Conducting Film," U.S. Patent No. 3,405,271, Oct. 8, 1968 (see also Nos. 3,405,272 and 3,405,273).
13. Brattain, W. H., and J. A. Becker, "Thermistor Bolometers," *J. Opt. Soc. Amer.*, **36**, 354 (1946).
14. Becker, J. A., "Bolometric Thermistor," U.S. Patent No. 2,414,792, Jan. 28, 1947.
15. Wormser, E. M., and R. D. DeWaard, "Construction for Thermistor Bolometers," U.S. Patent No. 2,963,674, Dec. 6, 1960.
16. Wormser, E. M., "Bolometer," U.S. Patent No. 2,983,888, May 9, 1961.
17. Allen, C., et al., "Infrared-to-Millimeter, Broadband, Solid State Bolometer Detectors," *Appl. Optics*, **8**, 813 (1969).
18. Smith, M., "Optimization of Bias in Thermistor Radiation Detectors. I: Bias Conditions and Performance Figures," *Appl. Optics*, **8**, 1027 (1969); "II: Practical Bias Networks," **8**, 1213 (1969).
19. Chasmar, R. P., W. H. Mitchell, and A. Rennie, "Theory and Performance of Metal Bolometers," *J. Opt. Soc. Amer.*, **46**, 469 (1956).
20. Maserjian, J., "A Thin-Film Capacitive Bolometer," *Appl. Optics*, **9**, 307 (1970).
21. Bishop, S. G., and W. J. Moore, "Chalcogenide Glass Bolometers," *Appl. Optics*, **12**, 80 (1973).
22. Putley, E. H., "The Pyroelectric Detector," Chap. 6 in Willardson, R. K., and A. C. Beer (eds.), *Semiconductors and Semimetals*, Vol. 5, *Infrared Detectors*, Academic Press, New York (1970).
23. Putley, E. H., "The Pyroelectric Detector—An Update," Chap. 6 in Willardson, R. K., and A. C. Beer (eds.), *Semiconductors and Semimetals*, Vol. 10, *Infrared Detectors, Part 2*, Academic Press, New York (1975).
24. Astheimer, R. W., and F. Schwarz, "Thermal Imaging Using Pyroelectric Detectors," *Appl. Optics*, **7**, 1687 (1968).
25. Kimmitt, M. F., J. H. Ludlow, and E. H. Putley, "The Use of a Pyroelectric Detector to Measure Q-Switched CO_2 Laser Pulses," *Proc. IEEE*, **56**, 1250 (1968).
26. Beerman, H. P., "Improvement in the Pyroelectric Infrared Radiation Detector," *Ferroelectrics*, **2**, 123 (1971).
27. Blackburn, H., et al., "Pyroelectric Detector Arrays for Thermal Imaging," *Proc. Conf. Infra-Red Techniques, IERE Conf. Proc. 22*, London (1971), p. 81.
28. Gabriel, F. C., "Microphonically Balanced Pyroelectric Detectors," *Appl. Optics*, **13**, 1294 (1974).
29. Jansson, P. A., "Hot-Pressed TGS for Pyroelectric Detector Applications," *Appl. Optics*, **13**, 1293 (1974).
30. Yamaka, E., T. Hayashi, and M. Matsumoto, "$PbTiO_3$ Pyroelectric Infrared Detector," *Infrared Phys.*, **11**, 247 (1971).
31. Mahler, R. J., R. J. Phelan, Jr., and A. R. Cook, "High D*, Fast, Lead Zirconate Titanate Pyroelectric Detectors," *Infrared Phys.*, **12**, 57 (1972).
32. Glass, A. M., J. H. McFee, and J. G. Bergman, Jr., "Pyroelectric Properties of Polyvinylidene Fluoride and Use for Infrared Detection," *J. App. Phys.*, **42**, 5219 (1971).
33. Hudson, R. D., Jr., *Infrared System Engineering*, Wiley, New York (1969), p. 274.
34. Karlson, F. J., and J. V. Kiernan, "Black Pigment for the Blackening of Infrared Radiation Detectors," U.S. Patent No. 3,364,066, Jan. 16, 1968.
35. Takahashi, S., "Blackening for Sub-mm Wavelengths," *Infrared Phys.*, **13**, 301 (1973).
36. Betts, D. B., "The Spectral Response of Radiation Thermopiles," *J. Sci. Instr.*, **42**, 243 (1965).
37. Stair, R., et al., "Some Factors Affecting the Sensitivity and Spectral Response of Thermoelectric (Radiometric) Detectors," *Appl. Optics*, **4**, 703 (1965).
38. Blevin, W. R., and J. Geist, "Influence of Black Coatings on Pyroelectric Detectors," *Appl. Optics*, **13**, 1171 (1974).

Reprinted from *Proc. IRE*, **47**(9), 1508–1513 (1959)

Description and Properties of Various Thermal Detectors*

R. De WAARD† AND E. M. WORMSER†

RADIATION THERMOCOUPLES

IN principle a thermocouple offers a simple and direct means for measuring radiant energy. By blackening the junction of dissimilar metals, incident radiation is absorbed, causing a temperature rise at the junction and a resultant increment in the electromotive potential developed across the junction leads. To be useful in modern applications radiation thermocouples must both respond rapidly and offer receiver areas of controlled dimensions to match reasonably the image of a radiant source. For example, in spectroscopy, where monochromatic radiation levels are low, it is important that nearly all radiant energy defined by the slit falls on the sensitive area of the detector.

Most of the modern developments in radiation thermocouples have been aimed at achieving a combination of fast time response and good definition of receiver area. Techniques basically reduce to attempts to achieve minimum heat capacity in the receiver element and controlled heat loss from the element. These thermal characteristics, of course, must be commensurate with high thermoelectric power in the couple materials. In a recent book, Smith, Jones, and Chasmar[1] give a good account of both historical and modern developments in thermocouples, including theoretical treatments aimed at selecting materials with optimized electrical and thermal properties. Available commercial radiation thermocouples, in the main, fall into two categories. One type employs couple materials in bulk form[2]; the other type utilizes thin evaporated or sputtered films supported by a thin dielectric film.[3] In the first case a thin, low heat capacity gold foil, coated with an evaporated black, constitutes the actual receiver area; and couple materials are welded to the foil to form thermoelectric junctions. Thin evaporated coatings of gold or platinum black are also applied to the evaporated junctions to improve absorption of radiation.

Thermoelectric materials employed in commercial thermocouple devices are principally metals, *e.g.*, nickel, antimony, bismuth, or alloys of these. Metallic alloys with tellurium, a semiconductor, are also used. Much theory and experiment has gone into the choice of ma-

terials best suited for a given detector design. In general high thermoelectric power is desired, but thermal and electrical conductivity must also be considered. The latter should be high to minimize the resistance of the detector, and in some designs low thermal conductivity is desired to obtain high responsivity. Unfortunately, most materials with low electrical resistance tend to have low thermoelectric power. Intermetallic alloying of metals (including tellurium) produces materials which tend to utilize the best properties of individual components. A good tabulation of the pertinent properties of thermoelectric materials is given by Hornig.[2]

In the following sections several of the commercial radiation thermocouples will be briefly described and their properties and performance given.

Perkin-Elmer Pin-Type Thermocouple

This couple is designed[4] for fast response and ac operation in infrared spectrophotometric instrumentation. The receiver area is proportioned to match, as nearly as possible, the image (2×0.2 mm) of the exit slit in the monochromator. The receiver itself is a thin gold foil coated with gold black. The gold strip is welded at its ends on relatively massive wedge-tipped pins of special semiconducting thermoelectric materials. The combination of low thermal mass of the receiver strip, high thermoelectric power, and thermal conductivity in the junction materials is conducive to both fast response and high responsivity.

The evacuated detector is normally housed (Fig. 1)

Fig. 1—Spectrometer-type thermocouple detector.

* Original manuscript received by the IRE, June 29, 1959.
† Barnes Engineering Co., Stamford, Conn.
[1] R. A. Smith, F. E. Jones, and R. P. Chasmar, "The Detection and Measurement of Infrared Radiation," Oxford Press, New York, N. Y.; 1957.
[2] D. F. Hornig and B. J. O'Keefe, "The design of fast thermopiles and the ultimate sensitivity of thermal detectors," *Rev. Sci. Instr.*, vol. 18, pp. 474–482; July, 1947.
[3] L. Harris. *Phys. Rev.*, vol. 45, p. 635; 1934.

[4] M. O. Liston, *J. Opt. Soc. Am.*, vol. 37, p. 515A; 1947

behind a suitable window at the end of a stem-type mounting, the narrow stem construction being employed to minimize obscuration when the detector is located at the focal point of a fast mirror.

For ac operation 13 cps chopping of the radiation is considered a good compromise where couple responsivity has not been degraded much from its maximum at dc conditions and still provides a suitable frequency for amplification. The low impedance of the detector (10 ohms) calls for transformer coupling to a vacuum tube preamplifier. Short cabling is required to connect the detector and preamplifier in order to avoid introduction of electrical noise.

The nominal properties of this detector are listed below:

Receiver size $= 2 \times 0.2$ mm
DC resistance $= 10$ ohms
Responsivity at 13 cps chopping $= 2$ volts/watt
Noise (Johnson noise of dc resistance to 1.5 times Johnson noise) $= 4$ to 6×10^{-10} volt rms in 1 cps bandwidth
Noise equivalent power (P_n) $P_n = 2$ to 3×10^{-10} watt for 1 cps bandwidth.

The Hornig Thermocouple

The design[2] of this type of thermocouple detector is based upon physical construction features involving a thin, low-heat capacity, blackened, gold-foil receiver joined by short fine thermoelectric wires, whose combined heat capacity is small in comparison to the foil receiver area. Thermal conduction losses through the short leads coupled with radiation losses from the receiver, and the aggregate small-heat capacity provide for both fast response and high responsivity. This detector is operated in a vacuum.

The Hornig type thermocouple is available commercially[5] usually in a single receiver size. It is housed in a stem-type evacuated mounting, not unlike the previously described detector, again for application principally in spectroscopic instrumentation. A potassium bromide window affords uniform transmission of radiation from the visible out to 25 microns. The unit carries a charcoal trap connected by a semiflexible tube for maintenance of good vacuum.

The Farrand-Hornig thermocouple has the following average performance characteristics:

Receiver size $= 0.75 \times 0.75$ mm
dc resistance $= 6$-10 ohms
Absolute dc responsivity $= 7$-10 volts/watt
Responsivity at 5 cps chopping $= 3$-4 volts/watt
Responsivity at 10 cps chopping $= 1.5$-2.5 volts/watts
Noise (approximately Johnson noise of dc resistance) $= 5 \times 10^{-10}$ volt rms in 1 cps bandwidth
Noise equivalent power (P_n) $(P_n) = 2$ to 3×10^{-10} watts for 10 cps chopping frequency and 1 cps bandwidth.

[5] Farrand Optical Co., New York, N. Y.

Reeder Thermocouples

Thermocouples manufactured by the Charles M. Reeder Company[6] cover a wide range of receiver sizes and types of mountings. Models are available to fit commercial spectrometric instruments. In addition, special large area thermopiles designed for diversified laboratory applications are manufactured.

Extended development has gone into the design of these couples. Both pin type (Schwartz) and wire type (Hornig) and combinations of these basic constructions are employed in the Reeder detectors. Thermocouple materials used include bismuth, bismuth alloys, and alloys of tellurium. High responsivity is reported for these detectors and noise outputs are held substantially at the level of the Johnson noise of the thermocouple resistance.

Work is going on at the Reeder Laboratories on the development of spectrally selective thermopiles in the far infrared. Planned also are detectors extending into the 50 to 100 μ wavelength region.

PNEUMATIC TYPE DETECTORS

In its simplest form the pneumatic type detector is comprised of a small gas-filled chamber, equipped with an infrared transmitting window, some means for absorbing radiation admitted to the chamber, and finally a method for transposing pressure change in the chamber into measurable signal output, usually electrical or optical.

An early device of this type is described by Hayes.[7] A carbonized "fluff" material is sealed in the chamber to absorb radiation. One end of the radiation chamber is closed with a thin metal diaphragm which forms one plate of a condenser. Capacitance changes provide direct means for measuring the radiation input.

Zahl and Golay[8,9] are responsible for significant developments in the pneumatic detector. In effect, the carbonized fluff is replaced by a low-heat capacity-absorbing membrane, detection being accomplished by an optical amplification scheme.

Described briefly below are two commercially available pneumatic cells, along with some information on their performance.

The Golay Cell

The Golay type pneumatic cell, as manufactured by the Eppley Laboratory, Inc., is pictured schematically in Fig. 2. Operation is as follows. Radiation admitted through the infrared window (NaCl, K Br, KRS5 and others) is absorbed by a thin foil blackened receiver comprising one end of a small gas-filled chamber. Pres-

[6] Detroit 3, Mich.
[7] H. V. Hayes, "A new receiver of radiant energy," *Rev. Sci. Instr.*, vol. 7, pp. 202–205; May, 1936.
[8] H. A. Zahl and M. J. E. Golay, "Pneumatic heat detector," *Rev. Sci. Instr.*, vol. 17, pp. 511–515; 1946.
[9] M. J. E. Golay, "A pneumatic infrared detector," *Rev. Sci. Instr.*, vol. 18, pp. 357–362; 1947.

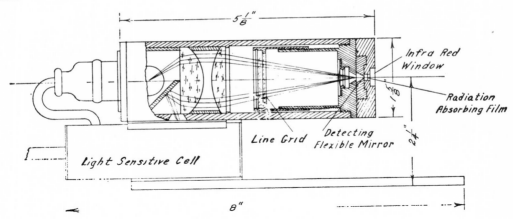

Fig. 2—Golay infrared detector.

sure changes distend a small flexible mirror located at the other end of the chamber. The optical system illustrated serves to detect distortions in the mirror. Light from an incandescent lamp, reflected by the flexible mirror, is imaged onto a photocell. A line grid is interposed to provide modulation of the light reaching the photocell. When the detecting mirror is undistorted the image of one-half the grid coincides with the other half in such a way that no light reaches the photocell. When the mirror membrane is distended the image defocuses and light energizes the photocell.

Drift common to all thermal detectors operated under dc conditions is eliminated by chopping the radiation and employing ac amplification. For spectrometric applications 10 cps chopping is suggested as optimum for the nominal detector time constant of about 15 milliseconds.

The standard receiver element has a 3/32-inch diameter absorbing disk, although other sizes and shapes can be obtained. Time constants can be made to range from about 2 to 30 milliseconds.

Operated in conjunction with a special synchronous-rectifier type amplifier, Noise Equivalent Input is quoted to be as small as 6×10^{-11} watt.

The Patterson-Moos Cell

In the Hayes and Golay type pneumatic cells the radiation absorbing elements are solid substances made as black as possible to absorb energy equally at all wavelengths. In another type of pneumatic cell the gas itself in the cell cavity absorbs radiation resulting in temperature and pressure increase in the chamber. By filling the cavity with gases having selective absorption bands, one can achieve detection only in certain preselected wavelength intervals, since the detector will only respond at wavelengths where the gas absorbs.

The Patterson-Moos[10] cell is of this type. Over-all physical size is determined largely by the size of the sensing area or entrance port. Typical gross dimensions of a cell with a 20 mm² sensitive area are about 1 inch long by $\frac{1}{2}$ inch in diameter. The cell housing is metal and normally is hermetically sealed with a sapphire or germanium window. Capacitive or condenser type output serves to extract electrical signals. Time response is fast; hence these cells can be used with a radiation chopper up to 50 cps, or at higher frequencies with some loss in responsivity.

BOLOMETER TYPE DETECTORS

Bolometer type detectors consist of thin resistance elements exhibiting a large change of resistance with temperature. A dc polarizing voltage is applied across the bolometric strip to detect the change of resistance as a corresponding change in the polarizing voltage. The bolometer is hence seen not to be a transducer. Resistance change in the sensing element merely serves to modulate a steady potential applied across the bolometer.

Sensitive elements are either metals (nickel, platinum) or semiconductor (sintered metallic oxides). In the following paragraphs these two types of bolometers will be treated separately.

Metal Strip Bolometer

Metal bolometer elements have a positive temperature coefficient (α) of about 0.3 to 0.4 per cent per degree Centigrade. In order to make metal sensing elements both sensitive and fast in response, it is necessary to minimize their heat capacity per unit of area. In effect this means the sensing strip must be very thin.

[10] Jamaica 18, N. Y.

Other physical parameters importantly involved in making metal strip bolometers are heat conductivity in the element itself, thermal conduction losses from the element, and radiation exchange. Metal bolometer elements have been formed directly from the bulk materials. For example, Langton[11] has rolled small platinum wires into thin strips (less than one micron) and blackened them to form bolometer elements. The resistance of such bolometers is about 20 ohms and their time constant about 4 milliseconds. In order to reduce microphonics, the thin metal strip is mounted in a taut suspension under reduced pressure in hydrogen. Responsivity of 1 volt/watt is reported and noise is substantially the Johnson noise of the unit's resistance.

In order to reduce further the heat capacity of metal strip bolometers they have been formed by vacuum evaporation.[12] Evaporated films of nickel and bismuth in thicknesses less than 500 Angstroms were examined as bolometer elements. In the case of evaporated films, temperature coefficient and electrical conductivity deviate from those of the bulk metals, but useful bolometric elements are realized.

Because metal strip thermosensing elements have high reflectivity they must be coated with a black absorber to improve absorption of incident radiation. Coatings are usually thin evaporated layers of gold or platinum black. This coating treatment makes the metal bolometers substantially black from the visible to beyond 15 microns in the far infrared.

The inherent low impedance of metal strip bolometers makes it best to transformer-couple them to a vacuum tube amplifier. In modern practice chopping of the radiation beam accompanied with ac amplification is found best suited for maximum detectivity.

Metal bolometers, at present, are not available in a range of sizes and models. Baird-Atomic[13] markets a platinum strip bolometer (Model AT3) designed principally for infrared spectrometric application. This unit is housed in a steel-type mounting, equipped usually with a silver chloride window. The housing is permanently evacuated. The sensing element is a platinum ribbon coated with evaporated gold black. The standard receiver area is 7×0.3 mm. The resistance of the bolometer element is 40 ohms and it is reported to have flat response from 1 to 26 microns. Responsivity is quoted to be 4 rms volts per peak-to-peak watt of radiation. The thermal time constant is 16 milliseconds. Optimum beam chopping speed is 10 cps. Coupled to a high-gain transformer and a tuned narrow-band amplifier, detectivity is essentially limited by Johnson noise. Noise Equivalent Power (P_n) for this detector in a 1-cps noise bandwidth is of the order of 10^{-9} watt.

Thermistor Bolometers

The active element in thermistor type bolometers is a thin semiconductor film usually composed of oxidic mixtures of manganese, nickel, and cobalt. Unlike metal strip bolometers, the thermistor material itself is a good absorber of radiation, although some blackening agent is usually applied to improve absorption. Thermistors have a high negative temperature coefficient ($\alpha = 4$ per cent per degree Centigrade) tending to provide high responsivity when used as a bolometric element. In common with the metal bolometer, the thermistor detector also operates by virtue of resistance change produced by incident radiation. In order to achieve fast time response, thermistor films or "flakes" are attached to good heat conducting thermal sinks. The normal bolometer bridge circuit comprises two identical thermistor elements mounted on the same base to effect compensation for changes in ambient temperature. One element, the active receiver, is exposed to radiation while the other is shielded from radiation. This circuit arrangement is ideally suited for ac operation, the signal junction being the common terminal between the two thermistors.

Initial development of thermistor detectors was carried out at the Bell Telephone Laboratories.[14] Manufacturing techniques were worked out and operating characteristics studied.[15] Later several organizations[16] continued manufacture and development of thermistor detectors.[17] Recent developments on thermistor detectors and their properties are described in a Navy Department Report.[18]

Two typical thermistor detector assemblies are pictured in Fig. 3. Bolometer elements are housed in a small (about 5/8 inch diameter $\times 3/8$ inch long) evacuated capsule. Window materials are coated silver chloride, KRS5, or other infrared optical windows, depending upon the spectral range of interest.

Fig. 4 represents the performance of a family of modern high-quality thermistor detectors. They can be made in a range of time constants from less than one millisecond to the order of 15 milliseconds. This kind of behavior is typical of Class II detectors as defined by Jones.[19] Because they all have the same quality (Jones' $M_2 = 0.3$) one can trade responsivity for time constant and choose a detector best suited to the chopping speed associated with a particular detection problem. This

[14] W. H. Brattain and J. A. Becker, "Thermistor bolometers," *J. Opt. Soc. Am.*, vol. 46, p. 354; 1946.
[15] J. A. Becker, *et al.*, "Final Report on Development and Operating Characteristics of Thermistor Bolometers," OSRD 5991.
[16] Thermistor detectors are currently available from Barnes Engineering Co., Stamford, Conn., and Servo Corp. of America, New Hyde Park, N. Y. The properties of thermistors described in this paper refer to those manufactured by Barnes Engineering Co.
[17] E. M. Wormser, "Properties of thermistor infrared detectors," *J. Opt. Soc. Am.*, vol. 43, pp. 15–21; January, 1953.
[18] R. DeWaard and E. M. Wormser, "Thermistor Infrared Detectors: Part I—Properties and Developments," Navy Bureau of Ordnance, NAVORD Rept. 5495; April, 1958.
[19] R. C. Jones, "Phenomenological description of the response and detecting ability of radiation detectors," paper 4.1.1, this issue, p. 1495.

[11] W. G. Langton, "A fast sensitive metal bolometer," *J. Opt. Soc. Am.*, vol. 36, p. 355; 1946.
[12] B. H. Billings, W. L. Hyde, and E. E. Barr, "An investigation of the properties of evaporated metal bolometers," *J. Opt. Soc. Am.*, vol. 37, pp. 123–132; 1947.
[13] Baird-Atomic, Inc., Cambridge, Mass.

Fig. 3—Thermistor detector assemblies.

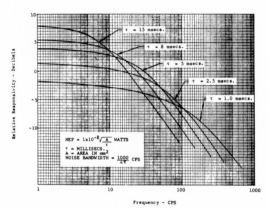

SAPPHIRE-BACKED CONTROLLED TIME CONSTANT DETECTORS

Fig. 4—Family of thermistor detectors.

family of detectors can be represented by a single simple relationship, given in Fig. 4, which states the noise equivalent power (NEP) as a function of detector area and time constant.

These detectors are available in a variety of sizes and rectangular shapes of the sensing element. Thermistors as small as 0.1×0.1 mm are being produced and it is possible to make them with linear dimensions as large as 10 mm. In the detection of small targets, however, it is advisable to keep the detector size as small as possible, since the larger detectors tend to decrease in detectivity.

The resistance of standard thermistor bolometers is high (1 to 5 megohms), making it convenient to couple them directly to vacuum tube amplifiers. Noise is substantially that attributable to the Johnson noise of the bolometer resistance. Noise levels in normal usage are of the order of 1 microvolt rms or less, hence it is advisable to provide close physical spacing between the bolometer and a vacuum tube preamplifier to avoid noise pickup.

Developments in Thermistor Bolometers

As a result of development work[20] carried on at Barnes Engineering Company in the past several years, thermistor detectors of new types and improved performance have evolved, and are now commercially available. These are briefly treated under separate headings below.

Optical Immersion Techniques

The detectivity of thermistor bolometers, in common with many other detectors, increases with decreasing receiver area. There are limits to the extent of optical gain one can practically achieve with conventional optical systems in the infrared. In the far infrared it is necessary to use reflecting optics exclusively and these become both expensive and bulky in large sizes. It is possible by means of optical immersion of the detector element to achieve substantial optical gain directly at the detector.

Thermistors have been optically attached to concentric germanium hemispheres to effect a large gain in detectivity. Fig. 5 illustrates the image reduction afforded by the high index ($N=4$) germanium optical element. The image at the plane face of the germanium element is reduced by a factor of 4 in linear dimensions from that of the image in space. This amounts to a reduction of 16 times in the area of the receiver and, since thermistor detectivity varies inversely with the square root of its area, a gain of a factor of 4 in detectivity results.

The thermistor is mounted centrally on the plane surface of the germanium hemisphere separated by a thin film which is transparent to infrared and electrically insulating. The insulating film has the important dual role of providing good optical contact and electrical insulation. Detector time constant is also determined by controlling the thickness of this layer. The germanium immersed bolometer is suitable for use in optical systems having an effective optical speed less than $f/1.0$. A typical immersed bolometer is shown in Fig. 6. Gross dimensions are about 5/8 inch diameter × 3/8 inch long. The bolometer carries a flange mounting base to facilitate precise location in an optical system.

Mosaic Techniques

A thermistor manufacturing process has also been worked out whereby it is possible to lay down a mosaic or matrix of a great number of small thermistor elements. Multiple thermistors are "printed" in desired patterns on ceramic substrates. Electrical junctions are likewise printed to provide signal and bias contacts. In Fig. 7 is pictured a printed assembly containing 100 thermistor elements arranged in a 10×10 matrix. Linear arrays of thermistor elements have also been produced by this technique. Individual bolometer element properties are comparable to those of the standard

[20] This work, in part, has been sponsored by the Navy Dept., Bureau of Ordnance.

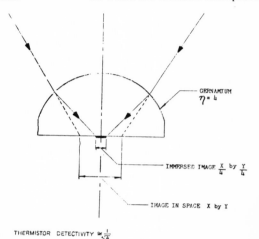

THERMISTOR DETECTIVITY $\approx \frac{1}{\sqrt{A}}$

Fig. 5—Illustrating image reduction with germanium immersion.

Fig. 6—Immersed bolometer.

Fig. 7—100-element mosaic.

thermistor detectors described above. At the present time, time constant cannot be varied much in the printing technique, 2 milliseconds being the nominal time constant obtained.

A feature worth mentioning of the printed-type thermistor is its adaptability to high-temperature operation. Because no organic cements or other temperature limiting materials are involved in construction, they can be exposed and used at high ambient temperatures.

Selective Wavelength Detectors

As already stated thermistor bolometers are uniformly black at all wavelengths in the infrared. Through special techniques it is now possible to make thermistor detectors spectrally selective; *i.e.*, they can be made to respond to radiation in certain specific wavelength regions. In brief, this is accomplished by first making the thermistor element completely reflective (overlay of evaporated gold). To achieve spectral selection the resulting blind thermistor element is coated with a material having appropriate absorption spectra. As coatings or sensitizers both organic films and powdered slurries of inorganic compounds have been employed. It is possible by this means to measure the radiant energy from a material by coating the detector with a thin film of the same material. In this case the detector senses only radiation emitted in characteristic absorption bands. Using an inorganic mineral as a sensitizer, a detector has been made with sensitivity at 15 microns extending one micron on either side, matching the strong CO_2 absorption band. Everywhere else in the infrared this detector is substantially insensitive. Selective detectors for other spectral regions can be similarly produced.

Thermistors for Matching Transistor Amplifiers

Most recently, spurred by developments in space technology and accompanying demands on small space and weight requirements, it has been desirable to employ transistor-type amplifiers with thermistor bolometers. The high impedance of the standard thermistor bolometer discussed above is not suitable for matching to transistor amplifiers. Theory and experiment show that the optimum resistance of thermistors for transistor coupling is of the order of 50K ohms. Suitable low impedance thermistors have been made and matched to transistorized preamplifiers. In general it has been possible to come within 6 db of theoretical detector noise (Johnson noise of the bolometer resistance).

24

Reprinted from *Rev. Sci. Instr.*, **18**(7), 474–482 (1947)

The Design of Fast Thermopiles and the Ultimate Sensitivity of Thermal Detectors

DONALD F. HORNIG, *Metcalf Research Laboratory, Brown University, Providence, Rhode Island*

AND

B. J. O'KEEFE, *Radiation Instruments Company, Boston, Massachusetts*

(Received March 13, 1947)

The design criteria for fast thermopiles of maximum sensitivity are obtained, with particular reference to their use with interrupted radiation in infra-red spectroscopy. It is shown that since the heat capacity of short, fine wires can be made small compared to that of the receiver, the conventional construction has certain advantages over evaporated thermopiles. Sources of noise are discussed and it is shown that in the cases treated the noise caused by the temperature fluctuations of the receiver is somewhat less than that resulting from Johnson noise. If Johnson noise is the chief factor, the minimum detectable power in the form of incident radiation is inversely proportional to a quantity $S = E/R^{\frac{1}{2}}$ (E = voltage sensitivity; R = thermopile resistance) defined as the absolute sensitivity. The optimum design of d.c. thermopiles is derived and a factor of merit for thermoelectric materials, $Q/(\kappa\rho)^{\frac{1}{2}}$ (Q = thermoelectric power; κ = thermal conductivity; ρ = electrical resistivity) is obtained. Physical constants and factors of merit for various materials are tabulated. Cd-Sb and Te alloys have the highest factor of merit but it is concluded

that their irreproducibility outweighs their slight advantages. General relations for the time constant are given in terms of the construction and it is shown that response times of less than 0.02 second are practicable. By using these relations the optimum design of thermopiles for use with interrupted radiation is obtained. Several concrete designs for use at a 5 c.p.s. interruption frequency are given. With a receiver area of 0.005 cm² the incident radiation, interrupted at 5 c.p.s., equal to the Johnson noise in a 1 c.p.s. frequency range is 5×10^{-5} μwatts. At this frequency the response is 90 percent of the equilibrium value. Approximately the same performance may be obtained at 10 c.p.s. Using the best available materials this sensitivity may be almost doubled. If ideal materials should ever be found there is still an ultimate limit to the sensitivity of any detector, thermopile or bolometer, which measures temperature, set by the temperature fluctuations. At 300°K this limit of detection is 4×10^{-6} μwatt for a detector of 0.005 cm² area and a measurement in a 1 c.p.s. frequency band.

INTRODUCTION

THE design of thermopile-galvanometer systems for maximum deflection sensitivity has been thoroughly discussed by many investigators.[1-4] However, when very high sensitivities are desired the inherent fluctuations in the signal must be taken into account and the proper criterion of sensitivity is the ratio of the signal to these fluctuations. Cartwright has given a quite complete treatment of the optimum design of both thermopiles and bolometers used with galvanometers when the Brownian motion of the suspension of a critically damped galvanometer is taken into account.[5] In all of these discussions, the response was assumed to be limited by the galvanometer period so the response time of the detector was unimportant.

In addition to the highest sensitivity one

would now also like to secure the most rapid possible response in order to reach either of two objectives: (1) to record at the highest possible speed, as on an oscilloscope; or (2) to be able to interrupt the radiation beam at a frequency high compared to the recording speed, thus eliminating long period drift which is usually a major problem. The development of suitable electronic amplifiers[6-8] has made it possible to modulate the radiation at a frequency limited only by the detector.

During the war a variety of detectors, both thermopiles and bolometers, was developed and some of them have been described in the literature.[9-13] With the exception of the supercon-

[1] E. S. Johanson, Ann. d. Physik **33**, 517 (1910).
[2] C. H. Cartwright, Rev. Sci. Inst. **1**, 592 (1930).
[3] F. A. Firestone, Rev. Sci. Inst. **1**, 630 (1930).
[4] John Strong, Rev. Sci. Inst. **3**, 65 (1932).
[5] C. H. Cartwright, Zeits. f. Physik **92**, 153 (1934).

[6] L. C. Roess, Rev. Sci. Inst. **16**, 172 (1945).
[7] M. D. Liston, C. E. Quinn, W. E. Sargeant, and G. G. Scott, Rev. Sci. Inst. **17**, 194 (1946).
[8] Western Electric Company, Inc., KS-10281 amplifier.
[9] L. C. Roess and E. Dacus, Rev. Sci. Inst. **16**, 164 (1945).
[10] I. Amdur and C. F. Glick, Rev. Sci. Inst. **16**, 117 (1945).
[11] F. G. Brockman, J. Opt. Soc. Am. **36**, 32 (1946).
[12] C. B. Aiken, W. H. Carter, and F. S. Phillips, Rev. Sci. Inst. **17**, 377 (1946).
[13] Louis Harris, J. Opt. Soc. Am. **36**, 597 (1946).

ducting bolometer developed by Andrews,[14,15] it appears that for spectroscopic purposes none of them has surpassed the prewar detectors in absolute sensitivity (measured in terms of the noise level) although they are all considerably faster than those previously available.

To obtain fast response thermopiles it is necessary to minimize their heat capacity. Until now this has usually been done by evaporating thin films of thermoelectric material which when blackened serve the dual roles of receiver for the radiation and generator of the signal.[9] This technique has been highly developed by Harris.[13] Several serious problems make this approach disadvantageous, among them: (1) very thin films of the materials used have a resistivity very much greater than that of the bulk metal;[16] (2) thin films usually show an abnormally low thermoelectric power; and (3) the thin films have a resistance which is much greater than optimum for the range of receiver sizes usually of interest.

We have found that if thermopiles are built by the older technique of attaching thermo-junctions in the form of wires to a thin, blackened receiver,[17] all of these difficulties are readily overcome. It is not difficult by using short thermoelectric wires (0.2–0.5 mm in length) to make the heat capacity of the wires small in comparison to the heat capacity of a very thin foil (0.1 micron gold), even when the foil area is less than 0.005 cm², so that times of response as short as those of evaporated piles or bolometers may be realized. The materials appear to retain their bulk electrical properties and optimum resistances are easily obtained.

The design criteria are obtained as follows. If the blackened receiver of either a bolometer or thermopile is exposed to a radiation flux P, its temperature increase at any time t after the illumination is started is given by

$$\theta = \theta_{\infty}[1 - \exp(-\lambda t)], \qquad (1)$$

where the equilibrium temperature increase, θ_{∞}, is given by

$$\theta_{\infty} = P/L, \qquad (2)$$

and λ, the reciprocal of the time constant, is

$$\lambda = L/C. \qquad (3)$$

In these equations L is the total heat loss to the surroundings per second per degree of heating and C is the total heat capacity of the system. In the case of thermopiles we have neglected the Peltier cooling at the hot junction which is negligible unless very many junctions or materials having a very high thermoelectric power are used.

SOURCES OF NOISE

It is this quantity, θ, which both bolometers and thermopiles measure. The measurement in both cases is theoretically limited by the fact that the temperature of the system fluctuates spontaneously. This problem has been excellently treated by Milatz and Van der Velden[18] who found that the frequency spectrum of these fluctuations is

$$\langle \theta_F^2 \rangle_{\text{Av}} = 4kT^2L^{-1}(1 + 4\pi^2\nu^2/\lambda^2)^{-1}\Delta\nu, \qquad (4)$$

where $\langle \theta_F^2 \rangle_{\text{Av}}$ is the mean square temperature fluctuation in the frequency range ν to $\nu + \Delta\nu$, k is Boltzmann's constant, and T is the absolute temperature. These fluctuations constitute a theoretical limiting factor for any detector which is thermal in nature. If a thermoelectric junction with thermoelectric power Q is attached to such a receiver, the random electrical signal generated by its temperature fluctuations is

$$\langle E_F^2 \rangle_{\text{Av}} = Q^2 \langle \theta_F^2 \rangle_{\text{Av}}. \qquad (5)$$

Another, and in most cases more important, source of fluctuation e.m.f. is the Johnson noise[19] resulting from the Brownian movement of electrons in the circuit resistance. This noise is quite accurately given by

$$\langle E_J^2 \rangle_{\text{Av}} = 4kTR\Delta\nu, \qquad (6)$$

where R is the total circuit resistance. If R is given in ohms this becomes, at 300°K,

$$E_J(\text{r.m.s.}) = 1.3 \times 10^{-4} R^{\frac{1}{2}} \Delta\nu^{\frac{1}{2}} \quad \mu \text{ volt}. \qquad (7)$$

For a typical thermopile of the type we shall

[14] D. H. Andrews, R. M. Milton, and W. De Sorbo, J. Opt. Soc. Am. **36**, 518 (1946).
[15] R. M. Milton, Chem. Rev. **39**, 419 (1946).
[16] L. Harris and A. C. Scholp, J. Opt. Soc. Am. **30**, 519 (1940).
[17] J. Strong, *Procedures in Experimental Physics* (Prentice-Hall, Inc., New York, 1939), Chapter 8.

[18] J. M. W. Milatz and H. A. Van der Velden, Physica **10**, 369 (1943).
[19] J. B. Johnson, Phys. Rev. **32**, 97 (1928).

consider, operated at room temperature, this noise is at least five times greater than that caused by temperature fluctuations. Under other circumstances, such as a low resistance detector operated in a gas, the temperature fluctuations may become more important than the Johnson noise.

There are still other sources of noise, all of which can be made small compared to the two already mentioned. Among these are: (1) the noise arising in the input stage of the amplifier, which is small compared to the previously mentioned noise sources if the thermopile is coupled to the input stage by a suitable input transformer; (2) contact noise in the thermopile circuit, which we have found negligible with metallic thermopiles but which might become important in making contact to semi-conducting materials; (3) gas microphonics, which can be eliminated by operating in a vacuum. We shall show that vacuum operation is desirable for other reasons as well.

Experimentally, thermopiles which we have built in accordance with the considerations here presented show a noise level equal to the Johnson noise limit to within our experimental error (20 percent).

MINIMUM DETECTABLE POWER

The power in the form of incident radiation which produces a signal equal to the root mean square value of the fluctuations can be conveniently defined as the minimum detectable power.

If the equilibrium (d.c.) voltage sensitivity is defined as the e.m.f. generated per microwatt of radiation falling on the receiver, we may write

$$E_\infty = Q\theta_\infty/P = Q/L. \qquad (8)$$

Then the voltage sensitivity at a time t after the illumination begins is

$$E = Q\theta/P = E_\infty[1 - \exp(-\lambda t)]. \qquad (9)$$

When an instrument with an exponential response is exposed to radiation modulated sinusoidally at a frequency ν, its equilibrium output is the sum of a d.c. component and a component at the frequency of modulation.

If we consider only the output component of frequency ν, we can further define the sensitivity

to modulated radiation as

$$E' = E_\infty[1 + 4\pi^2\nu^2/\lambda^2]^{-\frac{1}{2}}, \qquad (10)$$

where the term in brackets is the relative frequency response of the detector. In terms of any of these definitions of sensitivity, the minimum detectable power is

$$P_{\min} = (4kTR)^{\frac{1}{2}}\Delta\nu^{\frac{1}{2}}/E, \qquad (11)$$

where $\Delta\nu$ is the range of frequencies being measured. At 300°K this becomes

$$P_{\min} = 1.3 \times 10^{-4}\Delta\nu^{\frac{1}{2}}/S \quad \mu \text{ watt}, \qquad (12)$$

where

$$S = E/R^{\frac{1}{2}} \qquad (13)$$

is the only factor which depends on the design of the thermopile. We shall call S the absolute sensitivity corresponding to each of the voltage sensitivities previously defined.

In most applications, the frequency band is limited by the recording apparatus. Furthermore, in most electronic amplifiers, the resistance reflected into the primary of the input transformer can be neglected, so that the thermopile resistance can be taken as the total resistance of the circuit.

EQUILIBRIUM CHARACTERISTICS

If the thermopile response time is short compared to the time in which changes in the intensity of the incident radiation may occur, we should like to maximize S_∞. Let us now define the symbols:

A = total area of receivers.
κ = thermal conductivity of wire material.
ρ = resistivity of wire material.
a = cross-sectional area of wire.
l = length of individual wire.
σ = Stefan-Boltzmann constant.
L' = heat loss per second per degree C of heating other than by radiation or wire conduction.
n = number of equivalent series junctions.
m = number of equivalent parallel junctions.

The heat loss per second per degree of heating resulting from radiation is then $4\sigma A T^3$ and it will be convenient to define

$$\phi = 4\sigma A T^3 + L'. \qquad (14)$$

If we denote the two wire materials by subscripts 1 and 2, the total heat loss per degree tempera-

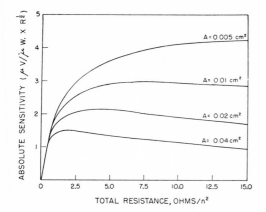

FIG. 1. Absolute sensitivity at equilibrium *vs.* total thermopile resistance/n^2, where n is the number of equivalent series junctions. The curves are plotted for thermopiles of various receiver areas built with 97 percent Bi–3 percent Sb and 95 percent Bi–5 percent Sn wires. All parallel thermopiles are given by $n=1$.

ture increase of all receivers is

$$L = \phi + n[\kappa_1(a_1/l_1) + \kappa_2(a_2/l_2)] \quad (15)$$

for both series and parallel thermopiles. The total resistance of the thermopile is

$$R = n[\rho_1(l_1/a_1) + \rho_2(l_2/a_2)] \quad (16)$$

for series thermopiles and

$$R = [\rho_1(l_1/a_1) + \rho_2(l_2/a_2)]/m \quad (17)$$

for parallel thermopiles.

If Eq. (13) is maximized with respect to a/l we obtain the optimum relation between the wire dimensions first found by Johansen[1]

$$(a_1/l_1)^2 \kappa_1/\rho_1 = (a_2/l_2)^2 \kappa_2/\rho_2. \quad (18)$$

When the thermopile is used so that its response time is not important and the wire dimensions are chosen according to Eq. (18), the absolute sensitivity for a series thermopile is given as a function of the total thermopile resistance by

$$S_\infty = nQ/[\phi + (n^2/R)\{(\kappa_1\rho_1)^{\frac{1}{2}} + (\kappa_2\rho_2)^{\frac{1}{2}}\}^2]R^{\frac{1}{2}}. \quad (19)$$

Parallel thermopiles are equivalent to the case $n=1$.

Since according to this equation the absolute sensitivity is a function of the parameter R/n^2, we can represent S_∞ for any number of junctions by a single curve. In Fig. 1, S_∞ is plotted against

R/n^2 for 97 percent Bi–3 percent Sb and 95 percent Bi–5 percent Sn wires. The abscissa is resistance directly for single junction and parallel thermopiles.

The optimum resistance, obtained by maximizing (19) with respect to R, is

$$R_{\mathrm{opt}} = n^2[(\kappa_1\rho_1)^{\frac{1}{2}} + (\kappa_2\rho_2)^{\frac{1}{2}}]^2/\phi. \quad (20)$$

The maximum sensitivity is then

$$S_\infty = Q/2\phi^{\frac{1}{2}}[(\kappa_1\rho_1)^{\frac{1}{2}} + (\kappa_2\rho_2)^{\frac{1}{2}}]. \quad (21)$$

This expression is independent of the number of junctions or whether they are arranged in series or in parallel. It is the maximum possible absolute sensitivity. It should be noted that it is inversely proportional to the square root of the total receiver area.

All of the preceding equations are valid for compensated thermopiles if twice the actual values of the resistivities are used.

THERMOELECTRIC MATERIALS

Equation (21) suggests as a factor of merit for thermoelectric materials the quantity

$$M = Q/(\kappa\rho)^{\frac{1}{2}}. \quad (22)$$

Several of the materials used in radiation thermopiles for which physical constants are available are compared on this basis in Table I.

In comparing pairs of materials we can use the exact factor of merit defined by

$$M' = (Q_1 - Q_2)/[(\kappa_1\rho_1)^{\frac{1}{2}} + (\kappa_2\rho_2)^{\frac{1}{2}}]. \quad (23)$$

Some of the combinations which have been used are compared in Table II.

It is apparent from the two tables that the most promising materials are semi-conductors. The physical constants of semi-conductors vary widely from specimen to specimen, being very sensitive to small amounts of impurities and depending on the previous treatment. The effect of small amounts of impurities on the resistivity of bismuth has been studied by Thompson[20] and it is well known that the thermoelectric power drops from $-60\mu v/°C$ to zero with the addition of 0.7 percent Sn.[21] The variations in the prop-

[20] N. Thompson, Proc. Roy. Soc. **A155**, 111 (1936).
[21] W. R. Thomas and E. J. Evans, Phil. Mag. **17**, 65 (1934).

TABLE I. Comparison of various thermoelectric materials.

Material	Q ($\mu v/°C$)	κ (watt cm^{-1} C$^{°-1}$)	$\rho \times 10^6$ (ohm cm)	Factor of merit $M \times 10^3$
Silver	+2.9	4.24	1.62	+1.11
Iron	+16	0.67	10.0	+6.2
Nickel	−19	0.59	7.8	−9.0
Antimony	+40	0.20	41.7	+14
Bismuth	−60	0.083	120	−19
Tellurium*				
a. (single crystal, ‖)	+436	0.018	35,000	+17
b. (poly-crystalline)	+376	0.015	30,000	+18
c. (poly-crystalline)	+372	0.010	200,000	+8
d. (Baker and Company)	+119	(0.02)	(3,100)	+15
Constantan	−38	0.212	49	−12
Chromel-p	+30	0.20	85	+7
95% Bi–5% Sn	+30	0.045	275	+9
97% Bi–3% Sb	−75	0.07	174	−21
90% Bi–10% Sb	−78	0.053	160	−27
99.6% Te–0.4% Bi**	+191	(0.02–0.032)	2,920	+(19–25)
99.1% Te–0.9% Sb**	+139	(0.02–0.033)	2,300	+(16–20)
98.5% Te–1.5% S***	+575	(0.015–0.030)	350,000	+(6–8)
65% Sb–35% Cd****	+106	(0.015–0.04)	790	+(19–31)
75% Sb–25% Cd****	+112	(0.015–0.04)	690	+(21–35)

* C. H. Cartwright, Ann. d. Physik **18**, 656 (1933).
** C. H. Cartwright and M. Haberfeld-Schwartz, Proc. Roy. Soc. **A148**, 648 (1935).
*** A. Petrikaln and K. Jacoby, Zeits. f. anorg. u. allgem. Chemie **210**, 195 (1933).
**** B. H. Volfson and V. H. Rozhdestvenskii, J. Exper. Theor. Phys. U.S.S.R. **3**, 447 (1933).

erties of Te and Te alloys are even greater, as may be seen from Table I. The polycrystalline material labelled (b) was prepared from the same Te as the single crystal (a). On the other hand the polycrystalline material (c) was from another source. Nevertheless, both batches of Te had less than 0.01 percent impurities. Other evidence indicated that really pure Te is a very poor thermoelectric material. Petrikaln and Jacoby (see third reference under Table I) found their purest Te to have an average thermoelectric power of +315$\mu v/°C$ but a resistivity of 0.35 ohm cm. Similarly, Cartwright (see second reference under Table I) using samples of Te even more pure than those listed in Table I found a thermoelectric power of +67$\mu v/°C$ and a resistivity of 0.35 ohm cm and higher. It is probable that the resistivity of Te-S alloys would be much lower and the factor of merit considerably higher than the values in Table I if they contained a few tenths of one percent Sb as an impurity.

Cd-Sb alloys are equally erratic. Samples of the same composition differ widely from one another and the properties fluctuate considerably as the composition is varied slightly (see last reference under Table I). Nevertheless, the factor of merit derived from Volfson and

Rozhdestvenskii's data is high throughout the range 55–80 percent Sb. Because of the considerably lower resistivities these alloys are probably more useful than the Te alloys.

Because they vary so widely the physical properties listed for semi-conductors must be taken only as a guide and must be measured on any particular specimen which is to be used.

RESPONSE TIME

In order to make fast response thermopiles we should like to maximize λ without decreasing S. One method of increasing λ (decreasing the time constant) is to increase the losses, L, as may be seen from Eq. (3). If A and R are fixed, the radiation and wire conduction losses are fixed. We can still increase L by adding the losses L'. An example of such a loss is the gas

TABLE II. Comparison of typical thermopile combinations.

Couple	M'
Chromel vs. constantan	9
Bismuth vs. silver	11
97% Bi–3% Sb vs. 95% Bi–5% Sn	13
90% Bi–10% Sb vs. antimony	20
90% Bi–10% Sb vs. 99.5% Te–0.5% Bi	21–25
90% Bi–10% Sb vs. 98.5% Te–1.5% S	7–9
90% Bi–10% Sb vs. 75% Sb–25% Cd	23–31

conduction if the thermopile is not evacuated. However, if we differentiate the response, Eqs. (9) and (13), we find that the initial slope

$$(dS/dt)_{t=0} = Q/CR^{\frac{1}{2}} \qquad (24)$$

of the response curve is independent of L'. Since S_∞ is inversely proportional to L it is apparent that the addition of any L' decreases the response at all later times. Hence an evacuated thermopile is more sensitive than a gas filled thermopile at any frequency of modulation of the incident radiation. Hereafter we consider only evacuated thermopiles for which $L'=0$.

The total heat capacity of the thermopile may be conveniently taken as the sum of three terms, the heat capacity of the foil from which the receivers are made, C_f, that of the material with which they are blackened, C_b, and that of the wires, C_w. If the thermocouple wire dimensions were picked according to Eq. (18) the inverse of the time constant is

$$\lambda = [\phi + (n^2/R)\{(\kappa_1\rho_1)^{\frac{1}{2}} + (\kappa_2\rho_2)^{\frac{1}{2}}\}^2] / [C_w + C_b + C_f] \qquad (25)$$

for series thermopiles. The same expression with $n=1$ applies to parallel thermopiles.

The straightforward way to increase λ is to decrease the heat capacity. If the receiver area has been fixed its heat capacity is minimized by constructing it from the thinnest possible foil. This process is limited by the fact that when the foil becomes too thin its thermal conductivity drops to the point where temperature gradients are established in the foil near the point of contact of the wires. When this occurs the sensitivity drops. With single junction thermocouples we have found no serious effect of this nature with gold receivers 0.1μ thick when the wire resistance was greater than two ohms. The effect can be minimized and thinner foils used by resorting to parallel construction, in which case the conduction by each wire is decreased and the heat is drawn from m points on the receiver (or from m receivers). The heat capacity of 0.1μ gold foil is 2.5 ergs/°C mm^2.

Similarly, the heat capacity of the black is reduced by using a minimum thickness of black consistent with complete absorption of the radiation. We have successfully used a mass of carbon

with a heat capacity estimated at 2.5 to 5.0 erg/°C mm^2 (based on the mass of carbon present).

In the case of the wires it is clear that the heat capacity is minimized if, for a given resistance, short, fine wires are used. If we define C_1 and C_2 as the heat capacity per unit volume of the wire materials and note that the average temperature increase of the wires is one-half that of the receiver, then the effective wire heat

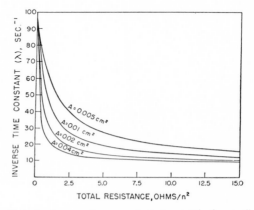

FIG. 2. Inverse time constant *vs.* total thermopile resistance/n^2, where n is the number of equivalent series junctions, for thermopiles built with 0.3 mm long wires of 97 percent Bi–3 percent Sb and 95 percent Bi–5 percent Sn. Curves are plotted for various areas of receivers made of 0.1-micron thick gold with the heat capacity of the blackening equivalent to 0.2-micron gold. All parallel thermopiles are given by $n=1$.

capacity for use in Eq. (25) is

$$C_w = n(a_1l_1C_1 + a_2l_2C_2)/2. \qquad (26)$$

If the wire dimensions have been chosen from Eq. (18) and the lengths of the two wires are set equal, this equation can be made more convenient. If the wire length is fixed by mounting considerations or has been made as short as is practicable for construction ($\sim$0.2 mm), then C_w is given as a function of R (and hence of wire diameter) by

$$C_w = l^2(\rho_1C_1 + \rho_2C_2)n^2/R. \qquad (27)$$

The same expression holds for parallel thermopiles if n is set equal to one. The inverse time constant for any number of junctions is then a function of R/n^2 for a given receiver area. In Fig. 2, λ is plotted from Eq. (25) in the case of a 0.1μ gold target, with blackening equivalent in heat capacity to 0.2μ gold, and BiSb and BiSn wires.

It is interesting to note in this case that the thermopile becomes faster as the wire diameter

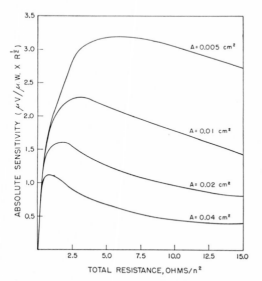

FIG. 3. Absolute sensitivity to radiation interrupted at 5 c.p.s. vs. total thermopile resistance/n^2, where n is the number of equivalent series junctions, for thermopiles built with 0.3 mm long wires of 97 percent Bi–3 percent Sb and 95 percent Bi–5 percent Sn. Curves are plotted for various areas of receivers made of 0.1-micron thick gold with the heat capacity of the blackening equivalent to 0.2-micron gold. All parallel thermopiles are given by $n=1$.

is *increased* (the resistance lowered). This follows from the fact that the effect of wire heat capacity is small until very low resistances are reached and the increase in C_w is outweighed by the increase in L. It is also noteworthy that λ reaches a limiting value at both low and high resistances. The low resistance (fast response) limit is independent of the target area but depends linearly on the length of wire. In the case plotted in Fig. 2 this limit is at $\lambda=98$ sec.$^{-1}$ ($\tau=0.01$ sec.) and it is perfectly practicable to reach a time constant of 0.02 sec. If the wire length is reduced to 0.2 mm, which we have successfully done, the time constant can be further reduced.

Frequently the design is fixed by the fact that wires of a fixed diameter are available and the resistance must be varied by changing the length. In this case the effective heat capacity of the

wire is given as a function of resistance by the expression

$$C_w = a_1^2 R(\rho_1 C_1 + \rho_2 C_2)/4\rho_1^2 \quad \text{(series)}$$
$$= m^2 a_1^2 R(\rho_1 C_1 + \rho_2 C_2)/4\rho_1^2 \quad \text{(parallel)}. \quad (28)$$

Here the wire heat capacity drops as the resistance is lowered so the rise in λ is more abrupt at low R. The limitation on the minimum practical length of wire (approximately 0.2 mm) means that λ can never be greater in this case than in the one considered in Eq. (27). In obtaining the expressions (27) and (28) for the wire heat capacity it was assumed that $(\kappa_1\rho_1)^{\frac{1}{2}}$ is equal to $(\kappa_2\rho_2)^{\frac{1}{2}}$. Considerable deviations from this approximation have little effect on λ and less on S.

THERMOPILES FOR INTERRUPTED RADIATION

Let us now consider the modifications which must be made in our previous design criteria if the thermopile is to be used with modulated radiation. In spectroscopic applications the radiation is chopped at some frequency ν, yielding a beam whose wave form is usually that of a square wave. If this is amplified, rectified, and recorded by a pen recorder, the band width is defined by the amplifier or recorder and we may still use the criterion S. After calculating the response to a square wave from Eq. (9), we obtain

$$S = S_\infty \tanh(-\lambda/4\nu). \quad (29)$$

We are now in a position to plot S for various possible thermopiles. Taking S_∞ from Fig. 1 and λ from Fig. 2 we plot S vs. R/n^2 at a 5-cycle interruption frequency for the case of fixed lead length (0.3 mm) in Fig. 3. It will be observed that the maxima are relatively broad and flat, the region within 20 percent of the peak encompassing a factor of ten in resistance. In each case the optimum 5-cycle resistance is about one-third of the optimum d.c. resistance. At higher frequencies the optimum resistance is lowered further but it can always be kept within a reasonable range by using a series thermopile.

In the case of fixed wire diameters the maxima are sharper because of the increasing heat capacity of the wires at higher resistances. In both cases the region of resistance lower than optimum corresponds to a higher percentage modulation and a lower S_∞. Resistances higher than the optimum yield a higher S_∞ but a lower percentage modulation.

The properties of several typical thermopiles built with 0.3 mm long leads of 97 percent Bi–3 percent Sb and 95 percent Bi–5 percent Sn are given in Table III. These thermopiles were designed for optimum operation at an interruption frequency of 5 cycles per sec. The data in the table may all be obtained from Figs. 1, 2, and 3 and the equations of the text, if allowance is made for a loss of approximately 20 percent because of reflection at the entrance window.

At 10 c.p.s. each of these thermopiles would have dropped about 25 percent in sensitivity. Thermopiles designed for optimum response at 10 c.p.s. would be somewhat improved and if 0.2 mm leads were used (which we have found quite practicable) it is possible to construct thermopiles which have essentially the same characteristics at 10 c.p.s. as those listed in Table III have at 5.

DISCUSSION

We have built a variety of thermopiles according to the designs presented here and have found that after allowing for reflection losses at the window (20 percent) they may be relied upon. The limitation is always in the physical constants of the thermoelectric materials. With the materials used in the figures we have found deviations of up to 25 percent.

Following these designs it is practicable to construct thermopiles to operate at an interruption frequency of 5 to 10 cycles per second with a sensitivity measured in terms of signal-to-noise ratio which is approximately as high as in d.c. operation. It is practicable to construct thermopiles which are more sensitive at any frequency than evaporated thermopiles which have been constructed heretofore. If short lead wires are used the chief limitation on response time is the heat capacity of the receiver.

Both the response time and the sensitivity are improved by using the smallest possible target. This is usually limited by the optics of the external system since it is practicable to construct receivers as narrow as 0.05 mm.

We may inquire as to how much improvement in thermoelectric materials is possible. From

Table I it is apparent that the absolute sensitivity may be increased approximately 50 percent by using 90 percent Bi–10 percent Sb vs. Sb instead of 97 percent Bi–3 percent Sb vs. 95 percent Bi–5 percent Sn. Another 30 percent improvement might be realized by using Te or Sb-Cd alloys but their lack of reproducibility makes this scarcely worth while.

THEORETICAL LIMIT OF SENSITIVITY

It is possible that materials with a very much improved factor of merit will be found. If sufficiently good materials become available the minimum detectable power will be lowered to the point where temperature fluctuations rather than Johnson noise become the limiting factor and in this case no further improvement is possible. From Eq. (5), the electrical signal of frequency between ν and $\nu+\Delta\nu$ generated by temperature fluctuations is

$$E_F(\text{r.m.s.}) = Q(4kT^2)^{\frac{1}{2}}L^{-\frac{1}{2}}(1+4\pi^2\nu^2/\lambda^2)^{-\frac{1}{4}}\Delta\nu^{\frac{1}{2}}. \quad (30)$$

The voltage generated by incident radiation

TABLE III. Characteristics of typical thermopiles built with 0.3 mm long leads of 97 percent Bi–3 percent Sb and 95 percent Bi–5 percent Sn.

	0.5	1.0	4.0
Total area of receivers (mm²)	0.5	1.0	4.0
Number of junctions	1	2	4
Resistance (ohms)	5	10	12.5
Response time (seconds)	0.036	0.036	0.041
Johnson noise in one-cycle frequency band at 300°K (microvolts) $\times 10^4$	2.9	4.0	4.6
Temperature fluctuation noise in one-cycle frequency range at 300°K (microvolts) $\times 10^4$	0.5	0.5	0.4
Equilibrium voltage sensitivity (microvolts/microwatts)	6.5	6.5	3.8
Response to 5 c.p.s. interrupted radiation (%)	88%	88%	84%
Absolute sensitivity (S) at 5 c.p.s. (μvolts/μwatt·$R^{\frac{1}{2}}$)	2.5	1.8	0.9
Minimum detectable power at 5 c.p.s. (microwatts) $\times 10^4$	0.5	0.7	1.4
Total heat capacity (ergs/°C)	4.8	9.5	36.3
Heat capacity of lead wires (ergs/°C)	1.0	2.0	6.3
Diameter of lead wires (cm) $\times 10^3$			
97% Bi–3% Sb	1.6	1.6	2.1
95% Bi–5% Sn	2.0	2.0	2.6

modulated at a frequency ν is obtained from Eq. (10). If we equate the two voltages we obtain the minimum detectable power

$$P_{\min} = (4kT^2L)^{\frac{1}{2}}\Delta\nu^{\frac{1}{2}}. \quad (31)$$

No matter how good the material is the losses can never be reduced below those caused by radiation. Hence the ultimate limit of detection can be written

$$P_{min} = 4(k\sigma A T^5 \Delta\nu)^{\frac{1}{2}}. \qquad (32)$$

Equations (31) and (32) may be applied to any type of detector which fundamentally measures temperature. At room temperature this limit has a value of $4\times10^{-6}\mu$watt for a receiver of 0.005 cm^2 area. This is only nine times better than can be attained with the best materials listed in the table. It is worth noting that this limit is also reduced if the target area is decreased.

A source of major improvement would be to operate with the cold junctions and surroundings at much lower temperatures. This has been done by Cartwright.[22] However, practically no data is available on the thermoelectric, electric, and thermal properties at low temperatures of materials which might be of interest.

[22] C. H. Cartwright, Rev. Sci. Inst. **4**, 382 (1933).

25

Copyright © 1964 by The Optical Society of America

Reprinted from *Appl. Optics*, **3**(4), 493–500 (1964)

Solid-Backed Evaporated Thermopile Radiation Detectors

Robert W. Astheimer and Seymour Weiner

Modern space-borne infrared systems require uncooled detectors sensitive to long wavelength radiation from 8 μ to 40 μ. The thermal detectors, such as thermistor and metal film bolometers, and radiation thermocouples are most practical. It is shown that the thermocouple detector has certain characteristics which make it particularly attractive for space-borne use. Chief among these is the fact that it has inherent dc stability and therefore does not require optical modulation. This permits systems to be designed with no moving parts. A technique is described for the construction of rugged thermocouple and thermopile detectors by vacuum deposition of evaporated materials through masks onto a solid substrate. This technique permits quantity production of complex detector element arrays that can survive and operate in severe military and space environments. A brief theoretical analysis of the performance to be expected from the new design is given and the properties of several configurations which have been fabricated are described. A discussion is included of some application considerations for use of these detectors.

I. Introduction

Modern infrared systems, particularly in military and space-borne applications, require infrared detectors which are sensitive to the long wavelength emission from objects at temperatures in the 100–300°K range. Detectors useful with such sources must respond to radiation of wavelengths from 8 μ to 40 μ. Photoconductive detectors which are highly sensitive in this region are available, such as copper- or zinc-doped germanium,[1] but unfortunately these must be cooled to temperatures in the neighborhood of liquid helium. Reliable cooling to these temperatures for long periods of time is presently impractical for space systems and often also for ground use because of logistic restrictions. The only suitable detectors are the thermal types, and of these the thermistor bolometer has found greatest use to date.

Although the thermistor bolometer[2] has proved very valuable over many years in a wide variety of applications, it has certain limitations associated primarily with the fact that a bias voltage is required which is many times greater than the signal to be amplified. A thermal detector which is free of this annoyance is the radiation thermocouple or thermopile. The radiation thermocouple has had a long history of development. An excellent description of the various constructions devised and a performance analysis is contained in ref. 3.

Radiation thermocouples have in the past been used primarily as detectors in laboratory-type instruments, such as infrared spectrometers, and have consequently been designed for maximum detectivity at low modulation frequencies. The ability to survive and operate under severe environmental conditions of shock, vibration, and temperature has not been given much attention. Typically, variations of configurations described by Schwarz,[3] and Hornig and O'Keefe[4] have been used. Although these produce detectors of excellent detectivity, they have relatively slow response and are rather fragile. It appeared that thermocouple detectors constructed along the lines of the so-called "solid-backed" thermistor bolometer[2] would have a number of merits. Although not inherently possessing as high a detectivity as the more fragile forms, it would be very rugged and reliable and would also have higher frequency response. Furthermore, such a construction lends itself to the use of thin film evaporation techniques which enable complex arrays of detector elements to be fabricated economically in quantity.

This paper gives a brief theoretical analysis of the performance to be expected from this design. The method of construction and the properties of several different thermopile detector configurations of this type which have been constructed are presented. It concludes with a discussion of application considerations for using these detectors.

It had been mentioned previously that the thermocouple detector requires no bias, which gives it certain advantages over detectors which do. It may be desirable at this point to amplify that statement, since, although if a bias supply can be omitted it simplifies the auxiliary equipment, this is by no means the major

The authors are with the Barnes Engineering Company, Stamford, Connecticut.

Received 21 May 1963.

advantage. The bias voltage will be many orders of magnitude greater than the signal to be detected; therefore any minute disturbances of the bias supply will produce spurious signals which may be large compared with the desired signal. Furthermore, thermistor bolometers undergo large changes in resistance with ambient temperature, and care must be taken to prevent excess bias current from damaging the detector at high temperatures where its resistance is low. Also, this change in resistance and bias current results in large changes in the detector responsivity and noise with ambient temperature change.

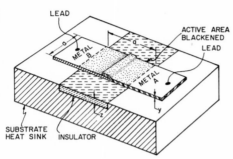

Fig. 1. Solid-backed thermocouple detector.

Another problem produced by the presence of bias current appears if the attempt is made to use a thermistor bolometer in an unmodulated optical system. Normally, the radiation to thermistor bolometers is chopped or scanned so that the desired fluctuating or ac signal can be separated from the dc bias by a capacitor. In an unmodulated optical system this cannot be done, and a balanced bridge technique must be resorted to. The best arrangement is to use a bridge made up of two identical thermistor elements, one of which is exposed to the radiation signal while the other is shielded from it. Even such a bridge will not remain in balance to the degree necessary in the presence of ambient temperature changes because of slight differences in the two elements and also because the bias current causes a self-heating temperature rise of several degrees, the exact amount depending upon the thermal impedance of the structure. The degree of the problem may be appreciated when one considers that the bias voltage will be of the order of 10 V and the signal of the order of 1 μV. The bridge balance must therefore be better than 1 part in 10^7. Furthermore, the thermistor elements composing the bridge are of necessity highly temperature dependent.

The electrical signal from a thermocouple detector will be between 100 and 1000 times lower than that from a thermistor bolometer for the same radiation signal. The electrical resistance and resulting Johnson noise power will be approximately proportionately lower. This increases the amplification requirements with the thermocouple detector; however, the only voltage appearing is the signal voltage itself. Therefore, we avoid the problem of identifying a very small change in a large bias voltage. For this reason dc or unchopped optical modes of operation are practical with thermocouple detectors where they would be completely out of the question with bolometers. This is of great advantage in space-borne instrumentation where power and life requirements preclude the use of mechanical scanning and chopping mechanisms.

II. Theory of Solid-Backed Thermocouple Radiation Detector

A solid-backed thermocouple radiation detector consists essentially of a pair of metallic junctions deposited onto a heat sink, one junction being in good thermal contact with the sink by conduction while the other is thermally isolated by a thin insulating layer. An idealized detector of this type is shown in Fig. 1. The area of the active junction is considered to be the portion over the insulator and is blackened, while the "cold" or reference junction is formed by the opposite ends of the metal strips which are in good conductive thermal contact with the sink. In this section expressions will be derived for the properties and optimization of thermopile detectors of this construction. The general treatment of thermal detectors given in ref. 3 will be followed.

Let us assume that the surface is irradiated with a radiation density of H (W/cm²). The junction will rise in temperature until the rate of heat loss by all processes equals the total radiant power incident. In the solid-backed configuration, the conduction losses to the heat sink through the insulating layer and the junction metals will be much greater than radiation or convection losses. The steady-state thermoelectric voltage generated by the junction will be:

$$V_s = \psi \Delta T = \psi Z H a^2,$$

where

ΔT = temperature rise of the active junction above the sink temperature,
ψ = combined thermoelectric power of the junction metals A and B,
a^2 = area of the detector ($a \times a$),
Z = thermal resistance of the active junction.

The responsivity, $\Re$, of the detector is:

$$\Re = V_s/Ha^2 = \psi Z. \tag{1}$$

The noise output, V_n, will be the thermal noise produced by the resistance of the junction

$$V_n = [4kT(\Delta f)R]^{1/2},$$

where

k = Boltzmann's constant,
T = absolute temperature,
(Δf) = bandwidth,
R = resistance of the junction.

Therefore, the noise equivalent power of the detector will be:

$$\text{NEP} = V_n/\Re = [4kT(\Delta f)]^{1/2}R^{1/2}/\psi Z. \qquad (2)$$

The time constant τ is given by:

$$\tau = ZC, \qquad (3)$$

where C is the heat capacitance of the active junction.

From Eq. (2) it is seen that to optimize detectivity we should minimize the ratio $R^{1/2}/Z$, which will depend upon the properties of the materials and the geometry as follows:

$$R = \rho_m a/ay = \rho_m/y \qquad (4)$$

$$Z = \frac{1}{\sigma_i a^2/z + 2\sigma_m ay/(a/4)} \quad \text{(approximately)}, \qquad (5)$$

where

ρ_m = average electrical resistivity of the junction metals,
σ_m = average thermal conductivity of the junction metals (assumed the same for both metals),
σ_i = thermal conductivity of insulating layer,
a, y, and z are as defined in Fig. 1.

The effective thermal conductivity length of the metal layers is assumed to be $a/4$ on each side. Substituting we have:

$$R^{1/2}/Z = \rho_m^{1/2}[(\sigma_i a^2/zy^{1/2}) + 8\sigma_m y^{1/2}].$$

The ratio will be minimized when

$$y = (\sigma_i/8\sigma_m)(a^2/z), \qquad (6)$$

which will be recognized as the case where the heat loss through the metals is equal to the loss through the insulating layer. In many practical applications, however, the resistance of the junction is not the noise limitation of the system. Therefore, it is desirable to thin down the metal layers somewhat more than Eq. (6) indicates to increase the responsivity at the expense of increased electrical resistance. This makes the heat loss through the insulating layer predominant, and Eqs. (1) and (5) can be simplified as follows:

$$Z = z/\sigma_i a^2, \qquad (7)$$

$$\Re = \psi(z/\sigma_i a^2). \qquad (8)$$

The heat capacitance of the junction is:

$$C = c_m a^2 y,$$

c_m = average heat capacity of the junction metals.

Thus from Eq. (3):

$$\tau = (c_m/\sigma_i)zy. \qquad (9)$$

We shall now compute the properties of a typical solid-backed thermocouple detector using these rela-tions. We shall use bismuth and antimony metals with a 0.025-mm mylar insulating layer and 1×1 mm active area.

ψ	= 100μV/°C,
a	= 0.1 cm,
z	= 0.0025 cm,
σ (bismuth)	= 0.08 W/cm °C,
σ (antimony)	= 0.20 W/cm °C,
σ_m (average)	= 0.14 W/cm °C,
σ_i (mylar)	= 0.00125 W/cm °C,
ρ (bismuth)	= $1.2(10)^{-4}$ Ω-cm,
ρ (antimony)	= $0.4(10)^{-4}$ Ω-cm,
ρ_m (average)	= $0.8(10)^{-4}$ Ω-cm,
c_m (bismuth and antimony)	= 1.3 J/cm³ °C.

From Eq. (6), $y = 0.0045$ cm. As has been mentioned previously, this will be reduced to about 0.002 cm.

From Eq. (8), $\Re = 0.02$ V/W. From Eq. (9), $\tau = 5$ msec. From Eq. (4), $R = 0.04$ Ω.

The noise in a 1-cps bandwidth is 25 pV. Therefore, from Eq. (2),

$$\text{NEP} = 25(10)^{-12}/0.02 = 1.25(10)^{-9} \text{ W}.$$

Thus, we see that the evaporated solid-backed thermocouple when optimized is inherently a very low impedance detector. Its NEP and response time are relatively good compared with other thermocouple designs. Actually the computed resistance is not realized, since the resistivity of thin evaporated layers is considerably higher than the bulk resistivity used in the computation. Furthermore, even if this low resistance were achieved, it would be difficult to construct an amplifier which would operate at the detector thermal noise level. The resistance, in practice, is about four times greater than computed from the bulk resistivity which increases the NEP by a factor of 2. The evaporation technique to be described permits easy production of thermopiles comprised of a hundred junctions or more which build up both responsivity and resistance.

It has been mentioned that these detectors are usable without optical chopping, and also that they may be arranged in multielement arrays which are electronically sampled. In such applications, frequency response is of less importance than responsivity. The responsivity can be increased by thermal signal storage at the expense, of course, of a longer time constant. This has been accomplished by relieving the heat sink under the active junction, as shown in Fig. 2, in order to leave a void. The insulating layer remains and now serves primarily to support the evaporated metal layers. This configuration is very similar to the evaporated thermopile structures described by Harris,[5] and Roess and Dacus.[6]

When this is done, the predominant heat loss is by conduction through the metal layers. In Eq. (5), the first term in the denominator now becomes negligibly small and the expression for the thermal resistance reduces to:

$$Z = \frac{1}{8}\sigma_m y. \qquad (10)$$

Inserting this value of the thermal resistance in Eqs. (1) and (3), we obtain the following responsivity and time constant for the example taken:

$$\Re = 0.045 \text{ V/W},$$

$$\tau = 12 \text{ msec.}$$

Further storage can be achieved by reducing the thickness y and width a of the metal layers after they leave the active area. This, of course, increases the electrical resistance.

In using solid-backed thermocouples and thermopiles, it is convenient to define a "specific responsivity", $\Re'$, which is the voltage developed by the detector per unit of radiation density or irradiance,

$$\Re' = \Re a^2 (\text{VW}^{-1}/\text{cm}^2).$$

From Eq. (8), we have $\Re' = \psi z/\sigma_1$, and we see that the specific responsivity is independent of area (unless the void design is used).

Suppose a detector area of 0.3×0.3 cm were desired. This could be made with one junction; however, if the area were broken up into a thermopile structure of 100 elements, each 0.3×0.3 mm and all connected in series, a specific responsivity 100 times greater would result. It is true that the resistance would increase also, but if the noise were limited elsewhere, or if the amplifier impedance could not be matched to the low value of a single element, there would be a direct increase in detectivity equal to the number of elements in the pile. The specific responsivity, $\Re'$, per junction in our numerical example is 200 μVW^{-1}/cm^2, and a 100-element thermopile comprising such elements would have an $\Re' = 0.02$ VW^{-1}/cm^2. Thermopile arrays can also be arranged in various patterns for specific detection and coding purposes.

III. Fabrication Techniques

The thermopiles being manufactured at the Barnes Engineering Company are fabricated by the deposition of evaporated materials through the apertures of metallic masks. The technique requires an integrated design, and accurate execution of masks, substrates, and alignment fixtures. A minimum of two materials is used in the thermopile. The geometrical layout for each material is governed by a mask that must also register properly with the substrate and all other masks.

The substrate is designed to support the thermopile configuration and in most of the existing arrangements must afford thermal isolation of the reference junction

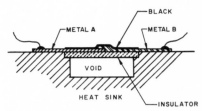

Fig. 2. Method of increasing responsivity.

from the active junction. Figures 1 and 2 show the methods used to provide thermal insulation for the "hot" junction.

The substrate material in Fig. 1 must have good electrical insulation properties to prevent the thermopile from being short-circuited. It must also have high thermal conductivity so that the reference junction will remain at the same temperature as the substrate. Substrate materials such as sapphire or anodized aluminum have been used successfully.

Epoxy resin is a reasonably good thermal insulator and has been used to insulate thermally the active thermopile junctions. A channel is cut into the substrate as is shown in Fig. 1. The resin, in liquid form, is introduced into the channel and then subjected to a curing treatment. When the resin is hard, the substrate is ready for the vacuum chamber.

The classical thermoelectric materials, antimony and bismuth, have been used in the Barnes Engineering thermopiles. These materials provide a thermoelectric voltage of approximately 100 μV per °C and are readily evaporated in the vacuum chamber. The substrate is aligned with the first mask, and all of the bismuth elements are deposited. The substrate is then aligned with the second mask, and all of the antimony elements are simultaneously deposited. Where interconnections are required in thermopile configurations, gold is used with an appropriate third mask. The junctions are usually formed with the deposition of the second thermoelectric material. It is obvious that the Bi and Sb portions may be interchanged with a resultant change in thermopile voltage polarity. Occasionally the elements of a reference junction are physically separated. Such a junction is completed with the deposition of the gold. The active junctions are blackened in a separate operation. Gold-black and various carbon blacks have been used.

IV. Typical Configurations

Four typical thermopile configurations are shown in Fig. 3. The first is a rectangular mosaic containing 120 active junctions and 120 reference junctions. The active junctions are arranged within a 1×8 mm area and are aligned over the thermal insulation in the substrate channels. It is important to note that the reference junctions may be exposed to radiation without degenerating the thermopile output voltage. The reference junctions are in close thermal contact with a heat sink and, in addition, are unblackened while the active junctions are blackened to improve their response to radiation.

The radial type thermopile is identical, in principle, with the rectangular mosaic, except that this arrangement permits the separation of the active and reference junctions. This separation is significant where it may be desirable to use an optical reflection filter, having good thermal conductivity, in contact with the thermopile. If such a filter were applied to the rectangular mosaic, the thermal conduction of the filter material would thermally short-circuit the active and reference junctions and thus degenerate the thermopile output voltage. Separation of the irradiated junctions from the reference junctions permits the use of filter materials regardless of thermal conductivity.

The mask-evaporation technique permits complex arrangements of detector elements for special applications to be easily produced. An example of such a special configuration is the "edge detector" shown in Fig. 3. Its principle of operation may be understood by reference to Fig. 4. Elements 1 and 2 are the two junctions of a thermocouple pair. In this case, both are blackened and thermally insulated from the sink and are therefore "active". Elements 3 and 4 are a similar pair of junctions, displaced laterally as shown and connected in series with reversed polarity.

Consider the image of a thermal edge or discontinuity moving across the detector from left to right, with the cold region preceding the hot region. The voltage generated by elements 1 and 2 will increase until element No. 1 is fully illuminated, and then decrease to zero when both are illuminated. The same signal, but of opposite polarity, will be produced by elements 3 and 4. When added, a linear slope passing through a null with a change of polarity will result, as shown dotted in Fig. 4(b). An important characteristic is that the position of the null point is independent of the magnitude of the thermal discontinuity. Such a linear

transfer function with a stable null is useful for horizon sensing or dimensional process control applications.

Another arrangement of the thermopile elements is used in the multiple layer thermopile. This device makes use of layers of thermopiles, stacked so that the reference and active junctions in each level lie over the respective junctions in the level below it. Thin coatings of electrically insulating material separate the thermopile layers. The thermopiles on each level are electrically connected in a series arrangement. The active junction area of the topmost layer is blackened. In the bottom level, the reference junctions are in thermal contact with the heat sink and the active junctions lie on the thermal insulator. In operation, radiation heats the active junctions of the top layer. This heat is conducted to the layers below and heats the active junctions in each level. The reference junctions are kept in close thermal contact with the heat sink. The upper layers of reference junctions are separated by films that are sufficiently thin to produce negligible thermal gradients.

The evaporated thermopile may be produced in an almost limitless number of configurations, of which the four described are but a sample.

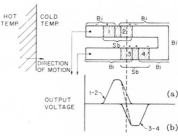

Fig. 4. Edge detector.

V. Properties of Evaporated Thermopiles

Table I lists the properties of four types of evaporated thermopiles. All of these are constructed with a void in the substrate, as shown in Fig. 2. Included for comparison is a typical 2 × 2 mm Schwarz-type thermocouple[3] and a 1 × 1 mm thermistor bolometer.

The Schwarz thermocouple offers better NEP and D* values, but it is fragile and cannot withstand the extreme treatment to which the evaporated thermopiles have been subjected.

The characteristics of the multiple layer radial thermopile are interesting in that the receiver area is identical with the area of the conventional radial unit, but the responsivity is twice that of the conventional unit. The D* is only slightly better than its single layer counterpart due to increased resistance.

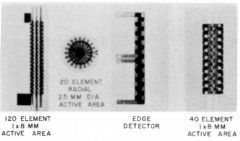

Fig. 3. Thermopile configurations.

Table I. Comparison of Long-Wavelength Detector Characteristics

Detector type	Receiver dimensions (mm)	Number of elements	DC responsivity $\mathcal{R}'$ (VW^{-1}/cm^2)	$\mathcal{R}$ (V/W)	τ (msec)	Resistance (Ω)	NEP (W)	D* 300°K, dc, 1 cps (cm/W)
Evaporated thermopile	1 × 8	40	0.02	0.25	8	30	2.8 × 10⁻⁹	1 × 10⁸
Evaporated thermopile	1 × 8	120	0.03	0.37	6	100	3.5 × 10⁻⁹	0.8 × 10⁸
Evaporated thermopile	2.5 diam	20	0.06	1.2	30	50	7.6 × 10⁻¹⁰	3 × 10⁸
Evaporated multiple layer thermopile	2.5 diam	40	0.12	2.4	40	150	6.6 × 10⁻¹⁰	3.4 × 10⁸
Schwarz-type thermocouple	2 × 2	1	0.20	5.0	35	10	8.2 × 10⁻¹¹	24 × 10⁸
Thermistor bolometer	1 × 1	1	14.0	1400	15	2.3M	2 × 10⁻¹⁰	5 × 10⁸

However, where the detector Johnson noise is not a limiting factor, the multiple layer unit offers twice the output voltage for the same irradiance.

A spectral response curve relative to a Golay cell is shown in Fig. 5 for the 2.5-mm diameter radial thermopile. The curve is typical of the spectral response that may be expected from any of these thermopiles, assuming the Golay cell to have a flat spectral response.

The ruggedness of the solid-backed evaporated thermopile was borne out by vibration and thermal shock tests. The vibration test was conducted in accordance with MIL-STD-810, dated 14 June 1962, method 514, Figure 514-3, curve F. The test consisted of one 30-min period of vibration normal to the thermopile surface and a second 30-min vibration period parallel to the thermopile surface. During each 30-min period, the frequency was swept at a constant octave rate from 5 to 2000 cps. A peak acceleration of 30 g was maintained between 99 and 2000 cps. The thermopiles experienced no change in physical or electrical characteristics.

The thermal shock test was performed on a thermopile in a sealed container with a small amount of silica gel. The container was lowered into a dewar of liquid nitrogen so that the bottom, on which the thermopile rested, was just above the liquid level. It was held in this position for 30 sec and then immersed in the liquid for 60 sec. After this, it was removed from the dewar and exposed to room temperature for 15 min. The container was next placed in a 100°C oven for 15 min, after which it was removed from the oven and permitted to come to room temperature. The thermopile was tested before and after the thermal shock test and showed no change in characteristics.

Operational tests have shown that the thermopiles continue to operate over the temperature range from −60°C to +100°C with no change in responsivity.

Certain applications are based on the difference in radiant energy received upon two detectors. Thermopile detectors, when used in this manner, are electrically connected in series opposition so that a zero or minimum output voltage indicates equal radiation on the two detectors. Although the difference may be small, each detector may be generating a large signal, and it is important that the responsivities of the two detectors remain balanced with ambient temperature changes.

This property was evaluated by a test in which two thermopiles of approximately equal responsivity were mounted on a common base and equally irradiated. The common base was then varied in temperature over the range of 25–128°C. With the thermopiles connected in series opposition, the change in net output voltage was no greater than 0.4% of the output from either detector.

VI. Application Considerations

Solid-backed thermopile detectors are most suitable in applications where their advantageous qualities are necessary and where their lower detectivity can be accepted. An area where they should find consider-

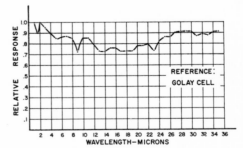

Fig. 5. BEC twenty-element radial thermopile.

230

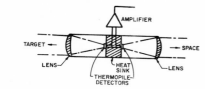

Fig. 6. dc comparison radiometer.

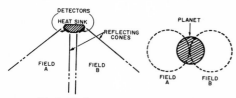

Fig. 7. Radiometric balance horizon sensor.

fields of view, is shown in Fig. 7. The two fields are arranged tangential to each other, and the instrument can, for example, be used to locate the centerline of a planet of uniform radiance by balancing the radiation received in each field. Four such cones and detectors would give two-axis information.

The detector output from such systems will be a dc voltage, probably in the range of 1 to 100 μV, and the problem arises as to how to amplify this signal to usable levels. Three satisfactory means are:

(1)   mechanical chopper amplifier,
(2)   magnetic amplifier,
(3)   photomodulator amplifier.

The mechanical chopper or breaker amplifier is well known,[3] and suitable vibrating type choppers are commercially available. These work well at very low signal levels, but the drive power, size, vibration sensitivity, and life are at times undesirable, particularly for space systems.

A photomodulation technique that has proved quite satisfactory is shown in Fig 8. A pair of neon bulbs is connected as a relaxation oscillator. Photosensitive resistors such as cadmium sulfide cells are placed in close proximity to the neon bulbs. These elements undergo a resistance change upon illumination of several orders of magnitude. A typical cell will have a dark resistance of $(10)^5$ Ω which will drop to 300 Ω when illuminated. One cell is connected in series with the detector and a step-up transformer, and the other shunts the transformer. As they oscillate they modulate the detector signal to the input of the step-up transformer. It has been found that such a modulator is quite free of spurious chopper induced signals or so-called "pedestal" and will permit ac conversions of signals down to about 0.5 μV.

The illuminated resistance of the photomodulator is usually appreciably greater than the resistance of the thermopile detector, and the electrical noise of the system is therefore determined by the thermal

able use is in space-borne instrumentation, such as radiometers for meteorological or planetary study, and infrared horizon sensors.

In the hard vacuum of space it is very difficult to lubricate properly optical-mechanical chopping and scanning mechanisms, and the power required to drive such mechanisms often produces a severe drain on the spacecraft power supply. Furthermore, space-borne systems usually must operate over a wide range of ambient temperatures and withstand a severe vibration environment during launch. The fact that thermopile detectors may be employed without optical modulation and, in addition, possess the other desirable qualities makes them very suitable for such applications.

A typical dc instrument is the simple comparison radiometer shown schematically in Fig. 6. Here two identical detectors and optical systems view separate fields of view. They are shown 180° apart in the figure, but any other angle could be used. The two detectors are connected in series opposition so that the dc output voltage represents the difference in radiation received from the two fields. The reference or "cold" junctions of both detectors are in thermal contact with the same heat sink, and their temperature is thus cancelled out. Care must also be taken in the design of the optical tube and lens structure to maintain an isothermal condition so that self-emission from the two lenses and housing will also cancel. If one field is directed toward space it will act as a stable reference at absolute zero for the other field, which is directed at the object whose radiance is to be measured. The radiometer signal will then be proportional to the absolute radiance of the target and be independent of the ambient temperature of the instrument.

Another comparison type of radiometer, which makes use of reflecting cones to define wide circular

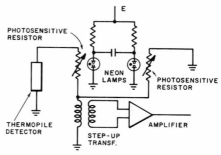

Fig. 8. Typical photomodulator circuit.

noise of the photomodulator. This is generally not of serious concern because the ultimate detectivity of systems which are not optically modulated is usually limited by thermal drift effects due to the self-emission of the optical elements.

Another means of converting the low level dc detector signal to an ac signal is by a magnetic modulator. Best results have been obtained with the second harmonic type of modulator.[7] This method is most suitable when an ac carrier power supply is available. Magnetic modulators have been found to be somewhat less stable, particularly with changes of ambient temperature, than either the mechanical or photoelectric modulators.

The photomodulating technique can be extended to sample sequentially arrays of thermopile detectors. One arrangement is shown in Fig. 9. Each thermopile in a linear array of detectors is connected through a photosensitive resistor to a common amplifier. Each photosensitive resistor has a neon lamp adjacent, only one of which is on at a time. A ring counter or similar device sequentially extinguishes the neon lamp that is on and lights the next one in the array. The low resistance of the photosensitive resistor which is illuminated connects its detector to the amplifier while all the others are isolated by the high resistance of the dark cells. Thus, each detector, in turn, is sequentially connected or "photoswitched" to the amplifier input. Such a system could be used for thermal imaging applications.

The ability of these detectors to operate without optical modulation often partially compensates for their low detectivity with respect to other thermal detectors, such as thermistors. This is because their frequency response or time constant can be traded off for higher responsivity by either increasing the thickness of the insulating layer or placing a void in the heat sink below the active junctions as has been described. The electronic modulation or sampling, of course, can be at any convenient frequency, since it is not dependent upon the thermal time constant of the detector. By this means, signal storage can be achieved in multielement mosaics of thermopile detectors which are electronically sampled.

One of the advantages of thermopile detectors is that they are extremely nonmicrophonic. This is partly due to their rugged construction, but even more because there is no bias voltage. In the presence of a large bias potential, vibration is likely to induce large spurious signals electrostatically unless great care is

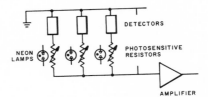

Fig. 9. Typical photosampling circuit.

taken. Because of their freedom from such effects, these detectors can be mounted directly onto rapidly moving or vibrating structures without special precautions. For example, an optical-mechanical chopping action has been produced by mounting a thermopile detector directly onto a vane which was caused to vibrate at 20 cps. By the use of an impedance-matching transformer, Johnson noise limited operation was attained without spurious signals from the vibrating motion.

In the foregoing, a brief discussion has been given of various means of applying thermopile detectors and amplifying and sampling their output signals. This has been done to illustrate some of the types of applications for which they are suitable and the signal processing techniques which have been developed.

To summarize, solid-backed evaporated thermopile detectors, although not the optimized configuration for maximum detectivity, have a number of attributes which makes them very suitable for use in certain applications. Of particular significance are their ruggedness, ability to operate without optical modulation, and the ease of producing large area detectors and complex arrays of thermopile elements. Photoswitching has proved to be a satisfactory all-electronic means of modulating the dc output signal for amplification or sampling a detector array.

References

1. R. F. Potter and W. L. Eisenmann, Appl. Opt. **1**, 567 (1962).
2. E. M. Wormser, J. Opt. Soc. Am. **43**, 15 (1953).
3. R. A. Smith, F. E. Jones, and R. P. Chasmar, *The Detection and Measurement of Infrared Radiation* (Clarendon Press, Oxford, 1957).
4. D. F. Hornig and B. J. O'Keefe, Rev. Sci. Instr. **18**, 474 (1947).
5. L. Harris, Phys. Rev. **45**, 635 (1934).
6. L. C. Roess and E. N. Dacus, Rev. Sci. Instr. **16**, 164 (1945).
7. A. G. Milnes, *Magnetic Amplifiers* (Macmillan, London St. Martin's Press, New York, 1957).

26

Reprinted from *J. Opt. Soc. Amer.*, **43**(1), 15–21 (1953)

Properties of Thermistor Infrared Detectors

ERIC M. WORMSER*

Servo Corporation of America, New Hyde Park, New York

(Received July 11, 1952)

Thermistor bolometers are fast, sensitive infrared detectors with good responsivity from 1 to 15 microns. The detectors consist of a thin flake of thermistor material mounted on a thermal sink. The time constant of the detectors is determined by the type of thermal sink on which the flakes are mounted. By cementing thermistor flakes to quartz or glass thermal sinks, detectors with time constants ranging from 3 to 5 and 5 to 8 milliseconds, respectively, are obtained.

Experimental results are given for quartz- and glass-backed thermistor detectors having sizes of sensitive areas ranging from 0.2 to 12.5 square millimeters. Performance criteria for these detectors are tabulated and plotted as a function of size of sensitive area. Typical frequency response and relative spectral responsivity data are given.

Detector flakes have to be mounted in sealed housings for noise-free operation. Two types of housings are described.

INTRODUCTION

THERMISTOR circuit elements and infrared detectors are *therm*ally sensitive re*sistors* whose resistance varies rapidly with temperature. They were originally developed by the Bell Telephone Laboratories.[1,2]

Thermistor infrared radiation detectors have properties comparable to rapid response metal bolometers, thermocouples and thermopiles. Thermistor type infrared detectors differ from other fast total radiation detectors in the following characteristics: (1) they have high impedance which permits direct coupling to an amplifying tube; (2) they can be manufactured with a wide variety of rectangular sensitive areas having linear dimensions ranging from 0.2 to 5 millimeters; (3) the solid backed types are extremely rugged and non-microphonic.

A large number of solid backed thermistor detectors have been manufactured and tested. It is the aim of this paper to briefly describe the general properties of these detectors and to present experimental data of interest in the application of those units.

GENERAL DESCRIPTION

The thermistor infrared detector or bolometer consists of a thin flake of sintered semiconductor material with a high negative coefficient of resistance. Two types of semiconductor materials are used for the manufacture of bolometers: (1) thermistor material No. 1, which is a mixture of the oxides of manganese and nickel having a resistivity of approximately 2500 ohm cm at 25°C, and (2) thermistor material No. 2, which is a mixture of the oxides of manganese, nickel and cobalt, having a resistivity of approximately 250 ohm cm at 25°C.

The resistance of detector flakes depends on their

thickness and the shape of their sensitive area. A standard flake thickness of 10 microns has been adopted. This has been found to be a good compromise between sensitivity and time constant on the one hand, and feasability to manufacture flat flakes on the other. Resistances of 10 micron thick flakes of various geometric shapes are shown in Fig. 1. As shown in this figure, it is possible to manufacture detectors with a wide variety of geometrical shapes of sensitive areas, maintaining resistances ranging from 1 to 10 megohms. This range of resistance is well suited to direct coupling to the grid of the first amplifying stage. It is desirable to maintain bolometer resistance above 1 megohm in order to have the noise in the first amplifying stage less than the limiting Johnson noise of the detector.

PRINCIPLES OF OPERATION

Heat energy impinging on the thermistor flake heats the flake and changes its resistance. The change in

* Present address: Olympic Development Company, Stamford, Connecticut.

[1] W. H. Brattain and J. A. Becker, J. Opt. Soc. Am. 36, 354 (1946).

[2] J. A. Becker *et al.*, OSRD 5991, Final Report on Development and Operating Characteristics of Thermistor Bolometers.

SHAPE	#1 THERMISTOR MATERIAL $\rho = 2500$ OHM CM	#2 THERMISTOR MATERIAL $\rho = 250$ OHM CM
SQUARE	2.5 MEGOHMS	0.25 MEGOHMS
RECTANGLE (4:1)	10 MEGOHMS	1.0 MEGOHMS
RECTANGLE (8:1)	20 MEGOHMS	2.0 MEGOHMS
RECTANGLE (12:1)	30 MEGOHMS	3.0 MEGOHMS

FIG. 1. Resistance of thermistor detector flakes as a function of shape of sensitive area (resistance at 25°C).

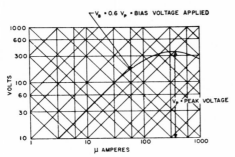

Fig. 2. Steady state voltage current curve of thermistor detector.

resistance is detected by applying a dc bias voltage across the detector flake. It is desirable to apply as large a biasing voltage as possible, since signal-to-noise ratio is proportional to bias voltage.

Voltage Current Characteristics

The amount of bias voltage which can be applied to a detector flake is determined by its steady state voltage-current characteristic.[3] A typical voltage-current curve is shown in Fig. 2. The peak voltage V_P of the voltage-current curve is determined by the equilibrium heat flow from the detector unit which is described by Eq. (1):

$$V_P{}^2/R_P = C_T(T_P - T_0),\qquad(1)$$

where V_P = peak voltage (volts), R_P = resistance of the thermistor flake at peak voltage (ohms), C_T = total or steady state rate of heat dissipation from the detector unit (watts/°C), T_P = temperature of the detector flake at peak voltage (°K), and T_0 = ambient temperature (°K). In order not to exceed peak voltage over a range of ambient temperatures, the detectors are usually operated with a bias voltage V_B which is 60 percent of the peak voltage at 25°C. The detector flake then operates at 5° to 10°C above the ambient temperature. At V_B the detector has 88 percent of its maximum responsivity.

Bolometer Bridge Circuit

When applied as infrared radiation detectors, thermistor elements are commonly used in bridge circuits as shown in Fig. 3. The bridge consists of an active thermistor flake exposed to the radiation to be measured, and an identical shielded compensating flake. Equal positive and negative bias voltages are applied. The use of a compensating thermistor maintains the signal junction S at dc ground potential for any variation in ambient temperature.

Effect of Bias Voltage on Signal-to-Noise Ratio

The limiting noise of a thermistor detector is the Johnson noise of its resistance. For bias voltage at 0.6

3 R. C. Jones, J. Opt. Soc. Am. 36, 448 (1946).

peak voltage, current noise is small compared to Johnson noise. The noise of the detector is determined by its resistance, while the signal is a function of the resistance and the bias voltage. The signal-to-noise ratio is thus proportional to the bias voltage which can be applied. Signal-to-noise ratios of large size thermistor bolometers have recently been considerably improved by increasing the steady state heat dissipation of detector assemblies, thus permitting use of increased bias voltages.

TYPES OF THERMISTOR DETECTORS

For most applications the detectors should have a fast response rate. This is achieved by providing a thermal sink in close proximity to the detector flake. Detectors with three types of thermal sinks are manufactured. For fastest response, thermistor flakes are cemented on Z-cut quartz which combines good thermal conductivity with high insulation resistance. For intermediate response rates flakes are cemented to glass. For slower response rates flakes are spaced by a small air-gap from a metal thermal sink. A comparison of the average characteristics of the three types of units is given in Table I.

For pulsed heat energy, the first time constant of the three types of detectors ranges, respectively, from 2 to 5 milliseconds for quartz, from 5 to 8 milliseconds for glass, and from 20 to 40 milliseconds for air-backed units. The increase in responsivity, as shown in Table I, is not as much as one might expect for the longer time constant units. This is due to the simultaneous decrease in the bias voltage which can be applied. Solid backed detectors are extremely rugged and nonmicrophonic, in addition to having fast response, which has led to their widespread application.

Response of Solid-Backed Detectors

When a pulse of radiant energy ΔW impinges on the detector flake, a fraction of the energy is absorbed and heats the flake. At the same time heat is conducted away from the flake to its thermal sink and the housing in which it is mounted. The voltage signal ΔV, as a function of time t, which is developed at the signal

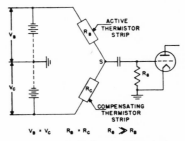

Fig. 3. Thermistor bolometer bridge circuit.

junction S is described by Eq. (2):

$$\frac{\Delta V}{\Delta W} = Fa\alpha V_B\left[\frac{1}{C_1}(1-\epsilon^{-t/\mathcal{T}_1})+\frac{1}{C_2}(1-\epsilon^{-t/\mathcal{T}_2})+\cdots\right], \quad (2)$$

where α = temperature coefficient of resistance, a = absorptivity of the blackened thermistor flake, F = constant of the bridge circuit (for active and compensating flakes of equal resistance $F=\frac{1}{2}$), C_1 = rate of heat dissipation from the flake to its quartz or glass thermal sink, C_2 = rate of heat dissipation from the quartz or glass thermal sink to the metal housing, C_3 = rate of heat dissipation from the metal housing to a large thermal sink at ambient temperature, and $\mathcal{T}_1$, $\mathcal{T}_2$, and $\mathcal{T}_3$ are the time constants of the respective heat flows where $\mathcal{T}_1 \ll \mathcal{T}_2 \ll \mathcal{T}_3$. For radiant energy pulses having a duration of the same order as the first time constant $\mathcal{T}_1$, only the heat flow to the thermal sink of the flake need be considered. Equation (2) can then be simplified to (2a):

$$\frac{\Delta V}{\Delta W} = Fa\alpha V_B\left[\frac{1}{C_1}(1-\epsilon^{-t/\mathcal{T}_1^{*}})\right], \quad (2a)$$

where $\mathcal{T}_1^{*}$ = effective first time constant of the detector, and $\mathcal{T}_1^{*} \simeq \mathcal{T}_1$. In describing the properties of thermistor detectors, the effective first time constant is used in this paper.

EXPERIMENTAL PROPERTIES OF SOLID-BACKED DETECTORS

From the point of view of application, the important properties of a thermistor bolometer are its resistance, the bias voltage which can be applied, its responsivity, its limiting noise level, and its effective first time constant. These properties can be combined into performance criteria as defined below.

Experimental Determination of Properties

To determine the bias voltage which can be applied to a thermistor detector, its steady state voltage-current characteristic, such as is shown in Fig. 2, is obtained. For this test, the housing containing the detector flakes is mounted to a large thermal mass at room temperature in a manner similar to its final installation. The values of current and voltage are recorded when equilibrium heat flow is reached at each point. In use the bias voltage is set at 0.6 of the peak voltage determined at room temperature.

To determine responsivity, the detector is exposed to infrared energy from a cavity blackbody source at a temperature of 393°K interrupted by a square wave chopping disk at a temperature of 300°K at a rate of 15 cps. The chopped radiation has its spectral peak at 7μ and 90 percent of its energy between 3μ and 20μ. Emissivities of the blackbody and chopping disk have been experimentally determined and the value of radiant energy incident on the detector corrected. The rms voltage output from the detector bridge, biased to 0.6 of its peak voltage, is measured by a thermal meter,

TABLE I. Comparison of three types of thermistor detectors. Size of sensitive area: 2.5×0.2=0.5 mm². No. 2 thermistor material, resistance 3 megohms at 25°C.

Thermal sink	Quartz	Glass	Air-spaced metal
Effective first time constant, seconds	0.002–0.005	0.005–0.008	0.020–0.040
Bias voltage, volts	212	130	81
Responsivity, volts rms/watts av at 15 cps	705	585	1210
Reference band width, cps	62.5	35.7	8.33
Johnson noise, microvolts	1.7	1.3	0.62
ENI, watts	2.3×10^{-9}	2×10^{-9}	5×10^{-10}
Detectivity, watts⁻¹	4.3×10^{8}	5×10^{8}	2×10^{9}

after amplification by an amplifier having a bandwidth of 2 to 1500 cps.

The values of responsivity tabulated are the rms volts output from the active thermistor strip for watts of average power input square wave interrupted at 15 cps (2×bridge circuit). Responsivity values are computed for unity total watts incident on the detector sensitive area and for unity watts/cm².

The noise level of detectors, when shielded from radiation, with operating bias voltage applied, is similarly measured. Under laboratory conditions, the operating noise level of acceptable detectors does not exceed 1.5 times the Johnson noise of their resistance at 25°C.

To determine the effective first time constant, a differentiating method is used.[2] The voltage output V and the differentiated output dV/dt resulting from a 15-cps square wave radiation pulse incident on the thermistor flake are connected to the horizontal and vertical deflection plates of a calibrated oscilloscope. The initial slope of the resulting parallelogram is a measure of the time constant.

Performance Criteria of Detector Units

In order to compare units having different time constants and sensitive areas of one particular kind of detector, and in an attempt to compare different kinds of detectors, it is desirable to establish general performance criteria for thermal detectors. One of the criteria which has been used in the literature to date is the equivalent noise input (ENI). The ENI will be here defined as the power in watts (using the 15-cps responsivity measured as described above) giving a signal equal to the Johnson noise in volts for a reference band width which is generally sufficient to pass signal pulses from the detector. The reference band width is here defined as $1/4\mathcal{T}$ in cycles per second, where $\mathcal{T}$ is the median value of first time constants experimentally determined for the type detector.[4]

[4] This reference band width is that used by R. C. Jones [J. Opt. Soc. Am. 39, 327–356 (1949)]; however, the value of ENI defined in this paper differs from that used by Jones, since responsivity at 15 cps (S_{15}) is used instead of computing zero frequency responsivity S_0. Jones' zero frequency responsivity, which is equivalent to Bell Telephone Laboratories' first time-constant, steady-state responsivity, is arrived at by substituting a single time-constant curve which is in contact with the experimental

TABLE II. Characteristics of quartz-backed thermistors over a range of sensitive areas.

No. of units	Sensitive area Length mm	Width mm	Area mm²	Thermistor material No.	Resistance at 27°C Megohms	Bias voltage Volts	Responsivity at 15 cps volts/watt/cm²	volts/watt	Time constant msec	For 62.5 cps band width ENI watts	Detectivity watts⁻¹
274	5.55	2.54	14.1	No. 2	0.48	242	5.8	41.0	4.56	1.0×10^{-8}	1.0×10^{8}
10	5.55	2.25	12.5	No. 2	0.55	260	4.8	38.7	2.5	1.9×10^{-8}	5.3×10^{7}
2	6.35	0.76	4.84	No. 2	1.54	207	5.35	110.5	3.5	1.2×10^{-8}	8.3×10^{7}
4	5.0	0.76	3.81	No. 2	1.46	176	3.12	82.5	5.1	1.5×10^{-8}	6.7×10^{7}
2	1.5	1.5	2.25	No. 1	2.37	300	5.35	238	4.0	5.0×10^{-9}	2.0×10^{8}
8	1.0	1.0	1.00	No. 1	2.07	184	2.84	284	4.8	5.2×10^{-9}	1.9×10^{8}
10	1.5	0.5	0.75	No. 2	0.70	100	0.52	69.6	2.6	1.2×10^{-8}	8.3×10^{7}
12	2.5	0.2	0.5	No. 2	3.02	212	3.52	705	4.0	2.3×10^{-9}	4.3×10^{8}
28	0.5	1.0	0.5	No. 1	1.2	125	2.56	510	2.7	2.0×10^{-9}	5.0×10^{8}
14	1.0	0.5	0.5	No. 1	4.65	216	5.35	1070	4.5	2.0×10^{-9}	5.0×10^{8}
14	0.75	0.6	0.45	No. 1	2.85	206	4.60	1025	3.1	1.7×10^{-9}	5.9×10^{8}
6	0.5	0.5	0.25	No. 1	2.1	170	2.73	1105	1.9	1.4×10^{-9}	7.1×10^{8}

TABLE III. Characteristics of glass-backed thermistors over a range of sensitive areas.

No. of units	Sensitive area Length mm	Width mm	Arae mm²	Thermistor material No.	Resistance at 27°C Megohms	Bias voltage Volts	Responsivity at 15 cps volts/watt/cm²	volts/watt	Time constant msec	For 35.7 cps band width ENI watts	Detectivity watts⁻¹
9	5.55	2.25	12.5	No. 2	0.60	145	4.5	35.5	5.8	1.6×10^{-8}	6.2×10^{7}
16	6.34	0.5	3.17	No. 2	2.96	282	6.0	189	6.4	6.5×10^{-9}	1.5×10^{8}
35	4.0	0.5	2.0	No. 2	1.55	110	2.85	142	6.06	6.7×10^{-9}	1.5×10^{8}
3	3.17	0.5	1.59	No. 1	12.8	416	14.0	876	6.2	3.0×10^{-9}	3.3×10^{8}
16	3.17	0.5	1.59	No. 2	1.31	132	3.9	245	5.8	3.5×10^{-9}	2.8×10^{8}
4	0.82	1.42	1.16	No. 1	1.04	80	3.0	152	5.3	2.8×10^{-9}	3.6×10^{8}
7	0.82	1.42	1.16	No. 2	0.11	31	1.0	85.5	6.2	3.0×10^{-9}	3.3×10^{8}
10	2.5	0.2	0.5	No. 2	3.0	130	2.9	585	5.9	2.2×10^{-9}	4.5×10^{8}
4	2.0	0.2	0.4	No. 2	2.5	102	2.66	665	6.0	1.8×10^{-9}	5.5×10^{8}
4	1.0	0.2	0.2	No. 2	1.25	54	1.5	750	6.3	1.2×10^{-9}	8.3×10^{8}

The criterion of "equivalent noise-input", however, suffers from the psychological defect that improved performance is associated with decreasing numerical values. A new performance criterion of "Detectivity" has recently been proposed by Jones,[5] which is here defined as the inverse of ENI. Detectivity provides positive correlation between improved performance and the associated numerical values.

EXPERIMENTAL RESULTS

Properties as a Function of Size of Sensitive Area

Solid-backed detectors having sizes of sensitive area ranging from 0.2 mm² to 25 mm² have been manufactured and tested. The properties of a range of quartz-backed and glass-backed units is summarized in Tables II and III. The properties listed for each size of detector are the average values for a representative sample of units manufactured. The reference band widths for determining ENI and detectivity for quartz- and glass-

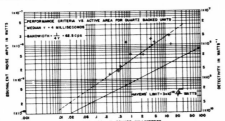

FIG. 4. Performance criteria vs active area for quartz-backed thermistor detectors.

curve at the point where the slope is 3 db. In order to arrive accurately at S_0 from the tabulated responsivity S_{15} an experimental frequency response curve has to be taken for each detector unit, since τ_2 and τ_3 vary from unit to unit.

Complete frequency response data are available only for a small number of the units tabulated. From these data we can approximately relate S_0 and S_{15} in general. From the foregoing data we find for quartz-backed units S_0 ranges from 1.0 to 1.25 times S_{15}.

[5] Research and Development Board, Natl. Bur. Standards, Symposium on Infrared Instrumentation, Houston, Texas, September 13–15, 1951.

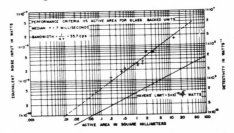

FIG. 5. Performance criteria vs active area for glass-backed thermistor detectors.

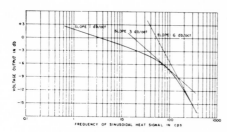

FIG. 6. Frequency response curve for quartz-backed
thermistor detector.

backed units are, respectively, 62.5 cps corresponding to
a median time constant of 4 milliseconds, and 35.7 cps
corresponding to 7 milliseconds.

In Figs. 4 and 5 the average performance criteria are
plotted as a function of size of sensitive area. For comparison, the engineering estimate of minimum detectable signals made by Havens in 1946[6] is shown in the
graphs.

Frequency Response of Solid-Backed Detectors

Sinusoidally varying infrared radiation from $\frac{1}{2}$ to
300 cps is generated through use of a special chopping
disk. The voltage output of detector flakes exposed to
the sinusoidal radiation input has been measured. In
Fig. 6 a typical frequency response curve of a quartz-
backed detector flake is shown. This flake has an effective first time constant of 2 milliseconds as measured by
the differentiating method. The first time constant
corresponds closely to the frequency at the point having
a slope of 3 db per octave on the frequency response
curve. At high frequencies this curve follows that of a
single time constant exponential device.

Variation of Responsivity Over the Detector Area

Variation of responsivity over a flake sensitive area of
3.07 mm×1 mm is shown in Fig. 7. This variation was
measured using a spot of radiation having 0.01 mm
diameter. An increase in responsivity is noted in regions
near the edges of the sensitive area. This variation is
assumed to be the result of higher specific resistance in
these regions, caused by a lower temperature difference
between the flake and its thermal sink. Loss of responsivity resulting from dissipation of the absorbed radiant
energy to the metallic lead areas is confined to the last
tenth of a millimeter in the length of the sensitive area.
Variation of responsivity as a function of angle of
incidence has been experimentally checked and found
to follow the cosine law.

Dynamic Range of Detectors

The responsivity of a detector, i.e., the relation of
voltage output to radiant energy input has been checked

[6] R. J. Havens, J. Opt. Soc. Am. **36**, 355 (1946).

for infrared radiation levels equal to Johnson noise, to
levels 1000 this value. Over this range the responsivity
has been found to be constant.

Variation of Responsivity with Wavelength

Radiation from a blackbody source at 1183°K, square
wave interrupted at 10 cps, was dispersed using a KBr
double monochromator having slit width compensation
giving uniform wavelength resolution. A quartz-backed
thermistor bolometer (0.6×0.75 mm) blackened with a
thin layer of Zapon lacquer and a fast vacuum
thermocouple (0.75×0.75 mm) blackened with gold
black were exposed to the dispersed radiation. A number
of comparison runs were made within a short time interval to avoid changes in atmospheric absorption. The results of a typical set of spectral recordings are shown in
Fig. 8. In order to compare the detectors, the signal
level measured is normalized by dividing it by the
Johnson noise of the respective detector.

The spectral responsivity curves (Figs. 8 and 9)
indicate that the thermistor detector tested has normalized responsivity amounting to 60–80 percent that of
the thermocouple from 2 to 7 microns, equal responsivity to that of the thermocouple from 8 to 10.5 microns.
Beyond 10.5 microns to 19 microns, the responsivity of
the thermistor ranges from 1 to 2 times that of the
thermocouple. This can probably be attributed to de-

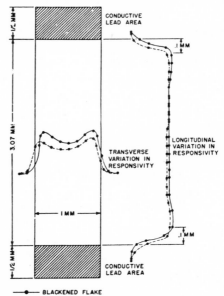

FIG. 7. Variation of responsivity over a thermistor flake
3.07×1 mm sensitive area.

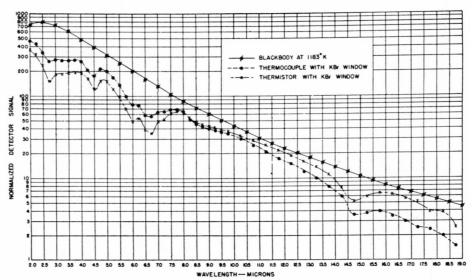

FIG. 8. Spectral responsivity curves of blackened thermistor detector (0.6×0.75 mm) and gold-blackened thermocouple (0.75×0.75 mm).

crease in the absorptivity of gold black in these regions.

A curve showing the ratio of normalized responsivity of the thermistor bolometer compared to that of the thermocouple is shown in Fig. 9. It should be pointed out that in view of the large quantity of data required, this ratio is based on measurements made on one detector of each kind.

THERMISTOR DETECTOR ASSEMBLIES

The minimum signal levels which can be detected by thermistor flakes are of the order of 10^{-8} to 10^{-9} watt. These radiation signals cause temperature changes of the order of 10^{-6} to 10^{-7} degree Centigrade in the flake and produce output signal levels of the order of 1 microvolt.

The low levels of temperature changes and electrical signals which can be detected by thermistor flakes, together with their high impedance, pose special requirements on the housings in which they are mounted.

Thermistor detector flakes have to be mounted in sealed housings to prevent noise caused by convection cooling. For units which are subjected to considerable mechanical vibration, it is further advantageous to evacuate the units to avoid swish noise which is caused by the movement of gas across the flakes.

Electrical contact with the detector flakes is made through glass-insulated lead wires sealed to the base of the housing. These leads must have insulation resistance of the order of 100 000 meghoms to avoid leakage currents which would otherwise contribute noise to the

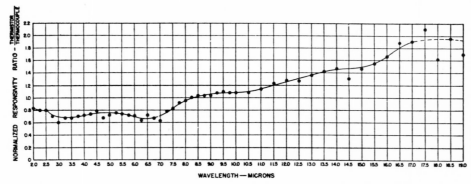

FIG. 9. Relative spectral responsivity of blackened thermistor detector as compared to gold-blackened thermocouple.

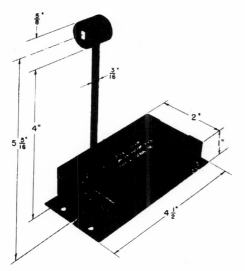

FIG. 10. Thermistor detector assembly mounted
on preamplifier unit.

detector output signal which is in the microvolt range. The connections from the detector housing to the grid of the first amplifying stage should be rigid to minimize capacity variation and should be as short as possible to avoid electrostatic pick-up.

Two standard types of housings have been developed which satisfy these requirements. The housing generally used for spectrographic and industrial application is shown mounted on a preamplifier unit in Fig. 10. This housing is suitable for compensated detector pairs having sensitive areas 2.5×0.5 mm or smaller. Windows

of silver chloride, rock salt, potassium bromide, KRS-5, or other infrared-transmitting substances, may be cemented into the small window aperture. The units are sealed but not evacuated. The signal lead and the bias leads to the matched detector pair are brought through the stem in glass covered wires to the preamplifier in the base of the unit.

For other applications and for larger sensitive areas a sealed capsule with a fused silver chloride window and a four-pin base is available, having an over-all diameter of $\frac{5}{8}$ in.

The small preamplifier unit shown in Fig. 10, using a low microphonic input tube, has been developed for use in conjunction with the detectors. This unit is mounted in close proximity to the detector. It provides a stabilized gain of 200 times, which amplifies the signals from the detectors to a level where use of ordinary flexible shielded cables is permissible.

Through application of the housings and techniques described, it has been possible to operate thermistor detectors at levels from 2 to 10 times their theoretical noise level under severe field conditions.

ACKNOWLEDGMENTS

The manufacturing and testing of thermistor detectors is being performed by a group of my co-workers at Servo Corporation of America, all of whom have contributed to the work reported in this paper.

The original manufacturing and testing procedures as well as the performance theory were, in a large measure, developed by personnel of the Bell Telephone Laboratories on contracts from OSRD and the Navy Department.

Thanks are due R. Clark Jones of the Polaroid Corporation for permission to use his new performance criterion of "Detectivity."

Reprinted from *Radio Electronic Engr.*, **42**(6), 260–264 (1972)

High Performance Pyroelectric Detectors

G. BAKER, B.Sc. *

D. E. CHARLTON, B.A. *

and

P. J. LOCK, B.Sc., Ph.D. †

Based on a paper presented at the Conference on "Infra-red Techniques" held in Reading from 21st to 23rd September 1971

SUMMARY

A survey of a number of possible materials for use in pyroelectric detectors shows that strontium barium niobate, lithium sulphate and triglycine sulphate (TGS) are the most interesting materials. Over a large range of frequencies and sensitive areas TGS and its derivatives offer the highest sensitivities. The major disadvantage of TGS, a tendency to depole, has been overcome by the addition of organic dopants. This can lead to biased hysteresis loops and improved dielectric performance. An increased range of operating working temperature is obtained by deuteration. A brief survey of some of the applications of pyroelectric detectors is given.

* *Mullard Southampton, Millbrook Industrial Estate, Southampton SO9 7BH.*
† *Formerly with Mullard Southampton, now with the Electrical Research Association.*

1. Introduction

Many years have passed since pyroelectricity was first observed but it is only within the last decade that any great interest has been shown in the use of this effect to detect infra-red radiation. Work up to 1969 has been described by Putley[1] and further developments have been reported by Schwarz and Poole.[2] This paper describes the present state of the art; in particular the influence of the material properties on the detector performance will be discussed.

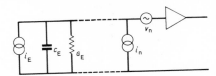

Fig. 1. Equivalent circuit of the detector.

The theory of the pyroelectric detector has been given initially by Cooper[3] and more completely in Putley's review.[1] It will be summarized here in order to establish the basis on which pyroelectric materials are judged for their suitability and the associated amplifier designed. Pyroelectric materials are those with a temperature-dependent spontaneous electrical polarization. Under equilibrium conditions this electrical asymmetry is compensated by the presence of free charges. If, however, the temperature of the material is changed at a rate faster than these compensating charges can redistribute themselves, an electrical signal can be observed. This means that the pyroelectric detector is an a.c. device, unlike other thermal detectors (such as thermistors) which detect temperature levels rather than temperature changes. It can be shown that at frequencies above that corresponding to the reciprocal thermal time-constant the current responsivity is constant; this leads to a good high frequency performance.

2. Noise Equivalent Power of Pyroelectric Detector

Noise in the system arises from the detecting element and the associated amplifier; in order to minimize it the device is used in the voltage mode. The equivalent circuit of the system is shown in Fig. 1. As the load on the signal current is capacitive above the electrical turnover frequency the voltage responsivity falls with increasing frequency. However, as the current noise

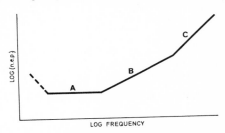

Fig. 2. Variation of noise equivalent power with frequency.

sources are affected in the same way, the signal/noise ratio is expected to be constant over a wide frequency range (up to the frequency where the voltage noise contribution dominates).

The noise equivalent power (n.e.p.) of the system can be shown to be given by

$$\text{n.e.p.}_{\text{(unit bandwidth)}} = \frac{1}{\eta}\sqrt{4kT\left\{\underbrace{TG_T}_{\substack{\text{thermal}\\\text{noise}}} + \frac{C_T^2+(G_T/\omega)^2}{\lambda^2 A^2}\underbrace{[G_E}_{\substack{\text{element}\\\text{noise}}} + \underbrace{G_N+R_N(G^2+\omega^2 C^2)]}_{\substack{\text{amplifier}\\\text{noise}}}\right\}}$$

where η = emissivity of the element

G_T = thermal conductance between the element and its surroundings

C_T = thermal capacity of the element

λ = pyroelectric coefficient

A = electrode area

G_E = electrical conductance of the element

G_N = noise conductance of the amplifier

$$\left[\frac{i_n^2}{\Delta f} = 4kT\,G_N\right]$$

R_N = noise resistance of the amplifier

$$\left[\frac{V_n^2}{\Delta_f} = 4kT\,R_N\right]$$

G = total conductance across the element

C = total capacitance across the element (including strays)

If the thermal noise contribution is disregarded and only frequencies $\omega > 1/\tau_T$ are considered we get

$$\text{n.e.p.} = \frac{xdC_P}{\eta\lambda}\sqrt{4kT\{\underbrace{G_E+G_N}_{\substack{\text{element}\\\text{noise}}}+\underbrace{R_N(G^2+\omega^2\,C^2)\}}_{\substack{\text{ampli-}\\\text{fier}\\\text{current}\\\text{noise}}\quad\substack{\text{amplifier}\\\text{voltage}\\\text{noise}}}}$$

where x = element thickness

d = density

C_p = specific heat.

The three noise contributions vary with frequency in different ways. The amplifier current limited case gives an n.e.p. independent of frequency (region A, Fig. 2). It is found that the conductance of the element varies with frequency in a similar way for most pyroelectric materials ($\sigma \propto f$); this leads to an n.e.p. proportional to $\sqrt{(\text{frequency})}$ (region B, Fig. 2). It is usual to use an f.e.t. as a high impedance amplifier input, and this device typically has a voltage noise behaviour leading to n.e.p. proportional to $\sqrt{f}$ at low frequencies and f at higher frequencies (regions B and C, Fig. 2).

3. Choice of Detector Material

It is apparent from the above that it is not possible to separate the effects of the pyroelectric material and the amplifier. Additionally, figures of merit which can be derived for different materials for each of the three types

of noise limited cases are of limited use as their individual applicability varies from frequency to frequency and detector area to detector area.

We have adopted the approach of computing the n.e.p. for a range of areas and frequencies for those pyroelectric materials for which sufficient data are available.

In order to do this, figures for the noise contribution of the amplifier were estimated. The best voltage noise achieved to date in a j-f.e.t. is $20\,\text{nV}/\sqrt{\text{Hz}}$ at 10 Hz dropping to $2\,\text{nV}/\sqrt{\text{Hz}}$ above 1 kHz. F.e.t.s with leakage currents down to $0\cdot5$ pA can also be obtained. Although these two specifications are not as yet available in one f.e.t. they have been used as it is unlikely that both will be equally important for the majority of situations and therefore one or other can be relaxed.

When the computation is carried out it is found that three materials are superior to all the others. These are triglycine sulphate, strontium barium niobate, and lithium sulphate (Figs. 3(a) to (d). In choosing which of these is likely to be most suitable for a certain application several factors must be considered. These are:

(1) bandwidth of operation,

(2) element area required,

(3) environmental conditions; e.g. temperature range.

Strontium barium niobate (SBN) is most suitable for use at low frequencies with small elements because of its high dielectric constant which leads to a relatively large voltage noise contribution from the f.e.t. Lithium sulphate is however most suitable for the larger area, higher frequency end of the range (because of its low dielectric constant and therefore a lower voltage noise contribution from the f.e.t.).

Over a large part of the frequency-area domain triglycine sulphate (TGS) is seen to give the best sensitivity. However, this material does have several disadvantages. These include a low Curie point (49°C) above which the pyroelectric effect is no longer present and, due to it being ferroelectric, a tendency to undergo partial reversal (depoling). These properties are also present in SBN, although somewhat higher Curie points can be achieved at the expense of some loss of sensitivity. Lithium sulphate has no Curie point and can be used up to higher temperatures; it is not ferroelectric and cannot therefore depole.

However, we have now developed a new material based on TGS which is completely resistant to depoling and has the further advantages of lower dielectric loss and dielectric constant.[4] Whe⋯⋯⋯⋯⋯ded to the solution from which the⋯⋯grown these crystals have their⋯⋯also their habits modified. The⋯⋯seen in the *P/E* hysteresis loo⋯seen that the doped material⋯the field axis. This behavio⋯

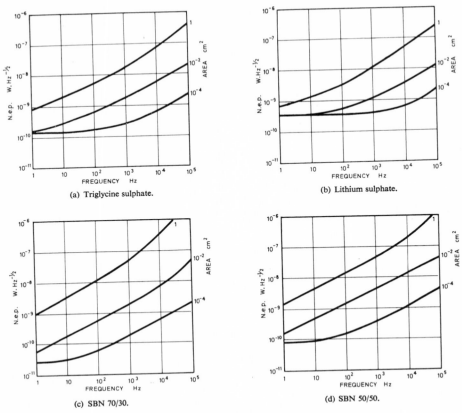

(a) Triglycine sulphate.

(b) Lithium sulphate.

(c) SBN 70/30.

(d) SBN 50/50.

Fig. 3. N.e.p. vs. frequency curves.

ferroelectric materials (e.g. colemanite[5]) and is known to arise from the presence of an internal biasing field. As can be seen from Fig. 4(b), the polarization in the absence of an external field can have only one value. Therefore, whenever the material is below its Curie point and no large external depoling fields are present, it is fully poled. This effect is permanent; repeated cycling through the Curie point, both with and without external field applied, has no effect. Detectors made from this material regain their performance rapidly on cooling to below the Curie point even after prolonged periods above it.

The lower dielectric loss of the alanine doped material (LATGS) can be seen from the measurements shown in Fig. 5. Although there are large spreads in the values, the conductivity of the doped material is about an order of magnitude lower than that of pure TGS. The dielectric constant is also lower; values of ε' down to about 15–20 have been observed at audio frequencies. As the Johnson noise due to the conductance of the pyroelectric element is the major factor limiting the performance of TGS,

this improvement in the resistivity is of considerable value. The lowering of the dielectric constant also leads to an improvement in performance, in this case at the higher frequency end of the range.

If the performance vs. frequency curve is recalculated for doped TGS and normalized for area by expressing as the specific detectivity D^*, the result is as shown in Fig. 6. This compares the predicted performance of the doped TGS pyroelectric detector with the performance found for other uncooled thermal detectors. Detectors have been made which fit this curve very well and in some cases surpass it; a 0·5 mm square element has been made which has a D^* of about $1·8 \times 10^9$ at a few hertz. It is apparent from the figure that above about 20 Hz the pyroelectric detector can give a higher sensitivity than any other uncooled thermal detector. Only the Golay cell is capable of a higher D^* and this only in the frequency range 5–20 Hz. With further developments in the fabrication of thinner elements it is probable that some further improvements are possible.

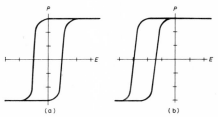

Fig. 4. Polarization *vs.* field hysteresis loops for

(a) TGS and (b) doped TGS.

x-axis; 3 kV.cm⁻¹ per division (*E*)

y-axis; 1 µC.cm⁻² per division (*P*)

To extend the possible applications we have also developed detectors with deuterated TGS elements. These have a working range up to about 56°C rather than about 45°C as for pure TGS and can also be made with non-depoling material.

4. Applications

The present state of the art on pyroelectric detectors can therefore be summarized as:

(1) $D^* > 10^9$ cm Hz$^{\frac{1}{2}}$ W⁻¹ at 10 Hz,
   $> 10^8$ cm Hz$^{\frac{1}{2}}$ W⁻¹ at 1 kHz.

(2) Detectors completely resistant to depoling.

(3) Temperature range up to 56°C.

Two fields in particular have now become accessible by virtue of these improvements. These are infra-red spectrophotometry and a variety of satellite applications.

Most infra-red spectrophotometers use either a Golay cell or a thermocouple-type detector, both of which are slow devices (5–20 Hz). The pyroelectric detector can be used as a direct replacement for either of these

devices but a further possibility also arises: that of much faster spectral scanning. An improvement of an order of magnitude in the scan rate is quite feasible. This would give a 2–15 µm spectrum in about 10 seconds. Process control uses are therefore quite possible and further increases in speed could be achieved with some loss in resolution.

Satellite instrumentation can require infra-red detection for several purposes, in particular horizon sensing and meteorological measurements such as the vertical temperature profile radiometer being developed for *Nimbus E*.[6] The prevention of depoling is of great value in these applications as alternative methods which have

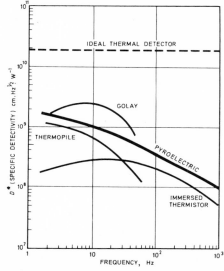

Fig. 6. D^* *vs.* frequency for thermal infra-red detectors.

been tried are less satisfactory and require more amplifier components. Another advantage of pyroelectrics is their very low power requirements (9 V, 5 mA), which is easily available in space systems.

Passive intruder alarms are another use to which the device can be put. Using a simple Cassegrainian optical system a range of detection in excess of 100 yards can be achieved for detection of persons crossing the field of view.

Laboratory applications include work with lasers, especially CO_2 (10·6 µm)[7] and HCN (337 µm). The detectors can be used for general problems such as power measurement and checking optical alignment. Bandwidths of operation out to megahertz and beyond can be simply achieved with some loss of performance or, by adding further electronic stages to the amplifier, retaining the high sensitivity if it is required. The possibility of heterodyne detection is also being investigated[8] but, ideally, lower dielectric loss materials are required for the high-frequency applications.

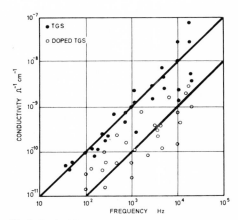

Fig. 5. Conductivity *vs.* frequency for pure and doped TGS.

In this paper we have described the advances made in pyroelectric detectors and pointed out some of their present uses. The state of the art is now such that these devices must be strongly considered on all future occasions when an uncooled thermal detector is capable of meeting a requirement.

5. Acknowledgments

The authors wish to thank the Directors of Mullard Limited and the Ministry of Defence for permission to publish this paper. They also wish to thank their colleagues who have contributed to this work. Particular thanks are due to Dr. B. E. Bartlett, Dr. F. A. Capocci, Mr. L. R. King and Mr. R. A. Lockett of Mullard Southampton, Mr. A. E. Jenkinson and Dr. E. T. Keve of Mullard Research Laboratories, and Dr. E. Putley of R.R.E., Malvern.

6. References

1. Putley, E. H., 'The pyroelectric detector', in 'Semiconductors and Semimetals', Vol. 5, pp. 259–285. Willardson, R. K. and Beer, A. C. (Ed.) (Academic Press, New York, 1970).

2. Schwarz, F. and Poole, R. R., 'Performance characteristics of a small TGS detector operated in the pyroelectric mode', *Appl. Optics*, **9**, pp. 1940–1, 1970.

3. Cooper, J., 'A fast-response pyroelectric thermal detector', *J. Sci. Instrum.*, **33**, pp. 467–72, 1962.
   Cooper, J., 'Minimum detectable power of a pyroelectric thermal receiver', *Rev. Sci. Instrum.*, **33**, p. 92, 1962.

4. Lock, P. J., 'Doped triglycine sulphate for pyroelectric applications', *Appl. Phys. Lett.*, **19**, pp. 390–1, 1971.

5. Wieder, H. H., 'Ferroelectric properties of colemanite', *J. Appl. Phys.*, **30**, pp. 1010–18, 1959.

6. Ellis, R. J., *et al.*, 'Infra-red atmospheric temperature sounding from satellites', *The Radio and Electronic Engineer*, **42**, pp. 155–61, April 1972.

7. Hadni, A., Thomas, R. and Perrin, J., 'Response of a triglycine sulphate pyroelectric detector to high frequencies (300 kHz), *J. Appl. Phys.*, **40**, pp. 2740–5, 1969.

8. Eng, T. and Gudmundsen, R. A., 'Theory of optical heterodyne detection using the pyroelectric effect', *Appl. Optics*, **3**, pp. 161–6, 1970.
   Leiba, E., 'Optical heterodyne system with pyroelectric detectors', *C.R. Acad. Sci. Paris*, B, **268**, pp. 31–3, 1969.

Manuscript first received by the Institution on 17th June 1971 and in final form on 4th April 1972. (Paper No. 1452/CC 131).

28

Reprinted from *J. Opt. Soc. Amer.*, **53**(6), 729–734 (1963)

Black Radiation Detector*

W. L. EISENMAN, R. L. BATES, AND J. D. MERRIAM

Infrared Division, U. S. Naval Ordnance Laboratory, Corona, California

(Received 30 July 1962)

A calorimeter in the form of a 15° cone has been constructed and evaluated for use as a "black" radiation detector. The internal surface of the cone is coated with colloidal graphite to enhance the absorptance. Spectral emissivity and reflectance measurements of the blackening material are used to estimate the effective absorptance of the conical receiver. To reduce many of the difficulties associated with a dc system, this detector is used with an ac amplifying system at a frequency of 1 cps. The spectral response of three commercial thermocouples is compared with that of the conical receiver.

INTRODUCTION

ONE of the most important properties of an infrared detector is the manner in which its output signal varies with the wavelength of the incident energy. This spectral-response characteristic is a primary consideration in determining the usefulness of the detector.

The radiation thermocouple has been used by most infrared laboratories as a standard detector against which the spectral characteristics of other detectors are compared. There has been a tendency among the laboratories to assume that the response of the thermocouple is constant over wide spectral regions. Though considerable effort has been spent on determining the optical characteristics of materials that may be used to enhance the absorptance of radiation receivers,[1] practically no data are available on the spectral characteristics of individual thermocouples.

The present interest in long-wavelength infrared detectors has raised a question as to the ability of thermocouples to function adequately as standard detectors in the long-wavelength region, and it therefore appeared advisable that a "black" detector be developed which could serve as a means for evaluating the current radiation standards.

Such a detector has been constructed, and a theoretical analysis of its emittance follows. Because of the complexity of a complete analysis, only two cases are presented—that of a perfectly diffusing reflecting wall and that of a perfectly specular reflecting wall. To evaluate the effective emissivity of this black detector, the theoretical results are weighted by experimental evidence. Its operating characteristics are described, and its spectral response is compared with that of several commercial thermocouples.

MAKING THE BLACK DETECTOR

The principal factor contributing to the absorptance of the receiver is its geometry; therefore, several shapes were considered. From a fabrication standpoint, the

conical cavity appeared to be the most attractive. A preliminary design requirement was that the response time of the detector be such that a modulated source could be used. (An air-conditioned laboratory with a large diurnal temperature cycle prohibits dc measurement under the signal conditions of interest.)

A feasible technique appeared to be the application of copper to a solid conical form made of material which could be dissolved, and it was also considered desirable that the fragile copper cone be mechanically supported on the form during the initial stages of fabrication. Therefore, a 15° solid conical form was machined from $\frac{1}{4}$-in. acrylic rod. The fragile tip of the form was finished by preliminary lathe-turning with fine emery paper and buffing with a counterrotating cloth wheel. Later it was found that its very sharp apex created a problem in applying the copper; consequently, a flat 0.003-in. area was left at the tip.

The polished plastic form was then placed in a vacuum evaporator in such fashion that the side of the cone could be rotated perpendicularly to a tantalum boat containing the copper. A thin coating of copper (approximately 0.5 μ) was evaporated onto the acrylic plastic form, and this evaporated copper film was then used as an electrode in an alkaline copper-plating bath. The coated cone was placed in the bath at the center of a cylindrical anode. During electroplating, the plastic form was continuously rotated to assure a uniform deposit of copper.

The copper-plated plastic cone was then mounted in a jeweler's lathe, and the copper was cut through according to the design dimensions, after which the cone was immersed in a container of trichloroethylene, which dissolved the plastic material. Finally, the remaining copper cone was rinsed in chloroform to remove any residual plastic material.

Originally, a minimum modulation frequency of 1 cps had been specified for the black detector. The electroplating technique used for depositing the copper film allowed accurate control of the film thickness. By controling the current density in the plating bath and the length of deposition time, the thermal mass of the receiver was reduced until a time constant of 0.17 sec was achieved for the response of the finished detector. A further reduction in response time could be achieved

---

* This work was performed by the Joint Service Infrared Sensitive Element Testing Program at the U. S. Naval Ordnance Laboratory, Corona, California. This program is supported by the Army, Navy, Air Force, and NASA.
[1] See R. A. Smith, F. E. Jones, R. P. Chasmar, *The Detection and Measurement of Infra-Red Radiation* (Clarendon Press, Oxford, England, 1957), pp. 89–91.

245

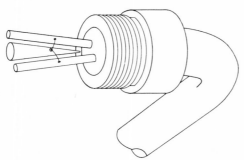

FIG. 1. Receiver assembly of the black detector. The length of the 15° conical receiver is 7.6 mm and the diameter at the base is 2 mm.

only at the expense of the mechanical integrity of the cone. However, preliminary measurements indicated that satisfactory sensitivity could be obtained when the detector was used with a narrow-band amplifier having a 1-cps center frequency.

The internal receiver surface was to be blackened by applying a material having suitable absorption characteristics throughout the spectral range of interest. Emissivity measurements were made on gold-black and on a colloidal graphite material (Aquadag) in the region of 2 to 15 μ by an emission method.[2] These materials were applied to flat polished-copper surfaces which duplicated the internal-surface texture of the cone.

The vacuum-deposition of gold-black requires particularly careful control in order to obtain coating with maximum absorptance characteristics.[3-5] Moreover, depositing the material into the sharp apex of the 15° cone was difficult. The colloidal graphite, however, could be applied readily to the inside of the cone, and it displayed adequate absorptance in the 2- to 15-μ region. Graphite material suspended in water was, therefore, applied to the cone by means of a fine camel's-hair brush.

Additional measurements were then made to determine whether the technique of applying the graphite, the thickness of the coat, or the substrate surface affected the emittance of the material. Emissivity curves for samples on different substrates, and with one, two, three, and four coats of graphite, proved to be uniform within 2%, which indicates that the emission characteristics are essentially independent of these considerations. The detector was subsequently given two coats of the graphite material, and examination of the apex of the cone after blackening indicated that the radius of curvature was approximately 0.0015 in.

[2] D. L. Stierwalt, "Infrared Spectral Emissivity of Optical Materials," NOLC Rept. 537, U. S. Naval Ordnance Laboratory, Corona, California, January 1961.
[3] L. Harris, R. T. McGinnies, and B. M. Siegel, J. Opt. Soc. Am. 38, 582 (1948).
[4] L. Harris and J. K. Beasley, J. Opt. Soc. Am. 42, 134 (1952).
[5] L. Harris, J. Opt. Soc. Am. 51, 80 (1961).

The thermocouple materials used were bismuth and bismuth with 5% tin. Although not optimum, this material had adequate sensitivity and contributed a sufficiently small thermal mass. The 1-mil thermocouple wire was soldered to heavy copper support wires with Wood's metal, by using zinc chloride as a flux. Then, in a similar manner, at a point midway between the base and the apex of the cone, the conical receiver was attached directly to the thermocouple junction.

The detector assembly was mounted inside a cylindrical brass enclosure on the end of a 3/16-in. stainless-steel tube. The tubing was fixed to a brass mounting base having the same dimensions as the base of a standard Perkin–Elmer or Reeder thermocouple. Before blackening, the mass of the conical receiver was 0.25 mg. The length of the cone was 7.6 mm and the diameter at the base was 2 mm. The completed detector, shown in Fig. 1, had a responsivity of 0.05 V/W.

EFFECTIVE ABSORPTANCE OF THE CONICAL RECEIVER

Gouffé[6] and others have analyzed the effective absorptance of the cavity of a conical receiver which has a perfectly diffusing reflecting wall and in which the

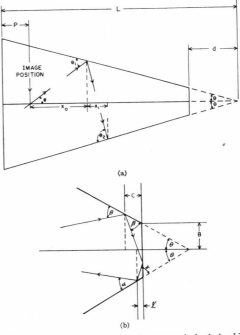

FIG. 2. Schematic drawings showing geometrical relationships between an incoming ray and the dimensions of the cone: (a) as ray travels toward apex; (b) as reflected from truncated end of cone.

[6] A. Gouffé, Rev. Optique 24, 1 (1945).

TABLE I. Number of reflections as a function of incoming angles.

ϕ^0	No. of reflections	ϕ^0	No. of reflections	ϕ^0	No. of reflections
0.0	1	4.0	11	18.0	9
0.33	1	5.0	11	20.0	9
0.50	2	6.0	11	22.5	9
1.0	4	7.0	11	24.0	9
1.5	4	8.0	11	26.0	8
2.0	6	10.0	11	28.0	8
2.5	10	12.0	10	30.0	8
3.0	11	15.0	10	32.0	8

incident energy enters in a collimated beam parallel to the axis of the cone. Although this case is important in other applications, these conditions could be met only partially by the black detector. Because of the low responsivity, it was necessary to focus the incident energy into the cone, with the result that rays entered at angles considerably off the axis. Since the surface reflection is assumed to be perfectly diffuse, this is of no great consequence. However, the small size of the particles of the graphite blackening material (approximately 2μ in diameter) indicated that a considerable amount of specular reflection could be expected at the longer wavelengths.

The other extreme case used for evaluation purposes is that of a perfectly specular reflecting wall, with the cone irradiated by a point image lying on the cone axis at a point (p) from the cone aperture. The constructed cone was truncated rather than perfect, and the incoming energy was considered to have a maximum divergence of 32° (this angle was defined by the optical system used with the black detector).

It can be seen in Fig. 2(a) that as a ray travels toward the apex of the cone we have the relation

$$\theta_n = \phi + (2n-1)\theta, \tag{1}$$

where n is the number of reflections made by the incoming ray. It can also be shown that when the ray is traveling toward the apex, the relation between the distance traveled into the cone and the number of reflections is

$$l = p + \sum_{i=0}^{n} X_i, \tag{2}$$

$$X_n = 2\left[(L-p) - \sum_{j=0}^{n-1} X_j\right]\tan\theta \Big/ [\tan\theta + \tan(\phi + 2n\theta)],$$

and

$$X_0 = (L-p)\tan\theta / (\tan\phi + \tan\theta), \tag{3}$$

where X_n is the distance traveled along the axis of the cone between (n)th and the ($n+1$)th reflection.

Where $\theta_n = \pi/2$, the ray will reverse its direction and emerge from the cone along the same path it entered. When $\theta_n > \pi/2$, the ray will reverse direction and emerge

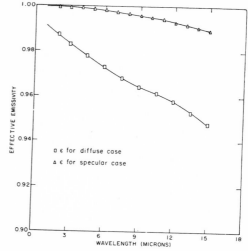

FIG. 3. The calculated spectral emissivity of the conical cavity, assuming the internal surface to be either a perfect specular reflector or a perfect diffusing reflector.

from the cone along a path different from that by which it entered the cone. Under these conditions, the relation between the number of reflections after the ray reverses direction and the axial distance traveled toward the aperture is

$$\zeta = (L-l) + \sum_{i=1}^{k} Z_i, \tag{4}$$

$$Z_n = 2\left[(L-l) + \sum_{j=1}^{k-1} Z_j\right]\tan\theta \Big/ \tan[\alpha - (2k-1)\theta] - \tan\theta,$$

and

$$Z_1 = 2(L-l)\tan\theta / [(\tan\alpha - \theta) - \tan\theta], \tag{5}$$

where $\alpha = \pi - (\theta_n + 2\theta)$. The ray will also reverse direction when $(\phi + 2n\theta) \geq \pi/2$.

If we have $\theta_n < \pi/2$ and $(\phi + 2n\theta) \geq \pi/2$, the distance traveled toward the aperture is given by

$$Z_0 = 2(L-l)\tan\theta / [\tan(\alpha+\theta) - \tan\theta]. \tag{6}$$

From the ($n+1$) reflection, Eq. (4) applies. All the Z_i are given here as positive.

If the ray reflects from the truncated end of the cone, as shown in Fig. 2(b), the axial distance from this end to the point of next reflection is given by

$$l' = B - C[\tan(\beta+\theta) - \tan\theta] / [\tan(\beta+\theta) - \tan\theta]. \tag{7}$$

From this point, Eq. (4) applies, with $\alpha = \beta$. If the ray reflects only from the truncated end, it emerges from the cone with no further reflections.

Table I, a tabulation of the number of reflections experienced by an incoming ray as a function of

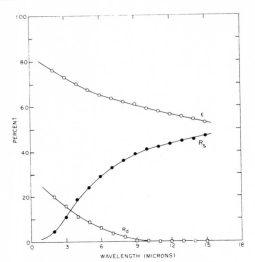

FIG. 4. Spectral emissivity and normal reflectance
of the graphite blackening material.

the entrance angle, is based on the equations given above.

The emissivity values of the blackening material and the data in Table I were used to calculate the effective emissivity of the conical receiver for the case of the perfectly specular reflecting wall, and the results were plotted in Fig. 3. For the calculation, it was assumed that any ray undergoing eight or more reflections was completely absorbed and that "p" equaled 1 mm. The effective emissivity for the case of the perfectly diffusing reflecting wall was calculated by using Gouffé's analysis. Although DeVos[7] states that Gouffé overcorrects the emissivity value, the latter's calculations were used as a conservative lower limit. The effective emissivity of the receiver for this case is also plotted as a function of wavelength in Fig. 3.

The curves for these two cases are taken as defining upper and lower limits for the effective absorptance of the conical receiver; but to estimate the absorptance of the receiver more accurately, it is necessary to know the character of the reflected energy. The spectral reflectance of the colloidal graphite was therefore measured on a Perkin–Elmer spectrophotometer, and in addition, the transmittance of the material when applied to a transparent substrate was measured. The samples exhibited no transmission in the 2- to 15-μ region. The specular reflectance and emissivity of the blackening material are shown in Fig. 4. Since the transmittance of the material was zero, the diffuse reflectance can be calculated from

$$R_d = 1 - (R_s + \epsilon). \qquad (8)$$

[7] J. C. DeVos, Physica **20**, 669 (1954); this paper contains an extensive bibliography.

The calculated diffuse reflectance is also plotted as a function of wavelength in Fig. 4. By using the diffuse and specular reflectance ratios, the calculated emissivity curves were proportionately adjusted so that a more realistic evaluation of the receiver absorptance was obtained. The adjusted emissivity of the black detector is shown in Fig. 5.

The above analysis is obviously based upon several assumptions. It was assumed that the specular absorptance of the blackening material is independent of the angle of incidence. Since the analysis is two dimensional, rays that do not cross the axis of the cone were not taken into account. Further, an isotropic energy distribution from a point source was assumed. It is difficult to analyze the extent to which the performance of the receiver was affected by these conditions. These assumptions, however, appear to have been justified by experimental evidence.

SPECTRAL RESPONSE MEASUREMENTS

The black detector was used as a unity reference at the output of a Leiss double monochromator equipped with sodium chloride prisms, and the spectral response of several commercial detectors was measured as a function of wavelength.

An optical bench constructed of aluminum plate was firmly attached to the monochromator base at the rear of the instrument. Two flat first-surface mirrors were mounted on the bench in line with the exit slit and so positioned that they reflected the energy into the condensing mirrors for the detectors at right angles. The first mirror was mounted so that it could be rotated out of the beam and thus allow the beam to be reflected by the second mirror. The optical alignment of the rotating mirror was regulated by a hardened-steel stop in such a way that the position of the mirror determined which detector would be irradiated by the energy from the monochromator.

The platinum-wire starting heater element from a Nernst glower was used as the energy source. A variable autotransformer adjusted the temperature of the source, and thus constant output energy from the monochromator over the spectral region from 1 to 14 μ was essentially maintained. This source provided nearly as much

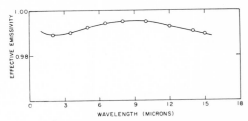

FIG. 5. Spectral emissivity of the black detector when the specular and diffuse reflectance of the blackening material is taken into account.

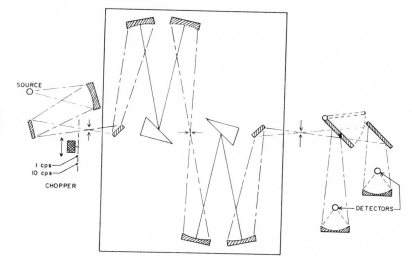

FIG. 6. Arrangement of the optical components used in the spectral-response measurements.

energy as a Nernst element and was much easier to control.

A single chopper disk, driven by a 60-rpm synchronous motor, was arranged to chop the entrance slit of the monochromator at either 1 or 10 cps. An amplifier with a bandpass of 0.5 cps and a center frequency of 1 cps was used with the black detector, and a 10-cps amplifier with a 1-cps bandpass was used for the channel of the detector undergoing evaluation. The chopper, which was mounted on a rail provided with mechanical stops to facilitate changing the chopping frequencies from the rear of the instrument, was controlled by means of a string loop. Figure 6 shows the relative positions of the optical components.

To determine experimentally the validity of the assumptions made in calculating the absorptance of the receiver, the condensing mirror for the black detector was adjusted so that the energy image fell at different positions (and was incident at different angles) within the cavity. The response of the black detector and of a radiation thermocouple were then measured as a function of wavelength for each position of the energy image. Any change in the response curve would have indicated a change in the spectral response of the black detector. Since no change in detector response was observed, it was concluded that the assumptions used in the calculations were valid within the observation error of 2%.

To measure the spectral-response curves of the commercial thermocouples, response data were taken in the 1- to 14-μ region for the black detector and the thermocouple, first in one position on the optical bench, and again with their positions reversed. Thus the relative spectral response of the thermocouple was obtained by taking the response of the black detector as unity. The results of the measurements made on two

Reeder thermocouples and one Perkin–Elmer thermocouple are shown in Fig. 7.

CONCLUSIONS

The black detector described has been used experimentally as a standard for the spectral calibration of other long-wavelength infrared detectors. The effective absorptance of the detector in the 1- to 15-μ region is

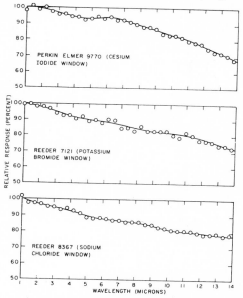

FIG. 7. Relative spectral response of three thermocouples.

constant within 2%. Such accuracy is considered satisfactory for most detector-calibration purposes.

Evaluation of the black detector in the 15- to 30-μ region has been started. Preliminary emissivity measurements of the colloidal graphite blackening material indicate that this material will not be satisfactory at 30 μ. However, recent emissivity measurements on commercial blackening materials indicate an absorption of greater than 90% out to 25 μ. No great difficulties are foreseen in applying these materials to the conical receiver. When applied to the black detector, such a material will yield an effective emissivity greater than 0.99.

The experimental data show that the relative spectral response of a commercial thermocouple may decrease as much as 30% in the 15-μ region. If the slope of the spectral-response curves remains constant, considerable loss of absorption will occur in the long-wavelength region. Since commercial thermocouples are generally used as spectral standards, significant errors may be expected in calibrations performed in the long-wavelength region if this decline in spectral response is not taken into account.

ACKNOWLEDGMENTS

It is a pleasure to thank D. L. Stierwalt and J. B. Bernstein for the spectral emittance measurements.

29

Reprinted from *J. Opt. Soc. Amer.*, **54**(10), 1280–1281 (1964)

Improved Black Radiation Detector*

W. L. EISENMAN AND R. L. BATES

*Infrared Division, Research Department, U. S. Naval Ordnance
Laboratory, Corona, California 91720*
(Received 20 April 1964)

BECAUSE of the lack of adequate standards, the assumption
has become prevalent in some infrared laboratories that the
response of thermal detectors is independent of wavelength. The
development of a black radiation detector for use as a standard
against which other infrared detectors could be evaluated was
reported last year.[1] An improved detector has since been developed
that minimizes some of the errors that existed in the original detector in the longer wavelength region.

The improved detector has an effective absorptance of 0.995 or
better out to 45 μ (as compared with 0.990 out to 15 μ in the earlier
model). Two important changes have been made: (1) Improved
fabrication techniques eliminated a 3-mil flat at the apex of the
receiver, thus forming a more nearly perfect conical cavity. (2)
Sicon,[2] a commercially available black paint, was selected as the
most promising blackening material for use in the receiver. Since
it was found that the vehicle used in the Sicon contributed little
to the black characteristics of the material but added greatly to
the mass of the paint film, the pigment was separated from the
paint by centrifuging and was then washed with acetone. A small
amount of the original vehicle was added to the centrifuged particles, which ranged in size from 2–10 μ, and this mixture was applied to the internal surface of the conical receiver with a finely
trimmed camel's-hair brush. These procedures have produced a
detector assembly that has essentially the same dimensions and
electrical characteristics as the original detector, except for the
improvements noted.

The absorptance and reflectance characteristics of the pigment–
vehicle mixture were determined in the 1–45-μ region (Fig. 1). To
analyze the effective absorptance of the conical receiver,[1] two extreme cases were considered; that of a diffusing internal reflecting
wall, and that of specular internal reflecting wall. In the specular
case, an analytical ray-tracing method was used to trace the rays
as they entered and passed down the conical cavity. The number
of reflections the ray underwent as a function of entrance angle
and the absorptance of the Sicon black were used to estimate
the effective absorptance of the cavity. This is shown as the upper
curve in Fig. 2. Gouffe[3] has analyzed the absorptance of a conical
cavity with perfect diffusing internal reflecting walls. His analysis

and the absorptance characteristics of the Sicon were used to
estimate the effective absorptance for the diffuse case, which is
shown as the bottom curve in Fig. 2. These two cases, then, may
be taken as an upper and lower limit for the effective absorptance
of the conical cavity. The ratio of the diffuse reflectance to the
specular reflectance of the Sicon was used to weight the upper and
lower limits calculated from the two cases. The solid center line in
Fig. 2 is the weighted effective absorptance curve for the Sicon-
blackened conical receiver.

Spectral response measurements out to 40 μ have been made for
several types of infrared detectors by comparing their output
as a function of wavelength to that of the black detector. The
measurements were made using a Leiss double monochromator
equipped with cesium iodide prisms and an external optical bench.
Chopping frequencies of 1 cps were employed for the black detector and 10 cps for the detector under evaluation. Spectral
responses of three commercial radiation thermocouples are given
in Fig. 3. These detectors were not selected for any particular
characteristic; they were in use in the laboratory and were thus
conveniently available for measurement.

Figure 4 shows the spectral response of a thermistor bolometer
and a Golay pneumatic detector. The spectral response curve of
the thermistor bolometer is typical of detectors of this type which

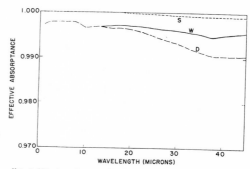

FIG. 2. Effective adsorptance of the conical receiver. S: specular case;
D: diffuse case; W: weighted case. The ratio of the diffuse reflectance to the
specular reflectance was used to weight the upper S and lower D limits.

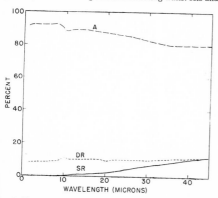

FIG. 1. Absorptance A and reflectance (DR; diffuse: Sr; specular)
of the "Sicon" pigment–vehicle mixture.

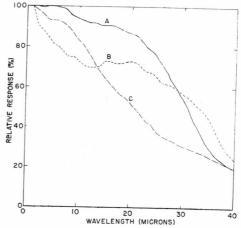

FIG. 3. Response as a function of wavelength for three commercial radiation thermocouples. A and B were equipped with KBr windows which have
some absorption at wavelengths longer than 30 μ. C had a CsI window.

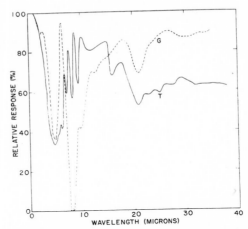

FIG. 4. Response as a function of wavelength for a Golay pneumatic detector G and a thermistor bolometer detector T.

have a fast response time. Because the response time is very dependent upon the amount of blackening material applied to the receiver's surface, it has been common practice to apply a very thin coat, or, in some cases, no blackening at all, to obtain time constants of the order of 1 msec. The Golay detector was equipped with a diamond window, which accounts for the large absorptance in the short wavelength region and the dip in response at approximately 21 μ. It can be seen that at longer wavelengths the response appears to be nearly independent of wavelength. The thermistor bolometer had a KRS-5 window.

The experimental data show that the response of commercial radiation thermocouples may be only one-third as much as 35 μ as at short wavelengths. The data also indicate considerable selectivity in the response of thermistor bolometer detectors. In general, the variations in the wavelength response are greater for bolometers having a short response time. Thermal detectors are, of necessity, used in a variety of radiometric experiments and equipment. The lack of blackness of the detectors can obviously lead to significant errors in the longer wavelength regions if this decrease in sensitivity is ignored.

* This work was performed under the Joint Services Infrared Sensitive Element Testing Program at the U. S. Naval Ordnance Laboratory, Corona, California. This program is supported by the U. S. Army, Navy, Air Force, and National Aeronautics and Space Administration.
¹ W. L. Eisenman, R. L. Bates, and J. D. Merriam, J. Opt. Soc. Am. **53**, 729 (1963).
² Midland Industrial Finishes Co., East Water Street, Waukegan, Illinois.
³ A. Gouffe, Rev. Opt. **24**, 1 (1945).

Editors' Comments on Papers 30 Through 33

30 **Andrews, Milton, and DeSorbo:** *A Fast Superconducting Bolometer*

31 **Boyle and Rodgers:** *Performance Characteristics of a New Low-Temperature Bolometer*

32 **Low:** *Low-Temperature Germanium Bolometer*

33 **Low and Hoffman:** *The Detectivity of Cryogenic Bolometers*

A study of Papers 1 and 2 shows that the thermal detectors having the highest detectivity are cooled, an observation that startles many who have mistakenly assumed that thermal detectors never need to be cooled. There are several reasons for cooling thermal detectors: cooling makes it possible to use temperature-dependent effects, such as superconductivity, that are not available at room temperature; cooling can reduce the thermal capacity of detector constructional materials and thus help to reduce the response time; and cooling in conjunction with radiation shielding can reduce the effects of background radiation. The papers in this section describe the characteristics of cooled thermal detectors.

A superconducting bolometer uses the very large change in resistance that occurs in certain metals and semiconductors during their transition from the normal to the superconducting state. Andrews, Milton, and De Sorbo, in Paper 30, describe the construction and performance of a superconducting bolometer made from a thin ribbon of columbium nitride. In order to keep the bolometer within its optimum operating range it was necessary to maintain its temperature within 0.01° of 14.36°K. Many readers will enjoy learning how this was accomplished. Additional information on superconducting detectors will be found in References 1 through 9.

Donald H. Andrews was, at the time of his death in 1973, Emeritus Distinguished Professor of Chemistry, Florida Atlantic University, Boca Raton, Florida.

In Paper 31, Boyle and Rodgers describe a bolometer that was developed for spectroscopic investigations in the far infrared. The bolometer consists of a slab cut from the core of a 56-ohm carbon composition resistor and operated at a temperature of 2.1°K. At these low temperatures, the carbon composition has an extremely large temperature coefficient of resistance and a very low specific heat. The result is a material that is very favorable for construction of a highly sensitive bolometer. Other work on cooled carbon bolometers is discussed in References 10 and 11.

W. S. Boyle is Executive Director, Pennsylvania Laboratories, Bell Telephone Laboratories, Inc., Allentown, Pennsylvania.

The performance of cooled thermal detectors has often been limited by some form of excess electrical noise. In Paper 32, Low describes a bolometer made from a single crystal of gallium-doped germanium, a particularly favorable choice since it is a material that seems to have negligible excess noise. The germanium bolometer is normally operated at a temperature of 2.1°K, although it is background-limited below 4.2°K. The performance of these bolometers continues to improve as the temperature is reduced, and they have been operated at temperatures as low as 0.37°K [12,

13]. In Paper 33, Low and Hoffman develop the theory describing the detectivity of cryogenic bolometers. The germanium bolometer, with Low showing the way, has become one of the most important detectors in astronomical research. Its high detectivity and extremely broad spectral response are ideal for astronomical investigations that may extend from the near infrared to the millimeter region of the spectrum. Additional information on germanium bolometers is contained in References 14 through 21.

Frank J. Low is a professor at the University of Arizona, Tucson, Arizona. In May 1974 he was honored by election to the National Academy of Sciences.

References

1. Martin, H. D., and D. Bloor, "The Application of Superconductivity to the Detection of Radiant Energy," *Cryogenics,* **1,** 159 (1961).
2. Andrews, D. H., R. M. Milton, and W. DeSorbo, "A Fast Superconducting Bolometer," *J. Opt. Soc. Amer.,* **36,** 353A (1946).
3. Fuson, N., "The Infra-Red Sensitivity of Superconducting Bolometers," *J. Opt. Soc. Amer.,* **38,** 845 (1948).
4. Dean, T. J., et al., "Superconducting Bolometers and Spectrometry in the Far Infra-Red," *Optica Acta,* **7,** 185 (1960).
5. Pankratov, N. A., and I. A. Khrebtov, "Conversion of the Temperature Fluctuation Spectrum of Liquid Helium into the Noise Spectrum of a Cryogenic Bolometer," *Sov. J. Opt. Tech.,* **36,** 378 (1969).
6. Pankratov, N. A., G. A. Zaytsev, and I. A. Khrebtov, "Stabilization of the Operating Temperature of a Superconducting Bolometer," *Sov. J. Opt. Tech.,* **36,** 521 (1969).
7. Pankratov, N. A., G. A. Zaitsev, and I. A. Khrebtov, "A Receiver–Amplifier System with Superconducting Bolometer," *Cryogenics,* **13,** 497 (1973).
8. Grimes, C. C., P. L. Richards, and S. Shapiro, "Josephson Effect Far Infrared Detector," *J. Appl. Phys.,* **39,** 3905 (1968).
9. Kamper, R. A., "The Josephson Effect," *IEEE Trans. Electron Devices,* **ED-16,** 840 (1969).
10. Tishchenko, T., "A Cooled Carbon Bolometer," *Cryogenics,* **11,** 404 (1971).
11. Corsi, S., et al., "Recent Advances in Carbon Bolometers as Far Infrared Detectors," *Infrared Phys.,* **13,** 253 (1973).
12. Drew, H. D., and A. J. Sievers, "A ³He-Cooled Bolometer for the Far Infrared," *Appl. Optics,* **8,** 2067 (1969).
13. Nolt, I. G., and T. Z. Martin, "An Adsorption Pumped ³He Cooled IR Detector," *Rev. Sci. Instr.,* **42,** 1031 (1971).
14. Jones, C. E., Jr., et al., "The Cooled Germanium Bolometer as a Far Infrared Detector," *Appl. Optics,* **4,** 683 (1965).
15. Zwerdling, S., R. A. Smith, and J. P. Theriault, "A Fast, High-Responsivity Bolometer Detector for the Very-Far Infrared," *Infrared Phys.,* **8,** 271 (1968).
16. Pankratov, N. A., and L. M. Vinogradova, "Calculation of Sensitivity and Noise of Cryogenic Germanium Bolometers," *Sov. J. Opt. Tech.,* **36,** 324 (1969).
17. Pankratov, N. A., and P. N. Nikiforov, "Cryogenic Germanium Bolometer Parameters as a Function of Impurity Concentration," *Sov. J. Opt. Tech.,* **36,** 797 (1969).
18. Deb, S., and M. K. Mukherjee, "Self-Cooled and Lead-Cooled Modes of Operation of a Semiconducting Thermal Detector," *Infrared Phys.,* **11,** 195 (1971).

19. Zwerdling, S., and J. P. Theriault, "Far Infrared Spectral Properties of Compensated Ge and Si," *Infrared Phys.,* **12,** 165 (1972).
20. Pankratov, N. A., and V. P. Korotkov, "High Input Resistance Tube Amplifier for Cooled Semiconductor Bolometers," *Sov. J. Opt. Tech.,* **35,** 563 (1968).
21. Zwerdling, S., and J. P. Theriault, "A Far Infrared Bolometer Pre-Amplifier with Low-Noise Performance at High Impedance," *Infrared Phys.,* **8,** 135 (1968).

30

Reprinted from *J. Opt. Soc. Amer.*, **36**(9), 518–524 (1946)

A Fast Superconducting Bolometer*

Donald H. Andrews, Robert M. Milton, and Warren DeSorbo

Chemistry Department, The Johns Hopkins University, Baltimore, Maryland

A superconducting bolometer has been developed for the fast detection of infra-red signals. It has a compound response in which the primary response time is about 5×10^{-4} sec. and the secondary response time about 5×10^{-2} sec. In the primary range a signal of $5 \times 10^{-4} \mu$ watt is equivalent to the noise of the apparatus. The frequency at which the response is down 6 db from the primary maximum value is 3000 c.p.s. and from the secondary maximum value about 140 c.p.s. The bolometer consists of a ribbon of CbN, $200 \times 10 \times 0.25$ mils, attached to the surface of a copper base-plate by means of a coating of Bakelite cement which covers the lower face of the ribbon and is about 0.1 mil thick. This unit is placed beneath a rocksalt window in the vacuum chamber of a cylindrical cryostat, $12''$ long $\times 6''$ diameter, which maintains operating temperature for about twelve hours when filled with 1000 cm³ each of liquid nitrogen, and liquid hydrogen kept at the triple point. For best operation this temperature must lie within a certain region of the superconducting transition zone at about 15°K where $dR/dT = 10$ ohm deg.$^{-1}$ and $R = 0.2$ ohm. The noise level varies with temperature in the transition zone and passes through several maxima and minima with variations in level of more than tenfold. Even at the minima it is at least ten times higher than in either the normal or superconducting states immediately above or below. For most sensitive operation, the temperature is kept constant at one of these minima by means of a precisely adjusted current which passes through a heating coil wound on the copper rod supporting the base plate. Radiation signals produce variations in resistance of the ribbon, observed by means of a bridge circuit, transformer, and amplifier.

I N the construction of receivers for the measurement of infra-red radiation, there have always been two objectives: sensitivity and speed. The nature of the measurement sometimes emphasizes sensitivity, sometimes speed, but both are always desirable. During the last few years a number of measuring devices have been built which have pushed sensitivity to the limit imposed by Brownian movement, that is, by inherent fluctuation in the instrument itself. If the sensitivity were to be increased beyond this limit, it would be necessary to lower the temperature of the essential parts of the receiving elements. Generally, such a lowering turns out to be a mixed blessing, because the reduction in thermal fluctuation is accompanied by even greater reduction in the response coefficient of the instrument, leaving one with a net decrease in sensitivity. For example, a single constantan-copper thermocouple has a response coefficient of about 40 microvolts per degree at room temperature, and this drops to about 3.7 microvolts per degree at 15°K. Over the same range the thermal fluctuation in terms of microvolts is reduced only by a factor of $4\frac{1}{2}$.

There are, however, certain phenomena for which the associated response coefficients increase markedly at low temperatures. Of these,

superconductivity is particularly suited for bolometric purposes. Thus, one finds that a coil of tantalum, with 0.7-ohm total resistance, has a response coefficient of 260 ohms per degree in the transition interval between the normal and the superconducting state.

This suggested to us the possibility of making bolometers of superior sensitivity by operating in the temperature range where superconducting transitions could be found. There were, of course, two disadvantages in such a procedure: first, the complexity of the apparatus necessary to produce and maintain a temperature near absolute zero, and, second, the difficulty in thermostating the superconducting bolometer so as to maintain it in its sensitive region within the superconducting transition zone. Nevertheless, the need for faster and more sensitive bolometers justified an investigation, and research[1] along these lines was started at The Johns Hopkins University in 1938. Before the first results[2] were published, the idea of the superconducting bolometer was suggested independently by A. Goetz.[3] The response characteristics of the first superconducting bolometer made of tantalum were so encouraging with regard to high sensitivity that a second

* Paper presented at the Winter Meeting of the Optical Society of America, Cleveland, Ohio, March 7–9, 1946.

[1] D. H. Andrews, *American Philosophical Yearbook* (1938), p. 132.
[2] D. H. Andrews, W. F. Brucksch, Jr., W. T. Ziegler, and E. R. Blanchard, Rev. Sci. Inst. **13**, 281 (1942).
[3] A. Goetz, Phys. Rev. **65**, 1270 (1939).

investigation was started in the fall of 1942 in order to see whether this high sensitivity could be coupled with high speed of response. Such bolometers are peculiarly suitable for high speed work because of the reduction in the heat capacity of all substances which takes place at low temperatures.

In planning this work we decided to try to overcome, insofar as possible, the two primary disadvantages of the superconducting bolometer, namely, the difficulty of producing low temperatures and of maintaining precisely thermostated surroundings. The discovery of the superconductivity of CbN by F. H. Horn, and subsequent studies by a group at Johns Hopkins,[4] pointed to the possibility of operating a superconducting bolometer in the region obtainable with liquid hydrogen alone, thus reducing the expense of maintaining the low temperature to perhaps $\frac{1}{10}$ that necessary to operate in the helium range. Horn and his co-workers had found that samples of powdered CbN, through changes in their magnetic properties, gave evidence of becoming superconducting in the region between 14–16°A. Horn nitrided his powdered columbium by heating in nitrogen and because of the difficulties associated with the high temperature of the process, had not prepared CbN in wire form. Following a suggestion by P.H. Emmett, we heated columbium wire in an atmosphere of ammonia and found that we were able to prepare samples which exhibited superconductivity in the range between 14–17°K. The temperature of the mid-point of the transition and the sharpness of the transition depended on the conditions of heating.

PREPARATION OF CbN

Nitriding both by ammonia and by nitrogen has been extensively studied in our laboratory during the last four years and will be reported in detail in subsequent publications, but it may be of interest to cite briefly a typical method of preparation here. A ten-inch piece of columbium wire (0.12 mm diameter) or of the ribbon rolled from it (25μ thick and 0.4 mm wide), is cut, and washed with carbon tetrachloride to remove any grease. This length is carefully folded in the

[4] F. H. Horn, W. F. Brucksch, Jr., W. T. Ziegler, and D. H. Andrews, Phys. Rev. **61**, 738 (1942).

middle to form a "V" and the two ends are attached by friction joints to electrical leads which come through the removable top of the Pyrex nitriding tube which is vertically suspended. A small weight is hung from the fold at the bottom of the V in order to keep the wire or ribbon taut. However, care must be taken not to put too much mechanical strain on the metal, as it may cause uneven heating, and as a consequence, inhomogeneous nitriding, thus affecting the shape and position of the superconducting transition.

The top of the nitriding tube with its suspended columbium wire is now fitted into the rest of the tube. A stream of ammonia, after passing through a mercury bubbler and a calcium chloride dryer, enters through the top of the nitriding tube. It passes out of the bottom and goes through a calcium chloride dryer, a safety trap, and is finally bubbled into water. When the air has been completely flushed away, a current is passed through the wire or ribbon, sufficient to raise the temperature to the desired level, 1200°–1400°C, which is maintained for 45 minutes. This gives a sample with the desired properties.

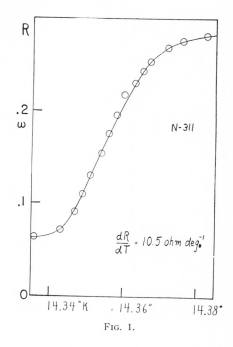

FIG. 1.

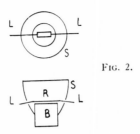

FIG. 2.

The change of resistance with temperature in the superconducting transition is shown in Fig. 1 for such a sample. The original columbium wire was obtained from Fansteel Metallurgical Corporation, Chicago, Illinois, and had specifications of impurity less than 1 percent.

MOUNTING OF CbN

The next problem was the mounting of this wire or ribbon in such a way that the radiation signal could be received and then withdrawn rapidly by thermal conduction so as to provide both a fast response and a fast restoration of initial temperature. A response time of the order of 1 millisecond was set as the objective.

The design of bolometer finally adopted is shown in Fig. 2. A piece (R) of CbN ribbon $0.5 \times 0.025 \times 0.0006$ cm is attached by Bakelite lacquer to the upper surface of a copper post (B) approximately 1 cm in diameter and 1 cm high. It is estimated that the thickness of the Bakelite lacquer between the ribbon and the post is about 0.001 cm thick. Since CbN is not wetted by solder directly, the ends of the ribbon are copper plated and tinned, and wires (LL) of No. 40 copper are soldered to them to provide electrical connections. A small half-sphere (S) spun from copper is soldered to the post and serves as a protection against mechanical shock. This is necessary because the ribbon, after nitriding, is exceedingly brittle.

BOLOMETER RESPONSE

Figure 3 shows a plot of signal response against frequency obtained with the bolometer for which the transition curve was given in Fig. 1. The form of signal used was a square-wave impulse of radiation obtained with a chopping wheel. The source of radiation was an iron plate with dulled surface, at a temperature of 128°C, the tem-

perature being measured with a calibrated thermocouple attached to the plate. The chopping wheel, which was placed about a centimeter in front of the radiation source, was made of heavy cardboard and was about 30 cm in diameter. It contained two holes located 1 cm in from the rim, diametrically opposite and 1.4 cm in diameter. The source was placed behind a hole in a plate back of the chopping wheel. Radiation from this source fell directly on the ribbon of CbN, without passing through any lens or mirror system, the only attenuation being caused by the rocksalt window. The ends of the ribbon were heavily tinned and thus not sensitive. It is estimated that the remaining area sensitive to radiation was approximately 0.01 cm². Thus, the energy falling on the bolometer was 0.08 erg per second.

The response is plotted in terms of the inches peak to peak, observed on the oscilloscope screen. In this particular observation, a current of 30 milliamperes was passed through the bolometer, which was connected to a bridge circuit and thence through a two-stage amplifier to an oscillograph, as shown in Fig. 4. The following items are included in the circuit: (R) bolometer, (D) dummy matching resistance, (BB) opposite bridge resistances, (T) transformer, (SS) secondary circuit to amplifier. A calibration indicated that a peak-to-peak voltage of 1 inch on the scope corresponded to a peak-to-peak voltage of 0.192 microvolt at the bolometer terminals. The corresponding noise level was 0.1 inch on the scope.

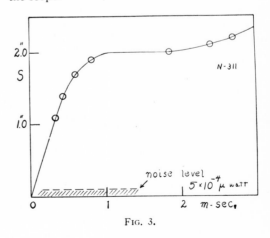

FIG. 3.

The signal rose lineally as the speed of chopping was decreased and leveled off for flashes in the range 1–2 milliseconds long. It then started to rise again. This behavior is explainable in terms of the construction of the bolometer which gives it a compound time constant. The shorter value corresponds to pulses of energy which are absorbed primarily in the CbN, but do not have time to penetrate through the layer of Bakelite into the copper. The longer time constant corresponds to a heating and cooling of parts of the upper surface layer of the copper rod. Because of the limitations of the amplifier with which we were working, we did not follow the response curve to the lower speeds of chopping.

Defining the time constant as that length of flash at which the response is equal to 0.632 of the response for a longer period sufficient to level off the curve, we see that this bolometer has a time constant of 5×10^{-4} sec. For a flash of this duration, the signal is about 15 times above the noise level. The limit of detection of the bolometer is thus about 2×10^{-6} erg. The noise level corresponds to a flux of 5×10^{-4} microwatt. This is in approximate agreement with the value of 6×10^{-4} microwatt reported by Bell, Buhl, Nielsen, and Nielsen,[5] in a study of bolometers of this type made by us.

Approximately 500 bolometers of various types were made during the course of our investigation, and detailed results will be reported in later publications. Some general comments may be of interest, however, on the effect of various factors on the behavior. For example, the thermal fluctuations were somewhat higher than might have been expected. The theoretical fluctuation level may be calculated[6] from the approximate formula $E^2 = 4kTR(f_1 - f_2)$, where k is the Boltzmann constant, T is absolute temperature, R is resistance in ohms, and $(f_1 - f_2)$ is the band width of frequency considered. Taking $R = 5$ and $(f_1 - f_2) = 1000$ we find for 300°K a value of E of 10^{-2} microvolt. At 15°K, $E = 2 \times 10^{-3}$ microvolt. The average noise level at the time of the observations shown in Fig. 3 was, however, about 10 times this value, or 2×10^{-2} microvolt.

[5] E. E. Bell, R. F. Buhl, A. H. Nielsen, and H. H. Nielsen, Abstract of Paper presented to The Optical Society of America, Cleveland Meeting, March 6–9, 1946.

[6] J. Strong, *Procedures in Experimental Physics* (Prentice Hall, Inc., New York, 1939), p. 436.

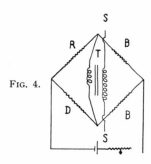

FIG. 4.

It appears that there is a special kind of fluctuation associated with superconductivity. This is illustrated in the photographs of the oscilloscope screen in Fig. 5. Figure 5A shows the noise when the element was completely normal. It is probable that this is amplifier noise. Figure 5B shows the noise at the point in the transition which yielded the best signal to noise ratio, a point lying approximately half way between the normal and the superconducting state. As may be seen, this noise is about 5 times the noise in the normal state. The noise when the columbium nitride is completely superconducting shrinks to a value lower than that observed in this part of the transition. However, over a region usually including about the upper 15 percent of the transition zone, the noise goes to much larger values, as shown in Figs. 5C, 5D, and 5E.

For values of the current through the bolometer between 5 and 50 milliamperes, the noise increases roughly proportionally to the current. Usually, in the region of around 50 milliamperes of current an instability develops in the bolometer with signs of violent fluctuation, ultimately resulting in a jump to the normal state where the bolometer remains until the current is cut off for a minute or two. This appears to be caused by the relation between resistance, current, and heat generated as discussed in the paper[2] on the tantalum superconducting bolometer. Unfortunately, at this stage of the development, there has not been sufficient time to work out in detail the relation between these factors and the shape and composition of the CbN ribbon, so that only a qualitative report can be given. The practical result is that usually a current of about 30 milliamperes gives best signal to noise ratio.

In various applications, it may be desirable

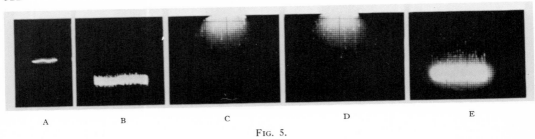

A   B   C   D   E

FIG. 5.

either to have a bolometer with as small a receiving area as possible in order to get sharp definition, or with a relatively large receiving area. For applications of the latter sort, we experimented with bolometers made with 2 to 10 ribbons laid parallel and connected in series. As a rule, the noise increases in proportion to the number of joints necessary in making such multiple ribbon bolometers. It thus appeared that some of the noise may be associated with the junction between the CbN and the lead wires. In fact, the greatest care was necessary in getting a well-adhering copper plate in order to avoid excessive noise.

With a very small current passing through the bolometer, it was found that the noise still persisted. In fact, superconductivity could be detected, even though no current were passing through the bolometer, merely by watching the changing characteristics of noise on the screen. Thus Figs. 5C, 5D, and 5E show the type of noise pattern in the upper 15 percent of the transition when no current was being passed through the bolometer at all. When these photographs were taken, the bolometer was merely coupled in series with the primary of the transformer in the amplifier, so that any fluctuating currents originating in the bolometer itself were transmitted to the amplifier, although all external sources of power were completely cut off. Under these conditions the bolometer exhibited a noise such as shown in Fig. 5A when either completely normal or completely superconducting as shown by reading of the resistance thermometer attached to the bolometer posts, and showed the fluctuation exhibited in Figs. 5C, 5D, and 5E in the upper part of the transition range. The base width of these peaks corresponded to a time interval of less than 10^{-4} second; it was impossible with the apparatus on

hand to make any accurate measurements. The height of the peaks corresponded to about 0.15 microvolt, and it appeared that an input of something like 10^{-5} erg would be required to produce such a signal. It does not appear that the cause of these fluctuations can be ascertained until further investigations are made. It is, perhaps, worth noting that bolometers of this sensitivity should respond to individual radioactive pulses.

In closing, we wish to make some brief comments on the apparatus with which the low temperature was maintained. It may be seen from Fig. 1 that in order to keep the bolometer in its sensitive range, it is necessary to maintain temperature constant to 2×10^{-2} degree and that within this range the rate of change in temperature must be relatively slow. By comparison with similar substances, we estimate that the heat capacity of CbN is 10^6 ergs degree^{-1} cm^{-3}, and that the bolometer ribbon has a heat capacity of 25 ergs degree^{-1}. Thus the minimum detectable radiation pulse of 2×10^{-6} erg yields a change in temperature of 10^{-7} degree. The bolometer is thus changing temperature at the rate of 10^{-4} degree per second when receiving this signal and fluctuation must be at a lower rate in order to avoid confusion.

This constancy of temperature was obtained with the help of a small metal cryostat made of cans of spun copper. The cryostat is shown schematically in Fig. 6. The main part of the cryostat is about 25 cm long and 12 cm in diameter. It consists of a series of nested copper cans with wall thickness 0.1 millimeter; the outer can serves as the vacuum jacket. The inner chamber, N, or "pot," contains the liquid hydrogen; this has a volume of approximately 1050 cc. A copper post, M ($\frac{3}{8}''$ O.D., $\frac{1}{4}''$ I.D., $\frac{1}{16}''$ wall tubing) soldered to the "hydrogen" pot, extends through

260

it from one end to the other and supports the superconducting bolometer base, S. Wound around the rod at M is a bolometer sensitizing coil (No. 40 constantan, approximately 500 ohms resistance). The purpose of this is to aid in controlling temperature as explained in a later section. The refrigerant is transferred into the pot through the tube, D, which is a super-nickel tube ($\frac{3}{16}''$ O.D., 0.010'' wall). The outer end of the tube has a screw cap, fitted with a rubber washer which holds the glass transfer tube through which the refrigerant is admitted. A supernickel tube ($\frac{1}{4}''$ O.D., 0.010'' wall parallel to D but not shown) acts as a vent tube while the pot is being filled; through it one evacuates the space above the surface of the liquid hydrogen to obtain temperature below the normal boiling point of hydrogen. "O" desig-nates the "nitrogen" pot having a volume of approximately 900 cm³. The purpose of this chamber is to provide a thermal dam to cut down the heat leak both by radiation and con-duction into the hydrogen pot. While in opera-tion, this contains liquid nitrogen whose normal boiling point is 77.4°K. The inlet and outlet tubes (both tubes $\frac{3}{16}''$ O.D., 0.010'' wall) for transferring the nitrogen refrigerant into this pot are similar to those on the hydrogen pot. In both cases the ends of the tubes come up to the top of the pot facing each other. Therefore, when the cryostat is almost full of refrigerant, i.e., liquid hydrogen in the hydrogen pot and liquid nitrogen in the nitrogen pot, it may be held either horizontally or vertically without spilling. A is the outer copper case which encloses the vacuum compartment surrounding both the hydrogen and nitrogen pot. Masonite spacers ($\frac{3}{16}''$ thick), L and W, are used as insulator rings between the hotter part which is at room temperature, and the colder inner pots, which in turn are similarly thermally insulated from one another. Q and R are radiation shields. These shields are made of copper sheet, and are each soldered to one of the refrigerant pots. The holes through the shields for the radiation beam are $\frac{15}{32}''$ and $\frac{13}{32}''$ in diam-eter, respectively. I is the window frame (having a $\frac{3}{4}''$ hole) which supports a rocksalt blank, G, which is about $\frac{3}{16}''$ thick and 1.0'' in diameter. A vacuum-tight connector is attached to the base of the cryostat. It consists of the metal base

of an ordinary 8-prong radio tube with the top cut off, and plate, grid, etc., removed. It is soldered into a hole in the cryostat. Eight copper leads (No. 32) in the cryostat are soldered to it. Four of the leads are connected to the bolometer (i.e., two potential and two power leads) and four corresponding ones are connected to the temperature regulator, M. Tube V (supernickel tube $\frac{1}{2}''$ O.D.) is used to evacuate the cryostat.

Two cylindrical charcoal containers (not shown) are attached to the nitrogen pot to neutralize any small vacuum leaks. A vacuum sufficient to provide good thermal insulation may be main-tained for several days without the aid of any external pumping, even though the vacuum compartment is cut off from the atmosphere only by an ordinary rubber tube and pinch clamp.

The weight of the cryostat and its dimensions are as follows:

Weight of copper cryostat, empty	10.7 lb.
Maximum diameter	5.2 in.

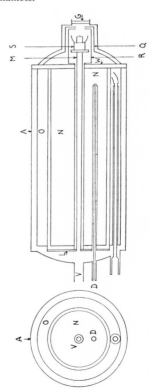

FIG. 6.

Over-all length tip of window frame to end of
   electrical plug   15 in.
Longest external tube projection (nitrogen inlet
   and exhaust tube)   1.0 in.
Length of electrical plug $1\frac{3}{8}$ in.
Weight of liquid nitrogen, 900 cc (density 0.80
   g/cc)   1.6 lb.
Weight of liquid hydrogen, 1050 cc (density
   0.07 g/cc)   0.16 lb.
Total weight of cryostat when completely filled   12.5 lb.

When the hydrogen is kept at the triple point, at 14.0°K and a current approximately 10 milliamperes is passed through the resistance coil of 500 ohms wound on a post, the bolometer is maintained in the middle of the transition zone with no need for adjustment for a period of 4 or more hours, depending on the type of operation being performed.

In order that the use of the bolometer might not be limited to the few laboratories in the country where large scale plants were available for making liquid hydrogen, we have developed a small portable liquid hydrogen unit. It consists of a liquefier of the Ahlberg Type, with a purifying unit which can be employed if the commercial hydrogen available contains impurities which may clog the liquefier. Operating from commercial tank hydrogen and blowing down the tank from 2500 to 1500 lb., it is possible to produce the liquid hydrogen necessary to charge the cryostat in about two hours. The liquefier stands about 1 meter high and is about 30×50 cm cross section, and weighs about 50 lb. Thus it can be transported readily.

Liquefier units of this type have been used at a dozen places in various parts of the country completely removed from any laboratory facilities, and have demonstrated their ruggedness and dependability over three years of operation.

Work is now under way to investigate the usefulness of superconducting bolometers in infra-red spectroscopy. Later publications will present further detail on various phases of construction and operation of the bolometer, the cryostat, and the refrigerating unit.

$\mathcal{31}$

Reprinted from *J. Opt. Soc. Amer.*, **49**(1), 66–69 (1959)

Performance Characteristics of a New Low-Temperature Bolometer

W. S. Boyle and K. F. Rodgers, Jr.
Bell Telephone Laboratories, Inc., Murray Hill, New Jersey
(Received April 16, 1957)

An infrared bolometer detector designed to operate at liquid helium temperatures has been constructed by using the core of a carbon composition resistor as the sensitive element. We have made use of the well-known fact that the temperature coefficient of resistance of certain commercial carbon resistors is extremely large at low temperatures. Since the specific heat at these temperatures is small, it has proven to be relatively simple to construct a detector with a time constant of the order of milliseconds and a signal-to-noise figure surpassing any other known thermal detector of radiation. Furthermore the resistivity of the material is such that most of the radiation striking the surface is absorbed even at frequencies in the far infrared. A brief discussion of the noise to be expected under nonequilibrium conditions is given. In practice it is found that current noise is about an order of magnitude larger than all other sources of noise.

INTRODUCTION

WE have recently been engaged in some low-temperature experiments at far infrared frequencies. As a matter of convenience we incorporated into this system a bolometer detector constructed from the core of a carbon resistor of the type that has been used extensively in low-temperature thermometry. Both the detectivity and figure of merit for this bolometer compare favorably with any other thermal infrared detector known to us. Although we have not carried out a detailed study to optimize the response character-

istics of the bolometer, we feel that in view of its performance a description is warranted at this time.

The rather unusual resistivity properties of carbon composition resistors have been described by Clement and Quinnell.[1] They give the following expression for the resistance as a function of temperature, $\log R + K/\log R = A + B/T$, where K, A, and B are parameters which vary with the room temperature value of the resistance,

[1] J. R. Clement and E. H. Quinnell, *Proceedings of the International Conference on Low Temperature Physics* (Oxford University Press, New York, 1951).

and T is the absolute temperature. In general, the parameter B, which determines the temperature coefficient of resistance in the helium range, increases with the room temperature value of the resistance. Typical values of these parameters for a 56-ohm resistor at room temperature are such that at $2°K$; $(1/R)dR/dT = \alpha = -2$. This large value of α, coupled with the low specific heat at these temperatures, enables one to construct a bolometer with an extremely high sensitivity; but not necessarily, of course, a correspondingly large signal to noise figure.

EXPERIMENTAL DETAILS

The bolometer was a flat slab cut from the core of a 56-ohm carbon resistor and then mounted directly on a copper backing, which was in good thermal contact with the helium bath. The bolometer was attached to the copper block with a 0.001-in. sheet of mylar interposed to provide electrical insulation. Formex was used to cement all connecting surfaces. When operating in the far infrared, the filtering system shown in Fig. 1 was used. This effectively removed all stray radiation below 40 microns.

The thermal relaxation time was measured and found to be independent of temperature in the helium range. In order to maximize the detectivity at the chopper frequency (13 cy/sec), the required relaxation time must be approximately 0.010 sec. This was achieved by using a slab 0.019 in. thick. Temperature fluctuations in the helium bath were at one point the main source of noise in the bolometer. This noise was effectively removed by pumping below the lambda point and also inserting an acoustic filter in series with the pumping line.

A measure of the sensitivity of the bolometer was obtained in the far infrared by using a Perkin Elmer spectrometer modified (in a similar fashion to McCubbin

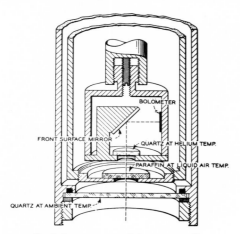

FIG. 1. Low-temperature bolometer assembly.

TABLE I. Operating characteristics of the bolometer detector.

Area		0.03 in.^{2}
Thickness		0.019 in.
Time constant		0.01 sec
Responsivity		
Measured	1.4×10^4 volts/watt.	
Calculated	1.8×10^4 volts/watt.	
Resistance		0.12×10^6 ohms
Noise for one cps band width		
Measured 1.6×10^{-7} volts.		
Photon $(10°K)$ 3×10^{-10} volts.		
Johnson 3.5×10^{-9} volts.		
Temperature 2×10^{-9} volts.		
Figure of merit		29
Noise equivalent signal		6×10^{-12} watts
(Band width $\frac{1}{3}$ cy/sec)		

and Lord[2]) to operate in this region of the spectrum. The energy flux from the spectrometer was measured with a standard calibrated thermocouple. Corrections for the absorption of the filtering system were applied to obtain the sensitivity figures quoted below.

When the helium bath temperature fluctuations were removed, the main source of noise in the bolometer was current noise. This was found in general to arise from the contacts. We have found that contacts made to the ends of the bolometer with an indium gallium amalgam gave the lowest noise figure. At this point it is difficult to say if the remaining current noise is inherent to the material or still coming from the contact area.

CALCULATED NOISE FIGURE

Of the fundamental sources of noise, Johnson noise is most readily calculated, and the value shown in Table I is obtained at the operating point.

Photon noise and temperature noise are quite distinct in our case since thermal contact is established by both conduction and radiation. An exact evaluation of photon noise would be difficult, since it would involve an accurate knowledge of the wavelength transmission characteristics of the filtering system. We may safely assume that the incident photon flux is much less than that corresponding to liquid air temperatures. Only by way of reference, therefore, we give in Table I the photon noise at $10°K$.

The usual expression for temperature fluctuation is not rigorously valid under nonequilibrium conditions. This is particularly troublesome at low temperatures where the difference in temperature between the two bodies may be comparable with the absolute temperature. We will present now a straightforward analysis of temperature or phonon noise in terms of the usual expressions for shot noise. This will give the temperature or power noise spectrum in both equilibrium and nonequilibrium conditions.

We shall use the following model for our calculations. The bodies, A and B, are connected by a thermally conducting path, which has zero specific heat and a

[2] R. C. Lord, T. K. McCubbin, J. Opt. Soc. Am. 47, 689 (1957).

thermal conductivity K. When a certain power, P, is being fed into A, the steady-state temperatures of A and B are T_1 and T_2, respectively.

We will assume that we are well below the Debye temperature and that the phonons make no collisions over the connecting path between the two bodies. If the transit time of a phonon from A to B is small compared to the period of the noise spectrum that we are concerned with, then the departure or arrival of a phonon can be taken as an instantaneous event. It is therefore permissible to use an analogous expression to shot noise in a vacuum tube to calculate the spectrum of fluctuations in power transfer. The expression for shot noise is $\omega_i(f) = 2e\bar{I}$ where e is the electronic charge, $\bar{I}$ the average current, and $\omega_i(f)$ is the square of the amplitude of the Fourier component of the fluctuation in current at a frequency f, or as it is usually called, the noise spectrum of the current fluctuations.

The density of phonons $N(\nu)$ in a frequency interval $d\nu$ is given by

$$N(\nu) = \frac{12\pi\nu^2}{c^3[\exp(h\nu/kT) - 1]} \quad (1)$$

where c is the appropriate mean velocity of the longitudinal and transverse modes, ν is the phonon frequency, h, k, and T have their usual significance.

A certain fraction δ of the phonons in A will leave per second. The noise spectrum in power leaving A for a part of the phonon spectrum $d\nu$ will be, $w_{1p}(f\nu) = 2N(\nu)\delta(h\nu)^2 d\nu$. Using the value of $N(\nu)$ given by (1) and extending the integration over ν to ∞ the total noise in power is given by

$$w_{1p}(f) \doteqdot \frac{24\pi\delta k^5 T_1^5 (4!)}{c^3 h^3}.$$

With a similar expression for $\omega_{2p}(f)$ and summing we obtain the total noise power as

$$w_p(f) \doteqdot \frac{24\pi\delta k^5 (4!)(T_1^5 + T_2^5)}{c^3 h^3}. \quad (2)$$

The parameter δ can be evaluated in terms of the heat transfer coefficient for a small temperature difference between A and B. The heat transfer W is given by

$$W = \int_0^\infty \frac{12\delta\pi\nu^2 h\nu d\nu}{c^3[\exp(h\nu/kT_1) - 1]}$$
$$- \int_0^\infty \frac{12\delta\pi\nu^2 h\nu d\nu}{c^3[\exp(h\nu/kT_2) - 1]} \doteqdot \frac{12\delta\pi k^4(3!)(T_1^4 - T_2^4)}{c^3 h^3};$$

for small temperature differences this reduces to

$$W = \frac{48\delta\pi k^4(3!)T^3\Delta T}{c^3 h^3}. \quad (3)$$

The expression for the noise $w_p(f)$ can be expressed in terms of a thermal conductivity $K = W/\Delta T$ and we find

$$w_p(f) = \frac{2kK(T_1^5 + T_2^5)}{T^3}. \quad (4)$$

We note that this expression should hold for any values of T_1 and T_2 provided we use a thermal conductivity which is appropriate to the temperature T and a small ΔT, even though a thermal conductivity cannot in general be defined.

The corresponding temperature fluctuation spectrum is readily evaluated. For some frequency, f, the amplitude of the temperature response to an harmonic input power, p, is $t = p/K[1 + (\omega\tau)^2]^{\frac{1}{2}}$, where C is the thermal capacity of A, assumed to be small compared to B, and τ is the relaxation time $= C/K$. The noise spectrum in temperature fluctuation is given therefore by

$$w_t(f) = \frac{w_p}{K^2[1 + (\omega\tau)^2]^2}$$
$$= \frac{2k(T_1^5 + T_2^5)}{K(1 + \omega^2\tau^2)^2 T^2}.$$

The foregoing expression reduces to the usual equilibrium one when $T_1 = T_2$.

We have made a crude measurement of the thermal conductivity coefficient, K, from values at the operating point of dP/dR and dR/dT. From these we estimate K to be 36 μw/°K. This sets the rms power for a 1-cycle band width at 10^{-13} watts for the operating point.

DISCUSSION

From the load line shown in Fig. 2, which was obtained at the operating temperature of 2.1°K, one may readily calculate the responsivity.

The responsivity,[3] S, in volts per watt at a frequency

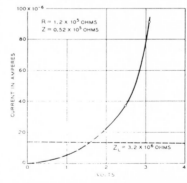

FIG. 2. Load line for the bolometer at 2.1°K.

[3] R. Clark Jones, J. Opt. Soc. Am. 43, 1 (1953).

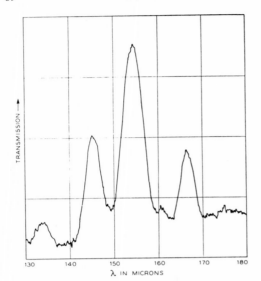

FIG. 3. Atmospheric absorption spectrum in the region of 150 microns.

f, is given by

$$S = \frac{Z_L(Z-R)}{2IR(Z+Z_L)(1+\omega^2\tau^2)^{\frac{1}{2}}},\qquad(5)$$

where Z_L is the load resistance, R the bolometer resistance, Z the incremental bolometer resistance at the operating point, and I the current through the bolometer.

This expression, when now applied to the responsivity to incident illumination, is not valid for a solid-backed bolometer where the temperature is not uniform throughout the element. In general, (5) will be smaller than the true responsivity (by just a factor of 2 if the temperature gradient is uniform throughout the detector). In the case discussed here, the current is not uniform and tends to flow on the side of the bolometer away from the heat sink. This sets the correction factor somewhat below 2 and, to be conservative, we use the responsivity calculated from (5) directly.

A comparison of this bolometer with other infrared detectors can be made in terms of the noise equivalent signal. As Jones[4] has pointed out, however, this is rather meaningless unless the operating conditions, i.e., band width, area, and time constant, are similar. We remark, therefore, only in passing, that the radiation power input, which is equal to the rms noise for a $\frac{1}{3}$-cycle band width, is 6×10^{-12} watts.

A more significant comparison can be made in terms of an expression introduced by Jones which he calls a figure of merit. The figure of merit applies to a general class of detectors in which the detectivity has the same functional dependence on the area, band width, and response time. In particular, he defines his figure of merit as $M_2 = 2.12 \times 10^{-10} A^{\frac{1}{2}} f^{\frac{1}{2}} D_{1M}(f)$, where A is the area of the detector, f the operating frequency, and $D_{1M}(f)$ the maximum detectivity in a one cycle band width. We refer the reader to Jones' article for further details and give here only a few typical values of figures of merit. The values are as follows: present bolometer ($M_2 = 29$), Golay cell ($M_2 = 2$), Perkin Elmer thermocouple ($M_2 = 0.8$).

The fraction of the incident energy absorbed by the bolometer should be large. The resistivity of 900 ohm cm at the operating temperature is such that the reflection coefficient will be determined almost entirely by the dielectric constant of the material. This resistivity is nevertheless large enough to set the optical absorption constant in the region of 0.01 cm^{-1}. Except in the very far infrared, when λ approaches the thickness of the bolometer, the efficiency of absorption should be high. The agreement between the calculated and measured responsivity is a confirmation of this.

We show, in Fig. 3, a spectrum of atmospheric absorption taken in the region of 150 microns with the modified Perkin Elmer instrument. These measurements were taken with a 100 lines per inch grating blazed at 27°. We believe that this exhibits more explicitly than the figures we have quoted above the performance characteristics of the detector.

ACKNOWLEDGMENT

The authors wish to acknowledge valuable discussion of the noise calculations with G. Wannier.

[4] R. Clark Jones, *Advances in Electronics* (Academic Press, Inc., New York, 1953), Vol. V, pp. 36 to 47.

32

Reprinted from *J. Opt. Soc. Amer.*, **51**(11), 1300–1304 (1961)

Low-Temperature Germanium Bolometer

FRANK J. LOW
Texas Instruments Incorporated, Dallas, Texas
(Received March 29, 1961)

A bolometer, using gallium-doped single crystal germanium as the temperature-sensitive resistive element, has been constructed and operated at 2°K with a noise equivalent power of 5×10^{-13} w and a time constant of 400 μsec. Sensitivities approaching the limits set by thermodynamics have been achieved, and it is shown that the background radiation limited or BLIP condition can be satisfied or 4.2°K. An approximate theory is developed which describes the performance of the device and aids in the design of bolometers with specific properties. The calculated noise equivalent power at 0.5°K, for a time constant of 10^{-3} sec, is 10^{-15} w. The detector is suitable for use in both infrared and microwave applications.

INTRODUCTION

THE advantages inherent in the operation of thermal detectors at low temperatures have led to the development of a number of cryogenic bolometers.[1-4] The sensitivities of these detectors have, in general, been limited by some form of excess electrical noise rather than by unavoidable thermal fluctuations.[5] By using single-crystal gallium-doped germanium[6] for bolometry in the liquid-helium temperature range, sensitivities have been obtained close to the theoretical limit. The time constant of the germanium bolometer at 2°K is variable from less than 10^{-5} sec to many seconds, and the responsivity can exceed 10^5 volts/watt.

Excess electrical noise in the *p*-type germanium used in the bolometer has been measured below 4.2°K and above 20 cps. It is negligible at current levels comparable to the required bias currents. The inherent detector noise, measured at small apertures, is divided almost equally between Johnson noise and phonon noise. Photon noise may predominate at larger apertures, depending upon the temperature of the background. For an aperture of 180° and a background temperature of 300°K, it is shown that the inherent detector noise at 4.2°K is much less than photon noise, thus the background limit can be achieved. By cooling the detector below 4.2°K the background limit can be achieved for even smaller apertures and lower background temperatures. The rapid response of the germanium bolometer has been obtained by utilizing the low thermal capacity and high thermal conductivity of germanium at low temperatures.

The theory which describes the performance of room-temperature bolometers[7,8] can be applied with certain modifications to the present low temperature device. The observed properties of several germanium bolometers agree well with the calculations.

EXPERIMENTAL DETAILS

Figure 1 shows the essential features of the experimental arrangement. The bolometer elements were cut from a suitably doped single crystal, and were lapped and etched to the desired thickness. Because of the piezoresistance of germanium at low temperature, it is necessary to mount the element in a strain-free manner. The element is supported in its evacuated housing by the two electrical leads. These leads provide the only appreciable thermal contact with the bath. This method of construction allows the thermal conductance to be varied several decades by changing the diameter, length, and composition of the leads. The thermal conductivity of the germanium is sufficiently high so that the temperature at the center of the element remains very nearly equal to that at the ends.

Absolute noise measurements were carried out with a low-noise, vacuum-tube amplifier,[9] a wave analyzer with

[1] D. H. Andrews, R. M. Milton, and W. de Sorbo, J. Opt. Soc. Am. **36**, 518 (1946).
[2] N. Fuson, J. Opt. Soc. Am. **38**, 845 (1948).
[3] J. A. Hulbert and G. O. Jones, Proc. Phys. Soc. (London) **B68**, 801 (1955).
[4] W. S. Boyle and K. F. Rodgers, J. Opt. Soc. Am. **49**, 66 (1959).
[5] The ultimate sensitivity of radiation detectors and their associated noise sources have been discussed by: R. Clark Jones, J. Opt. Soc. Am. **37**, 879 (1947), P. B. Fellgett, *ibid.* **39**, 970 (1949), and by the authors of reference 7.
[6] The resistance-versus-temperature characteristics of germanium doped with indium, arsenic and gallium have proven to be useful for resistance thermometry over the interval 1° to 40°K. See: I. Estermann, Phys. Rev. **78**, 83 (1950). S. A. Friedberg, *ibid.* **82**, 764 (1951). J. E. Kunzler, T. H. Geballe, and G. W. Hull, Rev. Sci. Instr. **28**, 96 (1957). F. J. Low, *Advances in Cryogenic Engineering* (Plenum Press, Inc., New York, 1961), Vol. 7.

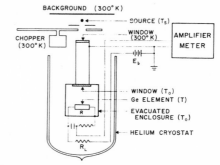

FIG. 1. Schematic diagram showing the arrangement of the apparatus and the temperatures of the various parts.

[7] R. A. Smith, F. E. Jones, and R. P. Chasmar, *The Detection and Measurement of Infra-Red Radiation* (Oxford University Press, New York, 1957).
[8] R. Clark Jones, J. Opt. Soc. Am. **43**, 1 (1953).
[9] J. J. Brophy, Rev. Sci. Instr. **26**, 1076 (1955).

an effective-noise bandwidth of 5.75 cps, and a true-rms voltmeter. The load resistor, grid resistor, and blocking capacitor were placed in the cryostat to reduce Johnson noise.

Blackening the surface of one of the bolometers allowed its responsivity to be measured optically. The 500°K blackbody radiation was admitted to the low-temperature, evacuated housing through a cooled sapphire window. The blackening was necessary for two reasons: (1) so that the emissivity could be taken to be approximately unity and (2) to avoid an intrinsic negative photoconductive effect which was observed. Because of this photoconductive effect, illumination with visible light causes an increase in resistance of the partially compensated gallium-doped germanium when it is at helium temperature. After the intrinsic radiation is removed the resistance returns to its original value in a time less than 10^{-3} sec.

THEORY

The following empirical relation for the resistance versus temperature of the bolometric material has been deduced from measurements between 1.1° and 4.2°K,

$$R(T) = R_0 (T_0/T)^A, \qquad (1)$$

where R_0 is the resistance at temperature T_0. The constant A is approximately equal to 4 but must be determined for each bolometer. The temperature coefficient of resistance α can thus be written as

$$\alpha(T) \equiv 1/R(dR/dT) = -A/T. \qquad (2)$$

Using these approximate relations, expressions will be developed for the responsivity S, the time constant τ, and the noise equivalent power NEP.

The following definitions will be employed:

R_L = load resistance
R = electrical resistance of bolometer element
E = voltage across the bolometer element
I = current through the bolometer element
$P = EI$ = electrical power dissipated in the bolometer element
T = temperature of the bolometer element
T_0 = temperature of the bath or heat sink
$\phi = T/T_0$ = reduced temperature
T_s = temperature of the radiation source
a = area of bolometer element
t = thickness of bolometer element
$v = at$ = volume of bolometer element
α = temperature coefficient of resistance
Q = background radiation power incident upon bolometer
ΔQ = applied radiation signal
ω = angular frequency of applied signal
$W = P + Q + \Delta Q$ = total power dissipated in bolometer
C = thermal capacity of bolometer element
G = thermal conductance between bolometer element and bath

τ = response time constant
τ' = C/G = thermal time constant
S = dE/dQ = responsivity
Z = dE/dI = dynamic resistance
NEP   = signal which produces unity signal-to-noise ratio for unity bandwidth
M_2 = Jones' figure of merit[10,11]
D^* = $a^{\frac{1}{2}}/$NEP = specific detectivity[11]

Referring to Fig. 1, we see that the incident radiation consists of two parts: the constant background Q and the alternating signal ΔQ. Although Q is generally much greater than ΔQ, the cooled window may, in some cases, be used as a filter to remove most of the background radiation so that $\Delta Q \cong Q$ at the bolometer. ΔQ is always much less than P. If $Q \gg P$ the bolometer cannot function in a useful manner, which leaves only the two cases $Q \ll P$ and $Q \cong P$.

With no signal applied, we have under steady-state conditions a balance between the total power dissipated in the bolometer and the power flowing to the bath,

$$G(T - T_0) = P + Q. \qquad (3)$$

If Q is not small compared to P we can define a temperature T_0' given by $G(T_0' - T_0) = Q$ so that

$$G(T - T_0') = P. \qquad (4)$$

The result is a degradation in performance arising from the higher effective bath temperature T_0'. The larger G the smaller the effect. If, on the other hand, Q is small compared to P, $T_0' \sim T_0$, and we can proceed with the usual small-signal analysis based on the steady state equation,

$$G(T - T_0) = P. \qquad (5)$$

We will assume throughout that G is independent of T, although not independent of T_0, and for simplicity take the load resistance to be very large, $R_L \gg R$.

The usual expressions[7,8] for the low-frequency responsivity can be written as

$$S = \alpha E/(G - \alpha P). \qquad (6)$$

For a bolometer with a single time constant,

$$\tau = C/(G - \alpha P). \qquad (7)$$

The parametric equations for the load curve are

$$E(T) = [GR(T - T_0)]^{\frac{1}{2}}, \qquad (8)$$

$$I(T) = [GR^{-1}(T - T_0)]^{\frac{1}{2}}. \qquad (9)$$

Substituting for E and using Eq. (5), S can be written as an explicit function of T,

$$S(T, T_0) = \frac{\alpha(T)(T - T_0)^{\frac{1}{2}}}{1 - \alpha(T)(T - T_0)} \left[\frac{R(T)}{G(T_0)} \right]^{\frac{1}{2}}. \qquad (10)$$

[10] R. Clark Jones, *Advances in Electronics* (Academic Press Inc., New York, 1953), Vol. V, p. 1.
[11] R. Clark Jones, Proc. IRE **47**, 1495 (1959).

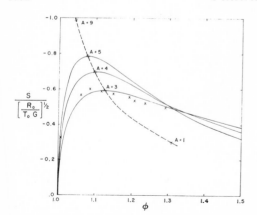

FIG. 2. Equation (12) is plotted for three values of A. The dashed line shows the position of the maximum for other values of A. Experimental points are shown for a bolometer with $A = 3.85$, $G = 183$ μwatt/$^\circ$K, $R_0 = 2 \times 10^4 \Omega$, and $T_0 = 2.15^\circ$K.

The response time constant becomes

$$\tau = \tau'/[1 - \alpha(T)(T - T_0)]. \qquad (11)$$

Introducing expressions (1) and (2) we have, where $\phi = T/T_0$,

$$S(\phi) = -\left\{\frac{A^2(\phi-1)}{[(A+1)\phi - A]^2\phi^A}\right\}^{\frac{1}{2}}\left[\frac{R_0}{T_0 G(T_0)}\right]^{\frac{1}{2}}. \qquad (12)$$

For a given bath temperature T_0, S has a maximum value which depends only on the constant A. In Fig. 2, Eq. (12) is plotted for three values of A; the dashed line shows the position of the maximum for other values of A. For $A = 4$, the optimum value of ϕ is 1.1, giving,

$$S_{max} = -0.7(R_0/T_0 G)^{\frac{1}{2}}, \qquad (13)$$

and

$$\tau = \tau'\phi/[\phi + A(\phi-1)] = 0.73\tau'. \qquad (14)$$

From Eq. (5), the optimum value of P is $0.1T_0 G$.

The Johnson noise and the phonon noise can be compared by using this approximate expression for S_{max}.

Equivalent Johnson noise power

$$= (4kTR)^{\frac{1}{2}}/S_{max} \cong 1.4(4kT_0^2 G)^{\frac{1}{2}}. \qquad (15)$$

Phonon noise power $= (4kT_0^2 G)^{\frac{1}{2}}$. $\qquad (16)$

The NEP, in the absence of photon noise, is the sum of these two nearly identical noise powers,

$$\text{NEP} \cong 4T_0(kG)^{\frac{1}{2}}. \qquad (17)$$

Thus the inherent noise equivalent power depends only on the bath temperature T_0, the value of G, and Boltzmann's constant k. Since G is the thermal conductance of the leads, the NEP is independent of the dimensions of the detector element. This has the important result that, unlike most detectors, the NEP does not vary as

the square root of the area. Therefore, the specific detectivity D^* cannot be used as a valid means of comparison with other detectors. For a given bath temperature and NEP, Eq. (17) gives the appropriate value of G. The area and thickness can be varied separately to obtain the minimum time constant.

The thermal time constant τ' can be calculated from the dimensions of the element using the measured value of G and the following relation for the thermal capacity[12]

$$C = 6.8 \times 10^{-6} T^3 v \text{ joules}/^\circ\text{K}. \qquad (18)$$

τ is then given by Eq. (14). τ can be measured by observing $S(\omega)$ versus ω. Using the measured load curve $R(T)$ and Eq. (5), the thermal conductance can be obtained with good accuracy. To determine the responsivity S at a given operating point, it is necessary to obtain R, Z, and E from the load curve. Jones[8] has given the following useful expression for S,

$$S = (Z - R)/2E. \qquad (19)$$

If the thickness is fixed, the product of NEP and $\tau^{\frac{1}{2}}$ does vary as the square root of the area. Therefore, using Jones' criteria,[10] the germanium bolometer qualifies as a class II detector. Jones' figure of merit for class II detectors is defined as

$$M_2 \equiv 2.12 \times 10^{-10} (af)^{\frac{1}{2}}(\text{NEP})^{-1}, \qquad (20)$$

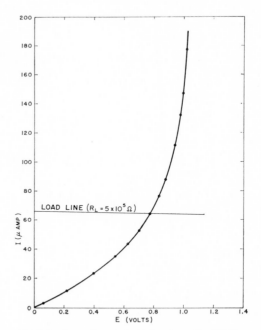

FIG. 3. Load curve for a typical bolometer at $T_0 = 2.15^\circ$K, with load line showing optimum operating point.

[12] P. H. Keesom and N. Pearlman, Phys. Rev. 91, 1347 (1957).

where $f = \omega/2\pi = 1/2\pi\tau$, and a is the area of the bolometer element in square centimeters. M_2 is dimensionless and independent of G and a. The following relation can be obtained by means of Eqs. (17) and (18),

$$M_2 \propto [tT_0{}^5]^{-\frac{1}{4}}. \tag{21}$$

In the small signal analysis given above the assumption was made that the background radiation Q is small compared to P. This is not the case if the detector views a hot background through a large aperture. Under these conditions the figure of merit is no longer independent of the magnitude of Q. It has already been noted that the constant heating produced by the background radiation is equivalent to an increase in the bath temperature T_0 to a higher value $T_0{}'$. Applying Eq. (21), the reduced value of the figure of merit $M_2{}'$ can be expressed as

$$M_2{}' = M_2 \left[\frac{GT_0}{Q + GT_0} \right]^{\frac{1}{4}}. \tag{22}$$

It is therefore possible to maintain a high figure of merit in the presence of large amounts of incident radiation by increasing the thermal contact with the bath. As G is increased the time constant decreases and the NEP increases.

If the detector noise is smaller than the photon noise generated by background radiation, then the detector is said to satisfy the BLIP condition.[13] Petritz[14] has pointed out that for room temperature thermal detectors the BLIP condition is approached more closely by increasing the time constant. It is interesting that when a thermal detector is operated with its heat sink at a temperature much lower than the temperature of the background, the BLIP condition can be satisfied by decreasing the time constant.

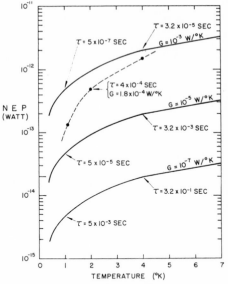

Fig. 4. Temperature variation of NEP given by Eq. (17). Measured values are included for a typical bolometer. The values of τ were calculated from Eqs. (14) and (21) by using $t = 10\,\mu$ and $a = 0.1$ cm^2.

EXPERIMENTAL RESULTS

By means of Eq. (20), the responsivity for different bias currents was determined from the load curve shown in Fig. 3. These data are plotted in Fig. 2 for comparison with the theoretical curves given by Eq. (12). The resistance versus temperature was measured at a power level below 10^{-7} w, and the results were fitted to Eq. (1) by the method of least squares, yielding $A = 3.85$ with a correlation coefficient of 0.9997. G was calculated from the load curve and the $R(T)$ data.

As a check on the electrical measurements, the responsivity of one bolometer was measured optically using a 500°K blackbody source. The agreement between the optical and electrical data was within the estimated error of about 20%. Under no conditions did the total incident power exceed 10^{-6} w.

Noise voltage measurements were carried out at 200 cps with an absolute accuracy of 10%. The integrating time of the rms voltmeter was sufficient to allow relative measurements to be made with an accuracy of 3%. With the amplifier input shorted, the rms noise voltage for a 1-cps bandwidth was 9.7×10^{-9} v. Since no change was observed when the bolometer was connected to the input of the amplifier, the total bolometer noise must have been less than 2×10^{-9} v. Excess current-dependent noise was observable only at higher currents and lower frequencies.

The characteristics of a typical Ge bolometer are summarized in Table I. Similar data are included for

TABLE I. Characteristics of a typical germanium bolometer and the carbon bolometer of Boyle and Rodgers.[4]

	Ge bolometer	C bolometer
T_0 (°K)	2.15	2.1
a (cm^2)	0.15	0.20
t (cm)	0.012	0.0076
R_L (Ω)	5.0×10^5	3.2×10^6
R (Ω)	1.20×10^4	1.2×10^6
Z (Ω)	5.00×10^3	5.2×10^4
S (volt/watt)	4.5×10^3	2.1×10^4
ϕ	0.12	0.13
τ (μsec)	400	1×10^4
f (cps)	200	13
G (μwatt/°K)	183	~36
Noise (volt)	2×10^{-9}	1.6×10^{-7}
Measured NEP (watt)	5×10^{-13}	1×10^{-11}
Calculated NEP (watt)	4.3×10^{-13}	1×10^{-13}
M_2	2.3×10^3	29
$D^* \left(\dfrac{\text{cm (cps)}^{\frac{1}{2}}}{\text{watt}} \right)$	8×10^{11}	4.5×10^{10}

[13] E. Burstein and G. S. Picus, "Background limited infrared detection," paper presented at IRIS, February 3, 1958.
[14] R. L. Petritz, Proc. IRE 47, 1458 (1959).

the carbon bolometer of Boyle and Rodgers.[4] Values for D^* are included but have limited significance. A more valid comparison with other detectors can be made by using Jones' figure of merit M_2.

Equation (17) gives the temperature variation of the NEP in the absence of photon noise. The smooth curves in Fig. 4 show this variation for three constant values of G. The experimental point at 2.15°K was taken from Table I. The other two points are for the same bolometer but, because of the thermal conductivity of the leads, the values of G are different. The values shown for τ were calculated for $t = 10$ μ and $a = 0.1$ cm².

By increasing R_0, responsivities greater than 10^5 volts/watt have been obtained at 2°K. By decreasing the thickness it has been possible to shorten the time constant to 10^{-5} sec without decreasing the area.

DISCUSSION

The germanium bolometer, by virtue of its low noise and fast response, possesses a figure of merit at 2°K two decades greater than any other thermal detector and three decades greater than the Golay cell. Sensitivity can be traded for speed to achieve optimum performance in given applications.

The BLIP condition for an aperture of 180° and a background temperature of 300°K is satisfied at 4.2°K, the boiling point of liquid helium, if the value of G is 10^{-3} watt/deg. For an area of 0.1 cm², Q is 1.5×10^{-3} watt and T_0' equals 5.70°K. The inherent detector-noise power taken from Fig. 4 is 2.7×10^{-12} w. This is to be compared with the room temperature radiation noise equivalent power of 1.8×10^{-11} w. The time constant under these conditions would be 50×10^{-6} sec.

In applications where radiation noise can be eliminated, there is much to be gained by operating at the lowest possible temperature. At 0.5°K, the practical lower limit for existing cryostats, a figure of merit of 10^6 should be attainable. Figure 4 indicates that at this temperature a time constant of 10^{-3} second and a noise equivalent power of 10^{-15} w would result if G were chosen to be 10^{-7} watt/deg., and if current noise should remain unimportant. For these conditions the electrical power P is 5×10^{-9} w, a value within the limits set by the cooling capacity of cryostats using liquid He³. These arguments indicate that it is possible to construct a thermal detector comparable in sensitivity to the photomultiplier and the radio receiver but capable of efficient operation in all parts of the spectrum.

ACKNOWLEDGMENTS

Valuable assistance in constructing and testing the bolometer was given by O. Hendricks and R. Stinedurf. Thanks are also due to Dr. G. K. Walters, Dr. J. Ross Macdonald, and Dr. R. L. Petritz, for their helpful discussions and suggestions concerning this work.

Reprinted from *Appl. Optics*, **2**(6), 649–650 (1963)

The Detectivity of Cryogenic Bolometers

F. J. Low and A. R. Hoffman

Texas Instruments Incorporated, Dallas 22, Texas. F. J. Low is now at National Radio Astronomy Observatory, Greenbank, West Virginia.
Received 25 September 1962.

The performance characteristics of germanium bolometers operating below 4.2°K have been reported previously.[1] It was shown experimentally that the limiting source of noise in these bolometers is the temperature fluctuation caused by the random flow of heat between the sensing element and the heat sink. In certain cases, however, this noise may be less important than the noise generated by random fluctuations in the intensity of incident or reradiated radiation. The magnitude of this radiation noise is calculated for cases of practical interest.

It is convenient to summarize all major contributions to the noise in a bolometer by writing the square of the noise equivalent power,* N.E.P., as the sum of five independent terms:

$$(\text{N.E.P.})^2 = \frac{4kTR}{S^2} + 4kT^2G + 8\epsilon\sigma kAT^5$$
$$+ 8\epsilon\sigma kAT_b^5 \sin^2\left(\frac{\phi}{2}\right) + \frac{Ci^\alpha}{f^\beta S^2}. \quad (1)$$

The first term is the square of the Johnson noise voltage across the bolometer of resistance R at temperature T, divided by the square of the responsivity, S. The Boltzmann constant is denoted by k. The second term is the square of the temperature noise power, where G† is the thermal conductance between the heat sink and sensing element. The third term arises from random fluctuations in the emission of the sensing element. This element, of area A, is assumed to radiate at all wavelengths with an average emissivity ϵ, and σ is the Stefan–Boltzmann constant. The fourth term represents the noise in the incident radiation originating from a blackbody background at temperature T_b. The detector is taken to be Lambertian, and the angle ϕ defines the size of the cone which accepts the background radiation. Furthermore, it is assumed that the emission and absorption of radiation are independent, that is, no stimulated emission is present. The last term is included to indicate the possibility of current noise of indefinite magnitude determined by the constant C. For our purposes we will exclude all excess noise from further consideration.

The term "ideal" bolometer refers to that case where S is very large and G is very small, leaving only the third and fourth terms of Eq. (1) as significant. In practice, however, the thermal conductance is never made negligibly small, so that the second term is always larger than the third and fourth terms when T is comparable to T_b. When T is much lower than T_b the third term vanishes and the background radiation noise can become domi-

---

* In the definition of N.E.P. and throughout this paper, we take the post-detection bandwidth to be unity.

† G refers to the flow of heat from the element strictly by conduction.

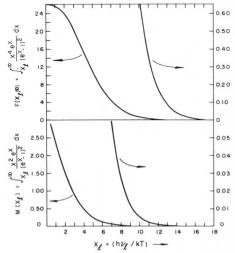

Fig. 1. $F(x_l, \infty)$ and $M(x_l)$ are both plotted as a function of x_l.

nant. We prefer the expression "perfect" bolometer for this latter case since the radiation power is carried away to the heat sink by a noiseless conductive heat current,* introducing no additional noise and leaving only the inherent temperature noise of the bolometer and the background radiation noise.

Using the results derived previously for the particular case of the germanium bolometer,[1] an approximate expression may be written for the N.E.P. of a nearly "perfect" bolometer†:

$$(\text{N.E.P.})^2 = 16kT^2G + 8\epsilon\sigma kAT_b^5 \sin^2\left(\frac{\phi}{2}\right). \quad (2)$$

Examination of this equation shows the conditions which are necessary for the two terms on the right to be of equal magnitude. The optimum bias power, P, has been shown[1] to be 0.1 TG, whereas the total radiation power, Q, is $\epsilon\sigma AT_b^4 \sin^2 \phi/2$. Therefore Eq. (2) may be written as,

$$(\text{N.E.P.})^2 = 8kT\left[20P + \left(\frac{T_b}{T}\right)Q\right]. \quad (3)$$

---

* The assumption here is that temperature noise is independent of the magnitude of the conductive heat current producing the temperature gradient between the sensing element and the heat sink.

† This result assumes that the temperature of the sensing element is only slightly greater than that of the heat sink. This is certainly true in most practical cases.

For cases of practical interest, (T_b/T) may be 100, and the bolometer becomes background limited when Q exceeds $P/5$.

Up to now we have only considered a flat spectral response for the bolometer. In practice, it is usually necessary to limit the spectral bandwidth by means of cooled filters. Assuming ideal filters, we will calculate the variation of specific detectivity, D^*, for a "perfect" thermal detector as a function of both short and long wavelength cutoffs, λ_s and λ_l, respectively.

Following Lewis[2] and Fellgett[3], we may write the exact expression for the noise associated with the radiation power received by the bolometer:

$$(\text{N.E.P.})^2 = \frac{4\pi A \epsilon (kT_b)^5 \sin^2\left(\dfrac{\phi}{2}\right)}{c^2 h^3} \int_{x_l}^{x_s} \frac{x^4 e^x}{(e^x - 1)^2}\, dx, \quad (4)$$

where $x \equiv (hc/\lambda kT_b)$. For a hemispherical field of view the specific detectivity is

$$D^*(x_l, x_s) \equiv \frac{A^{1/2}}{\text{N.E.P.}}$$

$$= 2.04 \times 10^{17}[\epsilon T_b{}^5 F(x_l, x_s)]^{-1/2}\left(\frac{\text{cm} - \text{cps}^{1/2}}{\text{watt}}\right), \quad (5)$$

where

$$F x_l, (x_s) = \int_{x_l}^{x_s} \frac{x^4 e^x}{(e^x - 1)^2}\, dx.^*$$

A plot of $F(x_l, \infty)$ is shown in Fig. 1. For arbitrary x_s and x_l, we may write $F(x_l, x_s) = F(x_l, \infty) - F(x_s, \infty)$.

The corresponding expression for a perfect photoconductor[3] is

$$D^*(x_l) = 3.00 \times 10^{27}[\epsilon\nu_l{}^2 T_b{}^3 M(x_l)]^{-1/2}\left(\frac{\text{cm} - \text{cps}^{1/2}}{\text{watt}}\right), \quad (6)$$

where $\nu_l = c/\lambda_l$, and

$$M(x_l) = \int_{x_l}^{\infty} \frac{x^2 e^x}{(e^x - 1)^2}\, dx.$$

A plot of this function is also shown in Fig. 1. In Fig. 2 we present a plot of D^* for a "perfect" cryogenic bolometer as a function of both short and long wavelength cutoffs, λ_s and λ_l.[†] The well-known curve for a perfect photoconductive detector as a function of long wave cutoff is included for comparison. The

---

* The transport integrals used here are conveniently tabulated in Natl. Bur. Std. (U.S.), Circ. No. 595.

† ϵ is assumed to be 1.

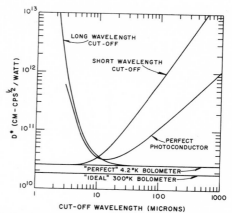

Fig. 2. The detectivity of a "perfect" bolometer is plotted as a function of both short and long wavelength cutoffs. Plots for a perfect photoconductor and two other cases are included for comparison. The background temperature is 300°K.

horizontal lines for a "perfect" bolometer with no cutoffs and an "ideal" 300°K bolometer are also included for comparison.

An example of increasing interest is that of a cryogenic bolometer operating at long wavelengths. For this case, the Rayleigh–Jeans approximation holds, and Eq. (5) reduces to

$$D_{R-J}{}^*(x_l, x_s)$$

$$= 2.04 \times 10^{17}(\epsilon T_b{}^2)^{-1/2}\left[\frac{\lambda_s{}^3}{1 - (\lambda_s/\lambda_l)^3}\right]^{1/2}\left(\frac{\text{cm} = \text{cps}^{1/2}}{\text{watt}}\right). \quad (7)$$

As λ_l/λ_s becomes large the power removed by the long wavelength filter becomes very small and the detectivity is independent of λ_l. On the other hand, large values of D^* are made possible by the use of cooled narrow-band filters, for which $\lambda_s/\lambda_l \cong 1$.

Finally, the variation of D^* with background temperature, T_b, can be shown to be identical for photoconductive and thermal detectors in the Rayleigh–Jeans limit. This is not the case at shorter wavelengths.

References

1. F. J. Low, J. Opt. Soc. Am. **51**, 1300 (1961).
2. W. B. Lewis, Proc. Phys. Soc. (London) **59**, 34 (1947).
3. P. B. Fellgett, J. Opt. Soc. Am. **39**, 970 (1949).

V
Ultimate Limits on the
Performance of Detectors

Editors' Comments on Papers 34 Through 39

The preceding papers have presented a wealth of information on the characteristics of a variety of infrared detectors. The performance of every one of these detectors will be limited by statistical fluctuations in some physical process. These fluctuations include, for instance, those associated with the photons in the incident signal flux, with the photons in the flux from the background, and with various noises arising in the detector itself. This group of papers delves deeply into these limits and provides a firm theoretical basis for the practical estimates of ultimate detector performance that were given in earlier papers. Paper 34 by Golay is, perhaps, more timely today than when it was written in 1952. It was written, from the viewpoint of a spectroscopist, at a time when there were no coherent sources available for use in the infrared. It discusses the penalty imposed by having to work with incoherent sources. Golay notes that the minimum detectable power using coherent radio detection techniques is about 4×10^{-20} watt while with an ideal thermal detector and incoherent radiation it is about 2.5×10^{-12} watt, or about eight orders of magnitude worse. Since Golay wrote this paper the laser has made the infrared coherent source a reality. A glance at the measurements reported in Papers 7, 8, and 25 shows that Golay's predictions have already been verified.

Marcel J. E. Golay is Senior Scientist, Perkin-Elmer Corporation, Norwalk, Connecticut, and is currently residing in Lutry, Switzerland.

Many of R. Clark Jones' immeasurably important contributions to the art of describing the performance of a detector are summarized in Papers 35, 36 and 37. In Paper 35, Jones derives an upper limit for the sensitivity (despite his strong feelings, expressed in Paper 3, against the use of the word sensitivity) of any radiation detector as well as expressions for the minimum detectable power. The results hold for any detector of size large compared with that of the radiation being detected and obeying Lambert's Law (interpreted in terms of a quantum efficiency for photon detectors). In the Papers 36 and 37, Jones gives the results of a study in which he introduces universally applicable means for describing the detectivity of any detector. He shows that all detectors can be placed in one of two classes, on the basis of the way that detectivity can be exchanged for speed of response. Figures of merit are introduced, one for Class I detectors and another for Class II detectors. Values of the figure of merit are computed for 49 thermocouples and one Golay cell, all of which were previously described in the literature. Values of the figure of merit for some more recent detectors are given in Section 5 of Paper 3. Two nearly book-length review articles by Jones provide further insight into his work [1, 2] and an excellent bibliography of the detector art up to 1959. In addition, see References 3 through 5.

Nudelman, in Paper 38, gives a review and an analysis of the factors that determine the radiation-detecting ability of semiconductor detectors. Signal and noise generation mechanisms are examined individually and then in terms of their contribution to D*, noise equivalent power, and Jones' figures of merit. Nudelman is to be commended for the lucid manner in which he shows how the characteristics of the newer detectors fit so well within the conceptual framework established by Jones (Papers 35 through 37). Other pertinent material will be found in References 6 through 9.

S. Nudelman is Professor of Radiology and Optical Sciences, University of Arizona, Tucson, Arizona.

Jacobs and Sargent, in Paper 39, have extended previous calculations of photon-noise-limited D* versus cutoff wavelength to include background temperatures as low as 1°K and wavelengths as long as 10,000 μm. Their results will be of increasing value for deep-space and far-infrared system applications. For additional material on this subject, see References 10 and 11.

S. F. Jacobs is a professor at the Optical Sciences Center, University of Arizona, Tucson, Arizona.

References

1. Jones, R. C., "Performance of Detectors for Visible and Infrared Radiation," in Marton, L. (ed.), *Advances in Electronics,* Vol. 5, Academic Press, New York (1953).
2. Jones, R. C., "Quantum Efficiency of Detectors for Visible and Infrared Radiation," in Marton, L. (ed.), *Advances in Electronics and Electron Physics,* Vol. 11, Academic Press, New York (1959).
3. Jones, R. C., "Energy Detectable by Radiation Detectors," *J. Opt. Soc. Amer.,* **50,** 883 (1960).
4. Fellgett, P. B., "On the Ultimate Sensitivity and Practical Performance of Radiation Detectors," *J. Opt. Soc. Amer.,* **39,** 970 (1949).
5. La Roche, U., "Entropy Efficiency of Detectors," *Infrared Phys.,* **13,** 273 (1973).
6. Long, D., "Generation–Recombination Noise Limited Detectivities of Impurity and Intrinsic Photoconductive 8–14 μ Infrared Detectors," *Infrared Phys.,* **7,** 121 (1967).
7. Williams, R. L., "Sensitivity Limits of 0.1 eV Intrinsic Photoconductors," *Infrared Phys.,* **8,** 337 (1968).
8. Williams, R. L., "Speed and Sensitivity Limitations of Extrinsic Photoconductors," *Infrared Phys.,* **9,** 37 (1969).
9. Borrello, S., "Detection Uncertainty," *Infrared Phys.,* **12,** 267 (1972).
10. Dereniak, E. L., and W. L. Wolfe, "A Comparison of the Theoretical Operation of High-Impedance and Low-Impedance Detectors," *Appl. Optics,* **9,** 2441 (1970).
11. Emmons, R. B., "The Frequency Response of Extrinsic Photoconductors with Reduced Background," *Infrared Phys.,* **10,** 63 (1970).

$$\mathcal{34}$$

Reprinted from *Proc. IRE*, **40**(10), 1161–1165 (1952)

Bridges Across the Infrared-Radio Gap*

MARCEL J. E. GOLAY†

Summary—The wide gap which exists between the generation and detection of coherent radiation on one hand, and the generation and detection of submillimeter incoherent radiation on the other hand, is discussed from the standpoint of spectroscopy.

While the resolving powers possible with microwave spectroscopy can be indefinitely extended with refinements in technique, the resolving powers possible with infrared spectroscopy are subject to an upper limit determined by the size of the external optics utilized. In the far infrared at wavelengths greater than 10 microns the resolving powers obtained in the past have been subject to further instrumental limitations, but the attainment of the optically allowable resolving powers appears possible now with recent developments in infrared spectroscopic instrumentation.

Introduction

IN THE PAST TEN years the radio engineer has succeeded in decreasing the wavelength of coherently generated radiation from around 10 cm down to a little over 1·mm. This hundred-fold extension of his mastery over the radio spectrum has been accompanied by the spectacular technical advances in communication and radar with which we are familiar. However, as we shall proceed to still shorter wavelengths, it is safe to predict that because of the increasing opacity of the atmosphere to such radiations, our interest in these will become more purely scientific.

We designate by the name of "microwave spectroscopy" the vigorous new science which has been made possible by the recent extension of radio techniques to the cm and mm spectral regions. Molecular-absorption spectroscopy forms one important branch of this new science, and represents the one scientific endeavor which is shared by the microwave spectroscopist and the infrared spectroscopist. Both want to know something about the frequencies at which molecules resonate, and, in order to find out, both must generate radiation of a known frequency and measure it after passage through an absorption cell filled with these molecules.

In the 6-mm wavelength region, where we know both how to generate and heterodyne coherent or cw radiation, the powers and sensitivities available for microwave absorption studies are overwhelmingly adequate.

In the region around 1 mm we know only how to generate pulsed, narrow-banded radiation; the microwave spectroscopist working in this region has borrowed the detector of the infrared spectroscopist until recently,

and is only now learning how to detect this radiation electrically.

At wavelengths less than 1 mm the molecular spectroscopist relies entirely on far infrared sources, on optical elements such as gratings, prisms, or interferometers, and on far infrared detectors. It may be illustrative to make an estimate of the number of orders of magnitude which separate radio generators from infrared generators and radio detectors from infrared detectors.

The power generated within a frequency band Δf by an infrared source at the absolute temperature T and having an area approximately $\frac{1}{2}$ wavelength square, which can be called a point source, is given by the expression

$$W = kT\Delta f. \tag{1}$$

For the purpose of comparison, assume a klystron generating 1 watt of coherent 3-cm radiation. Associate with this radiation a bandwidth determined by the uncertainty in our relative knowledge of this frequency with respect to a former measurement. One cps can be taken as a reachable figure for Δf if an excellent "flywheel," such as a quartz frequency standard, is utilized with loving care, as it should be in certain phases of microwave spectroscopy. Substituting these values for W and Δf in (1) and solving for T, we obtain the temperature needed for the same emission by an infrared point source. This temperature is $T = 7 \times 10^{22}$°C. Such a temperature does not exist any place in the world and is some 19 orders of magnitude greater than any practical laboratory temperature.

So much for generation. The infrared detection picture is equally rough. In the case of coherent radio detection, the least detectable power ΔW is also given by the second member of (1), except for the noise factor of the detector which should multiply this second member. Assuming 1 second for our measurement, or $\Delta f = 1$, and a detector at room temperature with a noise factor of 10, we obtain

$$\Delta W = 4 \times 10^{-20} \text{ watts.}$$

On the other hand, far infrared detectors are the so-called thermal detectors, and their greatest possible sensitivity can be calculated as though they were broadbanded radio receivers sensitive in the entire frequency spectrum. In this case we cannot use the formula above, which would give us infinity when Δf is infinite. The quantum conditions must be observed for the short

* Decimal classification: R111.2×535. Original manuscript received by the Institute, December 5, 1951; revised manuscript received, June 11, 1952.

† Signal Corps Engineering Laboratories, Fort Monmouth, N. J.

wavelengths of the spectrum, and when this is done we obtain for the square of the radiation fluctuations to and from the detector,

$$\overline{\Delta W^2} = 16 k k_1 T^5 S \Delta f = 4 k_1 T^4 S \cdot 4 k T \Delta f, \qquad (2)$$

where k_1 is the Stefan-Boltzmann radiation constant and S is the sensitive area of the detector. An analogy can be noted between this expression and the expression for the shot noise, $\overline{\Delta i^2} = 2 \, i \, e \Delta f$. The radiation flow to and from the detector, $k_1 T^4 S$, is analogous to the current i, and the analogue to the charge of the electron is $4 k T$, which represents, therefore, a measure of the average grain size in the quantized stream of radiation to *and* from the detector. At room temperature, for 1-mm wavelength for which we can assume a half-mm square for S, and for an observation time of 1 second or $\Delta f = 1$, the square root of this expression gives the smallest power detectable by an ideal thermal detector

$$\Delta W = 2.5 \times 10^{-12} \text{ watts.}$$

When this value is compared with the value obtained above for the case of radio detection, thermal detection is seen to be eight orders of magnitude less sensitive than radio detection. The presence of ΔW^2 as a square term implies ignorance of the nature of the radiation detected, and the radio spectroscopist who must resort to thermal detection of his high-quality coherent or nearly coherent radiation is rather in the position of a violinist who takes his Stradivarius to the pawnshop, to be given an estimate of its value based on the BTU content of the violin's wood. Yet, at submillimeter wavelengths, it may be the best he can do.[1]

With these premises, it will be assumed that the radio spectroscopist has a two-fold interest in infrared techniques. When he can generate, but not detect coherent radiation, he has a direct interest in knowing what sensitivity he can expect from practical thermal detectors. In the shorter wavelength region, in which he can neither generate nor detect coherent radiation, he has a physicist's interest in learning what the infrared spectroscopist can accomplish in the way of advance scouting of the spectrum in the region below 1-mm wavelength. Accordingly, some developments in the thermal detection of radio waves, and some other developments in purely infrared spectroscopic instrumentation will be reviewed in what follows.

Discussion

The two best known infrared detectors, the thermopile and the bolometer, come within one and a half orders of magnitude of reaching the sensitivity limit given by (2). However, the full realization of such a practical

[1] The concept of coherence or incoherence of electromagnetic radiation is utilized qualitatively only in this article, but mention should be made of the quantitative treatment of coherence which Gabor gives in his discussion, "Communication theory and physics," *Phil. Mag.*, vol. 41, pp. 1161–87; November, 1950, in which he utilizes the number of quanta in a unit cell of the frequency-time space as a measure of coherence.

sensitivity is predicated upon their being effectively blackened for complete absorption of incoming radiation. What looks like a good black in the visible region could be a very poor black in the 10-micron region. Actually, the effective blackening of the detectors in the 10-micron region is not difficult to accomplish. On the other hand, when these blacks must be deposited on a metallic sheet, as is the case for the thermopile and the bolometer, the condition that the electrical vector of the radiation be nearly zero at the surface of this sheet entails a corresponding decrease in the effective absorptivity of the black deposits for increasing wavelengths.

This difficulty does not exist in the pneumatic radiation detector, which operates on the old principle of the gas thermometer and comes within a half order of magnitude of the ultimate sensitivity permitted by (2).

The essential element of this detector is the radiation absorber which is placed in the center of a gas chamber. This radiation absorber consists of a broadbanded radio antenna in the form of a metallic sheet. The resistance of any square of this sheet matches approximately the parallel connected resistance of free space on both sides of it. This resistance of space, a full-fledged constant of nature, just like the speed of light, the charge of the electron, or Planck's constant, has the value $4\pi.30 = 377$ ohms, and its value is implied in the second Maxwell equation. Therefore, a metallic sheet of 188.5 ohms plays a role quite similar to that of a lumped resistance $R_0/2$ placed across a transmission line with a surge impedance R_0; one half of the energy reaching it will be absorbed; while a quarter each will be transmitted and reflected. (When this half resistance of free space is multiplied by the square of the electronic change, the quantity obtained is an action numerically equal to Planck's constant divided by 137, the famous number which, to this day, has been a major challenge to the theoretical physicists. It is noteworthy that the speed of light is *not* involved in this important relationship.)

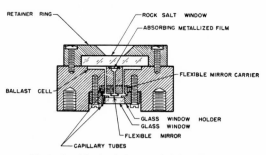

Fig. 1—Receiving head of pneumatic infrared detector.

Fig. 1 illustrates the principal elements of the pneumatic detector. The broadbanded radio antenna described above is the metallized film in the center of the small gas cell. It is heated by incoming radiation; the gas of the cell becomes heated, expands, and this expansion of the gas deflects the small flexible mirror, which has a surface tension of the order of that of a water bubble. For the detection of mm radio waves the large rocksalt window is replaced by a 0.1-mm fused quartz window, which permits a waveguide to approach the detecting antenna, encountering only a minimum of radiation leakage.

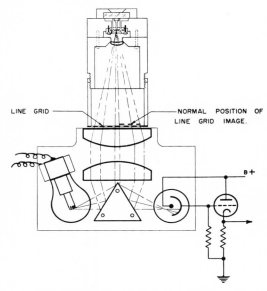

LINE GRID — NORMAL POSITION OF LINE GRID IMAGE.

B +

Fig. 2—General assembly of pneumatic infrared detector.

Fig. 2 illustrates the Schlieren-like optical system by means of which the deflections of the flexible mirror are converted into variations of light on a photocell. This optical detecting system permits the delivery of considerably more energy at the photocell output than is needed to cause the deflection of the flexible mirror, and therefore constitutes a preamplifier free of the flicker of thermionic tubes. Thus, the noise present in the photocell output represents the brownian motion of the flexible mirror, and can be verified to nearly disappear when the grid is displaced to a position of insensitivity, the residual noise being the shot noise of the photocell.

Because of the tight thermopneumatic coupling of this flexible mirror with the broadbanded antenna in the gas cell, a good part of the brownian noise of the flexible mirror reflects the fluctuations of the radiative interchange between this antenna and the background to which it is exposed. This is what permits this detector to approach within one-half order of magnitude the ultimate limit of broadbanded detection. In practice, it is used in connection with the so-called radiation chop-

ping method, and the ac signals obtained are amplified, rectified, filtered, and recorded.[2]

While this detector was developed for infrared applications, the use of a broadbanded antenna suggested that it would be equally sensitive to mm radio waves. This possibility was verified for wavelengths up to 3 mm by Townes,[3] who has realized with it a sensitivity of 10^{-11} watts for his molecular absorption studies with the $1\frac{1}{2}$-mm wavelength harmonics of a pulsed magnetron. Such weak powers cause deflections of the flexible mirror which are of the order of 20 trillionths of an inch.

Since Townes' original experiments at 1.5 mm, it has been found that these harmonics of a pulsed magnetron,[4] and even higher harmonics down to 1.1-mm wavelength,[5] can be detected with more sensitivity with a silicon detector, because advantage can be taken of the small duty factor of the magnetron in order to gate the detector during the emitting period only, and thus obtain a 30- to 33-db lower noise than for full-time sensitivity. For the detection of yet shorter radio waves, or of considerably longer trains of pulsed mm waves, thermal detection will probably remain a useful method while awaiting improvements in electrical detection.

The detector just described constitutes, therefore, a kind of infrared-radio link which the microwave spectroscopist can utilize when he works at the short wavelength frontier of the radio spectrum, where he can generate coherent or narrowbanded (pulsed) radiation, but cannot detect it electrically.

Were it possible to effect the complete thermal insulation of a tuned dipole, or of an array of tuned dipoles, so as to restrict the energy interchange between these dipoles and the radiative background to the spectral region to which they are tuned, and were some means available to measure sensitively the temperature of these dipoles, the arrangement so postulated would constitute a thermal detector *not* limited by the fluctuations of (2), but limited by the smaller fluctuations of energy interchange within the tuned spectral range. Unfortunately, considerable physical difficulties attend the realization of such tuned circuits, because of the high-frequency surface resistivity of available conductors, which does not vanish even at superconductive temperatures. Fine-wire bolometers with a linear resistance of the order of the impedance of space per wavelength might permit escape from the limitations of broadbanded detectors given by (2), but when it is considered that these wires are poorly coupled with the impedance of space on account of their inductance, that even if

[2] M. J. E. Golay, "Theoretical considerations in heat and infrared detection," *Rev. Sci. Inst.*, vol. 18, p. 347; 1947; "A pneumatic infrared detector," *ibid.*, p. 357; "The theoretical and practical sensitivity of the pneumatic infrared detector," *ibid.*, vol. 20, p. 816; 1949.

[3] J. H. N. Loubser and C. H. Townes, "Spectroscopy between 1.5 and 2 mm wavelength using magnetron harmonics," *Phys. Rev.*, vol. 76, p. 178; 1949.

[4] J. H. N. Loubser and J. A. Klein, "Absorption of mm waves in nd₃," *Phys. Rev.*, vol. 78, p. 348; 1950.

[5] J. A. Klein, J. H. N. Loubser, A. H. Nethercot, and C. H. Townes, "Magnetron harmonics at millimeter wavelengths," *Rev. Sci. Instr.*, vol. 23, p. 78; 1952.

placed in a good vacuum they have a thermal conductive path to ground through the terminating leads, and that bolometers have an inherently poor detecting sensitivity,[6] little appears to be gained in this direction.

Let us now examine what the infrared spectroscopist can accomplish with his over-all instrumentations, either with his classical instrumentation, or by means of recent developments and possible future developments. It will be clear, of course, that there is no instrumental long wavelength limit to the infrared technique, in the sense that there is an instrumental short wavelength limit to the radio technique. There is no point in working only with the infrared technique where the radio technique can be made to work.

It should be said at the outset that the comparison made before between the orders of magnitude involved in the generation and detection of radiation by the radio and infrared methods was slightly unfair to the infrared methods. The infrared spectroscopist has, in the past, done better than might be inferred from the foregoing, and there is room for further improvements. This is because the strength of the radio technique implies a weakness which the infrared technique does not have. Coherent generation requires that the essentially active elements of radio sources and detectors be not larger than, for example, a half wavelength, and also that they be monochromatic. That is, from an optical viewpoint, they must be point sources of a single-line spectrum and point detectors sensitive within a spectral band made as narrow as desired. As this limitation does not apply to infrared sources and detectors, the infrared spectroscopist can improve his lot by a few precious orders of magnitude, and he can do this in three steps, the benefits of which are, fortunately, cumulative.

The first step, which was taken almost intuitively at the birth of spectroscopy, consists of building spectrographs with elongated entrance slits instead of pinhole entrances, and the use of elongated entrance and exit slits in infrared monochromators followed as a matter of course. Thus, the radiative outputs of infrared monochromators are some two precious orders of magnitude higher than if pinholes were utilized, and this has permitted realizing up to the 10-micron wavelength the resolving powers inherent in the optical elements of these instruments, which is determined by the difference of path of the extreme rays in the collimated bundle.

The second step permitted the infrared spectroscopist consists in replacing both the single entrance and exit slits of his monochromator by two arrays of n slits each. With proper slit spacing, radiation of a specified wavelength—controlled by the position of the optical elements—will go in and out pairs of corresponding entrance and exit slits which have the same ordinal number in their respective arrays. Without any further precautions, the n-fold radiative output thus obtained for the specified wavelength range would be contami-

[6] Ref. 2, *loc. cit.*, pp. 351–352.

nated by the radiation of other wavelengths passing in and out noncorresponding slits; but this difficulty can be resolved by modulating the apertures of all the slits in accordance with functions of time which have orthogonal properties. This modulation can be effected by means of rotating discs with concentric slots of varying amplitude, and, in fact, a suitably selected portion of these slots can constitute the slits themselves.

With the arrangement just described, the radiation within a specified narrow spectral range passed by the monochromator, and no other, will be characterized by a specific time modulating, so that a measure of this radiation can be obtained. It must be noted that the old concept of a monochromator and a detector as two separate entities is being replaced here by that of a monochromator-detector combination, in which the "monochromator" does not yield a narrowbanded output which is then measured, but by means of which the measure of a narrowbanded component is obtained.

The choice of orthogonal functions deserves some discussion. An obvious choice would be sinusoidal functions. Thus, if the xth entrance and exit slits are sinusoidally modulated at the frequencies $f_1 + x\Delta f$ and $f_2 + x\Delta f$, with $f_2 - f_1 > n\Delta f$, the radiation of specified wavelength passed by all corresponding pairs of slits with the same ordinal number, and no other radiation, will be characterized by a sinusoidal modulation of frequency $f_2 - f_1$, and the problem of measuring the corresponding signal components in the detector output presents no difficulty.

Furthermore, every other spectral range will be likewise uniquely characterized by a sinusoidal modulation at the frequency $f_2 - f_1 + k\Delta f$, and individual selection and simultaneous measurement of these various radiations could be made by means of as many selective circuits. This simultaneous measure of many spectral elements constitutes the third step which is permitted an infrared spectroscopist, and just as the first two steps were permitted by the spatial extension of infrared sources, this step is permitted by spectral extension of these sources.

This third step still awaits realization, but by foregoing it, a wider choice of orthogonal functions becomes available for the second step. Thus, convenient two-value (1 and 0, open- and closed-slit) functions can be selected, and the experimental work done with slits modulated in accordance with such functions has demonstrated the general practicality of the system.[7] Mention can also be made of spectrometric systems with fixed and extended entrance- and exit-slit patterns, in which the measure of a specified narrowbanded range is obtained as the differential measure of two relatively large and broadbanded bundles of radiation passed by the exit-slit pattern of the monochromator.[8,9]

[7] M. J. E. Golay, "Multislit spectrometry," *Jour. Opt. Soc. Amer.*, vol. 39, p. 437; 1949.
[8] J. Strong, "Experimental infrared spectroscopy," *Physics Today*, vol. 4, p. 14; 1951.
[9] M. J. E. Golay, "Static multislit spectrometry," *Jour. Opt. Soc. Amer.*, vol. 41, p. 468; 1951.

Fig. 3 has been drawn to indicate the order of magnitude of increase in resolving power which the application of the second and third steps can be expected to yield. The conditions postulated for this were relatively conservative. First-order spectra from 2,000-line gratings were postulated for the whole range, which is

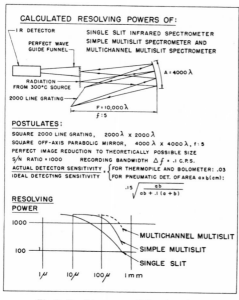

Fig. 3—Resolving power of infrared gratings.

CONCLUSION

When compared to radio sources, infrared sources have an extraordinarily small power per unit frequency. Likewise, infrared detectors have an extraordinarily small sensitivity when compared to radio detectors.

Since coherent detection is impossible in the infrared

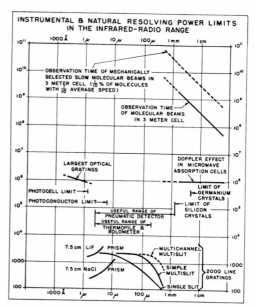

Fig. 4—Instrumental and natural resolving power limits in the infrared-radio range.

equivalent to the postulation of a maximum resolving power of 2,000, obtained with gratings ranging in length from 2 cm at 10-micron wavelength to 1 m at $\frac{1}{2}$-mm wavelength (resolving powers as high as 10,000 at 10 microns, and obtained with larger gratings, were reported at the last Ohio State Symposium on Molecular Structure by Peters, of Michigan University). If source temperatures higher than the postulated 300°C temperature are utilized, as well as larger time constants, and if the requirement of a S/N ratio of one thousand is relaxed, the use of the fairly convenient second step alone may go far in pushing to the mm region the wavelength at which the resolving power of the external optics is nearly realized, thus bridging what can be bridged of the spectroscopic gap between the infrared and the radio waves.

The handicaps under which the infrared spectroscopist must labor are illustrated by Fig. 4, in which the curves of Fig. 3 have been redrawn in the context of the resolving powers already obtained, or foreseeably obtainable in the two neighboring spectral regions.[10]

---

[10] Since the drawing of the dotted curve in the upper right corner, which was meant as an eventual target for the microwave spectros-

region, little basic progress is likely towards more sensitive detectors. The resolving powers to be expected will remain a function of the path difference of extreme collimated rays in the external optics used.[11] Also, spatial and spectral extension of infrared sources can be utilized more fully to develop spectrometric instrumentation permitting a near realization of the theoretical resolving power of the optical elements used, up to the shortest radio waves.

---

copy instrumentalist, Dicke and Newell have successfully demonstrated a new method for sharpening the resonance curve of a microwave absorption cell without resorting to the molecular-beam technique. With their method, the neighborhood of the lower left end of this dotted line appears reachable in the near future. (G. Newell, Jr. and R. H. Dicke, "A method for reducing the doppler breadth of microwave absorption lines," *Phys. Rev.*, vol. 83, p. 1064; August 15, 1951.)

[11] A few-fold increase in the resolving power of given dispersing elements can be obtained by an application of the multipass method, in which partially dispersed spectra are returned to these dispersing elements for further passes. Walsh has reported recently (*Jour. Opt. Soc. Am.*, vol. 42, p. 496; 1952) his successful application of this method. Multislit multipass spectrometers are discussed by the present author in a letter to the editor of the *Jour. Opt. Soc. Am.*, to be published shortly.

35

Reprinted from *J. Opt. Soc. Amer.*, **37**(11), 879–890 (1947)

The Ultimate Sensitivity of Radiation Detectors

R. Clark Jones

Research Laboratory, Polaroid Corporation, Cambridge, Massachusetts

(Received June 23, 1947)

An upper limit is obtained for the sensitivity of any radiation detector which satisfies Lambert's law and whose effective area is independent of the wave-length of the radiation. Expressions for the minimum detectable power are given by Eqs. (3.8) and (5.4).

The treatment given is based on thermodynamics, and makes no use of statistical mechanics. In the special case in which the fluctuations in the output of the detector are due entirely to fluctuations in the radiation, the results obtained agree with those obtained from a statistical mechanical treatment of the fluctuations in the radiation.

The treatment of the first five sections is confined to de-

tectors, called "temperature detectors", which operate by means of the change in their temperature brought about by the radiation. Such detectors include the thermocouple, the bolometer, and the radiometer. In Sections VI and VIII it is shown that the results hold for any detector satisfying the conditions in the first paragraph, provided that the emissivities are interpreted as quantum efficiencies. Section VII indicates that if the condition regarding Lambert's law is given up, then there exists no lower limit on the minimum detectable power. Section IX is devoted to numerical examples and a comparison with the reported sensitivities of bolometers.

INTRODUCTION

THE existence of statistical fluctuations sets an upper limit, which depends on the temperature, on the sensitivity of any device which detects electromagnetic radiation, including radiant heat.

The statistical fluctuations discussed here are those caused by temperature fluctuations of the detector and/or caused by statistical fluctuations in the radiation field. In actual radiation detectors, however, the voltage fluctuations which constitute Johnson noise frequently determine the limiting sensitivity. In these cases Johnson noise is the limiting factor only because the coupling between the radiation-detecting element and the electrical circuit is too weak. In the case of the bolometer, for example, the signal-to-Johnson noise ratio is proportional to the temperature coefficient of the resistance of the bolometer

element. Because there is no fundamental limitation on possible values of the temperature coefficient of resistance, it follows that there is no fundamental limit on the signal-to-Johnson noise ratio. Similarly, in thermocouple detectors there is no fundamental limitation on the Peltier coefficient, so that again the signal-to-Johnson noise ratio has no upper limit. In both of these cases, however, it will be shown in this paper that the fluctuations in the temperature of the heat-detecting element serve to provide a fundamental limit on the signal-to-noise ratio, which is not greatly in excess of the signal-to-Johnson noise ratios which are now obtainable.

Both the bolometer and the thermocouple are examples of "temperature detectors", defined here as detectors which operate through the mechanism of the change in their temperature produced by the radiation to be detected. Such detectors are also suitable for use as temperature-measuring devices. An example of a radiation detector which is not a temperature detector is the photoelectric cell, or the dipole antenna.

The results obtained in this paper hold for all types of radiation detectors whose size is large compared with the wave-lengths of the radiations being detected, and which obey Lambert's law. In the case of temperature detectors, the results apply to those which are conductively cooled, as well as to bolometer elements which are cooled primarily by radiation to their surroundings.

The theoretical value of the minimum detectable power will be computed first on the basis of the assumption that the detector is a temperature detector. In the case of such a temperature detector which has no conductive contact with its surroundings, and thus communicates with its environment only by electromagnetic radiation, the fluctuation in its temperature would be zero except for the fluctuation in the radiation itself. Accordingly, it may be concluded that the fluctuations in the output of such a detector are due solely to fluctuations in the radiation, and will accordingly be the same for any detector, whether or not the detector operates through measuring the change in its temperature.

Thus the results obtained in the paper hold not only for bolometers and thermocouples, but also for photoelectric cells, photo-conductive and photolytic cells, and any other detector whose dimensions are large compared with the wave-lengths being detected, and which obey Lambert's law.

In order to provide a clear distinction between Johnson noise and the noise at the output of the detector which is due to fluctuations in the radiation or to temperature fluctuations, the latter type of noise will be called temperature noise* throughout this paper.

Two different frequencies, denoted by ν and f, are used. The frequency ν is the frequency of the electromagnetic radiation itself, and f is the frequency of the modulation of the radiant power incident upon the detector.

---

* The term thermal noise would be preferred, but for the fact that this term is already in wide use as a synonym for Johnson noise.

I. ELEMENTARY BOLOMETER THEORY

This section contains no original material and serves only to introduce the notation and the framework of ideas which are used in the following sections.

Consider a material body of thermal capacity C which is in thermal equilibrium with a much larger body whose temperature is T_0. It is assumed that the dynamic behavior of the body may be specified in terms of a single time constant, so that the cooling constant κ of the material body with respect to its surroundings may be defined by the following differential equation for the approach to equilibrium with a large body, when the temperature of the small body is temporarily displaced for equilibrium:

$$C(d\theta/dt) + \kappa\theta = 0. \qquad (1.1)$$

In this expression t is the time and θ is the difference between the instantaneous absolute temperature T of the body and its equilibrium temperature T_0:

$$\theta \equiv T - T_0. \qquad (1.2)$$

Suppose now that because of a change in the environment of the body the flow of heat to the body (when its temperature is T_0) is increased by the amount H, where H may be a function of the time. The differential Eq. (1.1) then becomes

$$C(d\theta/dt) + \kappa\theta = H(t). \qquad (1.3)$$

Now suppose that the flow of heat H oscillates sinusoidally about the value zero with a frequency f:

$$H = H_0 \cos 2\pi f t. \qquad (1.4)$$

By Eq. (1.3), the amplitude of the steady-state oscillation of the temperature is then given by

$$\theta_0 = \frac{H_0/\kappa}{[1 + (2\pi f\tau)^2]^{\frac{1}{2}}}, \qquad (1.5)$$

where the time constant τ is defined by

$$\tau \equiv C/\kappa. \qquad (1.6)$$

In terms of τ the differential Eq. (1.3) may be rewritten in the following form:

$$\tau(d\theta/dt) + \theta = (H(t)/\kappa). \qquad (1.7)$$

II. TEMPERATURE FLUCTUATIONS

A thermodynamic treatment will now be used to obtain an expression for the mean-square fluctuation of the temperature of the material body with thermal capacity, C, about the temperature T_0 of the large body with which it is in thermal equilibrium.

Let l be any property of a body, and let $\bar{l}$ be its mean value at thermodynamic equilibrium. Then it may be shown[1] that

$$\langle (l-\bar{l})^2 \rangle_{Av} = -k \left/ \frac{\partial^2 S}{\partial l^2} \right., \tag{2.1}$$

where S is the entropy of the body and where k is Boltzmann's constant. Now let l be the temperature difference θ. Then when H is zero one finds from Eq. (2.1)

$$\theta_{r.m.s.}^2 = \langle \theta^2 \rangle_{Av} = -k \left/ \frac{\partial^2 S}{\partial T^2} \right.. \tag{2.2}$$

One has

$$\frac{\partial S}{\partial T} = \left(\frac{1}{T} \right) \left(\frac{\partial Q}{\partial T} \right) = \frac{C}{T}. \tag{2.3}$$

Upon differentiating (2.3) with respect to the temperature and substituting the result in (2.2), one obtains, finally,

$$\theta_{r.m.s.}^2 = kT^2/C. \tag{2.4}$$

It is perhaps worth while to pause in order to emphasize the generality of the result expressed in Eq. (2.4). This relation follows from the assumption that the body with thermal capacity C is in thermal equilibrium with a much larger body but makes no mention of the tightness of the coupling between the two bodies. If, for example, the two bodies were separated by a layer with poor conductivity, the frequency spectrum of the fluctuations would be quite different from the spectrum in the case of good conductivity. By itself, however, thermodynamics has nothing to say about the nature of the frequency spectrum of the fluctuations, and in order to obtain the frequency spectrum of the fluctuations further analysis is necessary.

Equation (2.4) may also be written in the convenient form

$$\theta_{r.m.s.}^2 = kT^2/(\kappa\tau). \tag{2.5}$$

[1] Paul S. Epstein, *Textbook of Thermodynamics* (John Wiley and Sons, Inc., New York, 1937), pp. 389–390.

III. MINIMUM DETECTABLE POWER OF A RADIATION DETECTOR

In this section the results derived in Section II are used to derive an expression for the minimum detectable power of any radiation detector, which operates by virtue of the change in its temperature which is induced by the signal which it is to detect.

Let the body whose thermal capacity is C be the detector. Furthermore, let H_m be the minimum detectable power of the detector, defined as the steady power *incident* upon the entire area of the detector which causes a change in its temperature equal to $\theta_{r.m.s.}$. This definition serves to introduce precision to a concept which is otherwise somewhat vague. In any event, however, the minimum detectable power just defined will be close to the value obtained by any reasonable criterion. Provided that the fluctuations in temperature are converted into changes in voltage by a proportionality factor which is independent of frequency, the definition of H_m just given means that the signal voltage will be equal to the r.m.s. voltage of the temperature noise. The case in which this condition is not satisfied is considered below in Section V.

Since the power *absorbed* by the detector is $\epsilon_s H_m$, where ϵ_s is the effective emissivity (absorption coefficient) of the surfaces of the detector for the signal in question, one finds from Eq. (1.3) that

$$H_m = \kappa\theta_{r.m.s.}/\epsilon_s. \tag{3.1}$$

By the use of (2.5), Eq. (3.1) may be written

$$H_m^2 = \kappa kT^2/(\epsilon_s^2\tau). \tag{3.2}$$

In order to make the minimum detectable power as small as possible at a given temperature, it follows from the last equation that it is desirable to make the time constant τ as long as possible, to make ϵ_s equal to unity, and to make the cooling constant κ as small as possible.

Equation (3.2) is the general expression for the sensitivity of a detector which operates by the mechanism of a change in its temperature. Consideration will now be given to the value of the cooling constant κ.

If H is the amount of heat absorbed per unit time by the detector, the cooling constant κ may be defined by

$$\kappa = dH/dT, \tag{3.3}$$

where in taking the derivative it is understood that the change in the temperature T of the detector takes place so slowly that the derivative has its maximum value.

The thermal radiation of a detector sets a lower limit to the value of the cooling constant κ. Provided that the detector satisfies Lambert's law, so that it is possible to define an emissivity which is independent of the angle of incidence of the radiation upon the detector, the quantity H defined in the last paragraph may be written

$$H = A \epsilon_r \sigma T^4, \qquad (3.4)$$

where A is the total area** of the detector, where σ is the Stefan-Boltzmann radiation constant, and where the quantity ϵ_r is the effective emissivity of the detector for thermal radiation of temperature T. Equation (3.4) thus yields the following explicit expression for the cooling constant κ:

$$\kappa = A\sigma \frac{d(\epsilon_r T^4)}{dT}, \qquad (3.5)$$

where it has been assumed that the area A is independent of the temperature.

It is convenient to define a third effective emissivity ϵ_u, defined by the relation

$$\kappa = 4A \epsilon_u \sigma T^3, \qquad (3.6)$$

so that ϵ_u may be written

$$\epsilon_u = \frac{1}{4T^3} \frac{d(\epsilon_r T^4)}{dT} = \frac{d(\epsilon_r T^4)}{dT^4}. \qquad (3.7)$$

The substitution of (3.6) in (3.2) then yields the following expression for the minimum detectable power of a detector whose cooling is purely radiative:

$$H_m^2 = 4A \cdot \sigma T^4 \cdot kT \cdot \epsilon_u/(\epsilon_s^2 \tau). \qquad (3.8)$$

It follows from this relation that in order to make the minimum detectable power at a given temperature as small as possible, the area of the bolometer should be made small, the time constant should be made long, and the ratio ϵ_u/ϵ_s^2 should be made as small as possible.

** If the bolometer is everywhere convex, then the area A is the geometrical area of the surface of the bolometer. If, however, the surface is such that not every point to it can be touched by a tangent plane, then the area A is the minimum area of an envelope enclosing the bolometer such that every point of the envelope can be touched by a tangent plane.

IV. EVALUATION OF THE EFFECTIVE EMISSIVITIES

The effective emissivity ϵ_s can be expressed in terms of the power spectrum of the signal and in terms of the frequency dependence of the emissivity. Let $H_s(\nu, r)$ be the power per unit frequency band width per unit area incident upon the detector at the point r at the frequency ν. The frequency ν is the frequency of the electromagnetic radiation. Similarly, let $\epsilon(\nu, r)$ be the emissivity of the detector at the point r at the frequency ν. Then the effective emissivity ϵ_s may be defined

$$\epsilon_s = \frac{\int dA \int H_s(\nu, r)\epsilon(\nu, r)d\nu}{\int dA \int H_s(\nu, r)d\nu}$$

$$= \frac{\int dA \int H_s(\nu, r)\epsilon(\nu, r)d\nu}{H_m}. \qquad (4.1)$$

In these expressions the integral with respect to A should be extended over the total area of the detector, and the limits on the integral with respect to the frequency ν are zero and infinity. Evidently, if $\epsilon(\nu, r)$ is independent of ν and r, then the effective emissivity ϵ_s is equal to ϵ.

The effective emissivity ϵ_r may be defined in terms of the power spectrum of the thermal radiation and the frequency dependence of the emissivity of the detector in exactly the same way employed in deriving Eq. (4.1). Let $J(\nu, T)$ be the power per unit frequency band of blackbody radiation of temperature T and frequency ν incident upon unit area of a surface, and let $\epsilon(\nu, r)$ be the same quantity previously defined. Then the effective emissivity ϵ_r may be defined by

$$\epsilon_r = \frac{\int dA \int J(\nu, T)\epsilon(\nu, r)d\nu}{\int dA \int J(\nu, T)d\nu}$$

$$= \frac{\int dA \int J(\nu, T)\epsilon(\nu, r)d\nu}{A\sigma T^4}, \qquad (4.2)$$

where, just as in the case of Eq. (4.1), the integral with respect to A is to be extended over the surface of the detector, and the integral with respect to frequency should extend from zero to infinity.

The substitution of (4.2) in (3.5) now yields the following expression for the cooling constant κ:

$$\kappa = \int dA \int \frac{\partial J(\nu, T)}{\partial T} \epsilon(\nu, r) d\nu, \qquad (4.3)$$

where it has been assumed that $\epsilon(\nu, r)$ is independent of the temperature of the detector.

By Eq. (3.6) the effective emissivity ϵ_u is now equal to

$$\epsilon_u = \frac{\int dA \int \frac{\partial J(\nu, T)}{\partial T} \epsilon(\nu, r) d\nu}{4 A \sigma T^3}. \qquad (4.4)$$

The theory of thermal radiation, as developed by Planck, yields the following expression[2] for the function $J(\nu, T)$:

$$J(\nu, T) = \frac{2\pi h \nu^3}{c^2(e^{h\nu/kT} - 1)}, \qquad (4.5)$$

where h is Planck's constant and c is the velocity of light in a vacuum. The derivative of J with respect to T is then

$$\frac{\partial J(\nu, T)}{\partial T} = \frac{2\pi h^2 \nu^4}{c^2 k T^2} \frac{e^{h\nu/kT}}{(e^{h\nu/kT} - 1)^2}. \qquad (4.6)$$

At this point it is convenient to introduce a dimensionless frequency defined by

$$\xi \equiv h\nu/kT. \qquad (4.7)$$

If, furthermore, $\epsilon(\nu, r)$ is expressed in terms of this new variable so that it becomes $\epsilon(\xi, r)$, the cooling constant κ as defined by (4.3) may now be written

$$\kappa = \frac{2\pi k}{c^2} \left(\frac{kT}{h}\right)^3 \int dA \int_0^\infty \frac{\xi^4 e^\xi}{(e^\xi - 1)^2} \epsilon(\xi, r) d\xi. \qquad (4.8)$$

Accordingly, the effective emissivity ϵ_u may be written

$$\epsilon_u = \frac{15}{4\pi^4 A} \int dA \int_0^\infty \frac{\xi^4 e^\xi}{(e^\xi - 1)^2} \epsilon(\xi; r) d\xi, \qquad (4.9)$$

with the help of the following relation between σ and the other constants:

$$\sigma \equiv \frac{2\pi^5}{15} \frac{k^4}{c^2 h^3}. \qquad (4.10)$$

The effective emissivities, ϵ_s and ϵ_u, which enter into the relation (3.7) have now been evaluated as explicitly as possible in the absence of more detailed information about the power spectrum of the signal $H_s(\nu, r)$ and the frequency dependence of the emissivity, $\epsilon(\nu, r)$. The results are expressed in Eqs. (4.1) and (4.9). In the special case in which $\epsilon(\nu, r)$ is independent of both the frequency ν and the position r, and is accordingly equal to a constant ϵ_0, both ϵ_s and ϵ_u become equal to ϵ_0. In the case of Eq. (4.1), this result follows by definition, but the result is not so immediately evident from Eq. (4.9). In this case, the equivalence of ϵ_u and ϵ_0 depends upon the following integral relation:

$$\int_0^\infty \frac{\xi^4 e^\xi}{(e^\xi - 1)^2} d\xi = 24\zeta(4) = \frac{4\pi^4}{15}, \qquad (4.11)$$

where $\zeta(z)$ is the Riemann zeta-function.[3]

V. ARBITRARY FREQUENCY RESPONSE

The relations developed so far have involved the assumption that the thermal characteristics of the detector may be expressed in terms of a single time constant; furthermore, in order that the definition of the minimum detectable power be a realistic one, it is necessary to suppose that the proportionality factor between the electrical output of the detector and the change in its temperature is independent of frequency. The possession of a single time constant, however, is usually the case only with detectors whose cooling is purely radiative. In the case of detectors which are cooled primarily by conduction as, for example, with the high speed bolometers developed during the war, the thermal behavior of the detecting element cannot be expressed in terms of a single time constant. Accordingly, it is desirable to generalize the expression (3.7) so that it holds for a detecting element which, in combination with its associated electrical amplifiers, has any specified frequency response.

[2] J. K. Roberts, Heat and Thermodynamics (Blackie and Son, Ltd., London, 1933), second edition, Chapter 19.

[3] E. T. Whittaker and G. N. Watson, Modern Analysis (Cambridge University Press, Teddington, England, 1935), fourth edition, Chapter 13.

The fact indicated by Eq. (2.5), that the mean-square temperature fluctuation is inversely proportional to the time constant τ, means that the power spectrum of the thermal fluctuations is independent of frequency, except as modified by the frequency response of the detector. Accordingly, if the frequency response of the detector plus its associated amplifiers is expressed by $Z(f)$ with the normalizing condition that the value of $Z(f)$ at zero frequency be unity, then the expression $1/\tau$, which occurs in (2.5) and in (3.7), should be replaced by

$$1/\tau \rightarrow 4 \int_0^\infty |Z(f)|^2 df. \qquad (5.1)$$

In the case of a detector with a single time constant, the function $Z(f)$ has the form

$$Z(f) = \frac{1}{1 + 2\pi i f \tau}, \qquad (5.2)$$

and in this case the expression on the right in (5.1) is equal to $1/\tau$.

The significance of the cooling constant κ remains unchanged; it is the constant effective in the limit of very low frequencies.

If the minimum detectable power of a sinusoidally oscillating signal, whose power varies sinusoidally, is defined by the condition that half of the peak-to-peak signal voltage is equal to the r.m.s. temperature-noise voltage, then the minimum detectable power of a sinusoidal signal with frequency f, as measured by half of its peak-to-peak amplitude, is given by

$$II_m{}^2 = \frac{4\kappa k T^2}{\epsilon_s{}^2} \frac{\int_0^\infty |Z(f)|^2 df}{|Z(f)|^2}. \qquad (5.3)$$

When κ has the minimum value given by Eq. (3.6), the last relation becomes

$$II_m{}^2 = 16A \cdot \sigma T^4 \cdot kT \cdot \frac{\epsilon_u}{\epsilon_s{}^2} \frac{\int_0^\infty |Z(f)|^2 df}{|Z(f)|^2}. \qquad (5.4)$$

In the last two equations the expressions are independent of the normalization of the function $Z(f)$, so that the equations are in a form suitable

for use with detector-amplifier systems with zero d.c. response.

The expressions (5.3) and (5.4) represent the minimum detectable power, as measured by half of its peak-to-peak amplitude, defined by the condition that the r.m.s. noise voltage is equal to the signal voltage, as measured by half of its peak-to-peak value. The same expressions are obtained, no matter what measure of the incident power and of the signal voltage are used, provided only that they be the same. On the other hand, however, it is essential that the noise voltage be measured by its r.m.s. value. Other measures of the noise voltage may be used only if corresponding changes in (5.3) and (5.4) are made.

It should perhaps be emphasized again that it is explicitly assumed that the level of Johnson noise is so much lower than that of the temperature noise that the effect of Johnson noise in raising the minimum detectable power may be ignored. In the event that the level of the Johnson noise may not be ignored, the r.m.s. Johnson-noise voltage should be computed independently, and the total r.m.s. noise voltage obtained by taking the square root of the sum of the mean-square Johnson-noise voltage and the mean-square temperature-noise voltage.

VI. THE RESULTS HOLD FOR ANY RADIATION DETECTOR

The general expressions (3.8) and (5.3) for the minimum detectable power of a radiation detector have been derived explicitly for a detector which operates through the mechanism of the change in its temperature brought about by the signal to be detected.

In the case of the detector which is cooled only by radiation, however, the fluctuations of temperature would necessarily be zero if there were no fluctuations in the transfer of heat by the radiation. Accordingly, it may be concluded that the fluctuations in the output of the detector are due to fluctuations in the radiation, and would, therefore, be the same for any radiation detector, no matter what its mechanism might be. This statement is qualified only by the assumptions which have been made in the derivation, the most important ones being the assumption that the effective area is independent of the frequency of the radiation and of the temperature, and that the absorption coefficient of the detector is independent of the angle of incidence.

The conclusions of the last paragraph are supported by the fact that the writer has been able to derive the expressions (3.8) and (5.4) by an altogether different method in which the fluctuations are attributed to statistical fluctuations in the absorption and emission of the light quanta of the radiation field. This derivation employs the statistical mechanics of Bose-Einstein assemblies and theory of power spectra employed by Rice[4] for the treatment of Johnson noise. This derivation will not be given here, because it yields no information not provided by the present treatment, and because it is less general in that it provides no information about the fluctuations of the detector which is cooled by other means than radiation.***

Since the fluctuations in a radiation-cooled detector are due entirely to the fluctuations in the incident radiation, it follows from the material presented that the mean-square fluctuation per unit frequency band width in the power incident upon a detector of area A is given by

$$16A \cdot \sigma T^4 \cdot \kappa T. \qquad (6.1)$$

Presumably this relation holds for all frequencies small compared with the reciprocal of the period required for the absorption or emission of the quanta.

VII. A FURTHER REDUCTION IN THE MINIMUM DETECTABLE POWER

It has been assumed up to this point that the quantum efficiency of the detector is independent of the direction of incidence of the radiation upon the detector. If this condition is given up, then it no longer is possible to provide an expression for the minimum detectable power which does not depend upon the degree of

---

[4] S. O. Rice, Bell. Syst. Tech. J. **23**, 282 (1944).
*** Just before the manuscript of this paper was sent to the editor, the writer's attention was called to a paper by W. B. Lewis, "Fluctuations in streams of thermal radiation," Proc. Phys. Soc. London **59**, 34 (1947). This paper contains essentially the treatment by statistical mechanics described above. Actually, the treatment in Lewis' paper does not consider the question of emissivities, and accordingly is restricted to detectors whose quantum efficiency is independent of the frequency of the radiation. This restriction is not fundamental, since the emissivities may very easily be inserted in Lewis' treatment, with the same result at that obtained here.

departure from Lambert's law. The signal-to-noise ratio of an array of antennas, for example, is proportional to the gain of the array, when the signal is incident from the direction of maximum response.

Even in the case of temperature detectors whose blackened surfaces provide a fairly close approximation to Lambert's law, it is possible by the use of radiation reflectors to produce a detector which exhibits a marked departure from Lambert's law, and thus permits one to obtain a lower minimum detectable power than that predicted by Eq. (5.4). A housing should be placed around the detector whose inner surface is as nearly as possible a perfect reflector for the frequencies to which the detector is sensitive; the housing must be incomplete in those directions from which radiation is to be detected. This housing will evidently reduce the cooling constant of the detector, because a fraction of the radiation which leaves the detector will return to the detector.

From the point of view used in the treatment by statistical mechanics, the reduction in the minimum detectable signal occurs because the photons which leave the detector and are reflected back by the reflector are absorbed and emitted at times so closely adjacent that the emitted and absorbed photons make no appreciable contribution to the low frequency components of the power spectrum of the radiation.

Evidently the degree by which the minimum detectable power may be reduced by the use of radiation reflectors will depend on the details of the geometry of the system and the degree to which the reflector approaches unity reflectance. These matters will not be discussed here.

The use of radiation shields yields necessarily a detector which does not obey Lambert's law.

VIII. SIGNIFICANCE OF THE EMISSIVITIES AS QUANTUM EFFICIENCIES

The reader may at first be disturbed by the fact that the general expression (3.8) for the minimum detectable power depends not only on the ratio, but also on the absolute values, of the emissivities. This situation occurs because the contribution to the signal voltage of the light quanta which constitute the signal is proportional to their number, whereas the temperature-noise voltage is proportional to the square root (aside from slight corrections because of the Bose statistics of the quanta) of the number of the quanta in the random radiation field. Thus if the number of quanta reaching and leaving the detector were all reduced by a constant factor, then the minimum detectable power would be increased.

Therefore, the significance of the emissivities for an arbitrary detector is that of the quantum efficiency of that detector. In the case of the temperature detector, the emissivity is simply the quantum efficiency for the absorption of incident light quanta. In the case of the photo-

electric cell, for example, the quantum efficiency is the ratio of the number of photoelectrons produced and utilized to the number of incident quanta of the frequency in question.

IX. NUMERICAL EXAMPLES

The Heat Detector

Consider first a temperature detector, such as a bolometer, whose emissivity is unity for all frequencies. Suppose that its total surface area is one square millimeter, that its temperature is room temperature (300 K), and that it is cooled entirely by radiation. Then if it is not equipped with radiation shields, but is exposed to the radiation field in every direction, so that it obeys Lambert's law, Eq. (3.8) yields the following expression for the minimum detectable power:

$$H_m = \frac{2.76 \times 10^{-12} \text{ watt}}{\tau^{\frac{1}{2}}}, \qquad (9.1)$$

where τ is the time constant in seconds, and where the following numerical values are used:

$$\sigma T^4 = 4.64 \times 10^{-2} \text{ watt/cm}^2,$$
$$kT = 4.11 \times 10^{-21} \text{ joule},$$
$$4A = 0.04 \text{ cm}^2.$$

Equation (9.1) is shown graphically in Fig. 1. For purposes of comparison, the sensitivity of a few modern bolometers is also shown on the plot. Because, however, of the present confused state with regard to the sensitivities of heat detectors, the writer has felt it necessary, in order to avoid adding to the confusion, to include in the Appendix a statement of exactly how the stated sensitivities were computed.

In examining Fig. 1 it must be remembered that the sensitivity of existing bolometers and thermocouples is limited not by the radiation fluctuations alone involved in Eq. (9.1), but rather by Johnson noise.

The dashed line, represented by the formula

$$H_m = \frac{3.0 \times 10^{-12} \text{ watt}}{\tau},$$

(where τ is in seconds) is the minimum detectable power of a bolometer operated at room temperature, as limited by Johnson noise, according to

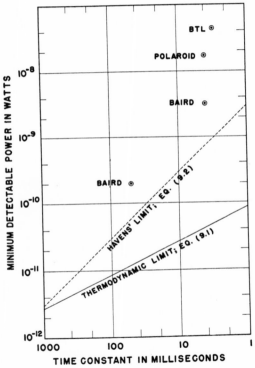

FIG. 1. The solid line shows the theoretical, minimum detectable power for a derector whose quantum efficiency is unity and whose effective area is one square millimeter at all wave-lengths, as predicted by Eqs. (3.8) and (5.4). The detector is assumed to have a temperature of 300°F, and to obey Lambert's law. The dashed·line shows the minimum detectable power, according to an engineering estimate made by R. J. Havens of what might be done in constructing bolometers with current materials and techniques. The four circles show the minimum detectable powers, computed on the basis specified in the Appendix from performance data supplied by their designers, of four bolometers constructed during or since the last war.

an estimate made by Havens[5] on the basis of current techniques and materials now available. This curve also refers to an area of one square millimeter. Havens' limit is by no means intended to be a fundamental limit which cannot be exceeded, but has rather the nature of an optimistic engineering estimate. It is evident from the crossing of the two curves that for time constants longer than about one second it may be possible to construct detectors whose sensitivity would be limited by temperature noise, rather than by Johnson noise.

It will be noted that Havens' limit depends on the time constant by the inverse first power of the time constant, rather than by the inverse one half-power. This difference arises because in temperature detectors limited by Johnson noise the coupling coefficient between the thermal and electrical effects is all important, and because in the case of the bolometer, for example, this coupling coefficient depends on the time constant; if, in a given bolometer, the cooling constant is increased with a corresponding decrease in the time constant, the voltage which may be placed across the element is increased, so that the coupling coefficient is also increased.

The Photoelectric Cell

Consider secondly a photoelectric cell with an area of one square millimeter and with a quantum efficiency of one percent for all frequencies greater than 3.0×10^{14} c.p.s., which is the frequency corresponding to a wave-length of one micron. The cell is to have zero quantum efficiency for lower frequencies.

Then, for any signal composed entirely of frequencies greater than 3.0×10^{14} c.p.s., the ratio ϵ_u/ϵ_s^2 which enters into (3.8) has the value

$$\frac{\epsilon_u}{\epsilon_s^2} = 100 \cdot \frac{15}{4\pi^2} \int_{\xi_0}^{\infty} \frac{\xi^4 e^{-\xi}}{(1-e^{-\xi})^2} d\xi \qquad (9.3)$$

by use of Eqs. (4.1) and (4.9). The quantity ξ_0 is defined by

$$\xi_0 \equiv h\nu_0/kT, \qquad (9.4)$$

where ν_0 is the cut-off frequency of the detector, taken here to be 3.0×10^{14} c.p.s. Thus one has

with $T = 300°K$:

$$\xi_0 = 48. \qquad (9.5)$$

For values of ξ_0 greater than 48 the exponential factor in the numerator integrand in (9.3) varies so much more rapidly than the other factors that the latter may be considered as constant; one thus has

$$\epsilon_u/\epsilon_s^2 = 3.85 \xi_0^4 e^{-\xi_0} \cong 3 \times 10^{-14}. \qquad (9.6)$$

The square root of this factor must now be appended to (9.1), because this factor was assumed to be unity in deriving this relation. One thus finds

$$H_m \cong \frac{5 \times 10^{-19} \text{ watt}}{\tau^{\frac{1}{2}}} \qquad (9.7)$$

for the minimum detectable power of the photoelectric cell in question.

For comparison with this relation, the power represented by one photon per second is 2×10^{-19} watt, when the photons all have a frequency of 3.0×10^{14} c.p.s. Thus the photo-cell in question, with a time constant of one second, would just comfortably be able to detect one photon per second.

Since the factor (9.6) depends very strongly on the cut-off frequency ν_0, changing by a factor of 100 when ν_0 changes by ten percent, it follows rather generally that photo-cells should be able to detect single photons whenever their cut-off wave-lengths are appreciably shorter than one micron.

Experimentally, however, commercial photomultiplier tubes with their peak sensitivities in the visible and with very sharp cut-offs toward the red are not able to detect individual photons unless they are cooled down to liquid-air temperatures.[6] Accordingly, it must be concluded that these tubes contain other sources of noise; the chief source of the additional noise is stated by Engstrom to be thermionic emission from the photo-cathode.

ACKNOWLEDGMENT

The writer is heavily indebted to the pioneer paper on this subject by Milatz and van der

[5] R. J. Havens, J. Opt. Soc. Am. 36, 355A (1946).

[6] Ralph W. Engstrom, J. Opt. Soc. Am. 37, 420 (1947).

Velden,[7] for the stimulation which led to work reported here. He is further indebted to Professor John Strong, of Johns Hopkins University, and to Dr. Bruce Billings, of Baird Associates, for helpful discussions about the matters discussed above.

APPENDIX

In order to inter-compare a number of bolometers whose properties have been measured in quite different manners, it is necessary to calculate the performance the bolometers would have if they were operated under a reference condition. The situation is complicated by the fact that with all actual bolometers, whose sensitivity is limited by Johnson noise, it is necessary to specify the frequency response of the amplifier, as well as of the bolometer.

The reference condition selected for the inter-comparison used here is as follows: The bolometer is assumed to be irradiated by a sinusoidal signal whose angular frequency is the reciprocal of the time constant τ of the bolometer. Thus the frequency f is related to τ by

$$f = 1/2\pi\tau. \qquad (9.2)$$

The amplifier is assumed to be flat at all frequencies, except for a simple RC high frequency cut-off whose time constant, defined by RC, is equal to the time constant of the bolometer. The incident power is measured by half of its peak-to-peak value, and the minimum detectable power is defined by the condition that half of the peak-to-peak signal voltage is to be equal to the r.m.s. noise voltage. (As pointed out in Section VI, however, the same numerical result for the minimum detectable power will be obtained if any measure of the incident power and of the signal voltage is used, provided only that the measures be the same.) The result is to be reduced to an area of one square millimeter on the basis that the minimum detectable power should be proportional to the square root of the area. (This dependence on the area is predicted not only by the thermodynamic theory given here, but also by more detailed theories of the operation of bolometers.)

The reference condition is an optimum one in two respects. First, if one wishes to detect a signal of frequency f and is able to vary the effective cooling constant of the bolometer, the greatest signal voltage is obtained if the cooling constant is adjusted so that the time constant is related to f by Eq. (9.2). Once this adjustment has been made, the greatest signal-to-noise ratio is obtained if the time constant of the high frequency cut-off of the amplifier is adjusted equal to that of the bolometer. The noise equivalent band width of such an amplifier characteristic is $\frac{1}{2}\pi f$.

The writer feels that the reference condition specified above has much to recommend it, since it has only one arbitrary numerical specification (area of one square millimeter) which is not uniquely specified by the behavior of the bolometer itself. Perhaps its greatest weakness is its restriction to bolometers which can be characterized by a single time constant. In any event, however, the evaluation of bolometers with multiple time constants will be essentially more complicated. Accordingly, the writer recommends its use in future reports of bolometer sensitivities, since the computed performance under the reference condition permits immediate comparison with the bolometers listed in Fig. 1.

It will be noted that the reference condition involves testing the bolometer with a sinusoidal signal with a frequency specified by Eq. (9.2), whereas the expression (3.8), which is used in plotting the line representing the thermodynamic limit in Fig. 1, refers to testing with an infinitely long pulse. It is evident that the use of a long pulse instead of the sinusoidal signal would reduce the minimum detectable power by the factor one-half. The use of the sinusoidal signal rather than a long pulse was chosen for the reference condition because the sinusoidal signal is much more closely related to the means of testing actually used. In comparing the experimental points in Fig. 1 with the line representing the thermodynamic limit, however, it should be borne in mind that the points should be lowered by a distance corresponding to a factor of two.

Until such time as bolometers are available whose sensitivity is limited by temperature noise rather than Johnson noise, bolometers of different time constants may be inter-compared on the

---

[7] J. M. W. Milatz and H. A. van der Velden, "Natural limit of measuring radiation with a bolometer," Physica **10**, 369 (1943).

basis that their minimum detectable power under the reference condition should be inversely proportional to their time constants, as suggested by Havens' limit.

The Baird Bolometers

The slower of the two Baird bolometers is reported† as follows:

$$A = 3.6 \text{ mm}^2, \quad \tau = 50 \text{ milliseconds.}$$

With a square-wave heat signal with a frequency of 10 c.p.s., the electrical response with a wide band (2–5000 c.p.s.) amplifier is 3.0 volts, read on a meter which measures the average absolute value of the voltage and which is calibrated to read the r.m.s. value of a sine wave, per peak-to-peak incident watt. The noise was measured separately in a band 8 c.p.s. wide (noise-equivalent band width) centered at 30 c.p.s.; the meter just described read 3.5×10^{-9} volt.

The observed response is equivalent†† to 2.7 average volts per peak-to-peak watt. If one ignores the small phase shift introduced by the low frequency amplifier cut-off of 2 c.p.s., it is easily computed that the 0.05-second time constant of the bolometer causes the average voltage to be only 0.164 of the peak-to-peak voltage obtained from a very low frequency square-wave signal. Thus at very low frequencies the response of the bolometer is 16.5 volts per watt, with any consistent measure of the power and of the voltage. In the reference condition, the effect of each of the two time constants is to reduce the response by the square root on one-half, so that the response in the reference condition is 8.2 volts per watt.

The measured noise voltage is equivalent††† to 3.96×10^{-9} r.m.s. volt in an 8-c.p.s. band, or to 3.13×10^{-9} r.m.s. volt in a 5-c.p.s. band, the band width corresponding to the reference condition.

It follows directly from the results of the last two paragraphs that the minimum detectable power of the bolometer in the reference condition would be 3.82×10^{-10} watt. Upon division by the square root of 3.6 in order to reduce the result to that for an area of one square millimeter, one obtains the result given in Fig. 1 of 2.01×10^{-10} watt. With a very long pulse instead of the sinusoidal signal, the minimum detectable power would be 1.00×10^{-10} watt.

The faster of the Baird bolometers is reported as follows:

$$A = 0.2 \text{ mm}^2, \quad \tau = 4.1 \text{ milliseconds.}$$

With a square-wave heat signal with a frequency of 10 c.p.s. the electrical response with a wide-band (2–5000 c.p.s.) amplifier is 2.75 volts, read on a meter which measures the average absolute value of the voltage and which is calibrated to read the r.m.s. value of a sine wave, per peak-to-peak incident watt. The noise was measured separately in a band 8 c.p.s. wide centered at 30 c.p.s.; the meter just described read 3.5×10^{-9} volt.

The observed response is equivalent to 2.47 average volts per peak-to-peak watt. Since the ratio of the bolometer time constant to the period of the square wave is the same as for the slower bolometer, the same factor of 0.164 should be used. One finds that at very low frequencies the response of the bolometer is 15.1 volts per watt, with any consistent measure of the power and of the voltage. In the reference condition the response is reduced to 7.6 volts per watt.

The measured noise voltage is equivalent to 3.96×10^{-9} r.m.s. volt in an 8-c.p.s. band, or to 1.09×10^{-8} r.m.s. volt in a 60-c.p.s. band, the band width corresponding to the reference condition.

It follows directly from the results of the last two paragraphs that the minimum detectable power of the bolometer in the reference condition would be 1.43×10^{-9} watt. Upon multiplying by the square root of 5 in order to reduce the result to that for an area of one square millimeter, one obtains the result given in Fig. 1 of 3.20×10^{-9} watt. With a very long pulse instead of the sinusoidal signal, the minimum detectable power would be 1.60×10^{-9} watt.

---

† Private communication from W. G. Langton, dated April 15, 1947.

†† A meter which measures the average value of a signal, and which is calibrated to read the r.m.s. value of a sine-wave signal, should have its indication multiplied by $2^{\frac{1}{2}}/\pi = 0.9003$ in order to obtain the average voltage.

††† A meter which measures the average value of a signal, and which is calibrated to read the r.m.s. value on a sine-wave signal, should have its indication multiplied by $2/\pi^{\frac{1}{2}} = 1.1284$ in order to obtain the r.m.s. voltage of a noise signal, or of any other voltage with a Gaussian distribution of amplitudes.

The Bell Telephone Laboratories Thermistor Bolometer

The thermistor bolometer is reported‡ as follows:

$$A = 0.6 \text{ mm}^2, \quad \tau = 3 \text{ milliseconds.}$$

A typical quartz-backed thermistor bolometer gives a signal which is equal to the noise when 2×10^{-8} watt of radiation falls on the bolometer for 0.003 second for a 30-c.p.s. pass band.

If one assumes that the "signal equal to the noise" means that the peak signal voltage is equal to the r.m.s. noise voltage, it is still not possible to make an exact reduction to the reference condition without making a detailed calculation of the transient wave form, and this cannot be done without more detailed information about the phase and gain characteristics of the amplifier. Accordingly, only an approximation reduction will be made.

It is estimated that if the pulse of radiation were infinitely prolonged, and if the amplifier passed direct current, the signal voltage would be roughly twice that actually measured. In order to obtain the minimum detectable power under the reference condition, the change in both the signal voltage and the noise voltage will now be computed for an amplifier which has the same maximum gain as that actually used, and which is flat except for a high frequency cut-off with a time constant of 0.003 second, and therefore at a frequency of 53.1 c.p.s. Under this reference condition the signal voltage will be one-half of its low frequency value, which change just annuls the increase estimated in the first sentence. Under this reference condition, the noise-equivalent band width is 83.3 c.p.s. instead of 30 c.p.s., so that the noise voltage is increased by the factor 1.67. Accordingly, the signal-to-noise ratio in the reference condition is reduced by the reciprocal of 1.67, so that the minimum detectable power under the reference condition is computed to be 3.34×10^{-8} watt. Upon dividing by the square root of 0.6 in order to reduce the result to that for an area of one square millimeter,

one obtains the result given in Fig. 1 of 4.31×10^{-8} watt. With a very long pulse instead of the sinusoidal signal, the minimum detectable power would be 2.15×10^{-8} watt.

The Polaroid Bolometer

The Polaroid metal bolometer is reported‡‡ as follows:

$$A = 6 \text{ mm}^2, \quad \tau = 4 \text{ milliseconds.}$$

With an amplifier with maximum gain at 30 c.p.s. and with a noise-equivalent band width of 100 c.p.s., the r.m.s. signal voltage is equal to the r.m.s. noise voltage when the r.m.s. signal power incident upon the bolometer at 30 c.p.s. is 3.3×10^{-8} watt.

Under the stated conditions, the minimum detectable power is thus 3.3×10^{-8} watt. In order to obtain the minimum detectable power under the reference condition, the change in both the signal voltage and the noise voltage will now be computed for an amplifier which has the same maximum gain as that actually used, and which is flat except for a high frequency cut-off with a time constant of 0.004 second, and therefore at a frequency of 39.8 c.p.s. Under this reference condition the frequency of the signal is 39.8 c.p.s. instead of 30 c.p.s., so that the response of the bolometer itself is reduced by the factor 0.882, and the response of the amplifier is reduced by the square root of one-half, so that the signal voltage is reduced by the factor 0.624. Under the reference condition the noise-equivalent band width is 62.5 c.p.s. instead of 100 c.p.s., so that the noise voltage is reduced by the factor 0.790. Accordingly, the signal-to-noise ratio in the reference condition is reduced by the factor 0.79, so that the minimum detectable power is raised to 4.18×10^{-8} watt. Upon dividing by the square root of 6 in order to reduce the result to that for an area of one square millimeter, one obtains the result given in Fig. 1 of 1.70×10^{-8} watt. With a very long pulse instead of the sinusoidal signal, the minimum detectable power would be 8.5×10^{-9} watt.

‡ W. H. Brattain and J. A. Becker, J. Opt. Soc. Am. **36**, 354A (1946).

‡‡ B. H. Billings, W. L. Hyde, and E. E. Barr, J. Opt. Soc. Am. **37**, 123 (1947).

Erratum: The Ultimate Sensitivity of Radiation Detectors

[J. Opt. Soc. Am. **37**, 879 (1947)]

R. Clark Jones

Research Laboratory, Polaroid Corporation, Cambridge, Massachusetts

THE writer has had the privilege of seeing in advance of publication a manuscript by Mr. P. B. Fellgett, Peterhouse, Cambridge, England. In the manuscript, Mr. Fellgett points out that the expression (3.8) in this paper holds less generally than is claimed in the paper. Although the expression holds for the case for which it was derived—namely, for detectors whose response is mediated by a change in temperature of the sensitive element, it is incorrect in application to a photo-cell for two reasons:

1. In the heat detector, the fluctuations are due equally to the radiation incident upon, and emitted by, the detector. But in a photo-cell, only the fluctuations in the incident radiation are effective. This change has the effect of multiplying the noise equivalent power by $2^{-\frac{1}{2}}$.

2. In the heat detector, the photons are weighted in accordance with their energy, whereas in the case of the photo-cell, the photons are weighted in accordance with their number. This change requires that for a given quantum efficiency, the weighting of photons of different frequency must be done differently. Evidently, however, if the fractional band width of the detector is small, this change is unimportant.

Mr. Fellgett concludes that his observations serve to modify the expression (3.8) by numerical factors whose order of magnitude is unity in all practical cases.

The writer wishes to thank Mr. Fellgett for communicating a copy of his manuscript in advance of publication and for approving the publication of these remarks.

The author wishes to take this opportunity to correct several misprints of minor importance in Paper I, all on page 886:

In the legend of Fig. 1, the temperature *300*°F should be *300*°K, and the word *derector* should be *detector*.

In column two, the designation *300*K should be *300*°K.

The last equation on page 886 should bear the number 9.2.

36

Reprinted from *J. Opt. Soc. Amer.*, **39**(5), 327–343 (1949)

A New Classification System for Radiation Detectors*

R. Clark Jones

Research Laboratory, Polaroid Corporation, Cambridge, Massachusetts

(Received February 4, 1949)

In Part I, a classification system for radiation detectors is proposed. The system is based upon the manner in which the noise equivalent power depends upon the time constant and the sensitive area, when the sensitivity is limited either by radiation fluctuations or by internally generated noise.

The major part of Part I is devoted to establishing the reference condition for the measuring of the noise equivalent power. In establishing the reference condition of measurement (Section 6), the output of the detector in the absence of a signal is first equalized so that the noise power per unit frequency band width is constant. The reference time constant is then defined in terms of the relative response as a function of signal modulation frequency (Section 7). The equalization is then modified by the addition of an RC low pass filter with a time constant equal to the reference time constant, and the noise equivalent power is measured. The power so obtained is termed the noise equivalent power in reference condition A, denoted by P_m. This procedure has the property that the noise equivalent power of the detector is measured in the presence of noise whose band width is equal to the band width of the detector.

In order to establish the reference condition of measurement it is necessary to consider the various sources of noise which are involved in detectors (Section 2), and the various types of time constants (Section 3). The important concept of responsivity-to-noise ratio is defined in Section 4. A general theorem involving the sensitive area and the responsivity-to-noise ratio is established in Section 5.

After the question of detectors with non-uniform spectral sensitivity is discussed in Section 8, the classification system is defined in Section 9: A detector is a Type I detector if its noise equivalent power depends upon its sensitive area A and its reference time constant τ in accordance with

$$P_m = A^{\frac{1}{2}}/k_1\tau^{\frac{1}{2}},$$

where k_1 is independent of A and τ. A Type II detector is defined by

$$P_m = A^{\frac{1}{2}}/k_2\tau,$$

and more generally, a Type n detector is defined by

$$P_m = A^{\frac{1}{2}}/k_n\tau^{\frac{1}{2}n}.$$

The usefulness of the proposed classification is illustrated in Section 10 by its application to sequential scanning systems.

In Part II the classification system proposed in Part I is used to determine the type number of each of eight different kinds of detectors. By a detailed analysis of each kind of detector, it is found that all of the detectors studied are either Type I or Type II detectors. The results of the analysis are summarized in Table I.

The detectors studied include bolometers (Section 2), thermocouples and thermopiles (Section 3), the Golay detector (Section 4), photographic plates and films (Section 5), vacuum and gas photo-tubes and photo-multiplier tubes (Section 6), and dipole antennas (Section 7). In the case of the dipole antenna it is shown that Johnson noise becomes equivalent to the noise produced by radiation fluctuations.

PART I. THE CLASSIFICATION SYSTEM

1. Introduction

AN ideal radiation detector might be defined as one which counts every one of the incident quanta which constitute the signal to be detected, and which gives no response when this signal is

---

* This is the second of a series of papers on radiation detectors, the second and third of which appear in this issue:
I. "The ultimate sensitivity of radiation detectors," J. Opt. Soc. Am. **37**, 879–890 (1947), referred to as Paper I.
II. This paper, referred to as Paper II.
III. "Factors of merit for radiation detectors," in this issue, referred to as Paper III.

absent. In general, no detector has this property, although some detectors for high energy photons, such as the proportional counter, are nearly ideal in this respect.

The difficulty is that all detectors have a background response of some kind. The zero-frequency component of the background response may be eliminated by a subtraction method, but the fluctuations in this background remain. No signal can be detected unless the amplitude of its response is comparable with or greater than the amplitude of the fluctuations which lie in the frequency band of the signal.

2. Types of Backgrounds

There are at least three different types of background to which consideration must be given in a comprehensive study of detectors.

Radiation Background.—Every detector must operate in the presence of the blackbody radiation field appropriate to its own temperature and to the temperature of its environment. The noise due to this background has been termed *temperature noise*, and is discussed in detail in the first paper of this series.† This background is particularly important for detectors, such as thermocouples and bolometers, which are responsive in the 5 to 15 micron range.

Internal Background.—Another type of background is that introduced within the detector itself. Examples of such backgrounds are thermionic emission (dark current) in photoelectric cells and photo-multiplier tubes, Johnson noise, current noise,[1] and perhaps semi-conductor noise[2] in bolometers and photo-conductive cells, Johnson noise in thermocouples, and density fluctuations in photographic plates.

Signal Background.—If one is concerned with detecting a small change in a steady signal, the statistical fluctuations in the signal itself may determine the smallest change which may be detected. An example of this type of background is the increased noise in the output of a photo-tube when daylight is incident upon the tube. The effect of this background on the absolute and relative performance of the human eye, television systems, and the photographic plate has been discussed comprehensively by Rose.[3]

3. Types of Time Constants

The purpose of this Part is to establish that many of the known radiation detectors as they are customarily operated may be classified usefully on the way that their noise equivalent power** depends on their time constant and sensitive area, when one takes into account the limitations introduced by the radiation and internal backgrounds.

Before defining the classification, however, it is necessary to define the concept of the time constant.

There are three different types of time constants used in the presentation.

There is first the *reference* time constant, defined in Section 7 on the basis of the response *versus* frequency curve obtained when the output has been equalized so that the noise power per unit band width is constant. This is probably the most fundamental definition of the time constant of a detector.

In general, any detector whose output is electrical may have its frequency response curve adjusted by electrical equalization so that the frequency response curve corresponds to any time constant. Any time constant obtained by using equalization different from that used in defining the reference time constant is termed the *effective* time constant.

Lastly, some detectors such as the bolometer possess one or more *physical* time constants. Examples of physical time constants are RC time constants, and the time constant obtained by taking the ratio of a heat capacity to a thermal conductivity coefficient.

In some cases, as for example the unequalized vacuum bolometer without current noise, these three differently defined time constants are all equal, but in other cases, they may be quite different.

4. Definitions Relating to the Responsivity

First, let the *responsivity* S of a detector with an electrical output be defined as the ratio of the output voltage to the power incident upon the detector. The responsivity S is conveniently expressed in volts per watt. The responsivity is a function of the modulation frequency of the radiation incident upon the detector, and in order to insure that the low frequency responsivity approaches the zero-frequency responsivity as a limit, it is necessary to postulate that the same measure of the amplitude be used for the radiation and for the voltage: for example, if the voltage is stated in terms of its root-mean-square value, then the incident power should be measured in terms of the root-mean-square value of the deviation of the incident power from its mean value. One may state, for example, that the responsivity is in r.m.s. volts per r.m.s. watt.

Secondly, let the *noise power* be defined as the mean square value of the noise voltage. Because the noise powers, rather than noise voltages, of non-overlapping frequency bands combine additively, it is convenient to deal with the noise power per unit band width, denoted by N.

Finally, the responsivity-to-noise ratio $\Re$ is defined as the ratio of the responsivity S to the square root of N:

$$\Re \equiv S/N^{\frac{1}{2}}. \qquad (4.1)$$

---

† Attention is called to the erratum on page 343 with reference to Paper I.

[1] C. J. Christensen and G. L. Pearson, Bell Sys. Tech. J. **15**, 197–223 (1936).

[2] B. Davydov and B. Gurevich, "Voltage fluctuations in semi-conductors," J. Phys., USSR, **7**, 138–140 (1943). In Russian.

[3] Albert Rose, J. Opt. Soc. Am. **38**, 196–208 (1948); J. Soc. Mot. Pict. Eng. **47**, 273–294 (1946).

** The quantity which is here termed the noise equivalent power was denoted by the phrase minimum detectable power in Paper I. This change is made in order that the phrase used should not imply more than is involved in its definition.

5. The Sensitive Area

The sensitive area A of a detector is defined as the area over which the detector responds to irradiation. It will be supposed in this section that the detector is equally responsive at every part of its sensitive area.

It will now be shown, under certain restrictive conditions, that the responsivity-to-noise ratio $\Re$ is inversely proportional to the square root of the sensitive area.

The restrictive assumptions are:

1. It is assumed that the output voltage of the detector is proportional to the radiant power incident upon the sensitive area, when the power has a constant spectral distribution.
2. It is assumed that the noise contributed by a given part of the sensitive area is statistically independent of the noise contributed by any other non-overlapping part of the sensitive area.

Now consider two separate cases: (a) in which a certain power P is incident upon a single detector, and (b) in which the same total power is incident upon n adjacent detectors, each of which is identical with the detector in case (a). The total sensitive area in case (b) is thus n times that in case (a). Suppose further that the detectors in case (b) are connected in series-aiding.

Then according to the first assumption, the output voltage will be the same in the two cases, since in case (b) each of the detectors will be irradiated by the power P/n. On the other hand, by the second assumption, the noise voltage in case (b) will be greater than that in case (a) by the factor $n^{\frac{1}{2}}$. It now follows directly that the responsivity-to-noise ratio of any detector satisfying the assumptions is inversely proportional to the square root of the sensitive area.

The same conclusion follows from any method of connecting together the detectors which permits each of the n units to contribute equally to the total output in case (b).

Accordingly, in order to intercompare detectors with different sensitive areas, the responsivity-to-noise ratios should be multiplied by a factor proportional to the square root of the sensitive area of each detector.

This result is of rather general applicability, and applies also to at least one detector which is not associated with an electrical amplifier: the photographic plate.

The proof is not applicable, however, to the dipole antenna, because the voltage output is proportional to the square root of the power falling upon it; thus the first assumption is not valid. The second assumption is also invalid, but it is substantially correct when the adjacent dipoles are separated by not less than one-half wave-length; this is usually the case with arrays of dipoles.

The result of the above proof may be applied to dipole antennas, however, if each of the individual dipoles is separately connected to a square-law rectifier, and if the outputs of the separate rectifiers are connected in series or parallel. It must further be assumed that the dipoles are separated by at least one-half wave-length, in order that the noise outputs be substantially independent in the statistical sense. See Part II, Section 7.

6. Reference Condition of Measurement

In order to permit a straightforward intercomparison of detectors with different time constants, it is desirable that their noise equivalent powers be stated for a suitable standard condition of measurement.

A reference condition of measurement was proposed in the first paper of the series. A reformulation of this proposal is given in this section.

The reference condition of measurement described below satisfies the following conditions:

1. The noise equivalent power of the detector is measured in the presence of the noise in a manner such that the band width of the noise is approximately equal to the band width of the detector.
2. The band width of the detector is measured after the amplifier gain has been equalized so that the noise spectrum is flat.

Let the detector be coupled to an amplifier, and let it be required that the amplifier be such and the method of coupling be such that the noise power introduced by the amplifier is small at every frequency compared with the noise power in the output of the detector.

The frequency response curve of the amplifier is now to be adjusted so that the noise power per unit band width is the same at all frequencies.[4] With this type of equalization the frequency response curve of the detector is measured, and the time constant of the detector is determined from the frequency response curve in a manner discussed in detail in the following section. The time constant so obtained is called the *reference time constant*.

After the reference time constant has been determined, the frequency response curve of the amplifier is changed by adding to the circuits of the amplifier a high frequency RC cut-off (series resistance, shunt capacity) with a time constant equal to the reference time constant of the detector.

One then determines the steady input power to the detector which produces a steady voltage*** at

---

[4] In order that this specification be unique, it is necessary to require that the complex gain *versus* frequency function of the amplifier be of the minimum-theta type. See Hendrik W. Bode, *Network Analysis and Feedback Amplifier Design* (D. Van Nostrand Company, Inc., New York, 1945).
*** There is no difficulty of principle in measuring a steady response equal to the r.m.s. noise voltage, since an arbitrarily long time is available for measuring the zero-frequency response before and after the incident power is initiated.

the output of the amplifier which is equal to the root-mean-square noise voltage at the output of the amplifier. The power so measured is termed the *noise equivalent power of the detector in the reference condition A* and is denoted by P_m.

A relation between the zero-frequency responsivity-to-noise ratio $\Re_0$ and the noise equivalent power P_m will now be derived. Evidently, the noise equivalent power is the ratio of the r.m.s. noise voltage E_N to the zero-frequency responsivity S_0 of the detector

$$P_m = E_N / S_0. \qquad (6.1)$$

If N_0 is the constant noise power per unit band width prevailing during the measurement of the reference time constant, then the noise power per unit band width during the measurement of the noise equivalent power is

$$N(f) = \frac{N_0}{1 + (2\pi f \tau)^2}, \qquad (6.2)$$

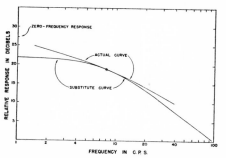

FIG. 1. Responsivity as a function of the frequency for the Schwarz thermopile, as measured by Nielsen. Since the noise is assumed to be Johnson noise only, the curve representing the reposnivity-to-noise ratio has the same shape. The curve slopes off more gradually at high frequencies than it does in the case of a detector with a single time constant. This characteristic defines a curve as a Class *I* curve.

the r.m.s. noise voltage E_N is then given by

$$E_N{}^2 = \int_0^\infty N(f) df = N_0 / 4\tau. \qquad (6.3)$$

By the definition of $\Re$, one has

$$S_0 = N_0{}^{\frac{1}{2}} \Re_0. \qquad (6.4)$$

Upon substituting Eqs. (6.4) and (6.3) in (6.1), one finds

$$P_m = 1 / (2\tau^{\frac{1}{2}} \Re_0). \qquad (6.5)$$

This is the desired relation.

Since it was established in Section 5 that the zero-frequency responsivity-to-noise ratio is in-

versely proportional to the square root of the sensitive area, it follows from the last equation that the noise equivalent power in reference condition A is proportional to the square root of the sensitive area. For this reason, it is convenient to define the *noise equivalent power in reference condition B*, denoted by H_m:

$$H_m \equiv P_m / A^{\frac{1}{2}}, \qquad (6.6)$$

where A is here the sensitive area expressed in square millimeters. As used in this equation, the area A is dimensionless, and might more accurately be defined as the ratio of the sensitive area to an area of one square millimeter.

The noise equivalent power in reference condition B is the noise equivalent power the detector would have if its sensitive area were changed to one square millimeter.

The measurement of the noise equivalent power in the reference condition involves using a band width of noise equal to $\pi/2\tau$ in radians per second, or to $1/4\tau$ in cycles per second. In practice, it will not usually be convenient to measure the noise equivalent power with exactly this band-width of noise. If, however, the noise is known to be flat, then a measurement of the noise equivalent power with any other band width may be used to calculate the noise equivalent power in the reference condition. Furthermore, it will not usually be convenient in practice to measure the noise equivalent power for a zero-frequency signal. If, however, the $\Re$ *versus* frequency curve of the detector is known, then again one may calculate the noise equivalent power in the reference condition. For the sake of specificity, the method of calculation is stated in the following paragraph.

Let the noise equivalent power be measured with an amplifier whose gain curve is such that the power spectrum of the noise is $N(f)$, and with a sinusoidal radiation signal of frequency f_s; let the noise equivalent power so obtained be denoted by P. Let the noise band width Δf of the amplifier be defined by

$$\Delta f = \frac{\displaystyle\int_0^\infty N(f) df}{N_{\max}}. \qquad (6.7)$$

Furthermore, let the zero-frequency value of $\Re$ be denoted as before by $\Re_0$, and let the value of $\Re$ at the signal frequency be denoted by $\Re_s$. Then the noise equivalent power in the reference condition A is obtained from P by

$$P_m = P \cdot \frac{\Re_s}{\Re_0} \cdot \left(\frac{1/4\tau}{\Delta f}\right)^{\frac{1}{2}} \cdot \left(\frac{N(f_s)}{N_{\max}}\right)^{\frac{1}{2}}, \qquad (6.8)$$

where τ is the reference time constant.

A somewhat different definition of the noise equivalent power in the reference condition was given in the first paper of this series. The definition given there involved the same specifications as those used here with regard to the noise spectra, but involved the use of a sinusoidal signal with the frequency $1/4\tau$ instead of the steady signal specified here. For detectors in accord with Eq. (7.1), the present noise equivalent power is one-half of that defined in Paper I. The extension of the definition of the noise equivalent power to detectors not in accord with Eq. (7.1), however, is accomplished

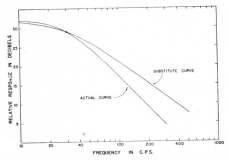

FIG. 2. Responsivity as a function of the frequency for the Strong bolometer, as measured by Nielsen. Since the noise is assumed to be Johnson noise only, the curve representing the responsivity-to-noise ratio has the same shape. The curve falls off more rapidly at high frequencies than it does in the case of a single time constant. This characteristic defines the curve as a Class II curve.

much more naturally with the present definition, and it is for this reason that the present definition was introduced.

Lest the reader receive from this section the impression that the noise equivalent power defined here is equal to the minimum detectable power under the same conditions, it should be pointed out that Rose[3] has found that in certain visual phenomena the minimum detectable stimulus is about five times the noise equivalent stimulus.

7. Determination of the Reference Time Constant

As specified in the last section, the reference time constant is to be determined from the shape of the responsivity-to-noise ratio *versus* frequency curve.

Consider first the case in which the shape of this curve corresponds to the existence of a single time constant. By this one means that the responsivity-to-noise ratio depends upon the frequency according to

$$\Re(f) = \frac{\Re_0}{(1+(2\pi f\tau)^2)^{\frac{1}{2}}}, \qquad (7.1)$$

where τ is the single time constant. This case offers

no difficulty; the reference time constant is set equal to the time constant τ.

When $\Re$ does not vary with frequency in accordance with Eq. (7.1), it is frequently stated that the detector "does not have a time constant," or that it "has more than one time constant." The latter method of expression has the following significance. If $\Re$ is expressed as a complex function that indicates both the amplitude and the phase of the response of the detector, then by a general mathematical theorem[5] relating to analytic functions without essential singularities, the function $\Re(f)$ can be completely specified in terms of its poles and zeros. Equation (7.1) corresponds to the existence of one zero at infinity and to one pole on the imaginary frequency axis. If further zeros and/or poles are involved, the dependence of $\Re$ on the frequency is more complicated.

Even in the case of complicated dependences of $\Re$ on the frequency, however, it is possible and convenient to define a single time constant for the detector. Perhaps the most simple, and at the same time the most generally useful definition of the reference time constant is

$$\tau = \frac{\frac{1}{4}\Re_{max}^2}{\int_0^\infty \Re^2 df}, \qquad (7.2)$$

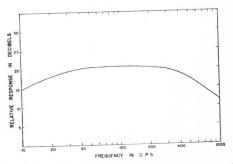

FIG. 3. Responsivity as a function of the frequency for the Golay pneumatic heat detector, as measured by Nielsen. Since the noise is assumed to be Johnson noise only, the curve representing the responsivity-to-noise ratio has the same shape. The defining characteristic of the Class III type of curve is that the responsivity-to-noise ratio falls off at low frequencies.

where $\Re_{max}$ is the maximum value of $\Re$, and where the numerical constant $\frac{1}{4}$ has been introduced in order that Eq. (7.2) should agree with Eq. (7.1).

In all of the specific detectors actually studied in the fifth paper of this series, however, the nature

[5] Liouville's Theorem. See E. T. Whittaker and G. N. Watson, *Modern Analysis*, fourth edition (Cambridge University Press, London, 1927), p. 105.

of the $\Re$ *versus* frequency curves is such that substantially the same result as that obtained from Eq. (7.2) can be obtained by procedures which substitute a curve representing a single time constant for the actual curve. The reference time constant and the effective value of the zero-frequency responsivity-to-noise ratio are both defined in terms of the substitute curve.

Three different classes of responsivity-to-noise ratio *versus* frequency curves were encountered which did not correspond to the existence of a single time constant. They are described as follows:

Class I

The most common type of departure from Eq. (7.1) is that illustrated in Fig. 1. This figure shows

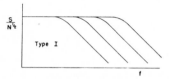

FIG. 4. Showing on logarithmic coordinate scales the relation between the responsivity-to-noise ratio $\Re$ and the frequency f for a Type I detector with a single time constant. The separate curves correspond to different values of the reference time constant. The curves all approach asymptotically the same horizontal line at low frequencies, in accordance with the fact that the zero-frequency responsivity-to-noise ratio is independent of the time constant for Type I detectors.

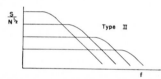

FIG. 5. Showing on logarithmic coordinate scales the relation between the responsivity-to-noise ratio $\Re$ and the frequency f for a Type II detector with a single time constant. The separate curves correspond to different values of the reference time constant. A straight line drawn tangent to the knees of the separate curves has a negative slope of 3 decibels per octave. These curves make very clear the fact that in order to maximize the responsivity-to-noise ratio for a sinusoidal signal with angular frequency ω, the time constant of a Type II detector should be set equal to $1/\omega$.

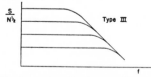

FIG. 6. Showing on logarithmic coordinate scales the relation between the responsivity-to-noise ratio $\Re$ and the frequency f for a Type I detector with a single time constant. The separate curves correspond to different values of the reference time constant. The curves all approach asymptotically the same straight line at high frequencies. This line has a negative slope of 6 decibels per octave.

the $\Re$ *versus* frequency curve measured by Nielsen[6] for the Schwarz No. 11 thermopile. The defining characteristic of this curve is that it slopes off more gradually as the frequency increases than it does in the case of a single time constant, even though the asymptotic slope may be as large as or even greater than 6 db per octave. For a Class I response curve it is here proposed that the actual curve be replaced by a substitute curve characterized by a single time constant in such a way that the substitute curve and the actual curve have in common the points at which they individually have slopes of 3 db per octave. Of all of the possible ways of making the two curves tangent to one another, the method just specified yields the the greatest factor of merit, M_2, as defined in Paper III.

Thus if the actual curve has a slope of 3 db per octave at the frequency f_0, at which point the responsivity-to-noise ratio is N decibels less than its zero-frequency value, then the reference constant is $1/(2\pi f_0)$, and the effective value of $\Re_0$ is $N-3$ decibels less than the actual zero-frequency responsivity-to-noise ratio.

Class II

This type of curve is illustrated in Fig. 2. This figure shows the curve measured by Nielsen for the Strong bolometer. The defining characteristic of this class of curves is that at high frequencies the curve falls off more rapidly than is the case with a single time constant. For such detectors the reference time constant may usefully be defined in terms of the frequency at which $\Re$ is 3 db less than its value at zero frequency. The effective zero-frequency responsivity-to-noise ratio is then to be taken equal to the actual value of $\Re_0$.

Class III

Class III is illustrated by the curve shown in Fig. 3. This figure represents the responsivity-to-noise ratio as measured by Nielsen for the Golay pneumatic heat detector. The defining characteristic of the class is that the behavior at high frequencies may either be that corresponding to Eq. (7.1) or be that corresponding to the behavior represented by a Class I or II detector, but that the curve falls off at low frequencies. In this case, provided the flat portion of the curve is several octaves wide, it is convenient and useful to extend the flat portion of the curve to zero frequency and to define the reference time constant and the effective zero-frequency responsivity-to-noise ratio in terms of the substitute curve so obtained.

[6] Harald H. Nielsen, *Comparative Testing of Thermal Detectors*, OSRD Report No. 5992, Contract OEMsr-1168 (October 31, 1945).

8. Detectors with Non-Uniform Spectral Response

Up to this point the nature of the radiation in terms of which the responsivity is to be measured has been left undefined. If the response is not the same for all wave-lengths of the radiation signal, then it is clearly necessary to introduce a parameter upon which the noise equivalent power must depend. In many cases it will be most convenient to take this parameter to be the wave-length of the signal, that is to say, the noise equivalent power must be expressed as a function of the wave-length of the radiation. If the factor of merit defined in Paper III is expressed in these terms, then the factor of merit for any given spectrum may be determined by averaging the factor of merit with respect to the spectrum.

In other cases, however, it will be more convenient to use the effective temperature of the radiation as the parameter. In this case the noise equivalent power and factor of merit are expressed as functions of the effective temperature of the radiation. This method of description has the disadvantage that when the absolute radiation temperature is not large compared with the temperature of the detector, the response then depends upon both of these temperatures. This difficulty may be avoided in part by stating the response the detector would have to radiation of a given absolute temperature if the detector were at the temperature absolute zero but had otherwise the same properties.

9. The Proposed Classification

On the basis of the foundation established in the preceding sections of this paper, it is now possible to present a simple statement of the proposed classification system for radiation detectors.

A detector is here defined to be a Type n (Type I, Type II, etc.) detector for a given range of reference time constants if over that range of reference time constants the zero-frequency responsivity-to-noise ratio $\Re_0$ depends upon the area A and the reference time constant τ in accordance with

$$\Re_0 = \frac{k_n \tau^{\frac{1}{2}(n-1)}}{2A^{\frac{1}{2}}}, \qquad (9.1)$$

where k_n is a constant which is independent of A and τ.

In view of the relation (6.5), this definition is equivalent to the statement that a detector is a Type n detector if over the given range of reference time constants the noise equivalent power in the reference condition A is related to the sensitive area A and the time constant τ according to

$$P_m = (A^{\frac{1}{2}}/k_n \tau^{\frac{1}{2}n}). \qquad (9.2)$$

Either Eq. (9.1) or (9.2) may be considered as the definition of a Type n detector.

By combining Eqs. (7.1) and (9.1) one finds

$$\Re(f) = \frac{k_n \tau^{\frac{1}{2}(n-1)}}{2A^{\frac{1}{2}}(1+(2\pi f\tau)^2)^{\frac{1}{2}}}. \qquad (9.3)$$

The relation defined by this equation is exhibited graphically in Figs. 4 to 6, with regard to the dependence of $\Re$ on f and τ. In each of these figures the separate curves correspond to different fixed values of the time constant. The figures utilize logarithmic coordinate scales.

In the case of the Type I detector shown in Fig. 4, the curves representing different time constants all approach asymptotically the same horizontal line, and in the case of the Type III detector shown in Fig. 6, the curves representing different time constants all approach asymptotically the same sloping line. In the intermediate case of the Type II detector, shown in Fig. 5, every curve crosses every other curve. Once one has clearly understood the significance of the curves in these figures, they represent the easiest way of remembering the distinction between the different types of detector behavior.

All of the detectors studied in Part II are either Type I or Type II detectors.

A list of the classification of various specific types of detectors is given in Table I of Part II.

With regard to the dependence on the reference time constant, the zero-frequency responsivity-to-noise ratio of a Type I detector is independent of the time constant, whereas with a Type II detector, $\Re_0$ is proportional to the square root of the time constant. Similarly, the noise equivalent power in the reference condition of a Type I detector is inversely proportional to the square root of the time constant, whereas with a Type II detector, P_m is inversely proportional to the time constant.

In considering the use of any given type of a detector for a specific application, it is important to know the type number of the detector because the optimum proportioning of the equipment with which the detector is associated depends on the type number of the detector. This situation is exemplified in the concluding section of this Part.

10. An Application to Sequential Scanning Systems

In this section a brief application of the classification system established in the preceding section will be presented. The chief purpose of this section is to illustrate the usefulness of the classification, but the problem studied is also of considerable interest in its own right.

The problem set is as follows. Consider a sequential scanning system whose instantaneous field of view covers the solid angle ω and which scans at such a rate that a total solid angle Ω is scanned in unit time. The field to be scanned contains no sources of a radiation of the type to be detected, except for one or two point targets. The only sources of noise are thus the thermodynamic and the internal backgrounds. The problem is to calculate the power of the target which yields a detector response equal to the r.m.s. noise of the detector.

The detecting system will be assumed to consist of a lens which focuses upon the detector the field of view at infinity. The detector responds to the total power incident upon its sensitive area. The scanning process will consist simply of a movement of the entire optical system in such a way that it views successively the various parts of the total field of view which is to be scanned.

The limitation on the sources of noise means that the present treatment will not apply, for example, to television systems as they are ordinarily employed. The limiting background in ordinary television systems is the third type of background discussed in Section 2.

The Stationary System

An expression will be derived in this section for the target strength which produces a steady signal voltage which is equal to the r.m.s. noise voltage, when the optical system is trained steadily in a given direction.

Let S be the strength of the target, as measured by the radiant flux which it produces at the location of the detecting system, in watts/cm². Then if the target is a point source, the power collected by the lens and thereby made to fall on the detector is equal to the product of the target strength S, the effective area of the collecting lens, and the energy transmission factor T:

$$P = \tfrac{1}{4}\pi D^2 T S, \tag{10.1}$$

where D is the effective diameter of the collecting lens.

The area of the sky which subtends the solid angle ω and which is focused upon the sensitive area A of the detector by the lens is defined by

$$\omega = A/f^2, \tag{10.2}$$

where f is the focal length of the lens. The humerical aperture N of the lens is defined by

$$N \equiv D/2f. \tag{10.3}$$

By combining these three relations, one finds

$$P = \tfrac{1}{2}\pi NDT(A/\omega)^{\frac{1}{2}}S. \tag{10.4}$$

If now in this expression the power P is set equal to the noise equivalent power P_m of the detector, then the target strength S becomes equal to the noise equivalent target strength S_m:

$$P_m = \tfrac{1}{2}\pi NDT(A/\omega)^{\frac{1}{2}}S_m. \tag{10.5}$$

According to the classification system developed in Section 9, the noise equivalent power P_m of a Type n detector depends upon its area A and its reference time constant τ according to

$$P_m = (A/\tau^n)^{\frac{1}{2}}/k_n, \tag{10.6}$$

where k_n is a constant independent of A and τ.

Upon eliminating P_m between the last two equations, and solving for S_m, one finds

$$1/S_m = \tfrac{1}{2}\pi k_n NDT(\tau^n/\omega)^{\frac{1}{2}}. \tag{10.7}$$

The discussion of the significance of this expression will be deferred until it has been modified by introducing the scanning process.

The Scanning Process

It will now be assumed that the optical system is moved continuously so that it scans over solid angle at the rate Ω.

It will further be assumed that the rate of scanning is such that each element of solid angle ω is surveyed for a length of time equal to the reference time constant τ. One then has

$$\Omega = \omega/\tau. \tag{10.8}$$

The expression (10.7) gives the noise equivalent target strength for a steady signal. If the signal has a duration of only τ, the noise equivalent target strength will be approximately doubled.**** If this correction is taken into account, one finds from Eqs. (10.7) and (10.8)

$$1/S_m = \tfrac{1}{4}\pi k_n NDT(\omega^{n-1}/\Omega^n)^{\frac{1}{2}}. \tag{10.9}$$

Discussion

All of the results of the present treatment are contained in the last equation.

Independent of the type of the detector, the sensitivity of the system, as measured by $1/S_m$, is

---

**** The factor is exactly two for a sinusoidal signal with the frequency $f = 1/2\pi\tau$. The signal voltage is decreased 3 decibels below its zero frequency value because of the decreased response of the detector itself, and 3 decibels more because of the decreased gain of the amplifier in the reference condition.

directly proportional to the constant k_n of the detector, to the numerical aperture N, to the linear size of the collecting lens, and to the over-all transmission of the optical system.

The last factor

$$(\omega^{n-1}/\Omega^n)^{\frac{1}{2}} \qquad (10.10)$$

of Eq. (10.9) is the only one which depends on the type number of the detector. If the detector is of Type I, this factor is simply

$$1/\Omega^{\frac{1}{2}}. \qquad (10.11)$$

Thus one finds the important result that with a Type I detector the noise equivalent target strength is independent of the resolution $1/\omega$ of the scanning system, and depends only on the rate at which solid angle is scanned.

For a Type II detector, the situation is quite different. In this case, the factor (10.10) becomes

$$\omega^{\frac{1}{2}}/\Omega. \qquad (10.12)$$

It follows that with a Type II detector the signal-to-noise ratio is adversely affected by an increase in resolution. It further follows that a heavier price must be paid for increasing the scanning rate $\dot{\Omega}$ than is paid with a Type I detector.

PART II. CLASSIFICATION OF EIGHT SPECIFIC KINDS OF DETECTORS

1. Introduction

This Part presents the considerations on the basis of which the type numbers of eight different kinds of radiation detectors have been assigned. The system of classification is that proposed in Part I, Section 9.

The conclusions of this Part are summarized in Table I. This table lists the Type I and Type II detectors separately, and states the range of reference time constants for which each classification holds.

For those kinds of detectors whose limiting noise is Johnson noise, the relation between the zero-frequency responsivity S_0 and the zero-frequency responsivity-to-noise ratio $\mathfrak{R}_0$ will be derived in this section for later use.

The relation will be derived on the basis of the assumption that the detector works into an open circuit. In this case, the noise power per unit frequency band width is given by

$$N_n = 4kTR, \qquad (1.1)$$

̩stant, T is the absolute electrical resistance of

TABLE I. Summary of the classification of the eight kinds of detectors studied in Part II. The range of attainable reference time constants for each of the detectors is also shown, as well as the section in which the detector is studied.

Detector	Range of τ in seconds	Section number
Type I detectors: $P_m \sim A^{\frac{1}{2}}/\tau^{\frac{1}{2}}$.		
Golay pneumatic heat detector	?	4
Vacuum phototubes limited by shot noise	$10^{-2} - \infty$	6
Gas photo-tubes limited by shot noise	$10^{-2} - \infty$	6
Photo-multiplier tubes limited by shot noise	$2.5 \times 10^{-10} - \infty$	6
Dipole antenna limited by temperature noise = Johnson noise	$10^{-9} - \infty$	7
Type II detectors: $P_m \sim A^{\frac{1}{2}}/\tau$		
Bolometers	$10^{-3} - 1$	2
Thermocouples and thermopiles	$10^{-2} - 1$	3
Photographic plates limited by grain structure	Reciprocity law	5

the detector. By the use of Eq. (4.1) of Part I, one then obtains

$$\mathfrak{R}_0 = S_0/(4kTR)^{\frac{1}{2}}. \qquad (1.2)$$

This is the desired relation for detectors whose sensitivity is limited by Johnson noise.

2. Bolometers

The discussion of bolometers is presented stepwise. It is first assumed that the only noise present is Johnson noise (internal background). It is then assumed that the limiting noise is temperature noise (radiation background). Finally, an expression is obtained which takes into account simultaneously both types of noise.

The zero-frequency responsivity of a bolometer may be written

$$S_0 = \alpha R I \Delta T/P, \qquad (2.1)$$

where α is the fractional change in the resistance per degree of temperature change, R is the resistance, I is the current through the bolometer, and ΔT is the rise in temperature of the bolometer when the radiant power P is incident upon its sensitive area. The change in temperature ΔT may be written

$$\Delta T = \epsilon P/\kappa A, \qquad (2.2)$$

where ϵ is the emissivity of the bolometer, κ is the steady-state heat flow from the bolometer to its surroundings per unit area per degree of temperature difference, and A is the sensitive area.

If C is the heat capacity of the bolometer per unit area, the physical time constant is

$$\tau = C/\kappa, \qquad (2.3)$$

and upon eliminating ΔT and κ by Eqs. (2.2) and (2.3), Eq. (2.1) becomes:

$$S_0 = \epsilon \alpha R I \tau/CA. \qquad (2.4)$$

Recognition will now be given to the fact that the permissible current I is usually limited by the fact that the current heats the bolometer, and that the maximum permissible temperature rise is limited. Let θ be the maximum permissible temperature rise. Then the maximum current is related to θ by

$$I^2R = \theta\kappa A = \theta CA/\tau. \tag{2.5}$$

Upon eliminating I between Eqs. (2.4) and (2.5), one finds

$$S_0 = \epsilon\alpha(\theta R\tau/CA)^{\frac{1}{2}} \tag{2.6}$$

If now the sensitivity of the bolometer is limited by the internal background of Johnson noise, the physical time constant is equal to the reference time constant, and Eq. (1.2) may be employed, with the result

$$\mathcal{R}_0 = \frac{\epsilon\alpha}{2} \cdot \left(\frac{\theta}{kTC}\right)^{\frac{1}{2}} \cdot \frac{\tau^{\frac{1}{2}}}{A^{\frac{1}{2}}}. \tag{2.7}$$

By comparison of this relation with Eq. (9.1) of Part I one concludes correctly that the bolometer is a Type II detector.

This conclusion is as yet upon a somewhat shaky basis, however, because if C is eliminated between (2.3) and (2.7), one obtains

$$\mathcal{R}_0 = \frac{\epsilon\alpha}{2} \cdot \left(\frac{\theta}{kT\kappa}\right)^{\frac{1}{2}} \cdot \frac{1}{A^{\frac{1}{2}}} \tag{2.8}$$

from which one might conclude that the bolometer were a Type I detector.

In order to resolve this situation, one must introduce the following practical consideration: In constructing a bolometer one first makes C as small as possible, because this shortens the time constant without affecting the zero-frequency responsivity. After this is done, κ is increased by various means until the time constant has been shortened to the desired value. This description is considerably oversimplified, but is substantially correct. The chief respect in which this description is incomplete is that most of the methods ordinarily used to increase the value of the constant κ do so in a way which leads to the introduction of further time constants; in other words, the quantity κ does not correspond to a simple resistance but to a two-terminal network containing reactive elements as well as resistances.

Since C is a constant which for a given type of bolometer is independent of the time constant, it follows from Eq. (2.7) that the bolometer limited by Johnson noise is a Type II detector for the range of reference time constants over which the time constant is adjustable by varying the cooling constant κ. By inspection of Figs. 1 and 2 of Paper III, this range is seen to be roughly from one millisecond to one second.

It is possible to make bolometers with reference time constants substantially longer than one second by increasing the value of the thermal capacity per unit area C. As indicated by Eq. (2.8), the bolometer is then a Type I detector. No advantage is gained from this procedure so that it is not used in practice. On the other hand, it was just this type of variation which was used in obtaining the maximum possible sensitivity of a bolometer or thermocouple in Paper I. In obtaining this maximum possible sensitivity it was assumed that the cooling constant κ had its minimum possible value, namely the value obtained when the only cooling is that due to radiation. Under these conditions the only way of varying the time constant is to vary the thermal capacity C. Although it is recognized that in each particular design of bolometer or thermocouple the minimum attainable value of C will always be used, it is nevertheless true that in different designs of bolometers or thermocouples the minimum attainable value of C will be different.

From Eq. (2.8) it may be found by the use of Eq. (6.5) of Part I that the noise equivalent power P_m is given by

$$P_m = \frac{(kT^2CA)^{\frac{1}{2}}}{\epsilon\tau} \cdot \left(\frac{1}{\alpha^2\theta T}\right)^{\frac{1}{2}}. \tag{2.9}$$

So far in this section it has been assumed that the only noise present is the Johnson noise associated with the electrical resistance of the bolometer. The noise equivalent power in the reference condition A will now be calculated by assuming that the only noise present is temperature noise.[††]

According to Eq. (2.4) of Paper I the r.m.s. temperature fluctuation $\langle\Delta T\rangle_{\text{r.m.s.}}$ of the bolometer due to temperature noise is given by

$$\langle\Delta T\rangle_{\text{r.m.s.}} = (kT^2/CA)^{\frac{1}{2}}. \tag{2.10}$$

When this relation is combined with Eqs. (2.2) and (2.3) one finds

$$P_m = \frac{(kT^2CA)^{\frac{1}{2}}}{\epsilon\tau}. \tag{2.11}$$

This equation expresses the noise equivalent power in reference condition A if the only source of noise is temperature noise.

Consider now the actual case in which both types of noise are simultaneously present. First let it be noted that the noise powers of the two noises will

†† The properties of temperature noise are discussed at length in Paper I of this series.

simply add because the two noise voltages are statistically independent. Since the noise equivalent power P_m is proportional to the r.m.s. noise voltage, it then follows that the noise equivalent power in the presence of both types of noise is equal to the square root of the sum of the squares of the noise equivalent powers defined by Eqs. (2.9) and (2.11):

$$P_m = \frac{(kT^2CA)^{\frac{1}{2}}}{\epsilon\tau}\left(1+\frac{1}{\alpha^2\theta T}\right)^{\frac{1}{2}}. \qquad (2.12)$$

If one compares this relation with Eq. (9.2) of Part I, one finds that the bolometer, whether it is limited by Johnson noise, by temperature noise, or by both, is a Type II detector over the range of time constants attainable by varying the constant κ.

In Eq. (2.12) the first term in the last factor represents the effect of temperature noise and the second term represents the effect of Johnson noise.

3. Thermocouples and Thermopiles

The zero-frequency responsivity of a thermocouple or a series-connected thermopile may be written

$$S_0 = nQ\Delta T/P, \qquad (3.1)$$

where n is the number of hot junctions, Q is the thermoelectric power of the pair of metals used, and ΔT is the change in the temperature of the hot junctions when the radiant power P is incident thereon. The change in temperature ΔT may be written

$$\Delta T = \frac{\epsilon P}{\kappa A + n(\sigma_1+\sigma_2)}, \qquad (3.2)$$

where ϵ is the emissivity, κ is the heat flow from the receiver attached to each of the hot junctions to the surroundings per unit area per unit temperature difference by all means except by the electrical conductors, σ is the heat flow per unit temperature difference from a single hot junction to a single cold junction through the electrical conductor, A is the total area of the receivers attached to the hot junctions, and the subscripts 1 and 2 refer to the two metals.

The physical time constant of the thermopile is then given by

$$\tau = \frac{CA}{\kappa A + n(\sigma_1+\sigma_2)}, \qquad (3.3)$$

where C is the heat capacity per unit area. The conductivities σ_1 and σ_2 may be written

$$\sigma_1 = k_1 a_1/l_1, \qquad (3.4)$$

$$\sigma_2 = k_2 a_2/l_2, \qquad (3.5)$$

where k_1 is the thermal conductivity of the electrical conductor of metal No. 1, a_1 is the effective cross-sectional area of this conductor, and l_1 is the length of this conductor. By combining Eqs. (3.2) and (3.3) one finds the following simple expression for the zero-frequency responsivity:

$$S_0 = nQ\epsilon\tau/CA. \qquad (3.6)$$

It is now necessary to derive the relation between the quantity n in the last equation and the resistance R of the thermocouple. If the couples are connected in series, the resistance R of the thermopile may be written

$$R = n(l_1\rho_1/a_1 + l_2\rho_2/a_2), \qquad (3.7)$$

where ρ is the electrical resistivity. By the use of Eqs. (3.4) and (3.5) the last relation may be written

$$R = n(k_1\rho_1/\sigma_1 + k_2\rho_2/\sigma_2). \qquad (3.8)$$

The resistance R will now be minimized, subject to the condition that $\sigma_1+\sigma_2$ has a constant value. This procedure leads to the following results:

$$\frac{\sigma_1}{\sigma_2} = \left(\frac{k_1\rho_1}{k_2\rho_2}\right)^{\frac{1}{2}}, \qquad (3.9)$$

$$\frac{a_1 l_2}{a_2 l_1} = \left(\frac{k_2\rho_1}{k_1\rho_2}\right)^{\frac{1}{2}}. \qquad (3.10)$$

By the use of Eq. (3.9), the resistance R may be written

$$R = \frac{n}{\sigma_1+\sigma_2}((k_1\rho_1)^{\frac{1}{2}} + (k_2\rho_2)^{\frac{1}{2}})^2, \qquad (3.11)$$

and upon combining Eqs. (3.3) and (3.11), one finds

$$R = \frac{n^2\tau}{CA}((k_1\rho_1)^{\frac{1}{2}} + (k_2\rho_2)^{\frac{1}{2}})^2\left(1+\frac{\kappa A}{n(\sigma_1+\sigma_2)}\right). \qquad (3.12)$$

By solving the last relation for n as it appears in the first factor of the last relation and substituting the result in Eq. (3.6), one finds the following expression for the zero-frequency responsivity of the thermopile:

$$S_0 = \frac{\epsilon Q}{(k_1\rho_1)^{\frac{1}{2}} + (k_2\rho_2)^{\frac{1}{2}}}$$

$$\cdot\left(1+\frac{\kappa A}{n(\sigma_1+\sigma_2)}\right)^{-\frac{1}{2}}\left(\frac{R\tau}{CA}\right)^{\frac{1}{2}}. \qquad (3.13)$$

If now it be assumed that the only noise is that due to the internal background of Johnson noise associated with the resistance R, one finds by the use of Eq. (1.2) that the noise equivalent power is given by

$$\mathcal{R}_0 = \frac{\epsilon Q}{(k_1\rho_1)^{\frac{1}{2}}+(k_2\rho_2)^{\frac{1}{2}}} \cdot \frac{1}{(kTC)^{\frac{1}{2}}}$$

$$\cdot \left(1+\frac{\kappa A}{n(\sigma_1+\sigma_2)}\right)^{-\frac{1}{2}} \cdot \frac{\tau^{\frac{1}{2}}}{A^{\frac{1}{2}}}. \quad (3.14)$$

It follows from the last equation that in an efficient design the cooling of the hot junction by conduction through the wires is made large compared with the cooling by all other causes, that is to say, one satisfies the condition

$$\kappa A \ll n(\sigma_1+\sigma_2). \quad (3.15)$$

By making use of the same practical consideration invoked in Section 2, it now follows from the last two equations that the thermocouple limited by Johnson noise is a Type II detector. By inspection of Figs. 1 and 2 of Paper III, one finds that this conclusion holds for time constants lying between about 10 milliseconds and one second.

By use of Eq. (6.5) in Part I and Eq. (3.14) in this section, the noise equivalent power P_m is given by

$$P_m = \frac{(kT^2CA)^{\frac{1}{2}}}{\epsilon \tau}$$

$$\cdot \frac{(k_1\rho_1)^{\frac{1}{2}}+(k_2\rho_2)^{\frac{1}{2}}}{2QT^{\frac{1}{2}}}\left(1+\frac{\kappa A}{n(\sigma_1+\sigma_2)}\right)^{\frac{1}{2}} \quad (3.15)$$

So far in this section it has been assumed that the only noise present is Johnson noise. By use of the same method employed in Section 2 and in particular by the use of Eq. (2.11), the effect of temperature noise may be introduced into the result (3.15). One finds

$$P_m = \frac{(kT^2CA)^{\frac{1}{2}}}{\epsilon \tau}$$

$$\cdot \left\{1+\frac{[(k_1\rho_1)^{\frac{1}{2}}+(k_2\rho_2)^{\frac{1}{2}}]^2}{4Q^2T}\left(1+\frac{\kappa A}{n(\sigma_1+\sigma_2)}\right)\right\}^{\frac{1}{2}}. \quad (3.16)$$

It follows from this equation that the thermocouple and thermopile, whether they are limited by Johnson noise, by temperature noise, or by both, are Type II detectors.

The conclusions of this section with regard to the classification of the thermopile are in no way changed if one assumes that the separate pairs of junctions are connected in parallel or in series-parallel.

4. The Golay Pneumatic Heat Detector

The theoretical basis of this detector has been described[7] at length by its inventor, and will not be repeated here.

The type number of this detector depends upon how one modifies the construction of the detector in order to change its time constant. The nature of the construction is such as to suggest that the constant κ is set in advance, and that the time constant can be adjusted only by changing the thermal capacity per unit area of the membrane. If this is true, the Golay pneumatic heat detector is a Type I detector, and it is provisionally so classified in Table I.

The theory indicates that the responsivity-to-noise ratio should be less by the factor $3^{\frac{1}{2}}$ than the theoretical maximum calculated in Paper I. The measurements reported in Paper III suggest an actual factor of 3.3.

5. The Photographic Plate

In this section the type number of the photographic plate will be determined on the basis of the assumption that the smallness of the minimum detectable energy is limited by the spatial fluctuations of the density within a uniformly exposed area (internal background).

Because the photographic plate does not yield an electrical voltage as its output, the reference condition of measurement defined in Part I does not apply to it. The chief problem accordingly lies in the definition of the reference time constant to be ascribed to the photographic plate. The procedure most closely analogous to that used with other detectors would be to define the time constant in terms of the characteristic times associated with the build-up and decay of the latent image. These time constants may be very long, of the order of days, months, or years.

But photographic plates are rarely used in a way such that the build-up and decay times of the latent image are the controlling ones. In detectors with an electrical output, the significance of the reference time constant τ is that if one uses the same equalization in the amplifier as that used in the reference condition of measurement, the noise equivalent power is increased only slightly if one

---

[7] Marcel J. E. Golay, Rev. Sci. Inst. 18, 347–362 (1947).

repeats the measurement every τ seconds. Accordingly, the writer postulates that the duration of the exposure of the plate is the reference time constant of the photographic plate, and that it is the time constant so defined that may most meaningfully be compared with the reference time constants of other detectors.

It is assumed that the plate has been given a uniform pre-exposure in order to bring the plate to the optimum point on its characteristic curve, and that the optimum method of development is employed. The optimum condition may be defined as that which maximizes the ratio k/κ, where k and κ are defined in Eqs. (5.1) and (5.3). Then, because of the distribution in the size of the developed grains, and because of their more or less random spatial distribution, there is a fluctuation of the density from one location on the plate to another. Furthermore, if one measures the density of a large number of different portions, each of area A, of the uniformly exposed region, there will be a fluctuation in the measured density. This fluctuation means that the average density of any given region of area A is not a unique function of the power incident upon the area, but contains also a random element. The fluctuation may be compared with the noise in electrical detectors, in that it serves to place a lower limit on the power which may be detected.

If $\langle \Delta D \rangle_{\text{r.m.s.}}$ is the r.m.s. fluctuation of the density in the area A, the fluctuation equivalent exposure of the plate is that exposure over the area A which produces a change in density equal to $\langle \Delta D \rangle_{\text{r.m.s.}}$.

Now consider the change in density ΔD produced on the plate by a certain power P incident upon the area A during the period τ. Because of the curvature of the characteristic curve of the photographic plate, the change in density ΔD will not be exactly proportional to the energy incident per unit area $P\tau/A$. Provided, however, that the fractional variation of the slope of the characteristic curve is small over the range of exposures considered, one has approximately

$$\Delta D = kP\tau/A, \qquad (5.1)$$

where k is a constant of proportionality.

It then follows from the last relation that the fluctuation equivalent power is

$$P_m = A \langle \Delta D \rangle_{\text{r.m.s.}}/k\tau \qquad (5.2)$$

subject to the restriction stated above.

The r.m.s. fluctuation of the density has been shown by a number of recent measurements[8] to be

proportional to the inverse square root of the area A

$$\langle \Delta D \rangle_{\text{r.m.s.}} = \kappa A^{-\frac{1}{2}}, \qquad (5.3)$$

where κ is another constant of proportionality. This relation is restricted by the condition that the area A must be so large that the average number of grains contained within it is large compared with unity. Equation (5.3) will usually hold[9] for areas not smaller than about 0.005 mm².

By combining the last two relations, one finds

$$P_m = (\kappa/k)(A^{\frac{1}{2}}/\tau). \qquad (5.4)$$

This relation indicates that the photographic plate is a Type II detector, subject to the two restrictions involved in Eqs. (5.1) and (5.3). Both of the restrictions have the effect of limiting the smallness of the area A.

The result (5.4) is limited in a further way, however, through the fact that the reciprocity law[10] was used in writing Eq. (5.1). Accordingly, the conclusion that the photographic plate is a Type II detector is limited also by the condition that the durations of the exposure must be within the range over which the reciprocity law holds.

Some of the Eastman plates intended for use in spectroscopy and astronomy obey the reciprocity law fairly closely over an exposure range of 10^7 to 1.

Presumably situations exist in which the time constants relating to the formation and fading of the latent image are the controlling ones. It has not been shown that the conclusions of this section apply to these cases.

The writer wishes to express his gratitude to Dr. David L. MacAdam of the Eastman Kodak Company for constructive criticism of several earlier drafts of this section.

6. Photoelectric Detectors

This category is intended to include both gas and vacuum photo-electric tubes, and photo-multiplier tubes.

The background which limits the sensitivity of these devices is the shot noise associated with the dark current of the photo-cathode. In order to obtain high sensitivity in the visible part of the spectrum, it is necessary to employ materials in the photo-cathodes which have a relatively low work function. Unfortunately, all such materials have a measurable thermionic emission at room temperature. Such emission may frequently be

---

[8] For a summary of the experimental information, See L. A. Jones and G. C. Higgins, J. Opt. Soc. Am. **36**, 203–227 (1946).

[9] Private communication from Dr. David L. MacAdam dated January 4, 1949.
[10] *Photographic Plates for Scientific and Technical Use*, sixth edition (Eastman Kodak Company, Rochester, 1948), p. 11.

reduced to a negligible amount by cooling the tube to dry ice or liquid air temperature.[11]

Vacuum Photo-Tube

Consider first a simple vacuum photo-tube. Since the thermionic emission from the photo-cathode is temperature limited, the mean square fluctuation in the cathode current is given by the usual expression[12] for shot noise:

$$\overline{i^2} = 2eI_c\Delta f, \tag{6.1}$$

where I_c is the cathode dark current, e is the charge of the positron, and Δf is the noise equivalent band width.

This noise combines additively with the Johnson noise current of the load resistance R of the photo-tube:

$$\overline{i^2} = (4kT/R)\Delta f. \tag{6.2}$$

The total noise current is thus given by

$$\overline{i^2} = (2eI_c + 4kT/R)\Delta f. \tag{6.3}$$

It is evident from this relation that the noise of the load resistor may always be made small compared with the shot noise by making the load resistance R large enough. The condition is

$$I_cR \gg (2kT/e) = 0.0518 \text{ volt at } T = 300°\text{K}. \tag{6.4}$$

Thus in order to prevent the Johnson noise of the load resistance from being greater than the intrinsic noise of the photo-tube, it is necessary that the load resistance be chosen so that voltage drop produced in it by the cathode current is large compared with 0.0518 volt.

The magnitude of the required resistance may be estimated as follows. The cathode dark current of a very good photo-tube is roughly 10^{-15} ampere.[13] In order that this current produce a voltage drop equal to 0.05 volt, the load resistor must be 5×10^7 megohms. If the anode capacity to ground has the value five micromicrofarads, the RC time constant is then 250 seconds. Since the anode capacity to ground cannot be less than the figure given, it follows that the physical time constant

of a good photo-cell cannot be made less than about 250 seconds when it is operated in such a way as to obtain the minimum noise power per unit band width.

The physical time constant in this case is not equal to the reference time constant, since the noise power per unit band width is inversely proportional to the frequency squared for angular frequencies greater than 0.004 radian/sec. Indeed, the reference time constant is infinitely short, since the responsivity squared and the noise power per unit band width both decrease in the same way as the frequency is increased.

Accordingly, it is again necessary to introduce a "practical consideration." First of all, because of the extremely high impedance of the detector at low frequencies, the only practical way of amplifying the output of the photo-tube is to connect it to the grid of a vacuum tube amplifier. Furthermore, in practice, amplifiers themselves possess a certain amount of noise; this noise may be defined in terms of the input resistance whose Johnson noise would produce the same noise level at the output of the amplifier; this noise equivalent resistance R_{eq} is of the order of 10^5 ohms for conventional low frequency amplifiers. From this consideration it follows that the reference time constant is shorter than the RC time constant by a ratio equal to the square root of R_{eq}/R:

$$\begin{aligned} \tau &= R_0C(R_{eq}/R_0)^{\frac{1}{2}}, \\ \tau &= (R_0R_{eq})^{\frac{1}{2}}C, \end{aligned} \tag{6.5}$$

where R_0 is the load resistance which makes Eq. (6.4) an equality, R_{eq} is the noise equivalent resistance of the amplifier, and C is the sum of the capacity of the photo-tube and of the input of the amplifier.

Equation (6.5) indicates the shortest reference time constant attainable with a simple photo-tube if one requires that the shot noise of the photo-tube itself should be dominant. With the values $R_0 = 5 \times 10^{13}$ ohm, $R_{eq} = 10^5$ ohm, and $C = 5 \times 10^{-12}$ farad, this equation yields $\tau = 0.01$ sec.

Provided that the second term on the right-hand side of Eq. (6.3) is small compared with the first— that is to say, provided that the reference time constant is longer than that defined by Eq. (6.5), the noise equivalent photo-current is given by

$$\begin{aligned} i_m &= (2eI_c\Delta f)^{\frac{1}{2}} \\ &= (eI_c/2\tau)^{\frac{1}{2}}. \end{aligned} \tag{6.6}$$

Since the dark current I_c is proportional to the area of the photo-cathode, and since for a given photo-emissive surface the noise equivalent power is proportional to the noise equivalent current, it follows that the simple photo-tube is a Type I

---

[11] Ralph W. Engstrom, J. Opt. Soc. Am. **37**, 420–431 (1947).

[12] E. B. Moullin, *Spontaneous fluctuations of voltage* (Clarendon Press, Oxford, 1938), Chapter 2.

[13] Dr. Ralph Engstrom in a private communication dated January 3, 1949, suggests that the cathode dark current of the very best 1P21 is about 10^{-15} ampere, and that the average cathode dark current for this tube is about 10^{-14} ampere. The HB-3 RCA Tube Manual states that the maximum anode dark current of the 1P21 is 10^{-7} ampere when the gain is 2×10^6; this corresponds to a maximum cathode dark current of 5×10^{-14} ampere.

detector for reference time constants longer than that given by Eq. (6.5).

Gas Photo-Tubes and Photo-Multiplier Tubes

The discussion so far in this section has brought out the two fundamental difficulties in obtaining high signal-to-noise ratios with simple photo-tubes:

(1) The values of the minimum load resistance are fantastically high: about 10^8 megohms.

· (2) The shortest time constant is about 0.01 second, which corresponds to a band width of 16 c.p.s.

The successful outcomes of the efforts to overcome these two difficulties were first, the gas photo-tube, and second, the photo-multiplier tube. The two are conveniently treated together.

Both contain internal amplification by means of which on the average G electrons reach the anode for each electron emitted by the photo-cathode. The amplification G is 10 for a typical gas photo-tube, and 10^6 for a typical photo-multiplier. The gas photo-tube suffers from the limitation that its amplification drops off above about 10^4 c.p.s., whereas the photo-multiplier maintains its gain up to about 10^9 c.p.s.,[14] at which point variations in electron transit time serve to reduce its response.

The mean square fluctuation in the anode current for these tubes is given by

$$\overline{i^2} = 2eI_cFG^2\Delta f$$
$$= 2eI_aFG\Delta f, \qquad (6.7)$$

where I_c is the cathode current, I_a is the anode current, and F is a numerical constant which takes account of the fact that the amplification process itself introduces noise which is over and above the amplified noise in the cathode current.

Let p_n be the probability that when one electron leaves the cathode, exactly n electrons reach the anode. Then F is defined by

$$F \equiv \frac{\sum_n n^2 p_n}{(\sum_n n p_n)^2}. \qquad (6.8)$$

It is evident from the form of this expression that F is not less than unity. Various reasonable assumptions about the relative probabilities p_n lead to values of F which lie between unity and two.[15] The factor F will be omitted in the remainder of this paper.

[14] R. D. Sard, J. App. Phys. **17**, 768–777 (1946).

[15] J. A. Rajchman and R. L. Snyder, Electronics **13**, 20–23, 58, 60, December (1940); W. Shockley and J. R. Pierce, Proc. Inst. Rad. Eng. **26**, 321–332 (1938); private communication dated August 17, 1948, from Dr. Hartland S. Snyder (Brookhaven National Laboratory).

The relation which now takes the place of Eq. (6.4) is

$$I_c R \gg 2kT/eG^2. \qquad (6.9)$$

With the gas photo-tube and its gain G of 10, it follows from the numerical assumptions used earlier in this section that the minimum value of the load resistor is 5×10^5 megohms, the RC time constant is 2.5 seconds, and the shortest reference time constant is 0.001 second, or a band width of 160 c.p.s. On the other hand, with a photo-multiplier and its gain of 10^6 or more, the minimum value of the load resistor as given by Eq. (6.9) is only 50 ohms! With a capacity of 5×10^{-12} farad, the RC time constant is 2.5×10^{-10} second, which corresponds to a band width of 6×10^8 c.p.s. Thus with a photo-multiplier it is possible to obtain a band width of 600 megacycles, which band width is of the same order as that of the widest amplifier yet constructed.

In summary, all three detectors are Type I detectors for time constants greater than a certain minimum value. The minimum time constants are roughly 0.01 second, 0.001 second, and 2.5×10^{-10} second, for vacuum photo-tubes, gas photo-tubes, and photo-multiplier tubes, respectively.

The writer would like to express his appreciation of very helpful criticism of an early draft of this section, rendered independently by: Dr. Ralph Engstrom, Dr. J. A. Rajchman, and Dr. Albert Rose, all of Radio Corporation of America, and Dr. Hartland S. Snyder, Brookhaven National Laboratory.

7. The Dipole Antenna

The dipole antenna is a somewhat different type of detector from the others which are considered in this paper, and the treatment of it is somewhat artificial from the present point of view.

There are three ways in which the performance of the dipole antenna differs from those treated so far in this paper:

1. The output voltage is not proportional to the power incident upon its effective area, but it is rather the output power which is proportional to the incident power.

2. The directivity pattern is different from that of the other detectors. More importantly, if one varies the area of the antenna by the only available method, namely by varying the number of dipoles in an array, the noise equivalent power is not proportional to the square root of the effective area, but is rather independent of the effective area. This fact is tied in with the fact that the directivity of the detector increases as the effective area increases. This difficulty may be avoided simply by introducing instead of the area, the product of the area and the reciprocal directivity index. This possibility has not been developed in these papers, however, because our primary interest lies in detectors which obey Lambert's law.

3. Temperature noise and Johnson noise become identical. From one point of view, the noise at the output of a dipole

antenna may be considered as the Johnson noise associated with the radiation resistance of the dipole, but from another point of view the noise may be considered as the response of the antenna to the thermal radiation field in which it is placed. The same result is obtained from either point of view. This situation is due to the fact that, except for ohmic losses, the dipole antenna is a completely reversible transducer with an efficiency of unity.

If, however, one wishes to force the dipole antenna into the mold established in Part I, it may be done as follows.

In an array of dipoles, let each dipole be connected separately through a matched transmission line to a square law rectifier, and let the outputs of the rectifiers be connected in series. It is supposed that the input of the rectifier has an impedance high compared with that of the dipole. Let ν denote the frequency of the voltage at the input of the rectifier, and let f denote the frequency at the output. Then the behavior of the square law rectifier may be written

$$V_f = \alpha V_\nu^2, \tag{7.1}$$

where V is the instantaneous voltage.

For the sake of simplicity of treatment, it is further supposed that the highest frequency f in the output of the rectifier is small compared with the lowest frequency ν at the input. It is further assumed that the band width Δf at the output of the rectifier is small compared with the band width $\Delta \nu$ of the dipole. Then it is easily shown that the noise power per unit band width N_f in the output of the detector is related to that at the input by

$$N_f = \alpha^2 \Delta \nu N_\nu^2, \tag{7.2}$$

where N_ν is assumed to be constant within the band width $\Delta \nu$, and zero otherwise. The noise power per unit band width N_ν is simply the Johnson noise associated with the radiation resistance Z of the dipole

$$N_\nu = 4kTZ, \tag{7.3}$$

whence

$$N_f^{\frac{1}{2}} = 4\alpha kTZ(\Delta\nu)^{\frac{1}{2}}. \tag{7.4}$$

Let the response to a signal be considered next. If P is the signal power incident upon the effective receiving area of the antenna, the open-circuit signal voltage at the input of the rectifier is given by

$$V_\nu^2 = ZP, \tag{7.5}$$

where Z is the radiation resistance of the dipole. One then has by Eq. (7.1)

$$V_f = \alpha ZP \tag{7.6}$$

or

$$S \equiv V_f/P = \alpha Z. \tag{7.7}$$

By combining Eqs. (7.4) and (7.7) one finds that the zero-frequency responsivity-to-noise ratio for a single dipole is

$$\Re_0 = \frac{S}{N_f^{\frac{1}{2}}} = \frac{1}{4kT(\Delta\nu)^{\frac{1}{2}}}. \tag{7.8}$$

If the dipoles in the array are separated by a distance of at least one-half wave-length, so that the noises at their outputs are nearly completely independent in the statistical sense, then one may use the argument given in Part I, Section 5, with the result that $\Re_0$ for the array is inversely proportional to the square root of the number of dipoles, and thus also inversely proportional to the square root of the total area of the array.

Thus, since $\Re_0$ is independent of the reference time constant, and is inversely proportional to the square root of the area, it follows that the array with the associated square law detectors is a Type I detector.

The argument given above is interesting in that no attempt was made to define a reference time constant. It was merely shown (Eq. (7.8)) that for frequencies f within the output pass-band of the rectifier, the responsivity-to-noise ratio is independent of the width of this band.

8. Other Detectors

The most important detector which has been omitted from discussion is the human eye. The eye, however, cannot be discussed fruitfully from the present point of view, because neither the area nor the time constant of the eye are adjustable over any substantial range. The eye is most usefully discussed on the basis of the third type of noise background defined in Section 2 of Part I. This treatment has been carried out by Albert Rose.[3]

Perhaps the next most important detector which has been omitted is the photo-conductive cell. In this case, the detector has not yet been developed from either the theoretical or the experimental aspect, to the point where one knows the dependence of the noise equivalent power on the reference time constant. The writer considers it highly probable that the photo-conductive cell is either a Type I or a Type II detector. Much the same statements may be made about the photovoltaic cell.

No effort has been made in this paper to consider the detectors for x-rays, γ-rays and high energy particles such as ionization chambers, proportional counters, Geiger counters, et cetera.

ACKNOWLEDGMENT

The writer wishes to express his appreciation to Dr. Marcel J. E. Golay of the Signal Corps, U. S. Army, for a helpful discussion of this method of classification.

37

Reprinted from *J. Opt. Soc. Amer.*, **39**(5), 344–356 (1949)

Factors of Merit for Radiation Detectors*

R. Clark Jones

Research Laboratory, Polaroid Corporation, Cambridge, Massachusetts

(Received February 4, 1949)

In Part I, the need for a suitably defined factor of merit for comparing the sensitivity of different radiation detectors is explained (Section 1). Both the advantages and the very important limitations of such a factor of merit are discussed.

In Section 2, separate factors of merit for Type *I* and Type *II* detectors are defined. The factor of merit for Type *I* detectors is defined as the ratio of the noise equivalent power given by the fundamental limit defined in Paper I to the noise equivalent power actually measured in the reference condition. The factor of merit for Type *II* detectors is similarly defined in terms of the actual noise equivalent power in the reference condition and the empirically determined minimum value for the noise equivalent power proposed by R. J. Havens. Section 3 contains a brief discussion of the significance of the factor of merit for the special case of the bolometer.

In Part II, the noise equivalent power in the reference condition, and the factor of merit is determined for about fifty thermocouples and bolometers. The results are presented in Tables I and II, and in Figs. 1 and 2. A description of each detector is given in Section 5, along with a statement of any special calculations which were necessary.

All of the results presented are based on measurements made by other persons; no measurements were performed by the writer. The sources of information are described in Section 3.

An Appendix is provided whose purpose is to call attention to the wide variety of possibilities in the specification of the relevant properties of radiation detectors, and thereby to encourage precision in the statement of these properties in the future periodical literature.

PART I. DEFINITION OF THE FACTORS OF MERIT

1. Introduction

THERE exists a need for a single, quantitative factor of merit for use in comparing the sensitivity of various radiation detectors. This criterion is needed for the intercomparison of similar detectors, such as an evaporated thermocouple and a wire thermocouple, and also for the intercomparison of dissimilar detectors, for example, a Golay pneumatic heat detector and a lead sulfide photoconductive cell.

Such a factor of merit should be capable of intercomparing units with sensitive areas of greatly different magnitude, with different speeds of response, and with different spectral response curves, when measured with an amplifier having any given frequency-response curve.

No such criterion has been developed in the past. In a few cases where the intercomparison of a number of detectors has been attempted, the "comparison" has consisted in the presentation of a table with a large number of entries for each detector, no one of which entries by itself permits a direct comparison of one detector with another.

Presumably the reasons for the failure to compare detectors in a way which, in however restricted a sense, permits a statement of a numerical factor of relative merit, are that

(1) comparisons are unpopular,

(2) comparisons carry with them a heavy load of responsibility, and

(3) a single factor of merit cannot take into account a number of important secondary considerations, such as ease of amplification, over-all physical size, need for auxiliary facilities, the physically attainable ranges of sensitive area and speed of response, and also price and delivery time.

Nevertheless, when it is necessary to choose a detector for a given task, a decision must be made. The writer proposes to assist in this decision by defining a factor of merit which depends upon, and only upon, the relative signal-to-noise ratios obtainable with the various detectors.

This paper contains the writer's proposal of a numerical factor of merit. The proposed factor of merit takes no account of the secondary considerations mentioned above as item 3, and the writer accepts the implications of items 1 and 2. It is offered not as a final judgment, but as a beginning, to be tempered and modified in the light of experience and opinion.

2. Definition of the Factor of Merit

The factor of merit proposed here is based on the reference condition of measurement defined in Paper II and, indeed, is almost implicit in the definition of the classification system. In Part I, Section 9 of Paper II a Type *n* detector is defined as one whose noise equivalent power P_m in the reference condition A depends upon the reference time constant τ and the sensitive area A according to

$$P_m = (A^{\frac{1}{2}}/k_n \tau^{\frac{1}{2}n}), \qquad (2.1)$$

where k_n is a parameter which is independent of A and τ, but which has different values for different detectors.

Since for a given area and time constant it is desired to make the noise equivalent power as small

* This is the third of a series of papers, the second and third of which appear in this issue. The work reported in this paper was supported in part by U. S. Navy Contract NObsr-42179.

as possible, it is evident that k_n is itself a factor of merit for a Type n detector. This parameter is of inconvenient magnitude for ordinary purposes, however, and one accordingly looks for a suitable way of defining a numerical multiple of k_n as the factor of merit.

For Type I detectors, a convenient multiple may be obtained by using Eq. (3.8) of Paper I. For a detector whose quantum efficiency is unity at every radiation wave-length, Eq. (3.8) states that the minimum value of the noise equivalent power is given by

$$P_m = (4\sigma T^4 kT)^{\frac{1}{2}}(A^{\frac{1}{2}}/\tau^{\frac{1}{2}}), \qquad (2.2)$$

where k is Boltzmann's constant, σ is the Stefan-Boltzmann radiation constant, and where T is the absolute temperature of the detector and of the surrounding radiation field. At the temperature $T = 300°K$ the last equation may be written

$$P_m = 2.76 \times 10^{-12}(A^{\frac{1}{2}}/\tau^{\frac{1}{2}}), \qquad (2.3)$$

where P_m is in watts, A in square millimeters, and τ is in seconds. If a detector satisfying Eq. (2.3) be considered to have a factor of merit equal to unity, then the factor of merit for any other Type I detector may be written

$$\begin{aligned} M_1 &= 2.76 \times 10^{-12} k_1, \\ &= 2.76 \times 10^{-12}(A^{\frac{1}{2}}/P_m\tau^{\frac{1}{2}}), \qquad (2.4) \\ &= 5.52 \times 10^{-12} \Re_0 A^{\frac{1}{2}}, \end{aligned}$$

where P_m is in watts, A is in square millimeters, τ is in seconds, k_1 is in the units resulting from using the units just mentioned in Eq. (2.1), and $\Re_0$ is in watt^{-1}-sec.$^{-\frac{1}{2}}$.

Equation (2.4) is the proposed factor of merit for a Type I detector.

For Type II detectors, a convenient multiple may be obtained by using Havens' limit. About three years ago, R. J. Havens[**] made an estimate of the minimum value of the noise equivalent power which could be obtained with thermocouples and bolometers with currently available materials and techniques, when the detector was at room temperature. His estimate was in no way intended to be a fundamental limit; it has the nature rather of an optimistic engineering estimate. His estimate has been very well confirmed in the sense that no bolometer or thermocouple known to the writer which operates at room temperature has better performance than is indicated by his estimate, although several detectors have approached it within a factor of two, as is shown in Part II.

Havens' limit is

$$P_m = 3 \times 10^{-12}(A^{\frac{1}{2}}/\tau), \qquad (2.5)$$

[**] R. J. Havens, J. Opt. Soc. Am. **36**, 355A (1946).

where P_m is in watts, A in square millimeters, and τ in seconds. If a detector which satisfies Eq. (2.5) be considered to have a factor of merit equal to unity, then the factor of merit M_2 for any other Type II detector may be written

$$\begin{aligned} M_2 &= 3.0 \times 10^{-12} k_2, \\ &= 3.0 \times 10^{-12}(A^{\frac{1}{2}}/P_m\tau), \qquad (2.6) \\ &= 6.0 \times 10^{-12}(\Re_0 A^{\frac{1}{2}}/\tau^{\frac{1}{2}}), \end{aligned}$$

where P_m is in watts, A is in square millimeters, τ is in seconds, k_2 is in the units resulting from using the units just mentioned in Eq. (2.1), and $\Re_0$ is in watt^{-1}-sec.$^{-\frac{1}{2}}$.

Equation (2.6) is the proposed factor of merit for a Type II detector.

The factor of merit M_2 for about fifty bolometers and thermocouples is plotted in Fig. 2.

3. Significance of the Factor of Merit

Because of the rather formal definition of the factors of merit given above, it seems appropriate at this point to discuss briefly the significance of the factor of merit.

In order to make the discussion as specific as possible, it will be supposed that the detector in question is a bolometer limited by Johnson noise. As shown in Part II, Section 2 of Paper II, the thermal capacity per unit area is to be held at its minimum attainable value (which value is supposed to be independent of the sensitive area) as the sensitive area is changed. The reference time constant is changed by varying the effective cooling constant κ, while again the thermal capacity per unit area is held constant at its minimum value. On the basis of this prescription, the factor of merit M_2 of the bolometer remains constant as the sensitive area and reference time constant are changed.

Consider now two bolometers with different sensitive areas and different time constants but with the same factor of merit. Because of the difference in their sensitive areas and time constants these two bolometers, in spite of the fact that their factors of merit are the same, will not in general yield equivalent results in any *particular* application. The fact that the two bolometers have the same factor of merit means, however, that if the two bolometers are reconstructed so that they each have the optimum sensitive area and optimum time constant for the particular application, then the performance of the two will be the same.

If the two bolometers have different factors of merit, then when they are each constructed so that they have the optimum sensitive area and time constant for a given application, the signal-to-noise ratios will be in proportion to their factors of merit.

PART II. FACTORS OF MERIT FOR FIFTY RADIATION DETECTORS

1. Introduction

This part contains the calculation of the factors of merit for about fifty Type *II* radiation detectors. The results of the calculations are shown in Tables I and II, and in Figs. 1 and 2.

As described in Section 3, a large number of references was consulted in connection with the collection of the data presented here. Unfortunately, a substantial number of the detectors described in the periodical literature are not described adequately for the present purpose, with the result that it was not possible to extract the information needed to calculate the factor of merit.

In some cases the missing information was obtained by correspondence with the authors, and in other cases the detector was omitted from consideration in this paper.

The most common omission in the published descriptions is information about the noise equivalent band width employed in making a measurement of the signal-to-noise ratio. In the older literature, information about the detector is often inextricably mixed with the properties of the associated galvanometer.

Because of the frequency of incomplete or vague description in the periodical literature, the writer has included in an Appendix to this paper a number of suggestions with regard to the presentation of information about measurements on detectors.

2. Assumed Operating Conditions

In computing the noise equivalent power in either of the reference conditions defined in the preceding paper, it is necessary to adopt several arbitrary conventions.

In the case of all detectors, it is assumed that they work into an open circuit, rather than into a load resistance equal in value to the resistance of the detector. In the latter case the factor of merit is less by the factor $2^{-\frac{1}{2}}$ than in the former case, if the only noise is the Johnson noise of the detector.

In the case of bolometers, it is assumed that the bolometer is operated with a series resistance large compared with the resistance of the bolometer strip, so that the responsivity of the detector in volts per watt is the product of the resistance change in ohms per watt multiplied by the maximum permissible current through the bolometer.

In the case of thermocouples and thermopiles, it is assumed that the construction of the hot and cold junctions is symmetrical, so that the resistances associated with the two types of junctions are equal. The Johnson noise is calculated for the series resistance of the hot and cold junctions.

In one sense, these conventions discriminate against the thermocouples in favor of the bolometers, since in principle one may use a cold junction of very low resistance, and thus obtain a gain in signal to Johnson noise ratio by the factor $2^{\frac{1}{2}}$ over that involved in the above convention with regard to thermocouples. On the other hand, however, in practice one nearly always uses a cold junction equal in resistance to that of the hot junction, whereas with bolometers one often uses a series resistance several times the bolometer resistance. Anyone who disagrees with the propriety of these conventions may divide the factor of merit of the bolometers by the factor $2^{\frac{1}{2}}$ on the basis that they should be used with a series resistance equal to the bolometer resistance, or he may increase the factor of merit of the thermocouples by the factor $2^{\frac{1}{2}}$ on the basis that they may be used with low resistance cold junctions.

3. Sources of Information

The data used in this report were obtained from three sources: the periodical literature, private communications, and Professor Harald H. Nielsen's comprehensive OSRD report[1] on the Ohio State tests of thermal detectors.

The available information was divided into two classes. Table I includes those detectors for which actual noise measurements were not available, and for which accordingly it was necessary to assume that the noise could be calculated on the basis that it was all Johnson noise. For the detectors in Table I the noise equivalent power and the factor of merit were calculated by Eqs. (4.5) and (4.7) below.

Table II includes those detectors for which actual noise measurements were available, so that it is possible to calculate the noise equivalent signal in the presence of the actual noise, rather than the noise calculated by assuming that the only noise is Johnson noise. For the detectors in Table II, the noise equivalent power and the factor of merit were calculated by Eqs. (4.9) and (4.10).

Although the actual noise in the output of thermocouple detectors is usually no higher than the computed Johnson noise, the actual noise of bolometers is frequently higher, sometimes much higher, than the computed Johnson noise. Accordingly, it should be recognized that the actual factors of merit of detectors 23 through 33 are probably somewhat less than those presented in Table I and plotted in Fig. 2. The degree to which they are less is not determinate on the basis of information now available.

There is a similar weakness in the case of some of the entries in Table II. In some cases, the noise was

---

[1] Harald H. Nielsen, *Comparative Testing of Thermal Detectors*, OSRD Report No. 5992, Contract OEMsr-1168 (October 31, 1945).

measured in a bandwidth which is much wider than that of the detector. This has the result that if the noise is high within the band width of the detector, but is otherwise lower, the high noise within the detector band width does not appear in the result of the measurement.***

Professor Nielsen's group actually measured the noise of the detectors studied by them. Unfortunately, however, as is emphasized in their report, the amplifiers used were not characterized by an unusually low noise level, with the result that the measured noise often reflected the noise of the amplifier rather than the noise of detectors. Since the relative importance of the noise added by the amplifiers varied from unit to unit, depending on the impedance of the detector, it was felt that the use of the measured noise would not permit an altogether equitable intercomparison of the different units. Accordingly, none of the noise measurements reported by Nielsen have been used, and all of the units tested by Nielsen are therefore included in Table I.

Although in a few cases (Numbers 22 and 46, for example) the data represent the typical performance of the manufacturer's production, most of the entries in Tables I and II represent measurements made on a single detector. Furthermore, in some cases (Numbers 3 and 23, for example), only a single sample of the detector has been tested. It is evident that conclusions based on such information should be accepted with some reserve.

Much of the information in the tables is not up-to-date. In particular, all of the tests reported by Nielsen[1] were completed before October 31, 1945. It is reasonable to suppose that in the meantime many of the detectors have been improved.

No information is given here about the spectrum of the radiation used to measure the noise equivalent power. This information is lacking in many of the references used, and is useful only if one knows also the response of the detector as a function of wave-length.

4. Derivation of Required Formulas

For those detectors listed in Table I, it is assumed that the only source of noise is the Johnson noise associated with the resistance of the detector. In this case simple expressions for the noise equivalent power and the factor of merit may be given in terms of the area A, the resistance R, the effective time constant τ, and the effective zero-frequency responsivity S_0.

***The writer wishes to acknowledge his indebtedness to Dr. J. A. Becker, Bell Telephone Laboratories, for pointing out the need for emphasizing the weakness in the data mentioned in this paragraph and the preceding one, and also for a helpful discussion of the sources of noise in bolometers.

The signal voltage E_s obtained from a given steady power P incident upon the detector is given by

$$E_s = S_0 P. \tag{4.1}$$

The r.m.s. Johnson noise voltage delivered by the detector is given by

$$E_N = (4kTR\Delta f)^{\frac{1}{2}} = (kTR/\tau)^{\frac{1}{2}}, \tag{4.2}$$

where k is the Boltzmann gas constant. The noise equivalent power P_m in the reference condition A is now to be obtained by equating these expressions. One finds

$$P_m = (kTR/\tau)^{\frac{1}{2}}/S_0. \tag{4.3}$$

Upon reducing the noise equivalent power to a sensitive area of 1 mm² by the use of Eq. (6.6) of Paper II, Part I, one finds that the noise equivalent power H_m in the reference condition B is given by

$$H_m = (kTR/A\tau)^{\frac{1}{2}}/S_0, \tag{4.4}$$

where A is the sensitive area in square millimeters. If the detector is at a temperature of 300°K, Eq. (4.4) becomes

$$H_m = 0.64 \times 10^{-10}(R/A\tau)^{\frac{1}{2}}/S_0, \tag{4.5}$$

where H_m is in watts, R is in ohms, A is in square millimeters, τ is in seconds, and S_0 is in volts/watt.

The factor of merit M_2 for Type II detectors now takes the form

$$M_2 = 3 \times 10^{-12} S_0 (A/kTR\tau)^{\frac{1}{2}}, \tag{4.6}$$

or, when the detector is at a temperature of 300°K,

$$M_2 = 0.0468 S_0 (A/R\tau)^{\frac{1}{2}}, \tag{4.7}$$

where the units are the same as in Eq. (4.5). Thus, in order to determine the factor of merit for a detector whose limiting sensitivity is set by Johnson noise alone, the factor of merit may be defined in terms of the effective zero-frequency responsivity S_0, the sensitive area A, the resistance R, and the reference time constant τ.

For those detectors which are listed in Table II, the noise was actually measured. The entries tabulated in Table II for each detector are defined as follows.

The power P_0 denotes the value of the steady incident power which produces a steady output voltage equal to the noise voltage under the actual condition of measurement. If the measurement was actually made with a sinusoidally modulated signal, the result tabulated was obtained by reducing the measured result to zero frequency by making use of the measured frequency response curve of the detector. Furthermore, if the detector does not possess a single time constant, the power P_0 is the effective value, as defined in Section 7, Part I, of Paper II.

The quantity Δf is the noise equivalent band width in the actual measurement, and A and τ denote the sensitive area and the reference time constant.

Then, since the noise equivalent band width in the reference condition is $1/(4\tau)$, the noise equivalent power P_m in the reference condition A is related to the noise equivalent power P_0 by

$$P_m = P_0/(4\tau\Delta f)^{\frac{1}{2}}. \qquad (4.8)$$

The noise equivalent power II_m in the reference condition B is

$$H_m = P_0/(4\tau A\,\Delta f)^{\frac{1}{2}}. \qquad (4.9)$$

From the last expression the factor of merit M_2 for Type II detectors is given by

$$M_2 = 6\times10^{-12}(A\Delta f/\tau)^{\frac{1}{2}}/P_0, \qquad (4.10)$$

where P_0 is in watts, A is in square millimeters, Δf is in cycles/second, and τ is in seconds.

5. Description of the Detectors

This section contains a brief description of each detector, and a statement of the source of information. If any intermediate computation was necessary, the nature of the calculation is stated.

The numbers in parentheses after the name of each detector give the reference numbers used in the tables and figures.

The letter N in parentheses indicates that the data were obtained from Nielsen's report.

The Weyrich Vacuum Thermocouple (1–2) (N)

This thermocouple has been used for many years by infra-red spectroscopists. It may be used at atmospheric pressure or it may be evacuated. The thermoelectric materials are two different antimony-bismuth alloys. Both of the junctions of the thermocouple are designed for use as radiation receivers. The values entered in Table I refer to the area of a single receiver and the resistance of the two junctions in series. The time constants were obtained with the use of a ballistic galvanometer.

The Eppley (Farrand) Thermocouple (3) (N)

This thermocouple was developed by the Eppley Laboratory and is of the type supplied for use in the Farrand heat detector. As with the Weyrich thermocouple, both of the junctions are fitted with radiation receivers. The information in Table I refers to the area of a single receiver and to the resistance of the two junctions in series. The time constant was determined from the frequency response curve, since it is of the form which corresponds to a single time constant.

The Schwarz Thermopiles (4–5) (N)

These thermopiles are manufactured by Adam Hilger, Ltd. of London, and employ a pin type of construction. The thermopiles contain two hot receivers and two cold receivers, each 2 mm² in area. The information in Table I refers to the area of two receivers and to the resistance of the four receivers in series. The measured zero-frequency responsivities of the lower pair of receivers for the three units listed in Table I are 0.75, 1.00, and 1.2 volt/watt.

Information with regard to the frequency response of these thermopiles is given only for unit No. 11. The frequency response curve of these units is of Class I, defined in Section 7, Part I of Paper II. The curve has a slope of 3 db per octave at 8 c.p.s., at which point the responsivity is 8.2 db less than it is at zero frequency. Accordingly, the reference time constant is 20 milliseconds and the effective zero-frequency responsivity is 5.2 db less than the measured zero-frequency responsivity.

If one assumes that the relative frequency response curve of units No. 9 and No. 12 is the same as that of No. 11, one finds that the effective zero-frequency responsivities are 0.41, 0.55, and 0.66 volt/watt, respectively.

The Buck-Harris Evaporated Thermopiles (7–8) (N)

These thermopiles were developed by Willard E. Buck and Dr. Louis Harris at the Massachusetts Institute of Technology. The units were prepared by an evaporation technique and were of the folded type. The case is not evacuated. The values entered in Table I refer to the area of the hot receivers and to the resistance of the entire thermopile.

The measured zero-frequency responsivity of the two units listed in Table I are 0.345 and 0.375 volt/watt.

The frequency response curve of these detectors is of Class I, defined in Part I of Paper II. Unit No. 1 has a slope of 3 db per octave at 12 c.p.s., at which point the response is 4.8 db less than its zero-frequency value; the reference time constant and effective zero-frequency responsivity are thus 13.3 milliseconds and 0.280 volt/watt. The frequency response curve of unit No. 2 has a slope of 3 db per octave at a frequency of 12 c.p.s., at which point the response is 4.0 db less than at zero-frequency. Accordingly, the reference time constant is 13.3 milliseconds and the effective value of the zero-frequency responsivity is 0.334 volt/watt.

The Eppley (Emerson) Thermopile (9) (N)

This thermopile was manufactured by Eppley Laboratory for the Emerson Radio and Phonograph Corporation. The thermopile contains 12 hot junctions and 12 cold junctions arranged so that the 12 hot receivers cover an area $\frac{1}{4}'' \times \frac{1}{2}''$, and so that the cold receivers fill a similar adjacent area, with the result that the receivers cover an area $\frac{1}{2}'' \times \frac{1}{2}''$. The information in Table I refers to the area of the hot junctions and to the resistance of all of the junctions. The time constant was determined by using a ballistic galvanometer.

The Harris Thermocouples (10–11)

These two thermocouples are described by Louis Harris, Phys. Rev. **45**, 635–640 (1934). Each is evacuated, and is backed with a cellulose film. In both cases, the effective area of the hot junction is 1.2 mm². The bismuth-tellurium couple has a resistance of 150,000 ohms, a time constant of 212 milliseconds, and a zero-frequency responsivity of 7.6×10^{-5} volt/(44 calories/(cm²/sec.)) = 34.2 volts watt. The bismuth-antimony couple has a resistance of 25,000 ohms, a time constant of 90 milliseconds, and a zero-frequency responsivity of 8.8×10^{-6} volt/(44 calories/(cm²/sec.)) = 3.96 volts/watt. The time constants are calculated from the fact that the attenuation factors at 30 c.p.s. are 0.025 and 0.059, respectively.

The Harris and Scholp Thermocouples (12–16)

These evacuated thermocouples are described by Louis Harris and Alvin C. Scholp, J. Opt. Soc. Am. **30**, 519–522 (1940); **31**, 25 (1941), supplemented by a private communication from Professor Harris dated May 28, 1948, and were made by the sputtering process. The units specified in Table I as items 12 through 16 are the blackened units designated in Table II of the reference cited above as couples 3, 5, 7, 8, and 2, respectively. All but No. 7 have an effective area of 1.2 mm², whereas No. 7 has an effective area of 4.8 mm². The first three couples employ bismuth and antimony, whereas the last two employ bismuth and tellurium. The authors state the zero-frequency responsivity in units of 10^{-5} volt, with an irradiation of 10^{-4} watt/cm²; their values have been converted to volts per watt by making use of the stated areas.

The Hornig and O'Keefe Thermopiles (17–19)

These are wire thermocouples employing a bismuth-antimony alloy as one conductor and a bismuth tin alloy as the other. They are described by D. F. Hornig and B. J. O'Keefe, Rev. Sci. Inst. **18**, 474–482 (1947). The information in Table I is taken directly from the authors' Table III.

The Perkin-Elmer Thermocouples (20–22)

These are evacuated thermocouples employing unspecified alloys assembled in a pin type of construction. The thermocouples were developed in the laboratories of the Perkin-Elmer Corporation. The thermocouples are described by Max D. Liston, J. Opt. Soc. Am. **37**, 515(A) (1947), and in the New Instruments Section, Rev. Sci. Inst. **18**, 373 (1947). The area of the receiver attached to the hot junction is 0.4 mm².

The unit described in the first reference has a resistance of 10 ohms, a zero-frequency responsivity of 5 volts/watt, and a speed such that the peak response to a square wave signal of 5 c.p.s. is 70 percent of the response to a steady signal.

In the second reference the resistance is specified as 20 ohms, the zero-frequency responsivity as 10 volts/watt, and a speed such that the peak response to a square wave signal of 5 c.p.s. is 80 percent of the response to a steady signal.

In a private communication to the writer dated June 1, 1948, Mr. Liston stated that the current production thermocouples had the same resistance and sensitivity as specified in the second reference but with an increased speed such that the peak response to a square wave signal of 15 c.p.s. is 75 percent of the response to a steady signal.

By use of the formula

$$\text{Attenuation Factor} = \tanh 1/(4\Delta f)$$

appropriate for peak response to square wave irradiation, one finds that the time constants to be associated with the three units are 58 milliseconds, 45 milliseconds, and 17 milliseconds, respectively.

The Strong Bolometer (23) (N)

This bolometer was constructed by Dr. John Strong in the laboratories of Harvard University. It contains six unbacked nickel strips operating in a hydrogen atmosphere at a pressure of 1 mm of mercury. The sensitivity is stated to be 2.54 ohms per incident watt, with a current of 0.24 ampere. This corresponds to a measured zero-frequency responsivity of 0.61 volt/watt.****

The frequency response curve of this detector is of Class II, defined in Paper II. The response at 30 c.p.s. is 3 db less than at zero frequency. The reference time constant is accordingly 5.3 milliseconds.

**** This is incorrectly given as 0.3 volt/watt in reference 1. (Confirmed by a private communication from Dr. E. T. Bell dated June 25, 1948.)

TABLE I. Data for the 33 detectors assumed to have a noise output equal to the calculated Johnson noise.

Reference number	Alternate identification	Time constant τ milliseconds	Zero-frequency responsivity S_0 volts/watt	Sensitive area A mm²	Resistance R ohms	Noise equivalent power H_m 10^{-10} watts	Factor of merit $1/M_2$	M_2
The Weyrich Vacuum Thermocouple (N)								
1	in air	68	0.29	2.0	20	26.7	60.5	0.0165
2	evacuated	1000	4.35	2.0	20	0.465	15.5	0.0645
The Eppley (Farrand) Thermocouple (N)								
3		90	0.375	1.0	5.8	13.7	41.1	0.0241
The Schwarz Thermopiles (N)								
4	#B4772/9	20	0.41	4.0	23.3	26.7	17.8	0.0562
5	#B4772/11	20	0.55	4.0	51.0	29.3	19.5	0.0513
6	#B4772/12	20	0.66	4.0	20.7	15.6	10.4	0.0962
The Harris Evaporated Thermopiles (N)								
7	#1	13.3	0.280	11.0	100	59.7	26.5	0.0377
8	#2	13.3	0.334	11.0	100	50.0	22.2	0.0450
The Eppley (Emerson) Thermopile (N)								
9		250	0.1	80	18	5.76	48.0	0.0208
The Harris Thermocouples								
10	Bi-Te	212	34.2	1.2	150,000	14.4	98.9	0.0101
11	Bi-Sb	90	3.96	1.2	25,000	77.7	233	0.00429
The Harris and Scholp Thermocouples								
12	#3	177	50	1.2	1480	1.07	6.31	0.166
13	#5	110	47.5	1.2	1430	1.40	5.13	0.195
14	#7	134	60.2	4.8	4900	0.928	4.15	0.241
15	#8	345	406	1.2	183,000	1.05	12.1	0.0826
16	#2	175	400	1.2	90,000	1.05	6.13	0.163
The Hornig and O'Keefe Thermopiles								
17		36	6.5	0.5	5	1.64	1.97	0.508
18		36	6.5	1.0	10	1.64	1.97	0.508
19		41	3.8	4.0	12.5	1.47	2.01	0.498
The Perkin-Elmer Thermocouples								
20	J. Opt. Soc. Am.	58	5	0.4	10	2.66	5.14	0.195
21	Rev. Sci. Inst.	45	10	0.4	20	2.13	3.20	0.313
22	letter	17	10	0.4	20	3.47	1.97	0.508
The Strong Bolometer (N)								
23		5.3	0.61	5.7	4.2	12.4	2.19	0.457
The Felix Bolometers (N)								
24	2A	20.2	1.09	17.2	16	3.99	2.69	0.371
25	15	21.2	1.41	17.2	16	3.01	2.13	0.469
26	11A	21.5	1.14	17.2	16	3.69	2.64	0.379
27	19	21.2	2.26	17.2	16	1.88	1.33	0.752
The Polaroid Bolometers (N)								
28	Ni 324	4.7	1.46	4.5	64	24.1	3.78	0.265
29	Ni 347	7.2	1.60	4.5	44	14.7	3.53	0.283
30	Ni 350	3.6	1.61	4.5	128	35.3	4.24	0.236
The Thermistor Bolometers (N)								
31	S-19	5.9	730	0.60	3×10^6	25.5	5.02	0.199
32	S-20	1.27	378	0.62	3×10^6	105.0	4.45	0.225
33	XB-108	135	3460	0.58	3×10^6	1.14	5.13	0.195

The Felix Bolometers (24–27) (N)

The Felix bolometers were developed by the Heat Research Laboratory of Massachusetts Institute of Technology under the direction of Alan C. Bemis. The bolometer contains four unbacked nickel strips operating in an atmosphere of hydrogen at a pressure of 4 mm of mercury. The measured zero-frequency sensitivities of the four bolometers tested, in the order used in Table I, are 10.3, 13.3, 10.7, and 21.2 ohms/watt. Since the current was 0.106 ampere,

the measured zero-frequency responsivities are 1.09, 1.41, 1.14, and 2.26 volts/watt. The time constants were determined from the frequency response curves, since they are of the type which corresponds to a single time constant.

The Polaroid Bolometers (28–30) (N)

The Polaroid bolometers were developed in the laboratories of the Polaroid Corporation under the direction of Dr. Bruce Billings. These bolometers

TABLE II. Data for 15 detectors whose noise level was directly measured.

Reference number	Alternate identification	Time constant τ milliseconds	Noise equivalent power P_0 10^{-10} watt	Sensitive area A mm²	Band width Δf c.p.s.	Noise equivalent power H_m 10^{-10} watt	Factor of merit $1/M_2$	M_2
The Superconducting Bolometers								
36	22–k–¼	10.0	3.67	1.8	4000	0.216	0.0720	13.9
37	25–k–¼	1.2	14.5	1.2	4000	3.02	0.121	8.28
38	8–f–1	0.9	18.5	0.7	4000	5.83	0.175	5.71
39	11–f–1	1.7	14.6	0.4	4000	4.43	0.251	3.98
40	24–k–¼	0.7	35.3	0.7	4000	12.6	0.294	3.40
41	10–f–1	1.5	20.6	0.4	4000	6.65	0.333	3.00
42	5–d–1	0.8	90.0	1.6	4000	19.9	0.531	1.88
43	13–g–1	4.2	115.0	6.4	4000	5.54	0.776	1.29
The Baird Bolometers								
44	slow	50	60	3.6	5000	1.00	1.67	0.600
45	fast	4.1	64.8	0.2	5000	16.0	2.19	0.457
The Thermistor Bolometer								
46	A	3.0	100	0.6	30	215	21.5	0.0465
The Polaroid Bolometer								
47	A	4.0	263	6.0	100	85	11.3	0.0884
The Aiken Bolometers								
48	G 165	3.8	185	27.5	1.85	210	26.6	0.0376
49	G 22	6.9	232	27.5	1.85	196	45.1	0.0222
The Golay Pneumatic Heat Detector								
50		5.0	0.48	7.0	1.0	1.28	(0.213)	(4.69)

were prepared by evaporating nickel upon a nitro-cellulose film about 800A thick. The three reported in Table I contain four sensitive strips in the form of a cross. The measured zero-frequency responsivities of the three bolometers in the order specified in Table I are 118.5, 130, and 144.5 ohms/watt. Since the current was 15 milliamperes, the corresponding zero-frequency responsivities are 1.78, 1.95, and 2.17 volts/watt.

The frequency response curves of the three units are all of the type specified in Part I of Paper II as Class I. The frequency response curve of unit 324 has a slope of 3 db per octave at 34 c.p.s., at which point the response is 4.7 db less than its zero-frequency value. The reference time constant and the effective zero-frequency responsivity are thus 4.69 milliseconds and 1.46 volts/watt.

The frequency response curve of unit 347 has a slope of 3 db per octave at 22 c.p.s., at which point the response is 4.7 db less than its zero-frequency value. The reference time constant and the effective zero-frequency responsivity are thus 7.24 milliseconds and 1.60 volts/watt.

The frequency response curve of unit 350 has a slope of 3 db per octave at 44 c.p.s., at which point the response is 5.6 db less than its zero-frequency value. The reference time constant and the effective zero-frequency responsivity are thus 3.62 milliseconds and 1.61 volts/watt.

The Thermistor Bolometers (31–33) (N)

These bolometers were developed by Bell Telephone Laboratories under the direction of Dr. J. A. Becker. The sensitive element is a polycrystalline semi-conductor. Unit S-19 is glass-backed, unit S-20 is quartz-backed, and unit XB-108 is air-backed. The measured responsivities of the three units listed in Table I at a frequency of 15 c.p.s. are 675, 450, and 275 r.m.s. volts/r.m.s. watt.

The frequency response curves of units S-19 and S-20 are of the type designated in Part I of Paper II as Class I. The frequency response curve of unit S-19 has a slope of 3 db per octave at a frequency of 27 c.p.s., at which point the response is 2.3 db less than the response at 15 c.p.s. Accordingly, the reference time constant is 5.9 milliseconds and the effective zero-frequency responsivity is 0.7 db greater than its value at 15 c.p.s.: 730 volts/watt.

The frequency response curve of unit S-20 has a slope of 3 db per octave at 125 c.p.s., at which frequency the response is 4.5 db less than the response at 15 c.p.s. Accordingly, the reference time constant is 1.27 milliseconds and the effective zero-frequency responsivity is 1.5 db less than its 15 c.p.s. value: 378 volts/watt.

The frequency response curve of unit XB-108 corresponds to a single time constant of 135 milliseconds. Accordingly, the zero-frequency respon-

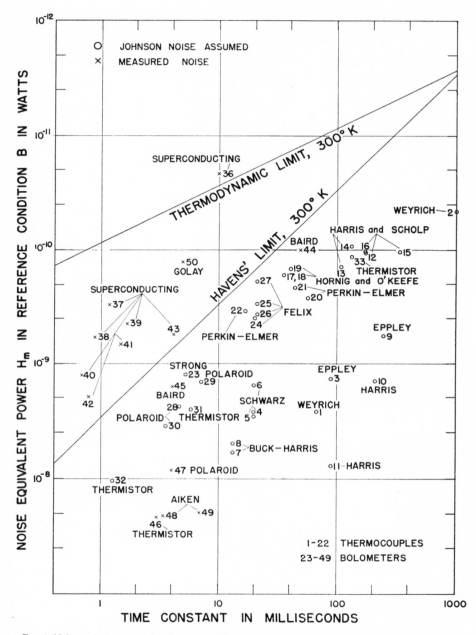

FIG. 1. Noise equivalent power in reference condition B as a function of the reference time constant for the approximately fifty Type II detectors listed in Tables I and II.

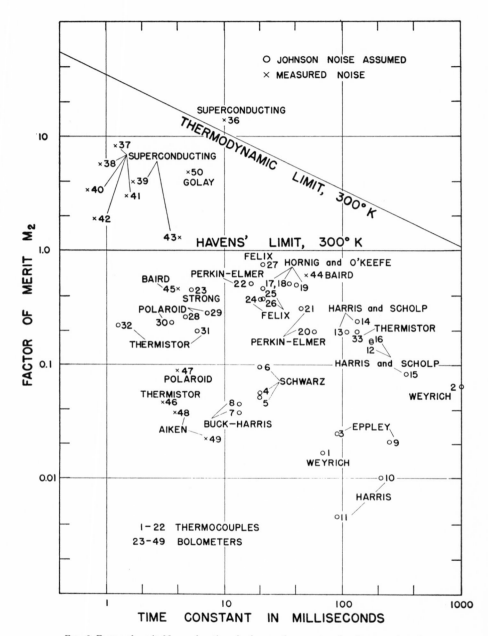

FIG. 2. Factor of merit M_2 as a function of reference time constant for the approximately fifty Type *II* detectors listed in Tables I and II.

sivity is 12.7 times the responsivity at 15 c.p.s., or 3500 volts/watt.

The Superconducting Bolometers (36–43)

The results for these bolometers are copied directly from the article by Nelson Fuson.[2]

In the article cited, the author states that the band width employed to measure the noise (4000 c.p.s.) is much wider than the band width of the detectors (16 c.p.s. to 230 c.p.s.), and states further that preliminary tests on two bolometers with a narrow band width amplifier indicate factors of merit several times smaller than those listed in Table II.

The Baird Bolometers (44–45)

The information in Table II is copied directly from Paper I.

The Thermistor Bolometer (46)

The information shown in Table II is copied directly from Paper I.

The Polaroid Bolometer (47)

The information shown in Table II for this bolometer has been taken directly from Paper I.

The Aiken Bolometers (48–49)

These evaporated gold bolometers were developed in the laboratory of the Electromechanical Research, Inc. by a group under the direction of Dr. Charles B. Aiken. The units discussed here are described in C. B. Aiken, W. H. Carter, and F. S. Phillips, Rev. Sci. Inst. 17, 377–385 (1946) and in private communications from Dr. Aiken to the writer dated June 7, 1948, and June 11, 1948. The time constants and sensitive areas are stated in the 1946 article and the remaining information is contained in the private communications. The zero-frequency responsivities in the secondary of the transformers associated with the detectors listed in Table II are 8.66 and 7.44 volts/watt. The r.m.s. noise voltage measured at the same point in a band 1.85 c.p.s. wide were found to be 0.16 and 0.18 microvolt. From these measurements the values of P_0 reported in Table II were calculated.

The Golay Pneumatic Heat Detector (50)

The values stated for this detector are taken directly from the article by Nelson Fuson, J. Opt. Soc. Am. 38, 845–853 (1948), who in turn obtained

[2] Nelson Fuson, J. Opt. Soc. Am. 38, 845–853 (1948). The writer is indebted to Mr. Fuson for his kindness in sending a copy of his manuscript prior to its publication.

the information through a private communication dated April 26, 1948, from Mr. Roy Anderson of the Eppley Laboratory.[†]

Since this detector is a Type I detector, it is not permissible to state the factor of merit M_2 for this detector. For this reason, the values of M_2 and $1/M_2$ entered in Table II are enclosed in parentheses. The value of M_2 has been calculated and displayed in Fig. 2 in order that the Golay pneumatic heat detector may be compared with other detectors of similar reference time constant. In this case, however, there is no implication that the factor of merit M_2 for other Golay detectors with different time constants will be the same. It is rather the factor of merit M_1 that has this property for the Golay detector. Accordingly, the points representing different Golay detectors should lie along a line parallel with the line marked "Thermodynamic Limit, 300°K" in Figs. 1 and 2. The factor of merit M_1 is 0.30.

6. Discussion

Inspection of Fig. 2 indicates clearly that no thermopile or bolometer operating at room temperature has a factor of merit as large as unity. Only the superconducting bolometers (36–43) and the Golay pneumatic heat detector (50) have factors of merit greater than unity. The Golay detector operates at room temperature.

Inspection of Fig. 2 indicates further that the maximum factors of merit attained at room temperature are substantially the same for bolometers and thermocouples.

The suitability of the definition of the factor of merit M_2 proposed in the preceding paper is borne out with particular nicety by the results on the thermistor bolometers (31–33) shown in Fig. 2. The three units use the same type of sensitive element, but they employ greatly different coefficients of thermal conductivity between the sensitive element and its surroundings. In spite of the 100:1 range of time constant, the largest factor of merit is only 15 percent greater than the smallest.

ACKNOWLEDGMENT

The writer wishes to thank the following individuals for their kindness in communicating information on specific detectors: Dr. Charles B. Aiken, Electromechanical Research, Inc.; Dr. E. T. Bell, Ohio State University; F. G. Brockman, Phillips Laboratories, Inc.; Nelson Fuson, The Johns Hopkins University; Professor Louis Harris, Massachusetts Institute of Technology; William Langton, Baird Associates; and Max D. Liston, Perkin-Elmer Corporation.

[†] See also Eplab Bulletin No. 10.

APPENDIX

The purpose of this section is to call attention to the wide variety of possibilities in the specification of the relevant properties of radiation detectors, and thereby to encourage precision in the statement of these properties in the periodical literature. In a letter to the writer from Nelson Fuson dated June 16, 1948, the suggestion is made that the present paper "will be particularly helpful to a lot of us who are working on one particular detector if you emphasize the need for stating *all* the experimental conditions, which we so often tend to overlook because they are, for our particular laboratory or test arrangement, kept constant with the result that we get careless and do not record them or think of them as variables."

Many of the matters mentioned below may seem too obvious to require specific mention, but they are included because of specific instances of failure to observe them in the literature.

Necessary Information

In the case of those detectors, such as the thermopile, for which it is reasonable and convenient to assume that the only source of noise is the Johnson noise associated with their resistance, it is sufficient to state the following information about the detector:

1. The electrical resistance.
2. The sensitive area.
3. A curve of relative responsivity *versus* frequency. If the frequency response may be characterized by a single time constant, a statement of its value is sufficient.
4. The relative response to different radiation wave-lengths.
5. The responsivity (in volts/watt) at a single frequency with a specified spectral energy distribution.

One may, of course, combine Items 3 and 5 by stating the responsivity in absolute terms as a function of frequency.

In the case of detectors for which it is not reasonable to assume that the only noise is Johnson noise, it is sufficient to state the following quantities:

1. The sensitive area.
2. A curve of relative responsivity *versus* frequency.
3. A curve of the relative noise power per unit band width *versus* frequency, under the same conditions used to determine Item 2.
4. The relative response to different radiation wave-lengths.
5. A single measurement of the signal-to-noise ratio under fully defined conditions.

Item 5 may be omitted if Items 2, 3, and 4 are on an absolute basis.

Sensitive Area

In stating the sensitive area of the detector in the case of thermocouples and thermopiles, it should be made clear just what area is involved. One may state the area of the receiver associated with a single junction, with all of the hot junctions, or with all of the hot *and* cold junctions. With that type of bolometer in which one strip is the load resistor of the other, one should state whether the specified area is that of one strip or of both strips.

Responsivity

The responsivity may be stated for a steady signal, for a sinusoidally modulated signal, for a square wave signal, or for isolated pulses of given shape. The magnitude of the radiation signal may be specified in terms of its peak-to-peak value, half of the peak-to-peak value, the average of the absolute value of the deviations from the mean (average value), the square root of the average of the square of the deviations from the mean (r.m.s. value), and each of these may refer to the complete signal, or only to the Fourier component with the fundamental frequency. The same comments apply to the specification of the resultant voltage response.

Thus the statement that the responsivity at a certain frequency is so many volts per watt contains considerable ambiguity; for example, the responsivity in peak-to-peak volts per r.m.s. watt is eight times the responsivity in r.m.s. volts per peak-to-peak watt, when the radiation is sinusoidally modulated.

Furthermore, it should be specified whether the responsivity is that corresponding to open circuit conditions, or if not, just what the impedance relations are. This problem is particularly acute with bolometers, because it is necessary to feed current through the bolometer from a load resistor; unless the load resistor has a resistance large compared with the bolometer resistance, it reduces the factor of merit for the bolometer.

The same uncertainties of specification that affect the sensitive area are also of concern here, since the responsivity in volts per watt is usually obtained from the responsivity in volts per watt per unit area, by multiplying the latter by the sensitive area.

Frequency Response

Even the curve of relative frequency response is not independent of the specification of the measurement of the voltage and of the radiation. Consider, for example, a detector with a single time constant equal to 15.9 milliseconds. Then the response of the detector is 3 db less than its low frequency value at 10 c.p.s. as measured by the r.m.s. response to a sinusoidal radiation signal, but the response as measured by the peak-to-peak response to square wave irradiation will not be 3 db less than its zero frequency value until the frequency rises to 17.8 c.p.s.

Thus even in specifying the relative frequency response curve, it is necessary to be careful to describe adequately the method of measurement.

Provided that the relative frequency response curve cannot be specified by a single time constant, it is desirable to exhibit the actual curve, preferably with logarithmic coordinates. If, however, for reasons of brevity this is not possible, then the minimum essentials are the specification of the frequency at which the response is down 3 db, and the specification of the frequency and attenuation factor for that point on the curve at which the slope is 3 decibels per octave, when the response is measured with sinusoidally modulated radiation.

Resistance

A specified resistance may refer to the resistance of a single junction, to the resistance of all of the hot junctions, or to the resistance of all of the hot *and* cold junctions.

With bolometers, the resistance may be that of the sensitive strip, or it may be that of the strip plus a load resistor; in some cases the latter may be another strip.

Noise Level

When measurements of the signal-to-noise ratio or the noise level of the detector are specified, it is essential to state the noise equivalent band width employed, and to state what property of the noise is measured: One may state the peak-to-peak, half of the peak-to-peak, average, or r.m.s. voltage of the noise.

In general, one should avoid specifying the noise level in terms of the peak value as observed on an oscilloscope, because a peak of any preassigned value may be observed if one waits long enough. Even if one is careful to measure the average maximum occurring in specified time periods, the relationship[3] of such measurements to the more direct measurements of the r.m.s. or average value is not a simple one, and depends on the power spectrum of the noise.

It is furthermore necessary to state the noise level at the same point in the circuit as that at which the responsivity is stated. For example, when a coupling transformer is used, it is not adequate to give the noise at the input grid of the amplifier, and the responsivity of the detector itself, unless further information is provided about the properties of the transformer.

Spectral Response

It has been customary to assume that the receiving area of the detector is equally black at all wave-lengths, so that the only relevant property of the radiation falling upon the detector is the total power incident upon the receiving area.

As the techniques of both construction and measurement improve, however, the error involved in this assumption will no longer be tolerable, and it will be necessary to specify the spectral response of the detector. Furthermore, in specifying the noise equivalent power in the reference condition, it will be necessary to specify the spectrum of the source used.

In the case of photo-conductive cells, this necessity has been recognized from the beginning, because of the large variation of sensitivity as a function of wave-length.

---

[3] For a detailed treatment of this relationship, see S. O. Rice, Bell Sys Tech. J. **24**, 46–156 (1945), especially p. 75.

Reprinted from *Appl. Optics*, **1**(5), 627–636 (1962)

The Detectivity of Infrared Photodetectors

S. Nudelman

This article presents a review and an analysis of the factors involved in determining the capability of semi-conducting photodetectors to detect radiation. The photoconductive detector is used as the vehicle for this analysis. Signal and noise generation are examined individually for their contribution to noise equivalent power (NEP), the Jones classifications of detectors, and detectivity-star (D^*). Equations for NEP are derived that permit direct examination of the influence of parameters such as bandwidth, area, and time constant in detector performance.

I. Introduction

The purpose of this article is to present a review and an analysis of the factors involved in determining the capability of semiconducting photodetectors to detect radiation.

By 1953, considerable information on thermal detectors, photoemissive devices, and photographic film was available, to the extent that Jones[1] was able to present a detailed accounting of the "state of the art." In addition, he was able to ascertain consistent patterns of detector behavior, to formulate a system of detector classification, and to propose figures of merit. Semiconducting photodetectors were in the early stages of their development, and it was possible to include only some of the preliminary information on lead-compound detectors in the report. Since that time, considerable effort has been expended toward the development of background-noise-limited infrared detectors, and to the understanding of the basic physical mechanisms responsible for their performance. G-R (generation-recombination) noise was uncovered as a limiting fundamental process, while the role of shot noise in p-n junction detectors was ascertained. Excess ($1/f$) noise received intensive study, and significant reduction of this noise was achieved in the development of most photodetectors. However, its properties still

remain to be organized and explained theoretically. The concept of detectivity-star (D^*) was introduced by Jones[2] to provide a numerical rating for a detector. This concept has proven useful, and on the basis of recent measurements appears to be fully applicable to semiconducting photodetectors. With the accumulation of this wealth of new information, it is now fruitful to present a re-examination of the mechanisms involved in photodetection and to derive expressions for noise equivalent power (NEP) for the various noise-limited conditions. An analysis of the meaning of D^* is also in order. The photoconductive detector is used as the vehicle for this analysis.[3]

II. The Photoconductive Detector

The solid-state quantum detector absorbs electromagnetic radiation (or photons) and generates from this absorption additional free charge carriers. The mechanisms of this process require that the quantum of energy associated with the photons ($E_{ph} = hc/\lambda$) be greater than some critical energy corresponding to allowed transitions between the conduction band, the valence band, and/or a discrete energy level in the energy-gap region of the semiconducting detector. This process is carried out without any significant temperature change. The additional carriers that are generated appear in a form suitable for measurement as a voltage or a current.

A. Signal Generation

The photoconductive detector, which is the only device that will be discussed here, is generally operated with the simple circuitry of Fig. 1. The voltage drop V_c across the detector is given by:

$$V_c = \frac{V}{r_L + r_c} \cdot r_c, \tag{1}$$

This paper was written while the author was at the Institute of Science and Technology, The University of Michigan, Ann Arbor, Michigan. His present address is: Armour Research Foundation, Chicago 16, Illinois.

Received 4 April 1962.

This paper is reprinted from the Proceedings of the Infrared Information Symposia, Vol. 6, No. 4, October 1961.

This work was supported in part by the Office of Naval Research and in part by the Aeronautical Systems Division, U.S. Air Force.

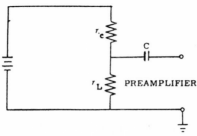

Fig. 1.

where

V = bias battery voltage,
r_L = load resistor,
r_c = resistance of the photoconductive detector.

The signal voltage is the variation in the voltage drop caused by the change in r_c, when the detector is exposed to signal radiation. Therefore:

$$\Delta V_s = I \cdot \Delta r_c + \Delta I \cdot r_c = I \cdot \Delta r_c, \qquad (2)$$

where

I = bias current = $V/(r_L + r_c)$.
$\Delta I = 0$,

which is applicable when photoconductive detectors are operated under constant-current conditions.

The change in resistance is due to the change in the number of carriers created by the absorption of signal photons. For the simplest kind of analysis, consider here the case of the intrinsic photoconductor whose hole and electron mobilities are assumed equal. The general case of unequal mobilities, and including the impurity photoconductive detector, results in an analysis requiring significantly extra arithmetic, and for the purposes of this paper leads to the same conclusions. Let the detector shape be a parallelepiped, of length l, width w, and thickness d. The quantity lw represents the surface exposed to radiation and defines the detector area A, with the electrodes placed across the wd end faces. The cell resistance is given by:

$$r_c = \rho \frac{l}{wd} = \frac{l}{\sigma wd} = \frac{l}{2ne\mu wd}, \qquad (3)$$

where

ρ = electric resistivity = $1/\sigma$,
σ = electric conductivity = $ne\mu_n + pe\mu_p = 2ne\mu$,
μ = mobility ($\mu_n = \mu_p$),
n = density of electrons,
p = density of holes.

When radiation is absorbed by the detector, it causes a change in the density of carriers Δn_s, and therefore a change in conductivity. The change in resistance is then given by:

$$\Delta r_c = \frac{dr_c}{dn} \cdot \Delta n = -\frac{\rho l}{wd} \cdot \frac{\Delta n_s}{n}. \qquad (4)$$

The signal voltage is then

$$V_s = I \cdot \Delta r_c = \frac{V}{r_L + r_c} \cdot \frac{\rho l}{wd} \cdot \frac{\Delta n_s}{n} \qquad (5)$$

or

$$V_s = \frac{V}{r_L + r_c} \cdot \frac{\rho l}{wd} \cdot \frac{\Delta N_s}{N} = Ir \frac{\Delta N_s}{N}, \qquad (6)$$

where

ΔN_s = total number of additional electrons given by $\Delta n_s wld$, equal to $N(t) - N$,
N = number of electrons in the absence of signal,
$N(t)$ = instantaneous number of electrons.

The photodetector small signal properties are governed by[4]

$$\frac{d}{dt} \Delta N_s = A\eta_\omega J_s - \frac{\Delta N_s}{\tau}, \qquad (7)$$

where

η_ω = responsive quantum efficiency,
τ = electron-hole lifetime,
J_s = incident number of photons per unit area per second.

The solution of this equation for a sinusoidal signal of modulation frequency $f = \omega/2\pi$ is

$$|\Delta N_s(f)| = \frac{A\eta_\omega J_s \tau}{\sqrt{1 + (\omega\tau)^2}}. \qquad (8)$$

The fractional change in conductivity is:

$$\frac{|\Delta N_s(f)|}{N} = \frac{\eta_\omega J_s \tau}{nd\sqrt{1 + (\omega\tau)^2}}. \qquad (9)$$

The signal voltage is now

$$V_s = \frac{V}{r_L + r_c} \cdot \frac{\rho l}{wd} \cdot \frac{\eta_\omega J_s \tau}{nd\sqrt{1 + (\omega\tau)^2}}. \qquad (10)$$

It is clear that signal response improves with longer lifetimes, improved quantum efficiencies, and decreasing equilibrium density of carriers.

B. Noise

There are five kinds of electrical noise that appear in the output of semiconducting photodetectors. They will be described below, briefly.

1. Johnson Noise

This type of noise is often referred to as Nyquist noise, since both Nyquist and Johnson treated this noise problem in 1928.[5,6] It can be expressed as:

$$\overline{v^2} = 4kTr\Delta f, \qquad (11)$$

where

k = Boltzmann's constant,
T = absolute temperature,
Δf = electrical bandwidth,
r = resistance,
v = voltage, that is, $[v(t) - v_{\text{av}}]$.

2. Generation-Recombination Noise

This type of noise is inherent in the electronic system of semiconducting materials. All atoms in a lattice vibrate in an organized manner to the extent that their vibrations are quantized and can be described in particle terminology as "phonons." Valence-band electrons are continually jostled by the vibrations of the lattice atoms. Every so often, the nature and phasing of vibrations between atoms is such that an electron in their midst is able to gain enough energy to be freed from its bound (valence-band) state, and to move about in the conduction band. The electron is said to have suffered a phonon collision, and to have absorbed energy in the process.

When electrons leave the valence band for the conduction band, charge carriers (electrons and holes) become available for the purposes of current flow. The equilibrium number of carriers created increases with a crystal's temperature, and for any one temperature will be greater for diminishing forbidden energy gap. The thermal (phonon) excitation process is statistical in nature, and the rate at which electrons are excited to the conduction band fluctuates.

In addition to the statistical pulses of generation, we find that a similar situation exists for the recombination of carriers. The electrons and holes wander about the crystal lattice with some thermal motion, and, during a "lifetime" characteristic of the semiconductor material, get close enough together to recombine directly, or through a recombination center. The lifetime is a statistically fluctuating quantity, as is the instantaneous number of electrons and holes. Current carrier fluctuations are therefore inherent in any semiconductor. When a sample is placed in a constant-current electric circuit, one should expect, and in fact one does observe, conductivity fluctuations causing electrical noise completely described by the generation-recombination process.

Since this noise is a bulk property of the crystal, and is due to conductivity fluctuations caused by carrier density changes, it follows from Eq. (4) that

$$\Delta r = \frac{dr}{dn} \Delta n = \frac{\rho l}{wd} \cdot \frac{\Delta n}{n} = -\frac{\rho l}{wd} \cdot \frac{\Delta N}{N} \quad (12)$$

and

$$\overline{v^2} = I^2 \cdot \overline{\Delta r^2}. \quad (13)$$

It can be shown that[7]:

$$\frac{\overline{(\Delta N)^2}}{N} = \frac{2\tau}{1 + (\omega\tau)^2} \cdot \Delta f. \quad (14)$$

Therefore,

$$\overline{v^2} = I^2 \cdot r^2 \cdot \frac{2\tau}{N[1 + (\omega\tau)^2]} \cdot \Delta f \quad (15)$$

or

$$\overline{i^2} = I^2 \cdot \frac{2\tau}{N[1 + (\omega\tau)^2]} \cdot \Delta f \quad (16)$$

as shown by Van Vliet.[8]

3. Current, 1/f, Modulation, and Excess Noise

All of the names above have been used at various times as names for the same kind of noise. It is a noise that appears commonly in photodetectors in addition to the noise already discussed, and is generally found to have characteristics that can be described by the expression (see ref. 6):

$$\overline{i^2} = C \frac{I^2}{fAd} \cdot \Delta f, \quad (17)$$

where

I = total current through the sample,
f = frequency,
C = constant.

Actually, Eq. (17) is not universal in that cases have been cited where the current exponent has been found as small as 1.25 and as large as 4,[9] while the frequency exponent has ranged from 1 to 3.[10] However, these extreme exponential variations are not particularly common, and can usually be associated with a specific material, or to some unique treatment and physical condition of a material.[9]

The physical mechanism of this noise is the least understood of all the noise found to date. Petritz[11] introduced the name "modulation" noise to identify the mechanism as something quite different from simple carrier density fluctuations.

The modulation suggestion offered by Petritz, however, can still be only a part of the picture. Brophy and Rostoker,[12] Brophy,[13] and Bess[14] found with direct experimental evidence that the noise fluctuations in Hall voltage followed the same frequency dependence pattern as the noise from a conductivity measurement. It must be concluded, therefore, that, like conductivity, 1/f noise also follows the fluctuation in the density of the majority current carriers.

The location of a major source of 1/f noise (at least in the case of germanium filaments), has been found to be the surface of the crystal. Recent studies by MacRae[15] indicated that a surface inversion layer (a *p*-type surface on an *n*-type crystal or vice versa) is necessary to generate significant 1/f noise.

Although the surface provides $1/f$ noise, it is quite clear that this kind of noise can also be generated in the bulk material. Brophy[16] finds that, by plastic deformation, he can create noise sources and cause an increase of excess noise by orders of magnitude. These sources are uniformly distributed throughout the crystal, and also contribute an unusual I^4 dependence. Bess[17] assumed that the noise was due to edge dislocations with impurities diffusion along the edge. Such distributions throughout the bulk can provide noise spectra similar to that obtained by inversion surface layers.[18] Even when surfaces are treated so that their noise contribution is minimized, and bulk sources are removed, residual $(1/f)$ noise may still exist. This noise has been associated with nonohmic contact regions, probably due to minority carrier drift across the contacts. Thus the sources of $1/f$ noise can be found at the surface, in the bulk, and at the electrical contacts of a detector.

4. Shot Noise

This noise is usually associated with vacuum tubes; it is described as the electrical noise that appears in the output of a vacuum tube when the grid, if any, is held at a fixed potential. However, it is more precisely described in what is usually called the temperature-limited condition. By temperature limited, it is meant that the anode voltage on the tube is sufficient to collect all of the electrons emitted from the cathode. Consider a temperature-limited diode connected to a resistance r. Because of the discreteness of the electronic charge, the number of electrons emitted at equal time intervals will fluctuate around an average value. The fluctuating current causes a fluctuating voltage across r which can be amplified and measured. The mean square current fluctuation turns out to be constant up to frequencies of approximately the reciprocal of the transient time, and can be described by

$$\overline{I^2} = 2eI\Delta f, \tag{18}$$

where e is the electronic charge and I is the bias current.[19]

Semiconductor photovoltaic detectors and back-biased p-n junctions operating in a photoconductive circuit also exhibit shot noise. In a p-n junction diode, a space charge region is developed across the barrier and an associated electric field. Electrons or holes created by phonons or the absorption of background photons, diffusing into the barrier region, are swept from the n-type material to the p-type material, or vice versa. This results in current pulses appearing across the diode with the same characteristics that are observed for the vacuum tube.

Petritz (see ref. 7) has developed a theory applicable to the lead chalcogenide (PbS, PbSe, PbTe) photoconductive films, which is suitable for use here. These films are composed of a system of tiny crystallites separated by intercrystalline barriers, where space charge regions exist. The barrier regions contribute both Johnson and shot noise, while the crystallites generate the usual Johnson noise. For thin barriers, compared to the carrier diffusion length, the barrier contribution to shot noise is given by

$$\overline{v^2} = \frac{2er_e^2 I}{n_l l} \cdot \Delta f, \tag{19}$$

where n_l is the number of crystallites per unit length, and r_e is the detector resistance.

5. Background-Radiation Noise

Background radiation can be thought of as a stream of photons originating from the detector environment. The cell walls surrounding the detector, its window, and the media viewed by the detector through the cell window all contribute to this radiation. The extent of the individual contributions is determined by their respective temperatures, emissivities, and geometry.

Photons originating from the background and impinging on the detector arrive and are absorbed in a statistically fluctuating manner. This results in a well-defined type of electrical noise. When this is the predominant noise, the photodetector is said to be "background radiation-noise limited." The generation of carriers by this process follows the same statistical law as the G-R noise. Thus the equilibrium density of carriers is given by $N_L = 2A\eta_L J_L \tau$, and the background-limited case is given by $N_B = 2A\eta_B J_B \tau$. The factor 2 arises from the generation of two carriers (an electron and a hole) for each photon, and

N_L, N_B are the equilibrium density of carriers generated by G-R, and background radiation, respectively;

η_L, η_B are the efficiency factors for the conversion of phonons and photons to carriers, respectively; and

J_L, J_B are the lattice flux and the radiation flux, respectively.

The complete expression that describes the fluctuation in carrier density, for the case of equilibrium where generation processes are equal to recombination processes, is given by Petritz[4,7] and Eq. (14) as:

$$\overline{\Delta N^2_{noise}} = \frac{4\tau^2 \cdot \Delta f \cdot A(\eta_B J_B + \eta_l J_l)}{1 + (\omega\tau^2)}. \tag{20}$$

III. Noise Equivalent Power

Consider an infrared detector exposed to some incident radiation in a circuit which provides an electrical signal voltage proportional to the radiant power. Assume also that the dominant electrical noise in the circuit is generated by the detector, not the preamplifier. Let the rms radiant power P be expressed as $H \cdot A$, where H is the rms irradiance and A is the area of the detector. The absorption of power P by the detector results in a signal voltage V_S in the circuit. When the power source is removed or masked, and the detector allowed to see only the background, then a noise voltage V_N is obtained. The predominant mechanism causing this noise might be internal to the detector, such as from Johnson, shot, and/or lattice-governed G-R noise sources. In the case of the best detectors, this noise might well be governed by the fluctuations in photon flux irradiating the detector, that is, the background radiation-noise limited condition. A simple proportionality can be set up, relating these voltage readings and the power P:

$$\frac{\text{NEP}}{V_N} = \frac{H\ A}{V_S}. \tag{21}$$

Thus if Johnson noise is responsible for V_N, the NEP represents the amount of power that would have to fall onto the detector to generate a voltage equivalent to V_N. This relationship is more usually observed in the form:

$$\text{NEP} = \frac{H \cdot A}{V_S/V_N} \tag{22}$$

Clearly, NEP is also the power exposure required by the detector to obtain a signal-to-noise voltage ratio of unity.

A relatively simple analysis can be first attempted for the purpose of providing some insight to the area dependence of NEP. Assume first a simple detector in a circuit, from which is obtained a signal voltage V_S and a noise voltage V_N. The noise voltage is assumed due to a uniform distribution of noise sources throughout the bulk of the detector. Consider two of these detectors placed adjacent to one another (end to end) and connected electrically in series. The total signal voltage is simply $V_{S1} + V_{S2}$, since the signals are coherent. The noise voltages add as their mean squares, since they are incoherent. The noise voltage of each unit is the rms value of the voltage fluctuation, or $V_n = \sqrt{\overline{v^2}}$, where

$$v = v(t) - r_{av} \tag{23}$$

and $v(t)$ is the instantaneous value of the noise voltage and v_{av} is the average value. The signal-to-noise ratio from the combined units is:

$$\frac{V_{ST}}{V_{NT}} = \frac{V_{S1} + V_{S2}}{(V_{N1}^2 + V_{N2}^2)^{1/2}} \tag{24}$$

If the two detectors are identical in area and thickness, $V_{S1} = V_{S2}$, and equal noise is generated by each of them. Then:

$$\frac{V_{ST}}{V_{NT}} = \frac{2V_{S1}}{\sqrt{2}\ V_{N1}} = \sqrt{2}\ \frac{V_{S1}}{V_{N1}}. \tag{25}$$

Suppose these two detectors are now placed side by side, and connected electrically in parallel. Since

currents can be added in a straightforward manner when dealing with parallel circuits, consider the noise sources as current generators. The total noise becomes (similarly to the noise voltage analysis):

$$I_{NT}^2 = I_{N1}^2 + I_{N2}^2 \tag{26}$$

and for identical detectors:

$$I_{NT} = \sqrt{2I_{N1}^2}. \tag{27}$$

The signal current is now:

$$I_{ST} = I_{S1} + I_{S2} = 2I_{S1}. \tag{28}$$

The signal-to-noise current ratio becomes

$$\frac{I_{ST}}{I_{NT}} = \frac{2I_{S1}}{\sqrt{2}\ I_{N1}} = \sqrt{2}\ \frac{I_{S1}}{I_{N1}} \tag{29}$$

The noise current can be expressed as a noise voltage by simply multiplying the total noise current by the effective resistance of the circuit. Therefore:

$$r_t = \frac{r_1 r_2}{r_1 + r_2} = \frac{r_1}{2} \tag{30}$$

for $r_1 = r_2$, and

$$V_{NT}^2 = I_{NT}^2 r_T^2 = \frac{I_{N1}^2 r_1^2}{2} \tag{31}$$

or

$$V_N = \frac{1}{\sqrt{2}} \cdot I_{N1} \cdot r_1.$$

The signal voltage is $V_{ST} = V_{S1} = V_{S2}$, since the circuit is now equivalent to two identical batteries placed in parallel. Then

$$\frac{V_{ST}}{V_{NT}} = \frac{I_{S1} \cdot r_1}{(1/\sqrt{2}) \cdot I_{N1} \cdot r_1} = \sqrt{2}\ \frac{V_{S1}}{V_{N1}}. \tag{32}$$

The analysis can readily be extended to three or more detector units. It is immediately apparent from the treatment with two units, however, that with either the series or parallel arrangement, the signal-to-noise ratio depends on the square root of the total detector area. This then indicates that the NEP of a detector should also depend on the square root of the area. The conclusions drawn are valid only for the special case of a uniform photoresponse and distribution of noise sources across the detector surface. It will also apply if the noise sources are uniformly distributed throughout the bulk of the detector. If a square-root areal dependence is not obtained for a particular detector type, then the generation of noise and/or signal is undoubtedly variable across the detector surface.

A detailed analysis of the dependence of NEP on area, for the various noise-limited cases, will be developed here starting with the definition of NEP, Eq. (22). The procedure followed will be to substitute appropriate expressions for the signal and noise voltages in this equation and to generate an equation which delineates the geometry, time constant, and bandwidth dependence of NEP. The factor for the incident power will be $H \cdot A$ where H is the irradiance, and the signal voltage is as given by Eqs. (2) and (10). Therefore:

$$NEP = HA \frac{V_N}{I \cdot \frac{\rho l}{wd} \cdot \frac{\eta_n J_s \tau}{nd \sqrt{1 + (\omega\tau)^2}}}, \qquad (33)$$

where V_N is the rms value of the noise fluctuation, that is, $V_N = \sqrt{\overline{v_N^2}}$.

A. NEP—Johnson-Noise Limited

Inserting Eq. (11) into Eq. (33), and using

$$
\begin{aligned}
r &= \rho l / wd, \\
I &= iwd, \\
i &= \sigma E, \\
E &= \text{electric field strength}, \\
\sigma &= 2ne\mu, \\
\mu &= D \cdot e / kT, \\
D &= L^2 / \tau,
\end{aligned}
$$

gives

$$NEP = K_1 \sqrt{\frac{A \cdot \Delta f \cdot (1 + \omega^2\tau^2)}{\tau}} \qquad (34)$$

and

$$K_1 = \frac{kT' \sqrt{2nd}}{e\eta_n J_s EL} \cdot H. \qquad (35)$$

Thus the relationship between NEP, area, time constant, and bandwidth is clearly spelled out for the Johnson-noise limited case by:

$$NEP \propto \sqrt{\frac{A \cdot \Delta f}{\tau}}. \qquad (36)$$

B. NEP—G-R and Background Noise Limited

In this particular case, it is informative to rewrite Eq. (33), using Eq. (6), to obtain:

$$NEP = HA \cdot \frac{V_N}{V_B} = HA \frac{Ir\sqrt{\Delta N^2_{noise}/N}}{Ir(\Delta N_{sig}/N)}, \qquad (37)$$

$$NEP = HA \frac{\sqrt{\Delta N^2_{noise}}}{\Delta N_{sig}}. \qquad (38)$$

Substituting Eq. (20) and Eq. (8) in Eq. (38) gives:

$$NEP = K_2 \sqrt{A \cdot \Delta f}, \qquad (39)$$

$$K_2 = 2H \frac{\sqrt{\eta_{Bn} J_B + \eta_l J_l}}{\eta_n J_s}. \qquad (40)$$

In this instance, the signal and noise voltages are identically dependent on bias current flow, so that NEP is independent of this factor. Here the square-root areal dependence of NEP arises from the averaging

process required in arriving at the value for the mean square fluctuation in carriers generated at random by the background photons and lattice phonons. Notice that NEP in this case does not depend upon the time constant. This is due to the behavior of the detector now being controlled by the same processes for signal and noise. Both the quantities ΔN_{sig} and $\sqrt{\Delta N^2_{noise}}$ are linearly dependent on τ. Their ratio being independent of τ means that the detector will respond equally well in frequency to background photons and signal photons.

C. NEP—Excess 1/f-Noise Limited

Inserting Eq. (17) into Eq. (33) provides:

$$NEP = \frac{K_3}{\tau} \cdot \sqrt{\frac{A \Delta f (1 + \omega^2\tau^2)}{f}} \qquad (41)$$

and

$$K_3 = \frac{nH\sqrt{Cd}}{\eta_n J_s}. \qquad (42)$$

In this instance, both the signal and noise are associated with bias currents. However, even though a complete understanding of the source of $1/f$ noise is not yet available, it is clear that, for a uniform generation of noise throughout the bulk and/or surface of the detector, the NEP will depend upon the square root of the detector area. In addition, NEP will vary as $f^{-1/2}$

D. NEP—Shot-Noise Limited

Substituting Eq. (19) into Eq. (33) and using

$$I = \sigma Ewd = 2ne\mu^* Ewd,$$

$$\mu^* = \frac{L^2}{kT} \frac{e}{\tau},$$

where u^* is an average reduced mobility for carriers in the detector film, and L is the carrier diffusion length, it follows that:

$$NEP = K_4 \sqrt{\frac{A \Delta f (1 + \omega^2\tau^2)}{\tau}} \qquad (43)$$

and

$$K_4 = \frac{H}{\eta_n J_s L} \sqrt{\frac{dnkT}{eEn_l}}. \qquad (44)$$

E. Summation

In all noise-limited conditions examined here, NEP is clearly dependent on the square root of the area (as expected from the geometrical arguments above) and on the square root of the bandwidth. The time constant and frequency dependencies of shot-noise-limited and Johnson-noise-limited detectors are identical.

This should be expected since both noise spectra are essentially flat. The factor in the NEP expression denoting this identical condition is

$$\sqrt{\frac{1 + (\omega\tau)^2}{\tau}}.$$

The case of $1/f$ noise provides a factor given by

$$\sqrt{\frac{1 + (\omega\tau)^2}{f} \cdot \frac{1}{\tau}},$$

while G-R noise leads to an expression independent of frequency and time constant. Finally, the background-noise-limited NEP does not contain any factors associated with a detector mechanism of operation. This should be expected since a detector measuring background noise is limited by that external noise, and thereby loses its identity. In other words, any number of different detectors operating with the same spectral characteristics could not be distinguished from one another in terms of NEP, if they were all limited by background noise. If the detectors are G-R phonon noise limited, then $J_L \gg J_B$. This condition can be identified by changing the detector's temperature and observing corresponding changes in noise. If the detector is background limited, then $J_B \gg J_L$, and changes in the background temperature would cause observable changes in the detector noise.

IV. Jones System of Classification

Jones[20] has contributed a substantial effort to the understanding and categorizing of infrared detectors. He suggested that the reciprocal of NEP, denoted as "detectivity," was a more suitable quantity for rating detectors. It represents primarily a psychological asset, in that larger detectivities, rather than smaller NEP's represent better detectors. Jones considers "detectivity" particularly desirable because it also avoids usage of the word "sensitivity," which has a variety of meanings in technical language. For the sake of completeness, a summary of the parts of Jones' paper (see ref. 1) pertinent to discussions here will be included below.

Jones observed that when data taken from thermal and photodetectors were examined, they could be distinguished or classified by the way in which their detectivity is related to time constant and frequency response. He noted that their behavior in one case is like the detectivity obtained for an ideal thermal detector, while in another, to an estimated best obtainable thermal detector. Accordingly, two classes of detectors were set up, the first based on the detectivity obtained for the ideal detector (capable of seeing background thermal or photon noise), while the second was based on the best obtainable heat detector from Havens' limit (see ref. 20).

In establishing this system of classification, a reference or detective time constant and a bandwidth are defined, such that $\Delta f = 1/4t$. These reference condi-

tions are to provide a measurement of the detectivity for the important special case where the bandwidth of the noise is the same as the bandwidth of the detector. Thus if a detector has a response out to some upper limit of frequency f, the bandwidth of the noise extends out to the same frequency. For the purposes of the ensuing discussion of the various noise-limited cases, it will also be assumed that the detector is limited only by the type of noise singled out for analysis in any one particular case, throughout the responsive bandwidth of the detector. The classifications established by Jones are:

Class I:
$$D = k_1 \frac{\sqrt{\tau}}{\sqrt{A}}, \qquad (45)$$

Class II:
$$D = k_2 \frac{\tau}{\sqrt{A}} \qquad (46)$$

Detectors limited by a flat noise spectrum such as Johnson noise have a detectivity in a unit bandwidth classified as:

Class Ia:
$$D_1(f) = \frac{k_1}{2\sqrt{A}} \frac{1}{\sqrt{1 + (\omega\tau)^2}}, \qquad (47)$$

Class IIa:
$$D_1(f) = \frac{k_2 \sqrt{\tau}}{2\sqrt{A}} \frac{1}{\sqrt{1 + (\omega\tau)^2}} \qquad (48)$$

In the case of the $1/f$-noise-limited detector, Jones finds it necessary to redefine the reference time constant, and in so doing the detectivities take the form:

Class Ib:
$$D_1(f) = \frac{2k_1 \sqrt{\pi} \sqrt{f\tau}}{\sqrt{A}} \frac{1}{\sqrt{1 + 16(\omega\tau)^2}}, \qquad (49)$$

Class IIb:
$$D_1(f) = \frac{2k_2 \sqrt{\pi} \sqrt{f \cdot \tau}}{\sqrt{A}} \frac{1}{\sqrt{1 + 16(\omega\tau)^2}}. \qquad (50)$$

In Section III, analytical expressions were derived for NEP in the various noise-limited conditions. They can now be used to develop Jones' system of classification. Furthermore, it will suffice to use here the detector's responsive time, rather than Jones' reference time constant.

A. Detectivity—Johnson-Noise Limited

This detectivity is derived from Eq. (34). Thus

$$D = \frac{1}{NEP} = \frac{1}{K_1} \frac{\sqrt{\tau}}{\sqrt{A \cdot \Delta f}} \cdot \frac{1}{\sqrt{1 + (\omega\tau)^2}} \qquad 51)$$

and, in a unit bandwidth, becomes

$$D_1 = k_1 \frac{\sqrt{\tau}}{\sqrt{A} \sqrt{1 + (\omega\tau)^2}} \qquad (52)$$

which is defined as a class IIa detector. In the bandwidth $\Delta f = 1/(4\tau)$, for $\omega\tau \ll 1$, the detectivity becomes

$$D = k_1' \frac{\tau}{\sqrt{A}} \qquad (53)$$

which is the form of the class II detector.

B. Detectivity—G-R-Noise Limited

In this case, both signal and noise depend on the frequency in the same manner. Thus the expression for NEP was found to be independent of the factor $\sqrt{1 + (\omega\tau)^2}$, and the detectivity is derived from Eq. (39) to be

$$D = \frac{1}{K_2} \cdot \frac{1}{\sqrt{A \cdot \Delta f}} \tag{54}$$

The detectivity in the reference bandwidth takes the form

$$D = k_2' \sqrt{\frac{\tau}{A}} \tag{55}$$

which is a class I detector.

C. Detectivity—Excess, 1/f-Noise Limited

This detectivity is derived from Eq. (41).

$$D = \frac{1}{\text{NEP}} = \frac{1}{K_3} \frac{\tau_p}{\sqrt{1 + (\omega\tau_p)^2}} \sqrt{\frac{f}{A \cdot \Delta f}} \tag{56}$$

and, for a unit bandwidth,

$$D_1 = \frac{1}{K_3} \frac{\tau_p}{\sqrt{1 + (\omega\tau_p)^2}} \sqrt{\frac{f}{A}} \tag{57}$$

The constant τ_p is the detector responsive time constant. Jones defines a reference time constant $\tau = \tau_p/4$. Using this time constant, it follows that:

$$D_1 = k_3 \cdot \frac{\tau\sqrt{f}}{\sqrt{1 + 16(\omega\tau)^2}} \cdot \frac{1}{\sqrt{A}} \tag{58}$$

which is the form of the class IIb detector.

D) Detectivity—Shot-Noise Limited

This detectivity is derived from Eq. (43).

$$D = \frac{1}{K_4} \frac{\sqrt{\tau}}{\sqrt{1 + (\omega\tau)^2}} \frac{1}{\sqrt{A \cdot \Delta f}} \tag{59}$$

Shot noise has a flat frequency spectrum, and provides a detectivity in a unit bandwidth similar to Johnson noise:

$$D_1 = \frac{k_4(\tau/A)^{1/2}}{\sqrt{1 + (\omega\tau)^2}} \tag{60}$$

and, in the responsive bandwidth,

$$D = \frac{k_4'}{\sqrt{1 + (\omega\tau)^2}} \frac{\tau}{\sqrt{A}} \tag{61}$$

and

$$D = k_4' \cdot \frac{\tau}{\sqrt{A}} \tag{62}$$

for $\omega\tau \ll 1$, which is the form of a class II detector.

E. Figures of Merit

By taking the ratio of the detectivity of class I detectors to that of the perfect thermal detector (limited by photon noise), Jones derives a figure of merit called M_1 (see ref. 1), expressed as:

$$M_1 = 2.76 \times 10^{-11} \times \frac{D \sqrt{A}}{\sqrt{\tau}} \tag{63}$$

A figure of merit for the class II detector was derived using Havens' limit as the reference detector.[21] Havens arrived at his estimate by treating such thermal detectors as heat engines, considering their operation from a theoretical analysis of the efficiency of such engines, and arriving at an ultimate limit of performance capability by including considerations of limiting noise, available materials, and techniques. On this basis, the class II figure of merit becomes:

$$M_2 = 3 \times 10^{-11} \times \frac{D \sqrt{A}}{\tau} \tag{64}$$

F. D-Star (D*)

The expressions for detectivity, for the different limiting-noise types, always contain the factor $1/\sqrt{A \cdot \Delta f}$. Therefore, they are a function of detector size and the electrical bandwidth of operation. A more useful number is one that establishes the performance of the detector independent of these quantities. Jones suggested that the quantity D^*, defined as

$$D^* = \frac{\sqrt{A \Delta f}}{\text{NEP}} = D\sqrt{A \cdot \Delta f}, \tag{65}$$

serves to characterize the detector in terms of the intrinsic properties of the material of which it is made (see ref. 2).

V. Discussion

The discussions carried out in Sections II to IV indicate that a considerable gain in understanding the noise properties of semiconducting photodetectors has been made since the Jones paper of 1953. The resultant analysis carried out in these sections indicates the following:

A. General

1. Class I detectors are limited by background and G-R noise.

2. Class II detectors are limited by all other sources of noise.

3. A class II detector can become a class I detector if its performance is improved to the point that background or G-R noise predominates over any other noise source. This can often be done by cooling a good detector. It appears that, when a detector becomes background limited, it loses its identity. The NEP

is descriptive of the rms fluctuation in background photons striking the detector. Thus the performance of the detectors is completely described by the condition of the background. If a number of different detectors, all having the same spectral response characteristics, were background limited, they could not be identified individually on the basis of NEP, D, D^*, or a class I designation. This would be so even though they might be made of different materials and/or have different time constants.

4. The significance of the figures of merit M_1 and M_2 is not clear, particularly in reference to Havens' limit. Since a detector can go from class II to class I, the reference might well be to the background-limited condition in both cases.

5. In a recent statistical analysis of the NEP's reported on lead-compound film detectors, Limperis and Wolfe[22] reported that these excess-noise-limited detectors follow an $(A)^{1/2}$ in accordance with the treatment of Section 3. Until about 1956, data did not clearly indicate a consistent geometry dependence for NEP. There were two factors which prevented obtaining consistent data:

(a) In bulk and film detectors, limiting-noise sources were not uniformly distributed through the detector. That is, most of the noise was generated at the electrodes around the detector periphery, localized regions on the crystal's surface, and similarly within the bulk.

(b) The photoresponse was generally nonuniform across a detector surface. The film detectors are particularly difficult to deal with in this regard.

It appears that detector technology has improved to the point where noise generation and signal response is sufficiently uniform to indicate a square-root areal dependence of NEP.

B. Units

The units of D^* have received considerable attention, and at one time were a matter of controversy. Jones' original definition of D^* appeared in the form

$$D^* = \sqrt{\frac{A}{1 \text{ cm}^2} \cdot \frac{\Delta f}{1 \text{ cps}} \cdot \frac{1}{\text{NEP}}}. \tag{66}$$

Thus it appeared that the definition normalized D^* to a unit area and bandwidth, and that the unit of D^* was reciprocal watts. This unit seemed appropriate since D^* was an intrinsic measure of a material's ability to detect radiation power. However, considerable opposition arose to the normalization procedure suggested by Jones, and the units ultimately accepted were those that appear in a straightforward examination of dimensions, namely $\text{cm} \cdot (\text{cps})^{1/2} \cdot \text{watts}^{-1}$. These units might appear somewhat peculiar in that D^* is supposed to provide a measure of the intrinsic ability of a detector material to respond to radiation power and yet contains the dimensions of size and frequency. However, a recent analysis of a substantial amount of data indicates that D^* appears to do its job. An explanation of this seeming contradiction comes from an examination of the equations derived earlier for NEP in the various noise-limiting cases.

Consider the background-noise-limited case where, from Eq. (39),

$$\text{NEP} = \frac{\sqrt{J_{b}\eta_{b}}}{\eta_{s}J_{s}} \cdot 2H \cdot \sqrt{A \cdot \Delta f}. \tag{67}$$

The factor $\sqrt{A \cdot \Delta f}$ cancels out in the computation of D^*, so that the remaining factors provide the units. The quantities η_{b} and η_{s} are efficiency factors. Therefore, the dimensional analysis is simplified to:

$$D^* = \frac{J_{s}}{\sqrt{J_{b}}} \cdot \frac{1}{H} = \frac{\text{No. photons}_{s}/\text{cm}^2 \text{ sec}}{\sqrt{\text{No. photons}_{b}/\text{cm}^2 \text{ sec}}} \cdot \frac{\text{cm}^2}{\text{watt}}$$

$$= \frac{\text{cm}}{\sqrt{\text{sec}} \cdot \text{watts}} = \frac{\text{cm}\sqrt{\text{cps}}}{\text{watts}} \tag{68}$$

since the numbers of photons are dimensionless, and the reciprocal of time is equivalent here to a frequency. The derived units of D^* becomes $\text{cm}(\text{cps})^{1/2} \text{ watts}^{-1}$, consistent with accepted usage.

An interpretation of the meaning of the units of D^* can now readily be put forth. In the measurement of NEP, radiation power from a blackbody illuminates a detector to generate a signal voltage V_S. A noise voltage V_N is obtained when the detector views an ambient black background. V_N may result from background fluctuations or noise sources internal to the detector. In the background-limited case, it was shown above that D^* is dependent only on the factors J_{s}, J_{b}, and H, while being explicitly independent of area and bandwidth. These are all factors in the measurement procedure, and the units of D^* can be attributed to J_{s}, J_{b}, and H. Since these quantities represent the numbers of photons and watts per unit area, it is clear that the measurement is inherently normalized. Thus if one wishes to perform a series of measurements in which the bandwidth and/or the area were variables, the experiment would be carried on at constant J_{s}, J_{b}, and H. The evaluation of D^* in this situation is truly independent of detector area. It appears, therefore, that Jones' judgment or the significance of D^* was correct.

It is fruitful to examine the units of D^* in the Johnson and shot-noise-limited cases. The appropriate expressions can be obtained by inserting Eqs. (34) and (43) into Eq. (65) respectively. They are: for D^* limited by Johnson noise:

$$D^* = \frac{\eta_{s}J_{s}eEL}{\sqrt{2nd\,kTH}} \cdot \sqrt{\frac{\tau}{1 + (\omega\tau)^2}}, \tag{69}$$

for D^* limited by shot noise:

$$D^* = \sqrt{\frac{L^2 e E n_t}{n d k t}} \cdot \frac{\eta_\omega J_s}{H} \sqrt{\frac{\tau}{1 + (\omega\tau)^2}}. \tag{70}$$

In the Johnson-noise case, the quantities eEL and kT represent electrical and thermal energies, and have the same basic units. Separating out these factors for the dimensional analysis, and omitting all numerical factors, results in

$$D^* = \frac{eEL}{kT} \cdot \frac{1}{\sqrt{nd}} \cdot \frac{J_s}{H} \cdot \sqrt{\tau}. \tag{71}$$

Substituting units provides:

$$D^* = \frac{\text{Energy}_{(\text{elect})}}{\text{Energy}_{(\text{therm})}} \cdot \frac{1}{\sqrt{1/\text{cm}^2}} \cdot \frac{\text{No. photons/area}\cdot\text{time}}{\text{watts/area}} \cdot \sqrt{\text{time}}$$

$$= \frac{\text{Energy}_{(\text{elect})}}{\text{Energy}_{(\text{therm})}} \cdot \frac{\text{cm}}{\text{watts}} \cdot \frac{1}{\sqrt{\text{time}}}, \tag{72}$$

$$D^* = \frac{\text{cm}\sqrt{\text{cps}}}{\text{watts}}. \tag{73}$$

In the case of shot noise, the quantity $n_t \cdot L$ is unitless, eEL is an electrical energy, and D^* can be expressed dimensionally as:

$$D^* = \sqrt{\frac{\text{Energy}_{(\text{elect})}}{\text{Energy}_{(\text{therm})}}} \cdot \text{cm} \cdot \frac{\text{No. photons/area}\cdot\text{time}}{\text{No. watts/area}} \cdot \sqrt{\text{time}}.$$

On simplifying:

$$D^* = \sqrt{\frac{\text{Energy}_{(\text{elect})}}{\text{Energy}_{(\text{therm})}}} \cdot \frac{\text{cm}}{\text{watts}} \cdot \frac{1}{\sqrt{\text{time}}}, \tag{74}$$

$$D^* = \frac{\text{cm}\sqrt{\text{cps}}}{\text{watts}} \tag{75}$$

It is clear that units are consistently maintained for D^* in the different noise-limited cases. An interesting feature of this examination reveals that the ratio of an electrical energy to a thermal energy appears as a factor for the Johnson-noise-limited case, and the square root of this quantity appears in the shot-noise case. This is consistent with different noise mechanisms responsible for the two cases. In both cases, an increasing bias voltage and a decreasing temperature should result in increasing D^*. The density of carriers n is also dependent on T, and decreases with reduced temperature. However, reduced T can also effect the magnitude of the energy gap, change the cutoff wavelength, and also affect the time constant τ. These apparently are the factors that are adjustable. In particular, when seeking the optimum bias for a photoconductive detector, it may well be a manner of raising the bias until an increased temperature from Joule heating results. Further biasing could cause adverse temperature effects, and therefore a reduced D^*.

The author wishes to thank T. Limperis and G. H. Suits for their helpful suggestions in reviewing the manuscript.

References

1. R. C. Jones, *Advances in Electronics* (Academic Press Inc., New York, 1953).
2. R. C. Jones, J. Opt. Soc. Am. **50**, 1058 (1960).
3. The photoconductive detector has also been treated in considerable detail by D. H. Roberts and B. L. H. Wilson, Brit. J. Appl. Phys. **9**, 291 (1958).
4. R. L. Petritz, Proc. IRE **47**, 1458 (1959).
5. J. B. Johnson, Phys. Rev. **32**, 97 (1928).
6. H. Nyquist, Phys. Rev. **33**, 110 (1928).
7. R. L. Petritz, Phys. Rev. **104**, 1508 (1956). See also ref. 4.
8. K. M. Van Vliet, Proc. IRE **46**, 1004 (1958).
9. J. J. Brophy, J. Appl. Phys. **27**, 1383 (1957).
10. T. G. Maple, L. Bess, and H. A. Gebbie, J. Appl. Phys. **26**, 490 (1955).
11. R. L. Petritz, Proc. IRE **40**, 1440 (1952).
12. J. J. Brophy and N. Rostoker, Phys. Rev. **100**, 754 (1955).
13. J. J. Brophy, Phys. Rev. **106**, 675 (1957).
14. L. Bess, J. Appl. Phys. **26**, 1377 (1955).
15. A. U. MacRae and H. Levinstein, Phys. Rev. **119**, 62 (1960).
16. J. J. Brophy, J. Appl. Phys. **27**, 1383 (1956).
17. L. Bess, Phys. Rev. **103**, 72 (1956).
18. J. R. Morrison, Phys. Rev. **104**, 619 (1956).
19. A. Van der Ziel, Proc. IRE **46**, 1019 (1958).
20. R. C. Jones, Proc. IRE **47**, 1495 (1958). See also ref. 1.
21. R. J. Havens, J. Opt. Soc. Am. **36**, 355(A) (1946).
22. T. Limperis and W. Wolfe, J. Opt. Soc. Am. **51**, 482 (1961).

Reprinted from *Infrared Phys.*, **10**(4), 233–235 (1970), with the permission of Microforms International Marketing Corporation as exclusive copyright licensee of Pergamon Press journal back files.

Photon Noise Limited *D** for Low Temperature Backgrounds and Long Wavelengths

S. F. JACOBS and M. SARGENT, III

Optical Sciences Center, The University of Arizona, Tucson, Arizona 85721, USA.

(*Received* 1 *September* 1970)

MODERN infrared detectors are, in many instances, limited in sensitivity by room temperature background photon noise. Where additional sensitivity is required in the laboratory the cooling of detector optics and field of view apertures reduces this limiting photon noise, while for space applications lower temperature backgrounds similarly result in reduced photon noise. The state-of-the-art detector is thus challenged to achieve even better sensitivities than it has thus far demonstrated. Furthermore, the exploration further into the infrared brings new interest in the low temperature background photon noise limit.

Recognizing these cold background and far infrared applications we have extended previous calculations of photon noise limited *D** vs. cutoff wavelength λ_c to include very low background temperatures and long wavelengths. We have used the formula[1]

$$D^*(T,\lambda_c) = \frac{c\eta^{1/2}}{2\pi^{1/2}h^{1/2}\nu^2 k^{1/2}T^{1/2}} \left\{ \sum_{m=1}^{\infty} \exp\left(-\frac{mh\nu}{kT}\right) \left[1 + \frac{2kT}{mh\nu} + 2\left(\frac{kT}{mh\nu}\right)^2\right] \right\}^{-\frac{1}{2}} \tag{1}$$

where η = detector quantum efficiency

h = Planck's constant

k = Boltzmann's constant

T = background temperature

c = speed of light

$\nu = c/\lambda_c$

This function can be closely approximated for $h\nu \gg kT$ by retaining only the first term

$$D^*(T,\lambda_c) = \frac{c\eta^{1/2}\exp(h\nu/2kT)}{2\pi^{1/2}h^{1/2}\nu^2 k^{1/2}T^{1/2}[1 + (2kT/h\nu) + 2(kT/h\nu)^2]^{1/2}} \tag{2}$$

However, for $T = 295°K$, λ_c must be less than $50\mu m$, which illustrates the inapplicability of (2) at long wavelengths and high temperatures.

Figure 1 shows a computer calculation of expression (1) for background temperatures down to 1°K and wavelengths out to 1 cm. This curve is identical for instance with Kruse[1] but additionally covers $T < 77°K$ and $\lambda_c > 50\mu m$.

We note that the curves are not independent, for from (1) $D^*(T,\lambda_c)$ is related to $D^*(T',\lambda_c')$ by the formula

$$D^*(T,\lambda_c) = \left(\frac{T'}{T}\right)^{5/2} D^*(T',\lambda_c') \tag{3}$$

where

$$\lambda_c' = \frac{T}{T'}\lambda_c \tag{4}$$

This relation is useful for determining values of $D^*(T',\lambda_c)$ which do not appear in Fig. 1 in terms of a value $D^*(T',\lambda_c')$ which does appear. For example, to find $D^*(1000°K, 4\mu m)$ from the 500°K curve,

$$D^*(1000, 4) = \left(\frac{500}{1000}\right)^{5/2} D^*\left(500, 4 \times \frac{1000}{500}\right) = 2\cdot3 \times 10^9 \text{cm Hz}^{1/2}/\text{W} \tag{5}$$

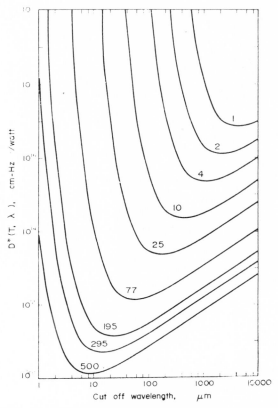

FIG. 1. Photon noise limited D^* at peak wavelength (assumed to be cutoff wavelength), for various background temperatures (°K) (Assumes 2π field of view and $\eta = 1$).

If higher accuracy is desired than can be determined from Fig. 1, one can use the formula above in combination with Table 1 which gives explicit values of D^* vs. λ_c for $T = 295°K$.

TABLE 1

λ (μm)	$D^*(295,\lambda_c)$	(μm)	$D^*(295,\lambda_c)$	λ (μm)	$D^*(295,\lambda_c)$	λ (μm)	$D^*(295,\lambda_c)$
1	$2 \cdot 19 \times 10^{18}$	10	$5 \cdot 35 \times 10^{10}$	100	$1 \cdot 67 \times 10^{11}$	1000	$1 \cdot 55 \times 10^{12}$
2	$4 \cdot 34 \times 10^{13}$	20	$5 \cdot 12 \times 10^{10}$	200	$3 \cdot 20 \times 10^{11}$	2000	$3 \cdot 10 \times 10^{12}$
3	$1 \cdot 64 \times 10^{12}$	30	$6 \cdot 29 \times 10^{10}$	300	$4 \cdot 74 \times 10^{11}$	3000	$4 \cdot 64 \times 10^{12}$
4	$3 \cdot 75 \times 10^{11}$	40	$7 \cdot 68 \times 10^{10}$	400	$6 \cdot 28 \times 10^{11}$	4000	$6 \cdot 19 \times 10^{12}$
5	$1 \cdot 70 \times 10^{11}$	50	$9 \cdot 13 \times 10^{10}$	500	$7 \cdot 82 \times 10^{11}$	5000	$7 \cdot 73 \times 10^{12}$
6	$1 \cdot 06 \times 10^{11}$	60	$1 \cdot 06 \times 10^{11}$	600	$9 \cdot 36 \times 10^{11}$	6000	$9 \cdot 28 \times 10^{12}$
7	$7 \cdot 93 \times 10^{10}$	70	$1 \cdot 21 \times 10^{11}$	700	$1 \cdot 09 \times 10^{12}$	7000	$1 \cdot 08 \times 10^{13}$
8	$6 \cdot 57 \times 10^{10}$	80	$1 \cdot 36 \times 10^{11}$	800	$1 \cdot 24 \times 10^{12}$	8000	$1 \cdot 24 \times 10^{13}$
9	$5 \cdot 80 \times 10^{10}$	90	$1 \cdot 52 \times 10^{11}$	900	$1 \cdot 40 \times 10^{12}$	9000	$1 \cdot 39 \times 10^{13}$

REFERENCE

1. KRUSE, P. W., L. D. McGLAUGHLIN and R. B. McQUISTAN, *Elements of Infrared Technology*, p. 360. John Wiley (1963).

VI

Techniques for Cooling
Infrared Detectors

Editors' Comments on Papers 40 Through 43

40 Crouch: *Cryogenic Cooling for Infrared*

41 Stephens: *Advanced Design of Joule–Thomson Coolers for Infra-Red Detectors*

42 Annable: *Radiant Cooling*

43 Daniels and du Pré: *Closed Cycle Cryogenic Refrigerators as Integrated Cold Sources for Infrared Detectors*

The rapid development of cooled photon detectors that started in the mid-1950s provided a strong impetus for the development of suitable cooling devices. The basic principles of such devices have been known for many years, but the requirements dictated by infrared systems placed a new premium on extreme miniaturization, minimum power consumption, simplicity of maintenance, and high reliability. In the limited space available here we can only include a brief sample of the many papers that have been published. We hope, however, that this sample will show how well the cryogenic community has met the challenge posed by the need to cool infrared detectors. Hudson has given an in-depth treatment of cooling techniques, a description of the engineering aspects of selecting and applying a cooling system to a specific application, and an extensive bibliography of the detector cooling literature up to 1969 [1]. Since this text is readily available, very little of its contents will be repeated here.

Paper 40, by Crouch, is a concise tutorial treatment of the art of cooling infrared detectors. He reviews the reasons why detectors must be cooled, notes the temperatures that are necessary, and makes a few well-chosen comments on each of the principal cooling techniques. The last half of the paper is an absorbing treatment of the systems approach as applied to the integration of a detector and a cooling system. Such information is usually jealously guarded by practitioners of the art and is rarely found in the literature. It is obvious that Crouch has learned his lessons in the real world of real hardware. Additional references include tutorial papers [2–7], liquid-transfer systems [8], solid-cryogen systems [9, 10], and cooled telescope assemblies [11].

J. N. Crouch, Jr., is an Engineering Specialist with E-Systems, Inc., Greenville Division, Greenville, Texas.

The Joule–Thomson cryostat (as it is called in the United States) is a gas liquefier that is small enough to be placed directly in the coolant chamber of a detector package. In Paper 41, Stephens notes that cryostats have the advantage of being quick-acting devices that are virtually free of microphonics, mechanical vibration, and acoustic noise. The high-speed units that he describes produce liquid in less than 2 seconds. Perhaps the most eloquent testimonial to the versatility of Stephens and his colleagues is the group of 14 photographs which show the variety of forms that the cryostat and its support equipment can assume. More information on cryostats is contained in References 12 through 17.

S. W. Stephens is Regional Sales Manager, Controls Division, Moog Limited, Cheltenham, England.

The early space-borne infrared systems used uncooled thermal detectors because there were no coolers that could meet the severe limitations imposed on weight and power consumption. The successful application of cooled photon detectors in space had to await the development of the radiative transfer cooler [18]. Annable, in Paper 42, describes a methodology for the design and analysis of such coolers and gives some very practical information on their construction. He describes coolers designed and built for satellite use that produce temperatures in the range 80 to 200°K. A later paper by Annable extends his analysis to include the effects of the absorption of radiant power from external sources [19].

R. V. Annable is a Senior Physicist, Optical Division, ITT Aerospace, Fort Wayne, Indiana.

The small, self-contained, mechanical refrigerator for cooling detectors was, for many years, high on the "wish list" of every infrared system designer. Prototypes of such coolers began to appear in the early 1960s and they soon became a commercially available item. In Paper 43, Daniels and du Pré describe a light-weight, low-power, Stirling-cycle refrigerator that provides a cooling capacity of 1.5 watts at a temperature of 77°K. The unit requires a power input of 45 watts and weighs 3.9 kg. Other units are described that produce lower temperatures and that are small enough to merit consideration for satellite applications. For more information on closed-cycle cooling systems, see References 20 through 26.

A. Daniels is Director, Mechanical Systems Research, Philips Laboratories, Briarcliff Manor, New York.

References

1. Hudson, R. D., Jr., *Infrared System Engineering*, Wiley, New York (1969), Chap. 11.
2. Beyen, W. J., and B. R. Pagel, "Cooling Requirements for Intrinsic Photoconductive Infrared Detectors," *Infrared Phys.*, **6,** 161 (1966).
3. Borrello, S. R., et al., "Cooling Requirements for Blip Performance of Intrinsic Photoconductors," *Infrared Phys.*, **11,** 225 (1971).
4. Gifford, W. E., "Refrigeration to Below 20K," *Cryogenics*, **10,** 23 (1970).
5. Klipping, G., "Open-Cycle Cryogenic Refrigeration Techniques," p. 359 in Hogan, W. H., and T. S. Moss (eds.), *Cryogenics and Infrared Detection*, Boston Technical Publishers, Inc., Cambridge, Mass. (1970).
6. Breckenridge, R. W., "Spaceborne Refrigeration Systems," p. 372 in Hogan, W. H., and T. S. Moss (eds.), *Cryogenics and Infrared Detection*, Boston Technical Publishers, Inc., Cambridge, Mass. (1970).
7. Williams, N., and A. A. Brooker, "A Practical Assessment of Refrigeration Systems for Cooling Infrared Detectors," p. 410 in Hogan, W. H., and T. S. Moss (eds.), *Cryogenics and Infrared Detection*, Boston Technical Publishers, Inc., Cambridge, Mass. (1970).
8. Nicholds, K. E., "Leidenfrost," *Cryogenics*, **10,** 45 (1970).
9. Gross, U. E., and A. I. Weinstein, "A Cryogenic-Solid Cooling System," *Infrared Phys.*, **4,** 161 (1964).
10. Kollodge, J. C., J. R. Thomas, and R. A. Weagant, "Nimbus Limb Radiometer, Apollo Fine Sun Sensor, and Skylab Multispectral Scanner," *Appl. Optics*, **11,** 2169 (1972).
11. McNutt, D. P., K. Shivanandan, and P. D. Feldman, "A Rocket-Borne Liquid Helium-Cooled Infrared Telescope. I: Dewar and Optics," *Appl. Optics*, **8,** 2199 (1969).

12. Smirnov, A. I., "The Use of Freons for Cooling Radiation Detectors," *Sov. J. Opt. Tech.*, **35,** 326 (1968).
13. Gross, S., "Infrared Sensor Cooling by the Joule–Thomson Effect," *Infrared Phys.*, **6,** 47 (1966).
14. Farmer, V. M., and D. P. Forse, "Improved Cooling Techniques for Detectors," *Infrared Phys.*, **8,** 37 (1968).
15. Drew, H. D., and A. J. Sievers, "A ³He-Cooled Bolometer for the Far Infrared," *Appl. Optics,* **8,** 2067 (1969).
16. Nicholds, K. E., "Joule–Thomson Cooling Systems for Military Infra-Red Systems," p. 326 in Hogan, W. H., and T. S. Moss (eds.), *Cryogenics and Infrared Detection,* Boston Technical Publishers, Inc., Cambridge, Mass. (1970).
17. Hughes, J. L., and K. C. Herr, "Mariner Mars 1969 Infrared Spectrometer: Gas Delivery System and Joule–Thomson Cryostat," *Cryogenics,* **13,** 513 (1973).
18. "IR Device Built to Detect any Vegetation on Mars," *Electronic Design,* **9,** 6 (Apr. 12, 1961).
19. Annable, R. V., "Radiant Cooling. 2: The Cone Wall Absorption of External Radiant Power," *Appl. Optics,* **11,** 1495 (1972).
20. Crawford, A. H., "Specifications of Cryogenic Refrigerators," *Cryogenics,* **10,** 28 (1970). (see also addendum in **10,** 429).
21. Maddocks, F. E., "Stirling Cycle Refrigerators for Airborne Infrared Detectors," p. 306 in Hogan, W. H., and T. S. Moss (eds.), *Cryogenics and Infrared Detection,* Boston Technical Publishers, Inc., Cambridge, Mass. (1970).
22. Prast, G., "The Vuilleumier Cycle," p. 283 in Hogan, W. H., and T. S. Moss (eds.), *Cryogenics and Infrared Detection,* Boston Technical Publishers, Inc., Cambridge, Mass. (1970).
23. Siciliano, S. G., "Evolution of a Closed-Cycle Airborne 20°K Refrigerator," p. 292 in Hogan, W. H., and T. S. Moss (eds.), *Cryogenics and Infrared Detection,* Boston Technical Publishers, Inc., Cambridge, Mass. (1970).
24. Daniels, A., "Cryogenics for Electro-Optical Systems," *Electro-Opt. Syst. Design,* **3,** 12 (July 1971).
25. Miller, B., "New Sensor Cooling Unit Developed," *Aviation Week,* **97,** 76 (Nov. 20, 1972).
26. Daniels, A., and F. K. du Pré, "Miniature Refrigerators for Electronic Devices," *Cryogenics,* **13,** 134 (1973).

40

Reprinted from *Electro-Technology*, **75**, 96–100 (May, 1965)

Cryogenic Cooling for Infrared

J. N. CROUCH
Apparatus Division
Texas Instruments Incorporated
Dallas, Texas

A NECESSARY GOAL in designing any infrared set is reducing noise sources for an acceptable noise level for adequate signal processing. Thermal or Johnson noise is considered the base level in photon detectors. This type of noise is directly dependent on absolute temperature; the lower the temperature, the less the Johnson noise. Generation-recombination noise also exhibits a temperature dependence. Current or $1/f$ noise seems to be unaffected.

Photon detectors with a large ambient-temperature noise figure are generally cooled to 90° K or below; the improved performance is often remarkable. Table I lists selected photon detectors at various operating temperatures. Notice the figure of merit D^*_{bb}, the black-body detectivity of the detector (independent of wavelength) normalized to unit area and unit bandwidth. Although the intent of Table I is to show only temperature dependence, care should be taken to consider all detector parameters when comparing values.

As shown, cooling the detector from ambient temperature to 90° K

or below can greatly enhance performance. Cooling the detector does not eliminate all noise sources, but the increased performance is sufficient to warrant the effort of providing cryogenic temperatures. (For the purposes of this article, cryogenic temperatures are those below 90° K, the approximate boiling point of liquid oxygen at atmospheric pressure.)

There are two types of cryogenic systems for infrared sets: open-cycle systems that expend the coolant after use and thus must be refilled, and closed-cycle systems that retain the coolant or working fluid and are thus continuously operable.

Open-Cycle Systems. In open-cycle cryogenic systems, the detector is cooled by transferring its heat to a liquefied gas that has a cryogenic boiling point. The heat of vaporization of the liquefied gas (cryogen) absorbs the heat, and the boiling point of the cryogen determines the detector temperature. A small range of detector temperatures is possible by varying the pressure over the liquid.

Although Table II lists seven cryogens along with their triple and boiling points, only five of them are commonly used in open-cycle systems.

Oxygen and fluorine have boiling points in the cryogenic range, but they are seldom used because of their hazardous chemical activity and toxicity.

Some open-cycle systems use the Joule-Thomson method to liquefy gases. A gas is expanded through a small orifice and works against its intermolecular forces to create a refrigeration potential. Highly efficient heat exchangers use this potential to lower the temperature of the gas regeneratively before it enters the orifice, and part of the gas is liquefied at the orifice. The refrigerant is supplied in gaseous form in high-pressure tanks. Although this type of system is simple in principle, it has contamination problems and a high weight ratio of storage tank to useful refrigerant.

Other open-cycle systems use a liquid-refill method in one of three forms.[1] One form is to use a liquid-filled, insulated Dewar with a pressure-buildup coil for supplying high-pressure gas to a Joule-Thomson

Table II — Boiling and Triple Points of Common Cryogens [2]

Cryogen	Boiling point at atmospheric pressure, °K	Triple point, °K
He	4.216	—
H₂	20.39	13.96
Ne	27.1	24.57
N₂	77.34	63.15
F₂	85.24	55.20
A	87.29	83.85
O₂	90.19	54.40

Note: Helium cannot be solidified under its own vapor pressure.

Table I — Operating Characteristics of Selected Detectors

Material	Operating temperature, °K	D^*_{bb} (500 °K, f, 1), cm—$\sqrt{\text{cps}}$/watt	Noise mechanism	Material	Operating temperature, °K	D^*_{bb} (500 °K, f, 1), cm—$\sqrt{\text{cps}}$/watt	Noise mechanism
PbS	77	40×10^8	Current	InSb	77	1×10^{10}	Current below 100 cps; GR above
PbS	295	4.5×10^8	Current				
PbSe	77	220×10^7	Current	InSb	295	†	—
PbSe	295	3×10^7	Current	Ge:Cu	20	1×10^{10}	Current below 1 kc; GR above
InSb	77	1200×10^7	Current				
InSb	295	1.4×10^7	Thermal	Ge:Cu	295	†	—
PbTe	77	3.8×10^8	Current	NbN	15	4.8×10^9	Unknown
PbTe	295	†	—	NbN	295	†	—
Ge:Au	65	1×10^{10}	Current below 40 cps; GR above	Ge:Cd	25	7×10^9	Current below 500 cps; GR above
Ge:Au	295	†	—	Ge:Cd	295	†	—
Ge:Zn	4.2	4×10^9	Current	Ge:Hg	45	9×10^9	Current below 500 cps; GR above
Ge:Zn	295	†	—	Ge:Hg	295	†	—

Notes: Mode of operation is photoconductive for all materials except InSb (photovoltaic) and NbN (superconducting bolometer); f= modulation frequency; GR=generation-recombination. †Detector cannot be operated at room temperature.

DESIGN TRENDS

orifice. Another system is the integrally mounted detector type. The detector, liquid-holding Dewar, and infrared-transmitting window are integrated into one package. There are no moving parts or valves, and the system is quite reliable.

The third type is the liquid-feed system. The cryogen is stored in a reservoir Dewar and fed through a line to the detector Dewar. The line may be either insulated or noninsulated. When noninsulated lines are used, a film of gas is established on the inside of the transfer line, and the liquid flows through the tube as small spheres which ride along the gas film. Liquid transfer using this two-phase flow is extremely adaptable because it allows the use of simple, flexible lines.

Closed-Cycle Systems. The desirable characteristics for an airborne, closed-cycle cryogenic system are high reliability to reduce maintenance and provide long trouble-free operation, and high efficiency to reduce weight, volume, and power requirements to a minimum. Some cooling methods that provide adequate refrigeration at the required temperature cannot provide reliable performance when miniaturized.

In general, there are four methods of closed-cycle cryogenic cooling that can provide adequate reliability when miniaturized: 1) the Joule-Thomson method, 2) the Claude method, 3) the Stirling method, and 4) the modified Stirling method.

The Joule-Thomson closed-cycle method uses a compressor and isenthalpic expansion of a gas through a small orifice to provide a refrigeration potential. Counter-current heat exchangers use this potential to reduce the temperature of the gas regeneratively. Liquefaction of part of the gas takes place at the orifice. The remaining gas and the boil-off gas are recycled to the compressor, closing the loop.

The Joule-Thomson method, although simple, has several disadvantages. First of all, it is an irreversible cycle and thus relatively inefficient. The second problem is contamination of the refrigerant. The small orifice to provide the throttling effect is generally 0.002 to 0.007 inch in diameter, and is also the coldest point in the system. Any condensable impurities in the system will tend to freeze out and block the orifice. This makes it necessary to use a nonlubricated compressor or elaborate clean-up techniques. High compression ratios (often 100 to 1) and small flows also put a strain on the compressor.

There are two major advantages of the method. First, there are no moving parts in the system in contact with the detector, and the likelihood of microphonics is minimized. Second, the compressor can be located quite a distance away from the cryostat to provide greater packaging freedom than offered by some other methods. The distance of separation for a compressor of a given capacity is limited by the pressure drop through the gas lines.

The Joule-Thomson method can be used only when the refrigerant to be liquefied is below its inversion point. The approximate inversion points of common refrigerants are: argon, 723° K; nitrogen, 621° K; neon, 220° K; hydrogen, 205° K; and helium, 50° K. Neon, hydrogen and helium must be precooled before they can be liquefied.

Methods involving isentropic expansion, such as the Claude cycle, are quite adequate for medium-scale refrigeration. In most cases, the gas is allowed to work against an expansion engine piston or a turbo-expander to provide a refrigeration potential. Work energy is transferred out of the cold area by a mechanical or electrical linkage to ambient.

The action of the expansion engine is essentially that of a steam engine. Warm gas at a high pressure and temperature enters the expansion area, works against the engine piston or turbo-expander, and exhausts at a lower temperature and pressure. The Claude cycle is more efficient than the Joule-Thomson cycle because it is reversible and approaches the Carnot efficiency. The use of expansion ratios of about five or six enhances compressor life. The gas (working fluid) does not have to be cooled below its inversion point. Care is taken to prevent liquefaction in the expansion engine because this would lower the efficiency. The Claude cycle is sometimes used to precool a gas before it enters a Joule-Thompson orifice for liquefaction.

Contamination of the gas by condensable impurities is a problem in the Claude cycle also, but it is not as serious as in the Joule-Thomson systems. The valves used in conjunction with the expansion engine can be blocked if enough impurities are present.

Two disadvantages inherent in the Claude cycle that limit its use in airborne infrared-detection sets are the poor reliability of the fast-acting parts in the cold zone, and the likelihood of microphonics introduced by moving parts near the detector. Long life in a cryogenic, unlubricated environment is difficult when miniature parts are used. Miniature turbo-expanders show promise, but their high rotating speeds (250,000 to 600,000 rpm) present difficulties.

The Stirling-cycle system can be used for cooling detectors to temperatures not easily reached by liquefied-gas systems. The basic parts used in the cycle are a compression chamber, an expansion chamber, and a thermal regenerator which is the counterpart of the counter-current heat exchangers used in the Claude and Joule-Thomson cycles.

The regenerator is constructed of a finely divided mass of high heat capacity. The gas flows alternately between the compression chamber and the expansion chamber through this regenerator. As the warm gas from the compression chamber flows through the regenerator, its heat is absorbed. The cold gas returning from the expansion chamber is warmed by the absorbed heat and returns approximately to ambient temperature. The regenerator heat capacity limits the lower temperature of the cycle. The specific heat of most regenerator materials decreases markedly with temperature. The minimum temperature obtainable with systems using thermal regenerators is generally 10° to 12° K.

The compression chamber and expansion chamber of the Stirling cycle are connected through the regenerator. The compression piston and expansion piston are operated roughly 90° out of phase. The close mechanical coupling of the compression and expansion chambers limits the use of the Stirling cycle to applications where space is available for the entire refrigerator. To eliminate this restriction, some Stirling-cycle refrigerators liquefy either air or nitrogen, and pump it using two-phase flow to the detector Dewar. The cycle is closed by collecting and reliquefying the boil-off-gases. The liquefied gas can be transferred approximately four to six feet.

The Stirling cycle depends on an ideal gas for efficient operation. Helium is used most often, although hydrogen could be used with some risk of explosion. Contamination is less of a problem with the Stirling

DESIGN TRENDS

cycle because of a purging action of the regenerator. Low pressure ratios also add to the long life of the mechanical parts. The efficiency of the cycle is quite high because of its isothermal expansion and compression, but input power increases as lower operating temperatures are required.

The modified Stirling cycle has many forms. The basic parts are the same as for the pure Stirling cycle: a compression chamber, an expansion chamber, and a thermal regenerator. The essential characteristic of all modified Stirling cycles is that the expansion and compression chambers are mechanically decoupled with little loss in performance; a compressor is used to supply pressurized gas, and a separate expansion engine produces the refrigeration potential allowing more flexibility than the Stirling cycle. Stirling and modified Stirling cycles are subject to the same minimum temperature range due to regenerator inefficiencies at low temperatures. The effect of contamination is about the same.

Integration

Problems and compromises associated with cryogenic-infrared interfaces that are described here are based on experience at Texas Instruments Inc. with several airborne infrared sets. Undoubtedly there are other solutions. In general, interface design coordination will include the consideration of problems and compromises associated with three broad areas: microphonics, vacuum, and cooling.

Microphonics. Numerous problems are covered by the word microphonics, which is too often used when a source of noise in the set cannot be located exactly. It is a real problem, however, and extremely important. Microphonics is defined as noise which is produced by one of three mechanisms: optical, thermal, or mechanical vibration. At times it is difficult to determine exactly which mechanism dominates because they are often interrelated.

Optical microphonics is the noise produced by a relative motion involving the detector and the associated optical system. This type of noise generally has a frequency proportional to the relative-motion fundamental frequency. The noise can be generated by the motion of the detector in relation to a fixed optical system, by the motion of the optical system in relation to a fixed detector, or by a combination of both effects. The first case is the most common.

This brings up the first compromise situation. The wall thickness of the cylinder on which the detector is mounted should be adequate to prevent any sway produced by the cooling system. The cylinder wall, however, is a heat leak from ambient and should be very thin to reduce heat conduction. If any sway can be allowed, the wall thickness should be reduced accordingly. If no sway can be allowed, the wall thickness should be increased to the required value, and the cooling system parameters adjusted to absorb the added heat load. Good mechanical mounting practice will eliminate the majority of optical microphonics after the detector mount is stabilized. The cooler-detector package can be hard-mounted to the same frame as the optics system if there is no objectionable sway in the detector mounting cylinder.

Thermal microphonics is the noise produced by fluctuations in the operating temperature of certain detectors. If a detector material exhibits a large change of resistance with temperature, thermal microphonics is more of a problem. Most detectors have a large rate of change of resistance with temperature above a certain point, but the resistance becomes nearly constant below this temperature. If the detector is operated above this critical temperature, slight variations about the operating temperature will produce thermal microphonics. The noise mechanism is associated with resistance changes in the detector due to temperature variations in the lattice. This is a lattice or phonon effect rather than a photon effect.

Thermal microphonics is particularly troublesome in systems where the detector is mounted in a closed-cycle cryo-engine device. The cooling power of such cryo-engines is cyclic, and this will produce fluctuations about the equilibrium temperature.

The thermal-microphonics mechanism produces noise with frequency components at the fundamental and harmonics of the cryo-engine cycling rate. In general, this type of noise is most evident at low frequencies. Even when the system bandwidth is narrow and centered at a high frequency, thermal microphonics may still be significant if the detector output is amplitude-modulated by the temperature fluctuation. Wide-band-pass systems with low frequency response should not be subjected to temperature fluctuations. Thermal filters, such as Teflon and lead-indium alloys, can be used in conjunction with the cooling system to attenuate the temperature fluctuations to an acceptable level.[3] Care should be taken not to load down the cooling system with too massive a thermal filter because additional thermal mass will increase the cooling time.

Mechanical microphonics is defined in this case as the noises produced either by detector assembly vibration or mechanical shock. Cooling systems with fast-acting parts near the detector are particularly prone to this type of noise. The most common case of moving parts near the detector is a piston operating in the cylinder upon which the detector is mounted. The vibration transmitted by this motion is dominant, although other inputs are often present. Careful attention to mounting the detector and routing the leads can help to alleviate the mechanical-microphonics problem.

The mounting of the detector depends on whether a glass or metal Dewar is used. With glass Dewars, the detector is mounted to a suitable metal mount with a low-melting-point solder, usually indium. The metal mount must be capable of forming a glass-to-metal seal with the glass Dewar. The thermal expansivity of the metal must closely match that of the glass so that rapid cooling will not cause the seal to crack. Kovar is a common detector-mount material used with glass Dewars. This type of mounting is permanent and has good rigidity. The temperature difference between the detector mount and the cold spot is very small and is not considered a problem. Cooling systems with glass Dewars use liquefied gases to provide the refrigeration. Joule-Thomson and Stirling-cycle liquid-transfer coolers sometimes use glass Dewars potted in metal housings.

A detector mounted on a closed-cycle cryo-engine involves another set of parameters. The detector mount is attached to the expansion cylinder in which a piston travels. The detector, attached to its mount with low-melting-point solder, is usually supplied by the detector manufacturer as a single component. Where possible, it is better to solder this mount to the cooling tip with a lower-melting-point solder. Often the operating or storage environment of the system eliminates the possibil-

DESIGN TRENDS

ity of fusion mounting so that the detector must be pressure-mounted to the cooling tip. Pressure mounting has the advantage of easier detector interchange.

Lead routing is a problem which is often overlooked because of its familiar nature. In systems using glass Dewars, the problem is greatly reduced by using painted leads. These, along with good glass-to-metal vacuum feedthroughs, tends to eliminate most noise due to lead vibration. Care should still be taken, however, to keep other leads as short as possible and to station them firmly to rigid parts.

In systems using metal Dewars or cryo-engines, the lead problem is more severe because painted leads cannot be used. The essential thing is to fix the lead wires firmly to rigid structural members. This can be done by using rigid connectors insulated from the metal wall or by using epoxy to fasten insulated wires to a mechanical support. However, a vacuum is harder to achieve if epoxy is present.

Noise due to vibrating leads is generally random and can be attributed to the lead sweep through the earth's magnetic field, the mechanical stress put on the lead, and the variation of the lead's position relative to another metal. Careful attention to good mechanical practice in supporting the leads can reduce this type of noise to an acceptable level.

Vacuum. One of the recurring problems of infrared sets is loss of vacuum in the cryogenic cooling system. This problem is so severe that some cooler manufacturers are using solid foam insulation. Foam or super-insulation is adequate at 77° K, but vacuum insulation is necessary at lower temperatures.

Degradation of vacuum occurs from leaks from the outside and internal outgassing. In glass Dewars, leaks from the outside must come through glass-to-metal, glass-to-IR-window, or glass-to-glass seals. The techniques of making these seals are well known, and the seals are quite reliable.

Leaks from the outside in metal Dewars can occur at several places. Permanent vacuum sealing at all points allows better control of leaks, but this is not always possible. The IR-window seal is a most likely spot for a leak. If the window is permanently sealed to the metal, a good vacuum seal can usually be achieved. If, however, the window is designed to be removable, the vacuum seal is much more vulnerable. Indium is used quite often for vacuum seals because of its good sealing and low outgassing qualities. The indium seal is kept in constant tension with a compressed O-ring which is not exposed to the vacuum.

The vacuum jacket around the cold spot is subject to leakage, especially if a removable seal is required; indium makes a good demountable seal. If demountable seals are used, an evacuation tube or a manual valve must be provided to allow repeated evacuation. Since the manual valve may also leak, it must be carefully designed and manufactured.

Internal outgassing is the release of entrapped gases from materials inside the vacuum. Cleanliness of the vacuum space is essential because grease, dust, or oil will cause serious problems when they begin to outgas at reduced pressure. Elastomeric O-rings and some forms of epoxy also outgas badly. In permanently sealed Dewars, all outgassing materials must be eliminated by stringent cleanup and handling techniques. Dewars with demountable seals must be cleaned thoroughly, and the outgassing materials must be eliminated or used carefully. Sealing O-rings should not be exposed to the vacuum, and epoxy should be used sparingly. An all-metal or all-glass system with no organic compounds is ideal.

Glasses and metals are prone to outgas; however, vacuum baking at elevated temperatures helps to eliminate some of the entrapped gases. In this technique, materials are heated in a high vacuum. The heat drives off the gases, which are swept into the vacuum pump. Careful attention to materials, vacuum baking, and stringent handling techniques will virtually eliminate the majority of outgassing problems.

Some infrared sets use detectors and other components that cannot stand high-temperature vacuum bakeout, and outgassing cannot be eliminated. Two common techniques are used to overcome this problem. The first is to provide the cooling system with enough capacity to freeze out the gases on the cold surface. This is called cryopumping; it works satisfactorily for a time on gases with a fusion point above the operating temperature. Some gases that have very low fusion points, such as helium and hydrogen, are unaffected by this technique. In addition, outgassing is accelerated at the low pressure levels produced by cryopumping. This results in a cumulative decay in the warm vacuum level. Eventually, the vacuum level is insufficient to allow cryopumping to take place.

The second technique employs dynamic pumping with small ion pumps. These pumps are electromagnetic in nature and effective for small vacuum spaces in miniature cooling systems. When this device is incorporated, the gases migrate from the vacuum into the ion pump where they are ionized by electrons emitted from the cathode. The ions are accelerated by a high potential and strike titanium surfaces with enough energy to sputter the titanium.[4] In addition to entrapping ions of noble gases, the fresh titanium, being an active getter material, completes the pumping action by forming stable, low-vapor-pressure compounds with active gases.

Cooling. In addition to microphonic and vacuum problems, other design areas need close coordination. For the most efficient performance, the cooling method should be chosen to do the job adequately; extra, unneeded capacity represents wasted power. The infrared engineer should estimate how much refrigerative power is required by the detector at the design temperature. A reasonable estimate can be obtained by considering three heat loads: radiant-energy input, bias power, and lead conduction.

Although the infrared engineer is interested in only a small segment of the electromagnetic spectrum, he must remember that all energy reaching the cold spot represents a heat load. Any optical filters or stops which can eliminate some of the excess energy should be used. If these stops require cryogenic temperatures, the added heat load and complexity will have to be weighed against the improved performance.

The bias power required for optimum detector operation is one of the largest heat loads imposed on the cooling system. The bias point should be chosen to obtain the least amount of power consistent with efficient detector operation.

The heat leaks introduced by conduction through the lead wires may also be significant. These can be minimized by using small-diameter wire with a low thermal conductivity. The heat leaks can be estimated by standard heat-transfer equations, although changes in thermal conductivity with temperature must be considered.[5]

DESIGN TRENDS

The cooling time is a parameter of concern to both cryogenic and infrared engineers since the less thermal mass there is to be cooled, the less cooling power is required for the proper cooling time. All design concepts that deliberately add thermal mass to the system should, therefore, be closely scrutinized. Fast cooling requires more power. Also, mechanical risk is high during cooling because the parts of the cooling mechanism are subjected to excessive thermal stresses during this period. ▲

Cited References

[1]. "Design Considerations for Cryogenic Liquid Refill Systems for Cooling Infrared Detection Cells," Haettinger, Skinner and Trentham, *Advances in Cryogenic Engineering*, Vol 6, Ed. by K. D. Timmerhaus, Plenum Press, Inc., New York (1961), pp 354-362.

[2]. *Cryogenic Engineering*, R. B. Scott, D. Van Nostrand Co., Inc., Princeton, N.J. (1959), p 268.

[3]. "Application for Long Wavelength Photoconductors in Engine-Refrigerated Cryostats," K. W. Cowans, 9th Infrared Information Symposium, 1963.

[4]. "Operation and Applications of Ion-Getter Pumps," C. V. Larson, 1961 AIEE Winter Meeting, Conference Paper No. 61-359.

[5]. "A Compendium of the Properties of Materials at Low Temperatures," WADD Technical Report 60-56, Wright Air Development Division, Wright-Patterson Air Force Base, Ohio.

Bibliography

Fundamentals of Infrared Technology, Holter, Nudelman, Suits, Wolfe and Zissis, The MacMillan Co., New York (1962). See Appendix C, p 395, for excellent sources of information about infrared technology.

Infrared Physics and Engineering, Jamieson, McFee, Plass, Grube and Richards, McGraw-Hill Book Co., Inc., New York (1963).

Elements of Infrared Technology, Kruse, McGlauchlin and McQuistan, John Wiley & Sons, Inc., New York (1962).

"Infrared Detectors," F. Schwartz, ELECTRO-TECHNOLOGY, November 1963, p 116.

"Low Temperature Physics," *Encyclopedia of Physics (Handbuch der Physik)*, Vols 14 and 15, Ed. by S. Flugge, Springer-Verlag, Berlin (1956).

Advances in Cryogenic Engineering, Ed. by K. D. Timmerhaus, Plenum Press, New York (1961).

Applied Cryogenic Engineering, Vance and Duke, John Wiley & Sons, Inc., New York (1962).

Scientific Foundations of Vacuum Techniques, S. Dushman, John Wiley & Sons, Inc., New York (1949).

Vacuum Symposium Transactions, sponsored by Committee on Vacuum Techniques, Inc., Box 1282, Boston 9, Mass.

Reprinted from *Infrared Phys.*, **8**(1), 25–35 (1968), with the permission of Microforms International Marketing Corporation as exclusive copyright licensee of Pergamon Press journal back files.

ADVANCED DESIGN OF JOULE–THOMSON COOLERS FOR INFRA-RED DETECTORS

S. W. STEPHENS

The Hymatic Engineering Company Limited, Glover Street, Redditch

Abstract—Many i.r. detectors require cryogenic cooling for optimum performance. Several methods of cooling are available, one of which is the use of Joule–Thomson liquefiers, which is considered in this paper.

INTRODUCTION

INFRA-RED detectors are widely used in such military applications[1] as missile guidance and fusing, airborne mapping and ground surveillance. Their use in civil applications,[2] in medicine and extraterrestrial studies is increasing.

The effect of infra-red radiation on detectors is to create free charge carriers in the material in addition to identical carriers which can be produced by thermal excitation. Thus, optimization of the sensitivity of the detector to incident radiation necessitates that the thermal excitation is reduced until it becomes negligible compared with that produced by the incident radiation. This is achieved by cooling the detector to a cryogenic temperature, which varies with coefficient of absorption of the material used.

Some typical detectors, the wavelength in which they operate, and the temperature to which they must be cooled to achieve background limited performance, are shown in Table 1.

TABLE 1

Detector material	Wavelength (μ)	Operating temperature (°K)
Indium antimonide	3–5	77
Mercury doped germanium	8–14	<40
Cadmium mercury telluride	8–14	77
Copper-doped germanium	2–25	4

The variation of detectivity, responsivity and noise with operating temperature, particularly for indium antimonide, is well known.[3]

POSSIBLE COOLING METHODS

Three basic physical effects are commonly used to achieve cryogenic cooling of i.r. detectors and these are (a) the evaporation of a liquid or solid. (b) The isentropic expansion of a gas, and (c) the isenthalpic expansion of a gas.

It is the isenthalpic expansion of a gas, known as the Joule–Thomson effect, which, is considered in more detail. A comparison of this, with the other processes, is discussed in a paper by Nicholds,[4] Technical Manager of The Hymatic Engineering Co. Ltd., presented at the Symposium on Infra Red Components, Techniques and Systems at the Ministry of Technology, Royal Radar Establishment, Malvern, April, 1967.

THE JOULE–THOMSON PROCESS

When a gas is expanded through an orifice a change in temperature occurs and this cooling effect is given by the formula

$$C_p \left(\frac{\mathrm{d}T}{\mathrm{d}p}\right)_H = T \left(\frac{\mathrm{d}v}{\mathrm{d}T}\right)_p - v$$

where $(\mathrm{d}T/\mathrm{d}p)_H$ is the cooling effect, and H is enthalpy.

For perfect gases the right hand side of the equation is always zero, but for real gases a change of sign may occur indicating that negative cooling, or heating, has taken place. The temperature at which this phenomenon occurs for any gas is known as the "Inversion Temperature". For example, hydrogen will not be cooled by this process unless the initial temperature is below 204°K.

For all real gases, the cooling obtained by this process is small and, using air, as an example, is approximately 0·3°C per atmosphere of pressure drop. To utilize this effect, therefore, it is necessary to make the effect cumulative by passing the expanded gas over a counter flow heat exchanger to pre-cool the supply gas.

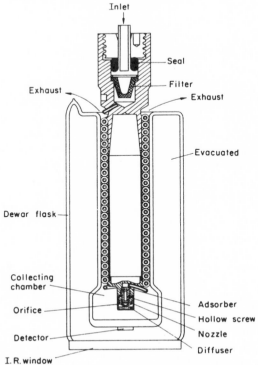

FIG. 1. Joule–Thomson liquefier/i.r. cell combination.

A general purpose Joule–Thomson liquefier, developed by the Hymatic Engineering Co. Ltd., and known as the "Minicooler" is shown mounted in a typical infra-red cell, in Fig. 1, and the thermodynamic cycle is illustrated in Fig. 2. High pressure gas, after entering the liquefier, is cooled at constant pressure in the heat exchanger, expanded at constant enthalpy through the orifice, or nozzle, and when exhausting gives up cold in the heat exchanger. It is during the nozzle expansion that a fraction of the gas is liquefied. This process is described in more detail in earlier papers.[5]

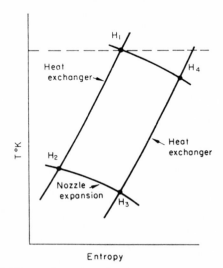

FIG. 2. Thermodynamic cycle of Joule–Thomson liquefier.

FIRST GENERATION COOLERS

The cryogenic activities of Hymatic commenced in 1955 when, as a natural application of the technology of fluid control, and precision engineering, the company undertook the development of a Miniature Joule–Thomson Liquefier, designed by Dr. D. H. Parkinson of the Royal Radar Establishment. The requirement was to cool an infra-red detector to 77°K within 2 min of initiation on an ambient temperature of +65°C and, also during this 2 min, to provide a reservoir of liquid to maintain the cryogenic temperature for 45 sec after termination of the gas supply. The Minicooler which resulted from this programme is shown in Fig. 3, and is that illustrated in cross section in Fig. 1.

A major part of the development programme was concerned with techniques for obtaining consistent and reliable performance. A problem normally associated with J.T. coolers is the blocking of the expansion nozzle due to contaminants in the gas supply which can solidify at temperatures above liquid air temperature. Solution of this problem was achieved by rigidly controlled standards in the selection of materials for the gas supply system, in the cleaning of the gas reservoirs and the gas control systems and in the purification of the gas itself. Additionally, an adsorber was introduced into the cooler close to the expansion nozzle,* thus avoiding loss of performance even if, in spite of the above precautions, a small amount of contaminant is present in the gas supply to the cooler.

* Covered by patents.

Several variants of this Minicooler have been produced against specific requirements, and some of these are shown in Fig. 4.

Comprehensive performance characteristics of these units have been recorded,[6] but a brief summary is contained in the graph at Fig. 5.

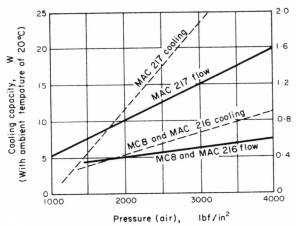

FIG. 5. Minicooler performance.

There are a number of approaches to the systems engineering relating to the supply of suitable gas and Hymatic have engineered a number of complete operational systems. These range from integrated "one-shot" systems for very short duration, as shown in Fig. 6, where the bottle may be a factory sealed unit, to long duration systems.

One such system uses a 400 atmosphere storage reservoir, as shown in Fig. 7 carried in an aircraft.

The charging of these replaceable reservoirs is carried out at the aircraft base, the Pure Air supply being obtained from a conventional, lubricated 6000 lbf/in² compressor and gas purification and monitoring plant specifically designed for the purpose. This plant without the compressor is shown in Fig. 8.

During operation the supply of gas from the reservoir to the Minicooler is controlled, as required by the flight plan, by a control pack as shown in Fig. 9.

RECENT DEVELOPMENTS

The wider applications of indium antimonide detectors and the availability of mercury–cadmium telluride operating in a longer wavelength has placed great emphasis on the requirement for cooling to 77°K. Whilst existing Minicoolers in large numbers have proved extremely reliable in service, there is a constant need to reduce cool-down time and overall system weight to an absolute minimum. The latter is particularly important in longer duration systems.

High speed cooling

Considering the rapid cool down requirement, it is obvious that to achieve shortest cool down time, the heat load on the cooler during this period must be minimal. The major heat load will be the thermal mass of the Minicooler itself. The heat exchanger surfaces, therefore, must have low thermal mass and a high coefficient of heat transfer. The configuration of coolers has been studied and the merits of different shapes and constructions analysed.

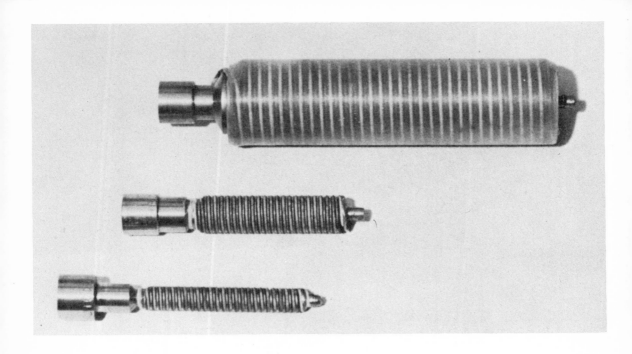

FIG. 3. Hymatic Minicooler Type MC8.

FIG. 4. Hymatic Minicoolers (approx. 1¼ × full size).
From left to right, MAC.216, MC8 and MAC.217.

352

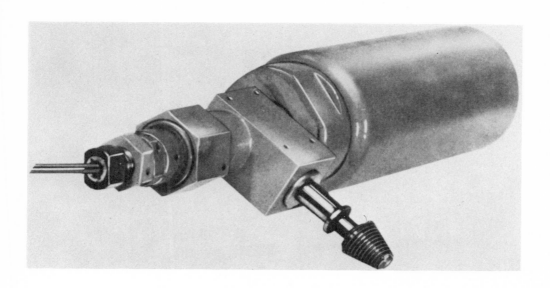

FIG. 6. One-shot system.

FIG. 7. Airborne storage reservoir.

353

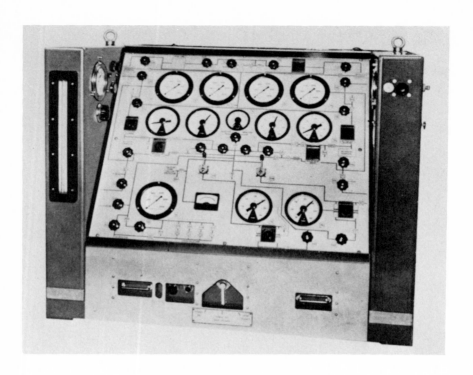

Fig. 8. Pure air charging equipment.

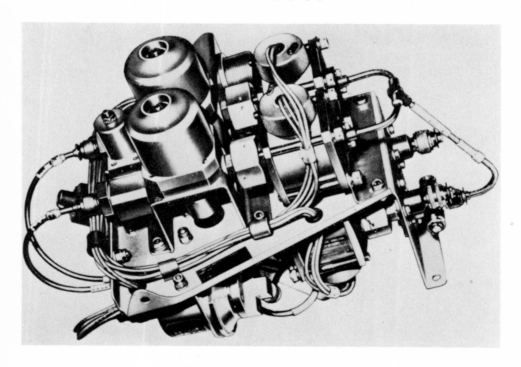

Fig. 9. Airborne control system.

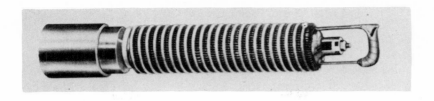

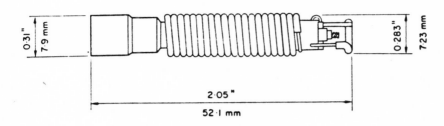

0.31"
7.9 mm

0.283"
7.23 mm

2.05"
52.1 mm

FIG. 15. Self-regulating Minicooler.

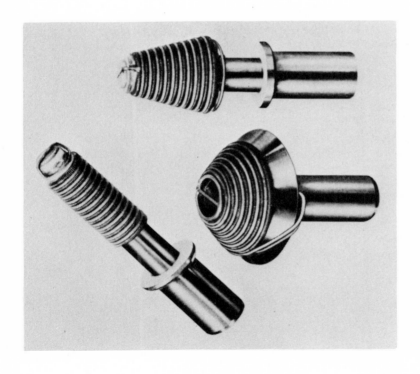

FIG. 10. High speed coolers.

Some of the configurations evaluated are shown in Fig. 10 and the best results are achieved using a conical cooler of 90° included angle. This unit* has a former of thin stainless steel and is shown in Fig. 11 in a possible detector arrangement where the flask is machined in Acrylic plastic.

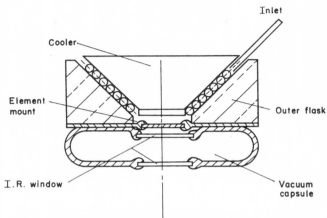

FIG. 11. High speed Minicooler/Detector arrangement.

The development of High Speed Minicoolers has reached the stage where it is relatively easy to produce liquid at the Minicooler nozzle in less than two seconds, but there is the added problem of transferring the cold to the detector. This places additional emphasis on the interface problem and necessitates an integrated approach to the design of the cooler/ cell combination. The shortest cool down times are achieved using Argon at 400 atmospheres and the curves in Figs. 12 and 13, reproduced from an earlier publication,[7] show respectively typical cool down times and illustrate the delay between achieving cold in the gas stream and achieving operating temperature at the detector itself.

Development now in hand is further reducing the cool down time of the cooler and is also aimed at reducing the delay in transferring the cold to the detector itself.

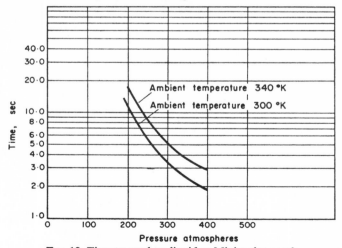

FIG. 12. Time to produce liquid at Minicooler nozzle.

* Covered by patents.

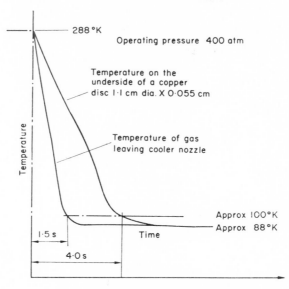

FIG. 13. Illustration of interface·effect.

Self-regulating Minicooler—Long duration systems

The greatest single limitation of Joule–Thomson coolers, until now, has been a system inefficiency which results from the need to work over a wide ambient temperature range, achieve short cool down time and run for many hours.

Until recently, expansion in Joule–Thomson liquefiers took place at a nozzle of fixed diameter, the size of which depends upon such parameters as:

Cooling power required

Operating pressure range

Ambient temperature range

and Cool down time.

The diameter is calculated to give the requisite cooling power under the most adverse combination of conditions. When the cooler operates under any other combination of conditions the gas consumption exceeds the demand, and means that excess gas storage capacity must be provided. The largest single condition affecting this inefficiency is ambient temperature and the curve reproduced[1] at Fig. 14 illustrates this point by comparing the actual gas flow with that which is necessary to maintain system operation. These curves are based on a constant heat load. A variable heat load further increases the difference between "actual" and "demand" consumption.

The most recent and significant advance in the design of Joule–Thomson liquefiers is the introduction, by Hymatic, of a Self-regulating Minicooler,* The fixed orifice is replaced by a slightly larger orifice with a needle which opens or closes the flow passages according to the temperature at the sensing point, whereby it is possible either to control the chamber at a given temperature or to maintain a constant level of liquid in the chamber, thus avoiding the excessive gas consumption which results if the level is allowed to rise and flood the Minicooler.

The self-regulating Minicooler maintains the desired temperature or liquid level irrespec-

* Covered by patents.

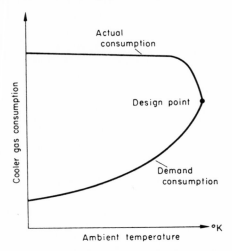

FIG. 14. Comparison of actual and necessary consumption.

tive of changes in supply pressure and ambient temperature, thus eliminating the need for additional control equipment (e.g. pressure reducing valves). The control is achieved entirely within the Minicooler.

At the moment of admission of gas to a Self-regulating Minicooler the nozzle is wide open, hence the maximum flow (for the particular ambient temperature and supply pressure) will occur and consequently a fast cool-down will result. Additionally, since there is no pressure reducing valve in the system, the full pressure available at the storage bottle is applied to the Minicooler and the maximum cooling power of the gas supply is available at the nozzle. (With a pressure reducing valve some of the cooling power is dissipated at the valve and is, therefore, wasted.) The unit is shown in Fig. 15.

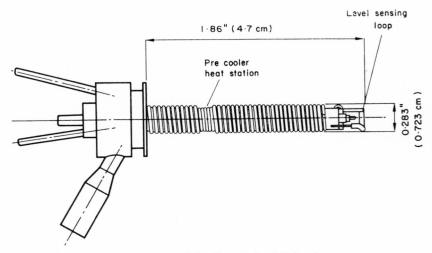

FIG. 16. Pre-cooled self-regulating Minicooler.

Pre-cooled Minicoolers

The cooling capacity of a gas when expanded through a Minicooler is dependent on its

IP—C

358

inlet and exhaust temperature, (normally ambient temperature). In high ambient temperature applications considerable benefit can be gained by cooling the gas before it is expanded
and this may conveniently be by adding to the Minicooler a pre-cooling stage employing
Freon. The flow of Freon, like that of the main cooling gas may be controlled according to
the load, by a technique similar to that, described earlier, for the Self-regulating Minicooler.
If a system is to operate only at low ambient temperatures, then the complication of pre-
cooling is unlikely to be justified, but if the system is required to operate over a wide
temperature range or at a sustained temperature above zero centigrade a considerable
saving in weight and bulk can be achieved, compared with a Self-regulating Minicooler
without pre-cooling.

The procedure for detail design of these coolers has been reported,[8] but it is worth
reproducing a comparison[9] of the system weights of each Minicooler variant against
probable specifications.

		Spec. A	Spec. B
Temperature required		77°–80°K	77°–80°K
Ambient temperature		+60° to −20°C	+60° to −20°C
Applied electrical load		1 W	0·2 W
Operating time		12 hr	2 hr
Cool down time		Not critical	3 min
Weight		16 kg max.	min. possible
Cost		low as possible	low as possible
Standard System			
Bottle capacity	litres	92·5	14·0
Bottle weight	kg	134·0	20·5
System weight	kg	170·0	28·5
Self-regulating Minicooler system			
Bottle capacity	litres	24·0	2·0
Bottle weight	kg	35·5	2·5
System weight	kg	45·5	5·5
Pre-cooled Self-regulating system			
Bottle capacities	litres	5·0 + 0·6	0·6 + 0·1
Bottle weights	kg	6·4 + 0·9	1·1 + 0·2
System weight	kg	13·6	4·5

INTERFACE PROBLEMS

One problem of the interface between cooler and detector has been described earlier.
Here the problem is essentially that of rapid transfer of cold from the J.T. gas stream to the
detector itself.

A further problem is that of microphony,[12] i.e. spurious signals produced in the detector. Microphony can be caused by mechanical vibration (producing strains within the
detector) by fluctuation of temperature or by minor disturbances on the detector mount
due to rapid boiling of the liquid in the dewar or variations of pressure within the dewar.
Compared with most mechanical systems, the J.T. cooler is, of course, free from mechanical
vibration, but careful design of the cooler/dewar/detector system is necessary to avoid
problems due to the boiling and pressure/temperature variations within the dewar. The
introduction of the Self-regulating cooler has provided a marked improvement in the
microphony problems of J.T. cooler/detector systems because the cooling effect is continuously matched to the demand thus providing a quiescent system.

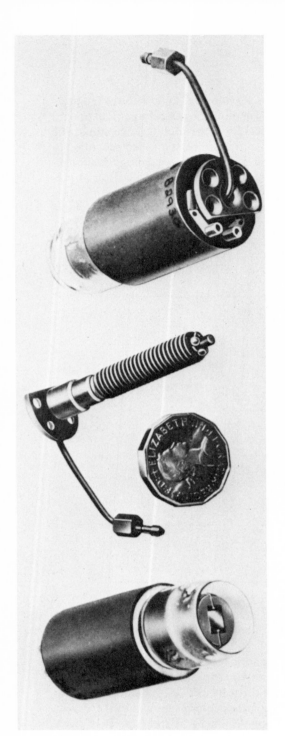

Fig. 17. Minicooler and infra-red detector.

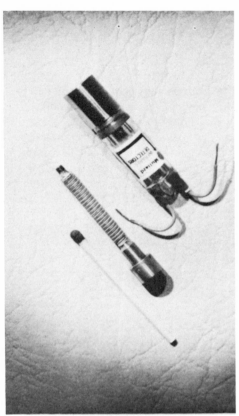

Fig. 18. MAC.216 Minicooler and Mullard infra-red detector.

360

FIG. 21. 77 °K miniature refrigerator.

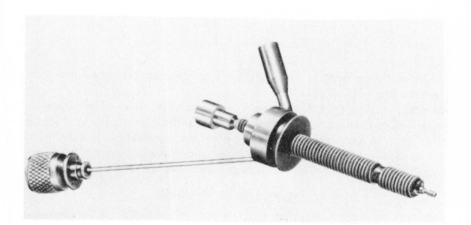

FIG. 19. Two-stage Minicooler.

For many years, Hymatic has been working in close collaboration with detector manufacturers on the integration problems and in alleviating the interface effect.

Figure 17 shows a Hymatic Minicooler type MC8 and Mullard[11] Indium Antimonide photo-conductive detector 4 mm × 4 mm element. Figure 18 shows the smaller MAC.216 also with an indium antimonide detector.

TWO-GAS LIQUEFIERS

Mercury doped germanium detectors require a temperature below 40°K for maximum sensitivity and the only suitable gases which have a boiling point below this value are neon, hydrogen and helium. Each of these gases has an inversion temperature well below normal ambient temperature and must, therefore, be pre-cooled before a Joule–Thomson liquefier will function. Neon and hydrogen can be liquefied in a two-stage liquefier, but helium requires three stages.

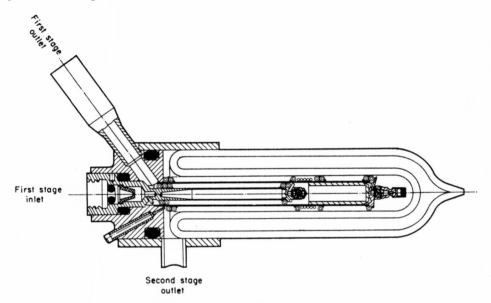

FIG. 20. Arrangement of two-stage Minicooler.

The solution to the problem of providing 40°K or lower for mercury-doped germanium detectors is shown in Figs. 19 and 20.

The first stage of the two gas Minicooler is the smallest Minicooler shown in Fig. 4 and the second stage is an increased length heat exchanger of the first unit discussed. This unit is being used to cool detectors and one current application is a balloon carried system for infra-red measurement in the upper atmosphere. The optimization of the cooler/i.r. cell combination has been discussed in an earlier paper.[10]

LABORATORY SYSTEMS

For laboratory work and feasibility trials, a convenient means of detector cooling to temperatures down to 77°K or 21°K is provided by the Hymatic Miniature Refrigerators SES.211 and SES.227 or SES.237 respectively. The units are complete with pressure control

FIG. 22. 21°K miniature refrigerator.

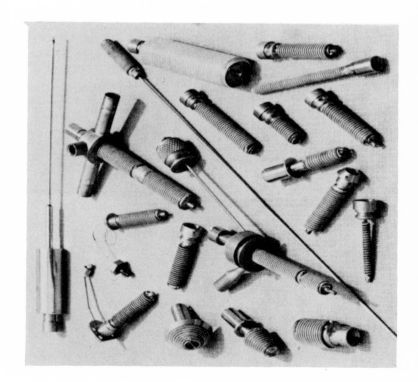

FIG. 23. Hymatic Minicoolers.

valves and instrumentation and the necessary gas cleaning equipment to permit their use with commercially available cylinders of high pressure gas. The systems are cheap, reliable and convenient and provide a reasonable simulation of an operational Joule–Thomson system thus enabling preliminary work to be carried out prior to the stage of designing special purpose systems.

The SES.211 shown in Fig. 21 requires only a supply of gaseous nitrogen. The unit is self-contained and allows localized cooling with single or multiple cooling heads and all of the Hymatic single stage Minicoolers are interchangeable on the unit.

A similar unit, for use with the two-gas Minicoolers, is shown in Fig. 22. This unit also requires a small supply of liquid nitrogen for the second stage gas cleaning circuit.

A simpler unit for laboratory use only is the SES.237[13] which uses liquid nitrogen pre-cooling and an MC8 Minicooler for final liquefaction.

CONCLUSION

This paper has been concerned with Joule–Thomson Liquefiers, or "Minicoolers", and whilst Fig. 23 shows a selection of currently available units it is emphasized that the optimum system performance is only achieved by close collaboration between the respective manufacturer of the detector and the liquefier and the overall systems engineer to derive the correct compromise between conflicting parameters. However, the system is not limited to the Minicooler itself.

Joule–Thomson liquefiers may be used in either open or closed circuit but the majority of current applications are open circuit. These systems have been discussed only briefly in this paper as the information is already readily available.[4, 14] It is sufficient to say that these systems exist and their reliability has been proved under service conditions.

First generation Minicoolers have always been competitive with alternative systems for many applications, and they have four major advantages:

> Rapid cool down.
> Low value of microphony.
> Freedom from mechanical vibration.
> Freedom from acoustic noise.

Second generation Minicoolers not only retain all these advantages but improved performance has made them even more competitive with alternative systems. The cool down time of high speed units has been reduced to less than 2 sec whilst microphony and system weight has been dramatically reduced by the introduction of self-regulating pre-cooled units. These units are more likely to be custom-designed against particular requirements to provide for optimization of various performance characteristics and for closer integrations with the detector to obviate interface problems.

This co-operation will not be limited to the Minicooler/Detector combination but will also necessarily include the systems engineer to determine the type of back up system and ground support equipment. This depends on the operational role and the logistics in service of supplying gas. Hymatic have provided the solution to many requirements and the answer may be anywhere between the factory-sealed reservoirs and on site Pure Air Charging Equipment as discussed.

Acknowledgements—The author wishes to thank the directors of The Hymatic Engineering Company Limited, and the Ministry of Technology for permission to publish this paper, and Associated Semiconductor Manufacturers Limited for permission to use photographs of their products.

REFERENCES

1. HAYWARD, H. T., Cryogenic cooling of electronic devices for military applications. *Electron. Compon.* June (1966).
2. SERGEANT, E. A., Cryogenics in industry, research and medicine. *Mod. Refrig. Air Control* October (1966).
3. MORTEN, F. D. and R. E. J. KING, Photoconductive indium antimonide detectors. *Appl. Optics* June (1966).
4. NICHOLDS, K. E., Cryogenic systems for infra-red detectors. *Symposium on Infra-red Components, Techniques and Systems*, Royal Radar Establishment, Malvern, April (1967).
5. HAYWARD, H. T., Technology at low temperatures. *Ind. Syst. Equipment.* September (1966).
6. McINROY, J., Minicooler performance, Hymatic.
7. HART, R. R., High speed minicooler systems, Hymatic.
8. GILES, B. A., Design of long duration 77°K cooling systems, Hymatic.
9. PIERSON, D. C., Characteristics of fixed orifice, self-regulating and pre-cooled Minicoolers, Hymatic.
10. HUGHES, W., 20°K Joule–Thomson cooling systems, Hymatic.
11. Associated Semiconductor Manufacturers, Mullard Southampton Works, Millbrook Trading Estate, Southampton.
12. GIBSON, C. R. G., Technical Memorandum TM.1138, Hymatic.

Reprinted from *Appl. Optics*, **9**(1), 185–193 (1970)

Radiant Cooling

R. V. Annable

An approach to the design and analysis of passive radiant coolers is developed. It is based on the use of a highly reflecting, specular cone channel to direct emission from a cooled volume (patch) to cold regions of space. The essence of the approach lies in the determination of the patch–cone radiative coupling, which is analyzed by means of the specular image technique. The optimum cone geometry depends on the parameter to be extremized as well as on the constraints imposed by limited dimensions on the cooler and a limited view of cold space. Specific equations for the calculation of the patch–cone radiative coupling are given for a cooler of rectangular geometry and employed in a sample cooler design. The over-all procedure based on the image technique has proved to be useful and accurate in the design and analysis of radiant coolers for spacecraft applications.

I. Introduction

The high vacuum of space and the low effective temperature of stellar regions provide the necessary environment for the passive radiant cooling of spacecraft components to cryogenic temperatures. A radiant cooler has an inherently long life, requires no moving parts or stored coolants, and consumes only the very small power needed for temperature regulation. Radiantly cooled fast, sensitive detectors can be used in a family of ir space instruments covering wavelengths from the visible to 16 μ. The atmospheric window between 3.4 μ and 4.2 μ can be covered by PbSe cooled to 200 K in a simple, single stage radiant cooler. The water band out to 6.4 μ can be covered by PbSe at 175 K or less in a more complex single stage cooler. Wavelengths out to at least 16 μ can be covered by a HgCdTe detector[1,2] cooled to the 100–80 K range. The lower temperature generally requires the use of a two-stage radiant cooler.

A radiant cooler can be designed for an earth (or planet) oriented spacecraft in a sun-synchronous or geostationary orbit. The general form is shown in Fig. 1. The patch (cold volume) is contained in an enclosure consisting of cold space and the cone walls to which it is physically attached. External sources such as the earth, sun, and spacecraft structure thermally load the cone and limit the view to cold space from the black patch. The outward sloping, specularly reflecting, low emissivity cone walls are usually de-

signed so that cold space is the only external source seen by the patch. The cone therefore acts as a crude collimator or directional antenna for patch radiation. The patch is then thermally coupled only to the cone walls and to cold space. The temperature of the cone is controlled by the addition of a low α/ϵ cone end which is thermally tied to the cone but not visible from the patch. The cone end offsets the high α/ϵ usually associated with the low emissivity cone walls and thereby keeps the cone temperature within bounds in the presence of direct and reflected sunlight. The cone itself may be thermally insulated from the spacecraft and partially shielded from external sources.

If the hemisphere above the patch is free of thermal sources or if the desired temperature is sufficiently high, no cone is required and the cone end becomes the patch by extending it to cover the area of the cone mouth.

A second stage of cooling may be added to the patch. It consists of a second, smaller cone which restricts the view of a second, smaller patch to cold space. The second stage external sources are usually limited to the cone walls of the first stage.

Ideally, the radiant cooler patch is made so large that the thermal inputs produced by the attachment of a detector or other component (joule heat, lead conduction, radiative input through the optical opening) are negligible compared with the power radiated by the patch at its equilibrium temperature. This permits flexibility in the optical and electrical designs and keeps the thermal design largely independent of optical and electrical requirements. In addition, the very large patch may be held in position by a caging mechanism during powered launch and by a weak, low conductance support during orbital flight. The thermal coupling between the patch and cone walls is then mostly radiative and is determined by the cone wall emissivity and

The author is with ITT Aerospace/Optical Division, Fort Wayne, Indiana 46803.
Received 27 February 1969.

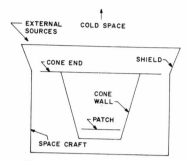

Fig. 1.   Generalized radiant cooler.

the cooler geometry. In any case, it is useful to determine the limiting performance of the cooler set by the radiative coupling between patch and cone. If the patch does not reach the desired temperature in this ideal case, it can never do so in the presence of additional thermal loads introduced by mechanical, optical, and electrical connections to the patch.

II. Radiative Coupling Between Patch and Cone

If f_n is the fraction of radiant flux from the patch which reaches space after n reflections in a perfectly reflecting ($\rho_c = 1$) cone, the effective reflectivity of the cone as seen from the patch is

$$\rho_{pc} = \Sigma f_n \rho_c{}^n = \Sigma f_n (1 - \epsilon_c)^n, \qquad (1)$$

where ρ_c is the reflectivity of the cone wall surface and ϵ_c its emissivity. The effective reflectivity of the cone is the fraction of radiant power emitted by the patch that reaches cold space. This concept is used in the study of cavities to express the fraction of incident flux reflected back out of the cavity.[3] In the case of the radiant cooler, the cavity is in the form of a truncated conical perforation. According to Kirchhoff's law, the effective emissivity of the cone walls as seen by the patch is then

$$\epsilon_{pc} = 1 - \rho_{pc} = 1 - \Sigma f_n (1 - \epsilon_c)^n. \qquad (2)$$

An expression for the effective patch-to-cone emissivity can also be obtained from the multiple images formed by specular reflections in the cone walls. A section through such a three-dimensional image array is shown in Fig. 2. The cone itself is shown by the solid lines of the central polygon, while the images are represented by the phantom lines. If, for example, the cone is a frustum of a right circular cone, the complete multiple images are generated by rotating Fig. 2 about the axis of the cone. On the other hand, if the cross section through the cone perpendicular to its axis is rectangular, the image pattern is more complex and two or more sections of the type shown in Fig. 2 are needed to describe its geometry completely.

The technique of multiple reflections has been used by Sparrow and others to determine heat exchange between surfaces with specular reflection.[4] It has also been employed by Williamson[5] and Hanel[6] in study of specularly reflecting cone channels. In his analysis, Sparrow uses a parameter called the exchange factor, which replaces the view (angle) factor used for diffusely reflecting surfaces. The exchange factor E_{ij} from an area A_i to an area A_j is the fraction of diffusely distributed flux from A_i that arrives at A_j both directly and by all possible intervening specular reflections. The exchange factor is also used by O'Brien and Sowell in their analysis of radiant transfer through specular tubes.[7] O'Brien and Sowell attribute the establishment of the exchange factor to Bobco.[8]

The exchange factor is related to the effective emissivity by

$$\epsilon_{ij} = \epsilon_i \epsilon_j E_{ij}. \qquad (3)$$

That is, the effective emissivity ϵ_{ij} in an enclosure of specularly reflecting surfaces (including black surfaces) is the emissivity of the diffusely emitting surface i times the fraction of its emission absorbed in the receiving area j. We therefore have the reciprocity relationship

$$A_i \epsilon_{ij} = A_j \epsilon_{ji} \qquad (4)$$

and the conservation equation[7]

$$\Sigma_j \epsilon_j E_{ij} = 1. \qquad (5)$$

The exchange factor can be employed to show that the patch temperature is independent of its emissivity when the patch is a graybody and the cone walls outward sloping and specularly reflecting.* The net radiant flux density leaving a gray patch of emissivity (absorptivity) ϵ_p is given by

$$(1/A_p)\Phi_{\text{net},p} = W_p - \epsilon_p H_p, \qquad (6)$$

where W_p is the radiant exitance of the patch and H_p the radiant incidence. If the patch is part of an en-

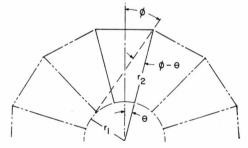

Fig. 2.   Multiple specular images of cone and patch.

* The general form of this proof was kindly supplied by one of the readers.

closure formed by itself, the specular cone, and black cold space, the radiant incidence is

$$H_p = E_{pp}W_p + E_{pc}W_c + E_{ps}W_s, \qquad (7)$$

and net flux density,

$$(1/A_p)\Phi_{\text{net},p} = W_p(1 - \epsilon_p E_{pp}) - \epsilon_p E_{pc}W_c - \epsilon_p E_{ps}W_s. \qquad (8)$$

Using

$$W_j = \epsilon_j \sigma T_j^4, \qquad (9)$$

the condition for zero net flux transfer (thermal equilibrium or steady state) yields

$$\frac{T_p^4}{\epsilon_c T_c^4} = \frac{E_{pc} + E_{ps}T_s^4/\epsilon_c T_c^4}{(1 - \epsilon_p E_{pp})}, \qquad (10)$$

The patch temperature is therefore independent of its emissivity when $E_{pp} = 0$, i.e., when the patch cannot see itself directly or by reflection. This condition is met by a flat or convex patch in a cone of outward sloping and specularly reflecting walls. In any real cooler, however, the temperature of the patch decreases with an increase in its emissivity because of thermal inputs from the supports, detector, and optics.

When $E_{pp} = 0$ and space is at zero temperature, the patch temperature for radiative coupling alone is given by

$$T_p = (\epsilon_c E_{pc})^{1/4} T_c. \qquad (11)$$

For a black patch, this may also be written as

$$T_p = \epsilon_{pc}^{1/4} T_c. \qquad (12)$$

A simple cone geometry that permits the direct calculation of the view factors f_n from a diffusely emitting patch is the frustum of a right circular cone (see Fig. 2). The view factor from the patch to cold space as seen by n reflections in the cone walls is then the difference in the view factors between two sets of circles with a common central normal[9]

$$f_n = F_{pa(n)} - F_{pa(n-1)}, \qquad (13)$$

where $F_{pa(n)}$ is the view factor from the patch circle to the cone mouth circle seen by n reflections. Another simple case is where no more than one cone wall reflection is needed for a ray from the patch to reach cold space. The effective patch-to-cone emissivity then becomes

$$\epsilon_{pc} = f_1 \epsilon_c = [1 - F_{pm(o)}]\epsilon_c, \qquad (14)$$

where $F_{pm(o)}$ is the direct view factor from the patch to the cone mouth.

III. Rectangular Cone Geometry

In a radiant cooler in which the cross section perpendicular to the cone axis is a rectangle, the boundaries of the cone mouth and of all its specular images in the cone walls are straight lines. Furthermore, the factor f_n [Eq. (2)] can be identified with the view factor $F_{p-m(n)}$ from the patch to the images of the cone mouth

(cold space) as seen by n reflections in the cone walls. As a result, the problem of determining the effective patch-to-come emissivity is reduced to one of calculating contour integrals about surfaces bounded by straight lines.

We can complete the rectangular geometry by assuming that the patch is also a rectangle whose normal lies along the cone axis. The view factor from an elemental patch area is then given by

$$F_{p-m(n)} = \frac{1}{\pi} \iint_{m(n)} \cos\vartheta \, \sin\vartheta d\vartheta d\varphi, \qquad (15)$$

where ϑ and φ are the spherical coordinates with respect to a normal to the patch element and the integration is over the angles subtended by the cone mouth as seen by n specular cone wall reflections. Integrating with respect to ϑ, this becomes a contour integral about the boundaries of the cone mouth images,

$$F_{p-m(n)} = \frac{1}{2\pi} \oint_{m(n)} \sin^2\vartheta(\varphi) d\varphi. \qquad (16)$$

For zero cone wall reflections, the integral is taken around the boundary of the cone mouth itself, so that we have

$$2\pi F_{p-m(0)} = \sum_{j=1}^{4} \int_{\varphi_{ii}}^{\varphi_{ik}} \sin^2\vartheta_j d\varphi, \qquad (17)$$

where φ_{ij} and φ_{ik} are the azimuth angles of the intersection of boundary line j with boundary lines i and k and the pole is normal to the surface of the patch. If the $\varphi = 0$, π plane is a section through the cone axis and perpendicular to two edges of the mouth, the boundary lines perpendicular to the plane have equations of the form

$$\sin^2\vartheta_i = (1 + \cos^2\varphi \tan^2\beta_i)^{-1}, \qquad (18)$$

where β_i is the elevation angle of the line above the surface of the patch in the $\varphi = 0$, π plane. The boundary lines perpendicular to the $\varphi = \pi/2$, $3\pi/2$ plane have equations of the form

$$\sin^2\vartheta_j = (1 + \sin^2\varphi \tan^2\beta_j)^{-1}, \qquad (19)$$

where β_j is the line elevation angle in the $\varphi = \pi/2$, $3\pi/2$ plane. The azimuth angle of the intersection of two boundary lines is obtained by setting $\vartheta_i = \vartheta_j$. Integration then yields

$$\int \sin^2\vartheta_i d\varphi = \cos\beta_i \arctan(\tan\varphi \cos\beta_i), \qquad (20)$$

$$\int \sin^2\vartheta_j d\varphi = \cos\beta_j \arctan(\tan\varphi/\cos\beta_j). \qquad (21)$$

These results assume that the plane of the cone mouth is perpendicular to the cone axis and therefore parallel to the patch.

To calculate the view factors to the specular images of the cone mouth, we also need the general equation of a straight line in spherical coordinates (image of a cone mouth boundary line),

$$\sin^2\vartheta_l = [1 + \cos^2(\varphi + \varphi_o)(\tan\beta - \tan\alpha \tan\langle\varphi + \varphi_o\rangle)^2]^{-1}. \qquad (22)$$

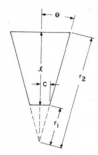

Fig. 3. Radiant cooler geometry.

The integration of $\sin^2 \vartheta_l$ is carried out by setting $\varphi' = \varphi + \varphi_o$ and using the result

$$\int \frac{d\varphi'}{1 + \cos^2\varphi'(\tan\beta - \tan\alpha \tan\varphi')^2} = \frac{2}{(4ac - b^2)^{1/2}}$$

$$\times \arctan\left[\frac{2a \tan\varphi' + b}{(4ac - b^2)^{1/2}}\right], \quad (23)$$

where

$$a = 1 + \tan^2\alpha,$$
$$b = -2 \tan\alpha \tan\beta,$$
$$c = 1 + \tan^2\beta.$$

IV. Limitations to Specular Model

The above model depends on specular reflectivity at the cone walls and neglects any directional or spectral variations in the emissivity. A cooler may be designed using a measured value for the hemispherical emissivity of a selected cone wall material. Any diffuse component of reflectance will then appear as an increase in the apparent specular emissivity. In addition, angular and spectral variations in emissivity may increase the thermal coupling between surfaces above that calculated from formulas using hemispherical emissivity.[10]

As a limiting example of the influence of diffuse reflection, consider a radiant cooler in the form of long coaxial cylinders or concentric spheres.[11] If the inner surface (patch) is black and has an area much less than the outer surface (cone wall), the effective emissivity for specular reflectivity at the outer surface is equal to the emissivity of the outer wall. On the other hand, the effective emissivity for diffuse reflectivity at the outer surface is equal to unity (the emissivity of the inner surface). The apparent specular emissivity for an outer surface of hemispherical emissivity ϵ at which a small fraction δ of the reflectivity is diffuse is then very nearly $\epsilon + \delta$.

V. Optimization of Cooler Geometry

As already mentioned, an ideal size for a radiant cooler patch is one in which the thermal inputs fixed in size (such as those introduced by a detector) are

negligible compared with the thermal equilibrium power emitted by the patch. It may then be desirable to place the patch in a cone of minimum length to conserve weight and space. On the other hand, the space available for the cone may be limited, so that it is desirable to obtain a patch of maximum size in a cooler of given length. This minimizes the influence of fixed thermal inputs to the patch. In either case, an optimum design for minimizing cone length or the influence of fixed inputs is obtained when the ratio of cone length to patch radius is a minimum. This minimum is subject to the constraint of a maximum patch view angle to cold space. The optimization then determines the angle of the cone wall with respect to the cone axis.

A simple case is illustrated in Fig. 3, which shows the cross section of a cooler that is rotationally symmetric about its axis (frustum of a right circular cone). This case arises, for example, in an earth-oriented geostationary orbit in which the cooler axis is parallel to the earth's rotational axis and the view to cold space is limited by the sun. The cone length l and patch radius c are given by

$$l = (r_2 - r_1) \cos\theta,$$
$$c = r_1 \sin\theta, \quad (24)$$

where r_1 is the apex to patch distance on the full cone, r_2 is the apex to mouth distance, and θ is the cone wall to cone axis angle. Now the maximum look angle ϕ of a spherical patch with respect to the axis is the solution to (see Fig. 2)

$$\sin(\phi - \theta) = r_1/r_2. \quad (25)$$

This relation is a good approximation for a flat patch when θ is small and a safe value for any θ, since ϕ for a flat patch is always less than or equal to ϕ for a spherical patch in the same cone. The angle ϕ is the maximum angle to the patch normal (cone axis) at which rays from the patch leave the cone mouth and the maximum angle to the cone axis of an external object visible from the patch. Thus we obtain

$$\frac{l}{c} = \frac{1 - \sin(\phi - \theta)}{\sin(\phi - \theta)} \cot\theta, \quad (26)$$

and the ratio of cone length to patch radius for a given ϕ is a function only of the cone wall angle θ.

The ratio of cone length to patch radius as a function of cone wall angle is shown in Fig. 4 for three values of maximum patch look angle. The curve is broad in the case of the larger look angles. A cone wall angle larger than optimum would then be selected in order to reduce other thermal inputs, such as the radiative coupling from the cone to the patch or the external irradiation absorbed in the cone walls. One would then like to find an optimization procedure related to thermal inputs other than the fixed patch inputs. For example, the geometrical parameters which influence radiative coupling between the cone and patch are θ and ϕ. The minimum coupling occurs at $\theta = \phi$. This means that $r_1/r_2 \to 0$ and $l/c \to \infty$, that is, either an infini-

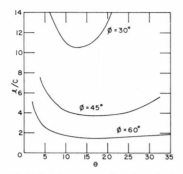

Fig. 4.   Relative length vs cone wall angle at fixed values of patch look angle.

tesimal patch or an infinitely long cone is required. Neither solution is practical, of course, and this approach to optimization produces no useful result.

VI.   Thermal Balance Equations

The thermal balance equations for both the patch and cone are needed in order to determine the equilibrium (steady state) temperature of the patch. Adding the conductive and detector thermal inputs to the radiative, we can obtain the thermal balance equation for a black, isothermal patch,

$$\sigma A_p T_p{}^4 = A_p \epsilon_{pc} \sigma T_c{}^4 + K_p(T_c - T_p) + \Phi_d, \qquad (27)$$

where

$\quad A_p$ = top (black) area of patch,
$\quad T_p$ = temperature of patch,
$\quad T_c$ = temperature of cone,
$\quad \epsilon_{pc}$ = effective patch-to-cone emissivity,
$\quad K_p$ = thermal conductance from cone to patch,
$\quad \Phi_d$ = detector thermal inputs.

The ir power emitted and absorbed by the low emissivity cone walls may, in general, be neglected compared with that emitted and absorbed by the attached high emissivity cone end. In this case, the thermal balance equation for the cone (assumed isothermal) is obtained by equating the radiative output (emission) of the cone end to the radiative and conductive inputs,

$$\epsilon_d A_d \sigma T_c{}^4 = \Phi_x + K_c(T_a - T_c), \qquad (28)$$

where

$\quad \epsilon_d$ = emissivity of cone end,
$\quad A_d$ = area of cone end,
$\quad \Phi_x$ = radiative power absorbed in cone end and cone walls,
$\quad K_c$ = thermal conductance from spacecraft to cone,
$\quad T_a$ = temperature of spacecraft.

The radiative inputs include the ir flux absorbed in the cone end and the visible flux (direct and reflected sunlight) absorbed in both the cone end and cone walls. As an example of a radiative source, consider the nighttime (ir) earth. The power absorbed in the cone on

an earth-oriented spacecraft in the absence of a shield (Fig. 1) is then

$$\Phi_x = \epsilon_d A_d F_{de} W_e, \qquad (29)$$

where

$\quad F_{de}$ = view factor from cone end to earth,
$\quad W_e$ = equivalent infrared exitance of earth.

The view factor from the cone end to earth is given by a contour integral about the visible boundary of the earth or its equivalent semicircle [see Eq. (16)]. For the nadir at $\varphi = 0$, the visible earth as seen from a vertical cone end is bounded by the plane $\vartheta = \pi/2$ and the semicircle

$$\sin^2\vartheta(\varphi) = \sec^2\varphi \, \cos^2\beta_e, \qquad (30)$$

where β_e is the angle from the nadir to a tangent line to the earth's surface. The view factor is then

$$F_{de} = \frac{1}{\pi} (\beta_e - \sin\beta_e \, \cos\beta_e). \qquad (31)$$

The equivalent ir exitance of the earth can be obtained by equating the power emitted by the earth with that absorbed from the sun. The result is

$$W = (\tfrac{1}{4}) \times S_o(1 - A) = 2.1 \times 10^{-2} \, \mathrm{W \, cm^{-2}}, \qquad (32)$$

where S_o is the solar constant (0.14 W cm^{-2}) and A the average albedo of the earth (0.4).

VII.   Sample Cooler Design

A radiant cooler designed for operation on an earth-oriented spacecraft in a near earth, sun-synchronous orbit is shown in Fig. 5. Figure 5(a) is a section in a

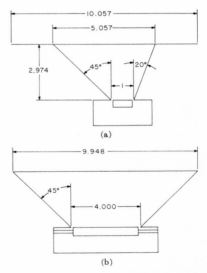

Fig. 5.   Basic design of a radiant cooler for use in a 6 o'clock near earth orbit (dimensions in units of the cone width at the patch top).   (a) Vertical plane.   (b) Horizontal plane.

vertical plane through the centers of the patch and earth; Fig. 5(b) is a section in a horizontal plane through the center of the patch and parallel to a plane tangent to the earth's surface at the satellite subpoint. The patch and cone cross sections are rectangular. Dimensions are shown in units of the cone width at the top of the patch. The cooler is designed for a 6 o'clock orbit (orbit normal to sun angle of zero) at an altitude of 1450 km. A 6 o'clock orbit may be used for the Nimbus E spacecraft. It results in maximum power from a set of fixed solar panels, but generally limits passive earth-oriented experiments to ir and microwave wavelengths. The patch normal is parallel to the orbit normal and points in an antisolar direction. The only external sources seen from the cone top (mouth and end) are the nighttime earth and a horizon sensor which begins at 54° above the patch normal in the vertical plane.

The horizon sensor cannot be seen from the patch and because of its small view factor (~3.5 × 10⁻³) has negligible influence on the cone temperature. The cone is insulated from the spacecraft (instrument housing) by multilayer insulation and supported by low conductance tubes. A maximum of two cone wall reflections is required for a ray from the top of the patch to reach cold space. In the vertical plane, patch rays leave the cone mouth at a maximum angle of 35° toward earth with respect to the cone axis and 53.2° away from earth. In the horizontal plane, the maximum angle is 66.9°. As a result, only the cone walls and cold space can be seen from the top of the patch. In addition, to minimize radiative coupling between the patch and cone, only the black top of the patch views the low emissivity cone walls. The sides and bottom of the patch are insulated by multilayer insulation. The entire patch is supported in orbit by low conductance tubes.

The radiative power absorbed by the cone is given by Eq. (29). At the equator, the radius of the earth is 6378 km. Adding an additional 17 km for the atmosphere (tropical tropopause), β_e from an altitude of 1450 km is 54.6°. The view factor F_{de} is then 0.1530.

For a cooler in which the width (minimum dimension) of the cone cross section at the patch top is 2.5 cm, the cone is adequately supported by eight 3.0 cm long fiber glass reinforced epoxy tubes of 0.65-cm o.d. and 0.50-cm i.d. The material has a thermal conductivity of 2.93 × 10⁻³ W cm⁻¹·K⁻¹ and the tubes a thermal conductance of 1.09 × 10⁻³ W K⁻¹. The multilayer insulation between the cone and spacecraft has an equivalent thermal conductivity of about 1 × 10⁻⁶ W cm⁻¹·K⁻¹ including the degradation produced by penetrations and other imperfections. The cone has an outer area of about 10³ cm², so that a 3.0-cm thickness of multilayer insulation adds a thermal conductance of 0.33 × 10⁻³ W K⁻¹, for a total of 1.42 × 10⁻³ W K⁻¹.

For F_{de} = 0.153, W_e = 2.1 × 10⁻² W cm⁻², T = 300 K, ϵ_d = 1, and A_d = 49.74 cm², the thermal balance equation for the cone [Eq. (28)] yields

$$\sigma T_c{}^4 + 2.85 \times 10^{-5} T_c = 1.18 \times 10^{-2} \text{ W cm}^{-2}.$$

The solution is T_c = 184 K. If the conductive input is reduced to zero, the cone temperature decreases to 154 K.

The equations needed to calculate the effective patch-to-cone emissivity are given in Sec. III. The maximum number of cone wall reflections from the patch-to-cone mouth (cold space) is two. If $F_{p-m(2)}$ is calculated, we can determine the remaining view factor from

$$F_{p-m(1)} = 1 - F_{p-m(0)} - F_{p-m(2)}. \tag{33}$$

If the cone mouth is reflected once in each of the four cone walls, it is found that the hemisphere above the patch is covered except for two triangular regions below the cone mouth corners formed by the intersection of the 20° cone wall with the two adjacent 45° cone walls. These regions are the cone mouth as seen from the patch by two cone wall reflections. Their boundaries are formed by the first cone wall reflections of the boundaries of the cone mouth and by the plane of the patch top ($\vartheta = \pi/2$). If the $\varphi = 0, \pi$ plane is the horizontal plane, the view factor to one of the triangular regions is given by the contour integral of $\sin^2\vartheta$ around its boundary,

$$2\pi F_{p-m(2)}{}^I = \int_{\varphi_1}^{\varphi_2} 1 \times d\varphi + \int_{\varphi_2}^{\varphi_1} \sin^2\vartheta_2 \times d\varphi. \tag{34}$$

The third boundary line has an equation of the form φ = constant and therefore does not contribute to the integral. The first integral is over a line in the $\vartheta = \pi/2$ plane. The equation for the remaining boundary line is given by Eq. (22) in which φ_0 is zero and α is twice the cone angle of the wall forming the image (40° in the sample design). The view factor to the second region of two cone wall reflections is given by a similar formula.

The view factors $F_{p-m(0)}$ and $F_{p-m(2)}$ were calculated for sixteen equal elemental areas over the top of the patch. By symmetry, there are eight different view factors. Since the elements are equal in size, the view factor for the entire patch is just the average view factor of all the elements. The results are listed in Table I. If the cone wall surface has an effective specular emissivity ϵ_c of 0.05, including degradations produced by nonspecular components, the effective patch-to-cone emissivity ϵ_{pc} is 0.0231 [Eq. (2)]. If the effective emissivity is increased to 0.06, ϵ_{pc} is increased to 0.0277.

The patch may be supported in orbit by two fiber glass reinforced epoxy tubes of 0.35-cm o.d., 0.25-cm i.d., and 7.0-cm length. The resultant thermal conductance is 3.94 × 10⁻⁵ W K⁻¹. Multilayer insulation of 2.5-cm thickness between the cone and sides and bottom of a 0.4-cm thick patch adds a conductance of 1.06 × 10⁻⁵ W K⁻¹, for a total of 5.00 × 10⁻⁵ W K⁻¹.

The patch black area in a cooler with a cone width of 2.5 cm at the patch top is about 17.6 cm² when allowance is made for a clearance between the patch and cone. If the detector thermal input (electrical and optical) is 3 × 10⁻³ W, the thermal balance equation of

Table I.   View Factors from Patch Elements to Space

Element	Number of cone wall reflections	
	0	2
A	0.5598	0.01695
B	0.5706	0.01544
C	0.5772	0.01414
D	0.5803	0.01375
E	0.5272	0.02073
F	0.5377	0.01836
G	0.5361	0.01670
H	0.5470	0.01590
Average	0.5545	0.01650

the patch [Eq. (27)] at a cone temperature of 184 K and an effective cone wall emissivity of 0.05 yields

$$9.98 \times 10^{-11}T_p{}^4 + 5.00 \times 10^{-5}T_p = 1.48 \times 10^{-2} \text{ W}.$$

The solution is $T_p = 100$ K. If the cone wall emissivity is increased by 20% (to 0.06), the patch temperature increases by only 1.2%. This shows that a highly accurate value of cone wall emissivity is not required for an accurate prediction of patch temperature.

VIII.   Optimization of Cooler Design

Minimizing the influence of fixed thermal inputs to the patch by means of the cone geometry may be undesirable because it also fixes the often larger radiative input from the cone walls. A better approach may be simply to increase the patch area to the point where the fixed inputs are relatively unimportant. The radiative input can be reduced by reducing the cone temperature as well as the patch–cone coupling factor.

In general, the cone end area is optimum and the cone temperature is a minimum when the cone end is sufficiently large that it determines the cone temperature. In practice, this usually means that the radiative power absorbed by the cone walls is a sufficiently small fraction of that absorbed by the cone end. The cavity formed by the cone walls is, by design, a weak emitter. In the sample design, a 10% increase in the power absorbed (e.g., in the cone walls) without additional emission would increase the cone temperature by only 1.9 K or 1%.

The cone temperature determined by the cone end can also be closely approached by a reasonably sized low α/ϵ cone end when both a daytime and nighttime earth are seen by the cone top as long as direct sunlight is excluded. For small solar elevations above the plane of the cone mouth, a relatively small shield (Fig. 1) can be used to accomplish this and incidently provide partial shielding of the cone top from the earth. As the sun elevation angle above the plane of the cone top increases, however, the shield tends either to restrict the patch's view to cold space or to increase greatly the size of the radiant cooler.

The optimum patch size may be defined as one sufficiently large that the increase in patch temperature produced by the introduction of detector thermal inputs has been reduced to an acceptably low value.

In the sample design, reduction of the thermal inputs to zero reduces the patch temperature by 8 K. In order to decrease this to 2 K, it would be necessary to increase the patch area by a factor of four, assuming that the thermal conductance of the supports increases in direct proportion to the patch area.

Suppose that the optimum cone end area and minimum cone temperature have been closely approached and that the patch has been made sufficiently large that the introduction of electrical and optical connections increases the patch temperature by an acceptably small amount. If the objective is then to obtain the minimum patch temperature, the design must minimize the radiative coupling between patch and cone (i.e., the effective patch-to-cone emissivity). As shown in Sec. V, this generally requires an infinitely large cooler in which the patch sees directly all the cold space in the hemisphere above it and the cone exactly shields all external sources. In the sample design, typical of operation from an earth-oriented spacecraft in a near earth orbit, the angular extent of cold space is limited in the vertical plane but covers a full 180° in the horizontal plane. The minimum radiative coupling between patch and cone is therefore approached as the ratio of patch dimension in the horizontal plane (length) to patch dimension in the vertical plane (width) is increased. In the limiting case in which the patch is confined to the horizontal plane, no cone at all is required and the coupling therefore goes to zero.

With a nonzero path width, however, it is necessary to provide cone walls on all four sides. The cone mouth is situated at the edge of the spacecraft, so that the patch must then be shielded from adjacent instruments and spacecraft structure. In addition, complete cone walls increase the mechanical rigidity of the cone and permit a simple protective cover to be placed over the mouth. The cover prevents contamination of the cooler during launch and shields against solar exposure prior to proper orientation of the spacecraft.

As the length-to-width ratio of the patch increases, the cone length decreases along with the radiative coupling. The ratio is usually limited by the horizontal dimension available for the radiant cooler and by the need for a patch width sufficient to house the components to be cooled. On the other hand, the change in patch temperature with respect to the radiative coupling between patch and cone is not large. If the patch temperature is determined only by radiative coupling to the cone walls, a 10% change in coupling produces only about a 2.5% change in patch temperature. The addition of other thermal inputs to the patch further reduces the change, as shown above for the sample design.

IX.   Radiant Coolers Used in Spacecraft

A variety of satellite-borne instruments utilize radiant coolers to obtain temperatures from 200 K to 80 K. The high resolution ir radiometer[12,13] (HRIR) flown on Nimbus I, II, and III (Fig. 6) contains a simple single-stage cooler that maintains a PbSe element at 200 K. The thermal conductance between the cone

Fig. 6. Nimbus high resolution ir radiometer.

and instrument housing is relatively high and results in a cone temperature of about 290 K. The patch has a black radiating area of 9.8 cm² and is radiatively insulated on its back side by a simple gold plating. The patch is held in place by six 3.20×10^{-2}-cm diam chromel AA or titanium alloy wires, which provide sufficient support to survive the Nimbus vibration environment. The gold coated cone restricts the patch viewing angle to ±30° in a vertical plane through the cooler axis to prevent coupling to the earth and spacecraft solar panels. The view angle in a horizontal plane of ±40° is determined for the most part by instrument design and available space.

The filter wedge spectrometer[14,15] (FWS) shown in Fig. 7 is designed for operation on the Nimbus D spacecraft. It contains a more complex single-stage radiant cooler that maintains a PbSe element at 175 K. The lower temperature extends the wavelength response of the detector to beyond 6 μ. The cooler has a large cone end of low α/ϵ material to permit operation at a maximum orbital plane to earth–sun line angle of 20°. The cone is supported from the instrument housing by a structure of fiber glass reinforced epoxy and is insulated by a blanket of multilayer insulation. The maximum cone temperature is about 245 K. The patch is supported by a fiber glass reinforced epoxy tube that also serves as a conduct for electrical leads to the patch. Patch rays leaving the cone mouth are restricted to ±28° in the vertical plane through the cooler axis to prevent coupling to the earth from a minimum spacecraft altitude of 925 km. The unsymmetrical angles in the horizontal plane (48° and 38.5°) are again largely determined by constraints on the over-all instrument design.

The feasibility model of a very high resolution radiometer (VHRR) for the improved TOS spacecraft is shown in Fig. 8. The HgCdTe detector in the ir channel is cooled to approximately 85 K by a two-stage radiant cooler. The cooler axis is tilted away from the earth in the vertical plane to obtain the maximum view to cold space. The view in this plane is limited by the

earth and by a horizon sensor scan mirror attached to the spacecraft. At an altitude of 1100 km, the total vertical view to cold space is 108°. The total view angle of 126° in an orthogonal plane through the cooler axis is set by the available space and by adjacent instrument structure. The first-stage patch is held in place by a four-pin caging mechanism during powered launch. The second-stage patch is not caged but supported from the first-stage patch by a fiber glass reinforced epoxy tube.

The feasibility of radiant cooling to 80 K under realistic thermal and mechanical conditions had previously been demonstrated by a laboratory test model of a two-stage cooler. The first-stage performance of this rather large (28.6 cm × 62.2 cm × 52.7 cm) cooler closely approximates the ideal in which the patch temperature is determined only by its radiative coupling to the low emissivity cone walls. Space chamber tests thermally simulated operation in a sun-synchronous orbit at an orbital plane to earth–sun line angle of 11° and an altitude of 1100 km. The first-stage cone attained a temperature of 202 K with a variation of ±2 K along its length. The first-stage patch (and second-stage cone) reached 108 K and the second-stage patch, 80 K. With only a single stage of cooling, the patch temperature decreased to 104 K.

X. Conclusions

An approach to the design of radiant coolers based on the specular image model of Secs. II and III has proven to be both suitable and accurate for the design and analysis of coolers in a variety of spacecraft instruments. To some extent, this may result from the fact that a highly accurate determination of the cone–patch radiative interchange is not necessary for an accurate prediction of patch temperature. This tends to reduce the influence of deviations from specular reflectivity at the cone walls.

The ultimate cooler performance requires an impractically large structure in which the thermal coupling between the patch and cone is purely radiative. The temperature of a graybody patch in such a cooler

Fig. 7. Filter wedge spectrometer for the Nimbus D spacecraft.

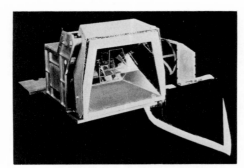

Fig. 8.   Feasibility model of a very high resolution radiometer for the improved TOS spacecraft.

is independent of its emissivity provided the cone walls are specularly reflecting and outward sloping. However, the presence of other thermal inputs to the patch in any real cooler dictates the use of a high emissivity patch.

The optimization of a cooler geometry to minimize the influence of fixed thermal loads on the patch determines the angle of the cone wall when the view to cold space is restricted to less than a hemisphere. However, this approach is often undesirable because it also fixes the radiative coupling factor between the patch and cone. Instead, one may simply select a patch size sufficiently large that the temperature increase produced by the introduction of the component to be cooled is reduced to an acceptable level.

Optimization of the cooler design (realization of the minimum patch temperature) may then be completed by minimizing the cone temperature and the radiative coupling between the cone and patch. In general, the cone temperature is a minimum when the cone end is sufficiently large that it determines the cone temperature. The absolute minimum of cone–patch radiative coupling requires an infinitely large cooler in which the cone exactly shields all external sources. In practice,

the optimization may be constrained not only by the limited space available but also by the patch width needed to house the cooled component. The minimum radiative coupling between the cone and patch subject to these additional constraints is generally realized in a cooler in which the patch dimension in a plane through the cone axis increases with the available look angle to cold space in that plane.

The development and application of radiant cooling as described in this paper have been supported by NASA contracts, beginning with the Nimbus HRIR program in September 1960. Initial momentum was supplied by R. Hanel, who suggested the use of a cone channel condenser in the reverse sense to direct patch radiation to space. Later encouragement and direction by I. L. Goldberg led to the extension of radiant cooling to near liquid nitrogen temperatures.

References

1. P. W. Kruse, Appl. Opt. **4**, 68 (1965).
2. C. Verié and J. Ayas, Appl. Phys. Lett. **10**, 241 (1967).
3. E. W. Treuenfels, J. Opt. Soc. Amer. **53**, 1162 (1963).
4. E. M. Sparrow and R. D. Cess, *Radiation Heat Transfer* (Brooks/Cole, Belmont, Calif., 1966), pp. 140–149.
5. D. E. Williamson, J. Opt. Soc. Amer. **42**, 712 (1952).
6. R. Hanel, ARS J. **31**, 246 (1961).
7. P. F. O'Brien and E. F. Sowell, J. Opt. Soc. Amer. **57**, 28 (1967).
8. R. P. Bobco, J. Heat Transfer **86**, 123 (1964).
9. M. Jakob, *Heat Transfer* (John Wiley & Sons, Inc., New York, 1957) Vol. 2, p. 14.
10. V. E. Holt, R. J. Grosh, and R. Geynet, Bell Syst. Tech. J. **41**, 1865 (1962).
11. R. B. Scott, *Cryogenic Engineering* (D. Van Nostrand Inc., Princeton, 1959), pp. 147–148.
12. I. L. Goldberg, L. Foshee, W. Nordberg, and C. E. Catoe, in *Proceedings of the Third Symposium on Remote Sensing of the Environment* (University of Michigan Press, Ann Arbor, 1964), pp. 141–151.
13. W. Nordberg, Science **150**, 559 (1965).
14. W. A. Hovis, Jr., W. A. Kley, and M. G. Strange, Appl. Opt. **6**, 1057 (1967).
15. W. A. Hovis, Jr., and M. Tobin, Appl. Opt. **6**, 1399 (1967).

Reprinted from *Appl. Optics*, **5**(9), 1457–1460 (1966)

Closed Cycle Cryogenic Refrigerators as Integrated Cold Sources for Infrared Detectors

A. Daniels and F. K. du Pré

Many ir detectors can only operate when cooled to low temperatures. A short description is given of an integrated refrigerator–detector combination in airborne use. Experimental results are presented for developmental refrigerators in which weight and power input have been reduced to a point where satellite-borne operation becomes feasible.

Background

Detectors in ir optical systems are often operated at low temperatures. For instance, airborne ir mapping systems that detect the natural ir radiation from the ground, occurring in the 8-μ to 14-μ atmospheric window, may use mercury-doped germanium[1] detectors at about 30°K or mercury-cadmium-telluride[2] detectors at 77°K.

Airborne ir warning systems that operate in the 3-μ to 5-μ band may use indium antimonide[3] detectors at 77°K; for extraterrestrial studies from satellites, a suitable detector might be gold-doped germanium[3] at 65°K, sensitive in the 2-μ to 9-μ region.

In each of the above detectors, the effect of the ir radiation is to create free charge carriers in the detector material. Since free carriers can also be produced by thermal excitation, optimizing the sensitivity of the detector requires that the thermal excitation be negligible. Therefore, the detector is cooled until the concentration of the thermally excited carriers becomes small compared to that produced by the background radiation from the target. In fact, without such cooling, the above detectors are almost useless because of the noise caused by the thermally excited carriers.

The temperature required to realize this background limited ir photodetection (BLIP) depends largely on the absorption coefficient of the material for the radiation. Thus, the temperature can be quite different for various materials.

For instance, in mercury cadmium telluride, all the incident radiation can be absorbed in a thin layer. In such a material, the concentration of charge carriers caused by the radiation is much greater than in a poorly absorbing material such as mercury-doped germanium.

The authors are with the Philips Laboratories, Briarcliff Manor, New York 10510.

Received 24 May 1966.

Therefore, we can tolerate a larger thermal concentration and thus a higher temperature in mercury cadmium telluride than in mercury-doped germanium. The reason for the poor absorption in the latter material is that only the impurity mercury atoms absorb the radiation instead of the lattice atoms, as in the case of mercury cadmium telluride. Details on the physics and chemistry of solid state ir detectors may be found in a review paper by Bratt *et al.*[4]

After this short discussion of the reasons why cooling of the above materials to their respective temperatures is essential, we now turn to the problem of choosing a suitable refrigeration method. The range of temperatures we have encountered, namely, from room temperature down to about 20°K, is the typical operating region of miniature, closed cycle, cryogenic refrigerators. Since there are several thermodynamic cycles that may be utilized for such refrigerators, we briefly discuss them in order to justify our choice—the Stirling cycle.

Refrigeration Methods

There are three physical effects used to produce low temperatures: the evaporation of a liquid, the isenthalpic expansion of a gas, and the isentropic expansion of a gas. A fourth physical effect, radiative exchange with cold interstellar space, has been used[5] to cool spaceborne ir detectors to 200°K and studied for application below 100°K. These effects are used in four of the best known refrigeration methods: the vapor-compression cycle, the Joule–Thomson process, the Stirling cycle, and the Claude process. The thermodynamic analysis of these effects and processes can be found in the literature[6]; however, the usefulness of a particular process to the designer, faced with having to choose a working refrigerator, depends on a combination of various properties, such as efficiency, weight, size, and maintainability. Therefore, we discuss the refrigeration processes cited, with emphasis on these parameters.

The *vapor-compression cycle*, the basis for the house-

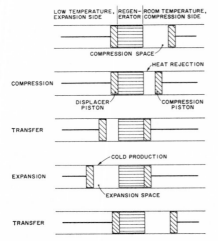

Fig. 1. Diagram showing the four steps of the Stirling cycle.

hold refrigerator, is reliable and efficient. However, its efficiency begins to decrease rapidly at −45°C; at this point a cascaded system is required in order to attain lower temperatures. Such a cascaded system would involve several interlocking refrigerant circuits, e.g., Freon, ethane and nitrogen, each with separate compressors, condensers, heat exchangers and evaporators. A system of this kind would be cumbersome and require a long cooldown time.

The *Joule–Thomson process* relies on the cooling effect obtained from the irreversible expansion of high pressure gas (at approximately 130 atm) through a throttle valve. The thermodynamic efficiency of this process is low; the reliability of high pressure gas compressors and the preservation of gas purity have always been serious operational problems.[7−9] Furthermore, in order to achieve low temperatures, two refrigerant circuits—e.g. nitrogen and hydrogen—are required. Thus, power consumption, size, and weight become unfavorable design characteristics.

The *Claude cycle* (and the similar Brayton cycle) produces refrigeration by expanding a gas through an engine and having the gas do work. This cycle, being reversible, has a good efficiency and has recently received considerable attention.[10] However, it requires the use of separate compression and expansion machines, thereby increasing size and operational complexity. In addition, intake and exhaust valves (in some designs operating at cryogenic temperatures) are required to control the gas flow; this results in an efficiency which is lower than that of the Stirling cycle.

The *Stirling cycle* is reversible and has a high thermodynamic efficiency. The operating efficiency is close to the theoretical since the same piston is used for both compression and expansion, thereby minimizing mechanical losses. The Stirling cycle is performed in a sin-

gle machine, using helium at low pressures (in the order of 6–8 atm); the compressor and expander spaces communicate through an open connection without valves. This cycle can be embodied in a small, relatively simple machine since the low operating pressure and low compression ratio employed are conducive to a lightweight design.

Since the Stirling cycle refrigerator is the only one that combines high efficiency with good reliability, small size, and light weight, we restrict our attention to it. And since there are excellent review papers[11], only a brief description of the Stirling cycle is given.

Stirling Cycle

Figure 1 illustrates the operation of the Stirling cycle. Helium gas is compressed in a compression space, and cold is subsequently produced by expansion of the gas in an expansion space. Characteristic of the Stirling cycle is the method of transfer of the gas between these spaces. The compression space, kept at ambient temperature, and the expansion space are in open connection via a short channel. This channel contains the regenerator, a porous mass of fine wires, one end of which is at room temperature and the other end at the low temperature—the temperature decreasing linearly from one end to the other. The regenerator permits the flow of helium gas from the warm to the cold space, and vice versa, without cold losses. On its way to the cold space, the helium gas is cooled by the regenerator mass which stores the heat contained in the gas. Returning from the cold space, the helium is warmed by the regenerator, thus returning the stored heat to the helium. Therefore, the helium, enters the compression or expansion space at the approximate temperature of that space; there is hardly any cold loss due to the

Fig. 2. Stirling cycle refrigerator for cooling mercury-doped germanium detectors.

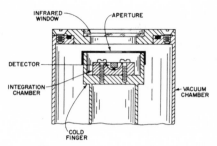

Fig. 3. Detector mounting method.

movement of gas from one area to the other. The necessary compression and expansion of the gas are performed by an out-of-phase motion of two pistons, thus eliminating the need for any valves.

In the following paragraphs, some practical miniature refrigerators based on this principle are discussed. We only consider the case where the detector can be integrated with the refrigerator to form a compact unit. Although there are circumstances where transfer of the cold from refrigerator to detector via a liquefied gas can be advantageous[12], direct integration is preferable since it results in a simpler, more compact instrument.

Miniature Stirling Cycle Refrigerator

Figure 2 shows a Stirling cycle refrigerator, presently in use, designed to cool mercury-doped germanium detectors to 25°K and below.* The power input is about 450 W with an operating speed of 1600 rpm; the working medium is helium gas at an average pressure of 8 atm.

The refrigerator proper is vertical and about 30 cm high. It can be operated in any attitude since only dry lubrication is used in the mechanism. The heat rejection system, consisting of a small radiator with a fan and a pump for circulating a cooling liquid, can be seen behind the refrigerator itself. Direct air cooling is also possible.

The cold cylinder, i.e., the cold finger, with the detector fastened to it is not visible since it is protected by a vacuum enclosure. A typical method for mounting the detector to the cold surface is shown in Fig. 3.

Though the refrigerator produces some vibration, no disturbing microphonics are produced as long as the leads to and from the detector are immobilized. When operating, the noise voltage is not more than 10% higher than experienced immediately after interrupting operation at low temperature.

Developmental Units

Figure 4 shows a lightweight, low power refrigerator designed to cool an InSb detector to 77°K or below.

---

* Manufactured by the North American Philips Company, Inc., Cryogenic Division, Ashton, R. I.

With a power input of 45 W to the dc motor, the cold production is 1.5 W. This means that even if a 1.5-W heater were placed on the cold spot, the temperature could still be maintained at 77°K. The total weight is 3.9 kg.

It should be noted that the low power input is partly due to the operation at 77°K instead of 25°K. According to the second law of thermodynamics, less power is needed to produce the same amount of cold at a higher temperature than at a lower temperature.

The detector, mounted directly on the cold spot, is housed in a permanently evacuated metal enclosure with a soldered-in sapphire window. The cooldown time is about 15 min. Since the power input is small, heat rejection is only by natural convection.

The performance of the cooler is shown in Fig. 5 which is a plot of net refrigerating power and input shaft power as a function of cold end temperature. The working pressure is 6 atm; the ambient temperature is +25°C.

The cooldown time of a refrigerator is a function of its cooling capacity in the intermediate temperature range and the mass it has to cool. With a conventional dewar-housed indium antimonide detector mounted on its cold finger, this unit attains 77°K in approximately 8 min.

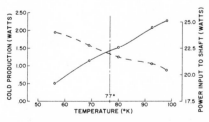

Fig. 4. Stirling cycle refrigerator for cooling indium antimonide detectors (77°K).

Fig. 5. Cold production and power input vs output temperature for miniature refrigerator.——cold production.----power input. Working pressure, 6 atm.

Fig. 6. Stirling cycle refrigerator suitable for satellite operation.

The cooler is capable of achieving the same temperature level with a lower working pressure and therefore less power input, but at the expense of the cooldown time.

Since the 1.5 W cold production of this unit is ample for many detector applications, one might want to reduce it further and thereby decrease the power input and the weight. It should be noted, however, that any drastic reduction in cold output will considerably increase the cooldown time. In situations where an extreme reduction in weight and power is essential and a short cooldown time is only of secondary importance, the unit shown in Fig. 6 should be of interest. The weight has been reduced to 2.3 kg, the power input to 25 W. The usable cold production is 0.5 W; the cooldown time is about 20 min. The weight and power input have been minimized to the extent where operation in a satellite can be considered for missions not exceeding 500 h of actual data-taking.

However, in addition to low weight and power, satellite-borne operation imposes stringent mechanical requirements. The operation must be vibration-free and start-up torques must be eliminated.

During the start-up of almost any motor-driven device, a torque is exerted on its base. This would set a satellite into a permanently spinning motion, which of course would have to be corrected. This effect would be especially disturbing if many starts and stops are needed. This effect can be completely overcome in the

Stirling cycle using a special drive mechanism.[11] This mechanism, the *rhombic drive*, uses two identical motors rotating in opposite directions. The drive can be balanced so no forces or torques are exerted by the refrigerator on its base—hence, the satellite.

In the developmental models discussed above, conventional crank mechanisms were used rather than rhombic drives. Until now, the rhombic drive has been used only in experimental models of somewhat larger capacity, but there is no reason to doubt that it can also be used in these low power units.

In conclusion, we can state that a miniature refrigerator integrated with an ir detector has been developed into a practical device for use in optical instruments, both ground-based and airborne.

References

1. S. R. Borello and H. Levinstein, J. Appl. Phys. **33**, 2947 (1962).
2. P. W. Kruse, L. D. McGlauchlin, and R. B. McQuistan, *Elements of Infrared Technology* (John Wiley & Sons, Inc., New York, 1962), p. 413. See also: Honeywell Research Center, Hopkins, Minn., Quart. Progr. Rept. 3, 20 October 1965, "Intrinsic Infrared Detector Technology"; ASTIA AD-473136 under a contract.
3. P. Bratt, W. Engeler, H. Levinstein, A. MacRae, and J. Pehek, Infrared Phys. **1**, 27 (1961).
4. P. Bratt, W. Engeler, H. Levinstein, A. MacRae, and J. Pehek, Syracuse University Tech. Rept. AFAL TR 64 343, 1 April, 1965, "Photoconductivity in Impurity Activated Germanium and Indium Antimonide"; ASTIA AD 459613 under a contract with Air Force Avionics Laboratory, WPAFB.
5. N. Fuschillo, C. Schultz, and R. Gibson, Trans. IEEE **AS-3**, 81 (1965), Suppl.
6. See, e.g., J. G. Daunt, in *Handbuch der Physik* (Springer, New York, 1956), Vol. 14, p. 1.
7. J. M. Geist and P. K. Lashmet, in *Advances in Cryogenic Engineering* (Plenum Press, New York, 1961), Vol. 6, p. 73.
8. J. S. Buller, Proc. 7th IRIS, p. 193 (1962).
9. A. S. Chapman, Proc. 8th IRIS, p. 67 (1963).
10. C. A. Schulte, A. A. Fowle, T. P. Henchling, and R. E. Kronauer, in *Advances in Cryogenic Engineering* (Plenum Press, New York, 1965), Vol. 10, p. 477.
11. J. W. L. Köhler, in *Progress in Cryogenics* (Heywood and Company, Ltd., London, 1960), Vol. 2, p. 41; and Sci. Am. **212**, 119 (1965).
12. K. W. Cowans and P. J. Walsh, in *Advances in Cryogenic Engineering* (Plenum Press, New York, 1965), Vol. 10, p. 468.

Author Citation Index

Abney, W. de W., 130
Abrams, R. L., 49, 117
Adams, H. D., 192
Aiken, C. B., 216
Akiyama, M., 50
Aldrich, N. C., 150
Alekseyev, A. M., 133
Alfieri, I., 12
Alkemade, C. T. J., 28
Allen, C., 209
Altemose, H., 29, 150
Amdur, I., 216
Ameurlaine, J., 150
Amon, W. F., Jr., 130
Anderson, L. K., 28, 56, 96, 115, 116
Anderson, N. C., 184
Andrews, D. H., 217, 254, 256, 257, 267
Annable, R. V., 342
Antonov, Y. I., 12
Arams, F. R., 13, 49, 56, 96, 115, 117, 184
Arnquist, W. N., 12
Asnis, L. N., 184
Astaf'yev, A. I., 133
Astheimer, R. W., 209
Autrey, E. A., 133
Avery, D. G., 13, 153, 166
Ayas, J., 96, 116, 374

Baertsch, R. D., 116
Bahr, A. J., 96, 115
Bailey, F., 29, 50
Baker, D., 80
Baker, G., 49, 184
Barr, E. S., 11, 12, 28, 50, 213, 294
Bartlett, B. E., 117, 166, 177
Bartlett, N. R., 130
Earyshev, N. S., 56
Bates, R. L., 29, 252
Beasley, J. K., 246
Beaupre, H. J., 150
Beck, J. D., 150
Becker, J. A., 12, 28, 209, 213, 233, 294

Beer, A. C., 13, 50, 115
Beerman, H. P., 209
Bell, E. E., 259, 318
Bell, R. L., 115
Belyakova, V. V., 133
Bemski, G., 74
Bené, R. W., 50
Benton, R. K., 49
Berdahl, C. M., 56
Bergman, J. G., Jr., 209
Bernamont, J., 79
Bernt, H., 29
Bess, L., 182, 335
Besson, J., 116, 174
Betts, D. B., 209
Betz, C. R., 137
Beyen, W. J., 17, 117, 135, 192, 341
Biard, J. R., 116
Biller, L. N., 133
Billings, B. H., 28, 213, 294
Bisbee, J., 133
Bishop, S. G., 49, 209
Blackburn, H., 209
Blackman, R. B., 77
Blanchard, E. R., 256
Blaney, T. G., 49
Blevin, W. R., 56, 209
Bloor, D., 29, 254
Blout, E. R., 130
Blum, A. N., 166
Bobco, R. P., 374
Bode, D. E., 117, 119, 133, 192
Bode, H. W., 298
Bogle, R. W., 81
Born, M., 96
Borrello, S., 117, 192, 202, 277, 341, 378
Bostick, H. A., 96, 117
Boyd, G. D., 115
Boyle, W. S., 267
Bradley, C. C., 49
Bradshaw, P. R., 56
Bratt, P., 17, 117, 135, 174, 193, 378

Subject Index

389